THE MATHEMATICS OF
FINITE ELEMENTS
AND APPLICATIONS

THE MATHEMATICS OF FINITE ELEMENTS AND APPLICATIONS
Highlights 1996

Edited by

J. R. Whiteman
Brunel, The University of West London, UK

JOHN WILEY & SONS

Chichester · New York · Weinheim · Brisbane · Singapore · Toronto

Other Wiley Editorial Offices

John Wiley & Sons, Inc., 605 Third Avenue,
New York, NY 10158–0012, USA

Jacaranda Wiley Ltd, 33 Park Road, Milton,
Queensland 4064, Australia

John Wiley & Sons (Canada) Ltd, 22 Worcester Road,
Rexdale, Ontario M9W 1L1, Canada

John Wiley & Sons (Asia) Pte Ltd, 2 Clementi Loop #02–01,
Jin Xing Distripark, Singapore 129809

VCH Verlagsgesellschaft mbH, Pappelallee 3,
D–69469 Weinheim, Germany

British Library Cataloguing in Publication Data

A catalogue record for this book is available from the British Library

ISBN 0–471–96270–8

Produced from camera-ready copy supplied by the editor
Printed and bound in Great Britain by Bookcraft Ltd, Midsomer Norton
This book is printed on acid-free paper responsibly manufactured from sustainable forestation,
for which at least two trees are planted for each one used for paper production.

CONTENTS

Preface

Contents

PREFACE

This book seeks to highlight certain aspects of the state-of-the-art of the theory and applications of finite element methods as of June 1996. All the papers result from presentations which were made at MAFELAP 1996, the Ninth Conference on the Mathematics of Finite Elements and Applications, which was organised by BICOM, the Brunel Institute of Computational Mathematics at Brunel University during the period 25-28 June, 1996. As with the previous eight conferences in the MAFELAP Series, the intention was to bring together mathematicians, engineers and other practitioners of finite elements at Brunel to discuss finite element techniques in as broad a context as possible.

On this occasion it was fitting that the third Zienkiewicz Lecture should be presented by Professor Ivo Babuska, in view of his vast contributions to the subject and in particular to adaptive finite element methods. The other invited speakers were Professors M. Feistauer, U. Langer, K. Morgan, J. T. Oden, D. J. Owen, J. N. Reddy, E. Stephan, P. Vassilevski, W. L. Wendland, M. F. Wheeler, J. R. Whiteman and O. C. Zienkiewicz. Papers by eleven of these authors are included in this book, together with a number of additional papers which have been chosen to highlight other current research areas in the finite element field. A total of over 180 presentations were made at the conference in lecture sessions, in mini-symposia and in poster-sessions. It has been possible to include the papers of only a fraction of these in this book.

A major innovation for MAFELAP 1996 was the inclusion of a large number of mini-symposia in the conference programme. These were organised by people who were either invited or who volunteered to undertake the task. These mini-symposia made a large contribution to the conference. In addition to these organisers the success of the conference depended on the programme committee, the persons who chaired the sessions, the speakers, the poster session authors and, as always on the extensive efforts of the BICOM Fellows, Research Students and Visiting Fellows. My sincere thanks go to all these people and especially to Mrs. Mary Reece, who undertook all the administration of the accommodation for the conference. I am pleased also to acknowledge financial support for the conference from the United States Army European Research Office. Once again Dr Michael Warby performed his wizardry in producing on the computer all the documentation for the conference, and by turning all manner of forms of LaTeX chapter manuscripts into the camera ready form of the manuscript of this book. It is a pleasure to express my thanks to him. Finally my heartful thanks go for the ninth time to my wife, Caroline, who not only produced the index for this book, but as always provided immense support and encouragement during the lead up period to the conference.

J. R. Whiteman
BICOM
Brunel University
September 1996

1

Trends and New Problems in Finite Element Methods

Ivo Babuška[1] and Barna Szabó[2]

[1]Texas Institute for Computational and Applied Mathematics
University of Texas
Austin, TX 78712 U.S.A.

[2]Center for Computational Mechanics
Washington University
St. Louis, MO 63130 U.S.A.

1.1 INTRODUCTION

Trends and problems in finite element analysis technology are discussed from the points of view of engineering practice and academic research. Ideally, these two areas are closely related. Experience has shown, however, that with few exceptions it takes ten or more years before the results of research become available for use in professional practice.

Research is continuing at a substantial intensity. One measure of the level of research activity is the number of papers published on various aspects of the finite element method and its applications. The number of these papers is about 4,000 annually [Mac86].

One measure of the level of industrial activity is the license fees collected by vendors of finite element software products. This is estimated to be 400 million dollars annually, with a growth rate of about 18 percent. This figure does not reflect levels of expenditure related to the industrial use of finite element analysis software products, which is many times larger but very difficult to measure. At present the fastest growing major finite element analysis software product is Pro/Mechanica Structure[1], which is based entirely on the p-version of the finite element method.

There is a declining trend in the development and use of in-house finite element analysis codes in favor of commercial codes and there is a strong demand for close integration among commercial finite element analysis codes and computer-aided

---

[1] Pro/Mechanica Structure is a trademark of Parametric Technology Corporation.

The Mathematics of Finite Elements and Applications
Edited by J. R. Whiteman © 1997 John Wiley & Sons Ltd.

design (CAD) software products. The proliferation of powerful computers makes the computation of larger and larger problems by an increasing number of users possible. The results of computation are typically presented in graphically very sophisticated and attractive ways, creating the usually false impression that the quality of the computed data is comparable to the quality of graphical presentation.

The prevailing trends and expectations were articulated recently by Mr. T. Curry, president of the MacNeal-Schwendler Corporation, as follows [Cur96]:

> *"Today, with immense business pressure to shorten product development cycles we find our customers' needs changing. Tools that require a significant investment in time and resources to master, let alone raw finite element technologies, are inconsistent with today's accelerated schedules."*

On one hand users demand software tools that do not require a significant investment in time and resources to master, on the other hand the consequences of wrong decisions can be very substantial. An example is the failure of the *Sleipner A-1* offshore platform. The finite element analysis, which was the basis of the design of the critical section, was inaccurate due to an inadequate mesh, elements of degree $p = 1$, and the absence of any accuracy control [Hol94], [RGA93], [JR94].

Clearly, engineers are responsible for their decisions based on computed information. They can exercise this responsibility well only with software tools that provide reliable a posteriori estimates of error in terms of all data of engineering interest. Given the competitive pressures noted by Mr. Curry, the demand for reliable a posteriori error estimation in terms of any data of interest is expected to increase.

The apparent economic value of finite element analysis technology based on the estimated levels of usage is very substantial. In this paper we address problems relating to the continued development and industrial use of finite element analysis technology. The discussion is restricted to solid mechanics. This restriction is made because the traditional applications are in solid mechanics where, relatively speaking, the finite element method (FEM) is most mature and has the widest range of applications. There is rapid progress in the application of FEM in computational fluid mechanics (CFD) as well.

1.2   FEM IN ENGINEERING PRACTICE

The type of usage of FEM in engineering practice falls into two broad categories: (a) *design* computations and (b) *certification* computations related to various safety considerations and/or regulations. Design and certification computations have different objectives and requirements.

1.2.1   DESIGN COMPUTATIONS.

The most important potential use of FEM is in preliminary and conceptual design. It has been estimated that in typical aerospace projects at the end of the conceptual design phase 66% of life cycle costs are committed and 60% of design change opportunities are lost. Figure 1 shows a typical relationship between time and growth of decisions that affect life cycle costs and the consequential decline in design opportunities.

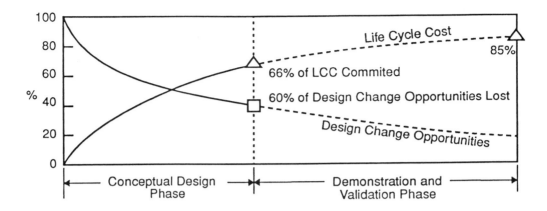

Figure 1   Typical relationship between decisions that affect life cycle costs and maturity of an aerospace project.

The conceptual phase of design is the most creative part of the design process. The objective is to investigate alternative design concepts from the points of view of feasibility, cost, expected performance, ease of manufacturing, and maintenance. Design considerations can change rapidly and widely as concepts are clarified and refined. The considerations are invariably multi-disciplinary in nature. This requires an effective framework for visual and quantitative communication among the various design groups. This role is performed by sophisticated CAD systems. An often quoted example is the design of the Boeing 777 aircraft. During the peak of the project, designers working on 7,000 workstations in 17 time zones managed to communicate effectively through the CAD system CATIA[2], [Pet95].

In general, finite element methods are either not employed in the conceptual design phase, or they are employed only for obtaining an estimate of the force distribution throughout a structure. Very coarse models, called 'stick' models, are created for this purpose. For example, an aircraft structure is typically modeled by very simple plate, beam and bar elements only. The expected output is a reliable distribution of forces acting on the major structural components. It is generally believed that the relative stiffnesses of the major components, and hence the load distribution, will be accurately captured by such models.

There are at least two reasons for the general under-utilization of FEA in the early part of the design cycle where it could provide the greatest benefit: First, a substantial amount of training is required for users of current FEA software products. Most designers lack the required expertise. Second, the time required for conducting finite element analyses is far too long in relation to the time available to designers for completing their work. Typically, design computations are based on traditional procedures. Many organizations have incorporated these procedures in computer programs to make them more convenient to use but have not changed the

---

[2] CATIA is a trademark of Dassault Systèmes.

underlying design methodology. An additional obstacle for many designers, especially for designers of aerospace structures, components and parts, is that design usually involves investigation of structural stability in the inelastic range which conventional finite element programs do not handle well.

The computational requirements for design are greatly varied from industry to industry, and even from department to department. For example, in automotive firms the requirements of car body designers are very different from those of power train designers. Present finite element software products attempt to meet all requirements across departments and industries. These software products are highly complex; difficult to use properly; slow in responding to new requirements, and often fail to meet specific engineering requirements.

Design computations generally involve parametric analysis. The objective is to select dimensions so that design criteria are met in a nearly optimal fashion. Designers usually focus on a few parameters they consider important and work with simple models in order to make parametric calculations possible in a short amount of time. Implied in this is the notion of hierarchic models: The simple model is understood as the lowest member of a family of models, capable of representing the real physical object with increasing reliability and detail.

Consider, for example, the problem of designing a reinforcement (boss) around a hole in a plate. The plate, made of ASTM-A36 steel (modulus of elasticity $E = 2.9 \times 10^7 \ lbf/in^2$; Poisson's ratio $\nu = 0.295$), is subjected to tension (Fig. 2).

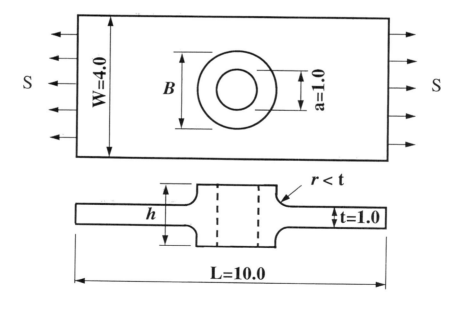

Figure 2   Plate with a reinforced circular hole.

Assume that the parameters to which numbers are assigned are fixed. The parameters representing the diameter of the boss (B), the thickness of the boss (h) and the radius of the fillet r ($r < t$) are to be selected by the designer such that the stress concentration factor denoted by Ktg and defined by $Ktg := \sigma_{\max}/S$ is not greater than 2.5 where S is the applied stress, see Figure 2.

In a case like this the designer would first investigate the parameters B and h because these parameters can be selected using a simple two-dimensional model, assuming plane stress conditions. The domain is defined parametrically and the mesh is associated with the domain so that when a parameter is changed, the mesh changes with it. In this case p-extension is the logical choice for controlling the errors of discretization. The reasons for this are discussed in Section 1.2.5. Of course, the two-dimensional model cannot account for the effects of the fillet, but is capable of approximating the maximum stress and its dependence on the dimensions B and h well. The computations were performed with Stress Check[3]. Varying one parameter at a time, the design curves shown in Fig. 3 were obtained. From the design curves the dimensions B and h were selected, so that the design conservatively satisfies the constraint $Ktg \leq 2.5$.

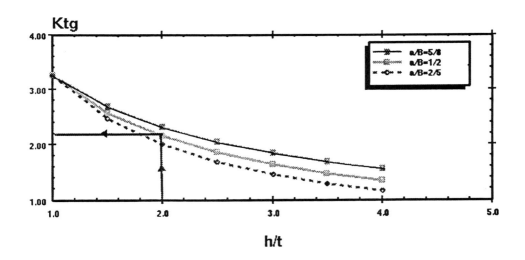

Figure 3   Variation of Ktg with respect to the design parameters h/t and a/B.

The next problem is to select the fillet radius such that the design criterion is met. This requires the use of a three-dimensional model which either represents the fillet or, alternatively, the program has to have a capability to compute the stress intensity factor along the edge to be filleted. It is possible to derive an asymptotic formula for the stress at the surface of the fillet as a function of the stress intensity factor and the radius of the fillet (see [AFBF95]). The stress concentration factor (or the stress intensity edge function) can be quickly estimated and the radius of the fillet selected

---

[3] Stress Check is a trademark of Engineering Software Research and Development, Inc.

such that the maximal stress in the fillet does not exceed the maximal stress on the surface of the hole.

Error estimation and control are an important part of analyses performed in support of the design decisions. This point is discussed next.

1.2.2 CERTIFICATION COMPUTATIONS

Computations performed to verify that a particular set of design criteria are met involve the use of finite element analysis programs together with the data that characterize the problem (e.g., domain, material, load, etc.). The goals of the computation are to determine maximal stresses, stress intensity factors, buckling stresses, etc.; compare them against allowable values, and report the results in terms of factors or margins of safety.

In general, a discretization can be considered adequate if the following four criteria are met:

1. the data of interest (in this case the maximum stress) are known to be finite;
2. the error in the energy norm is small;
3. there are no significant jumps in those stress contours which correspond to high stress levels at the interelement boundaries;
4. the data of interest (in this case the maximum normal stress) appear to have converged to a limit with respect to h-, p- or hp-extension, or a reliable a posteriori error estimator indicates that the error is small.

These rules work well in practice. Nevertheless it is possible to conceive counter-examples because the validity of the estimators is established for the asymptotic range but many engineering computations are performed in the pre-asymptotic range. Also, some estimators are not sufficiently robust.

These rules are satisfied in the case of the reinforced hole problem which was analysed using the p-version: Item 1 is satisfied, since there are no sharp re-entrant corners or other causes for stress singularities to occur. That item 2 is satisfied is seen from see Table 1 which shows the convergence of the potential energy with respect to increasing p; the estimated rate of convergence β, and the estimated relative error in energy norm (e_r), which is related to the root-mean-square error in stresses. For details we refer to [SB91]. Item 3 is satisfied, see Fig. 4, and item 4 is satisfied, see Fig. 5.

Recognizing the trends in industry toward abandoning the use of in-house programs, vendors of finite element analysis software products promote codes that satisfy needs previously met by in-house codes. The demand for a posteriori error analyses and adaptive approaches has increased substantially in the past few years. In Fig. 6 we present some results of a recently conducted survey of industrial users [Hal96] in comparison with a survey conducted in 1991 [HBD91].

The results of this survey show that approximately fifty percent of FEA users now employ some form of error estimation and adaptive methods.

Table 1 The reinforced hole problem. Three-dimensional model. Convergence of the potential energy.

p	N	$\Pi(u_{FE})$ (lbf in)	β	e_r %
1	69	$-3.424467054 \times 10^{-8}$	-	12.59
2	226	$-3.466245235 \times 10^{-8}$	0.60	6.21
3	383	$-3.473543257 \times 10^{-8}$	0.75	4.19
4	655	$-3.477141146 \times 10^{-8}$	0.83	2.68
5	1042	$-3.478937791 \times 10^{-8}$	1.36	1.43
6	1571	$-3.479397501 \times 10^{-8}$	1.27	0.85
7	2269	$-3.479574810 \times 10^{-8}$	1.69	0.46
8	3163	$-3.479623475 \times 10^{-8}$	1.69	0.26
∞	∞	$-3.479647034 \times 10^{-8}$	-	0.0

1.2.3 THE DANGER OF THE UNCRITICAL USE OF FEM.

Vendors of commercial finite element codes claim to provide for "painless" use of their code by nonspecialists as much as possible. Many users treat the code as a "black box". The results, made available in graphical form, often appear plausible for nonspecialists, even when the results are grossly inaccurate. An interesting case study was presented in [Kur95].

Some commercial codes provide a posteriori error estimation in terms of a percentage of the data of interest, but the method by which the estimation is obtained is not specified exactly. As seen from Fig. 6, there is an understanding of need for accuracy and hence information about the error (often in the energy norm or the similar L_2 norm) is interpreted as accuracy of *any* data of interest. Many users are not familiar with the notions of error in the energy norm and error in strain energy, or the difference between them. In addition, the notion of the error as the difference between the exact solution of the mathematical problem and the FE solution as its approximation is often not appreciated, because many users still think of the FEM as finite element modeling and not as a method for obtaining an approximate solution. This is illustrated by an example in the following.

The bracket problem shown in Fig. 7 was computed with various FEA codes. (Fig. 7 was generated by Stress Check, $p = 8$.) The analysts were asked to report the location and magnitude of the *maximal* equivalent (von Mises) stress. Inspection of the domain reveals that there is an edge (with an internal angle of about 270°) and hence the maximal stress corresponding to the linear theory of elasticity is infinity. Often it is argued that infinitely large stresses cannot exist and therefore the results are "correct". The authors were shown reports that gave some specific numerical value at the (reentrant) edge, where the exact stress is known to be infinity, yet it was stated that the error was 5% in maximal stresses.

The graphical representation looked very plausible to persons who do not view the results of finite element analysis as an approximation to a problem of elasticity but

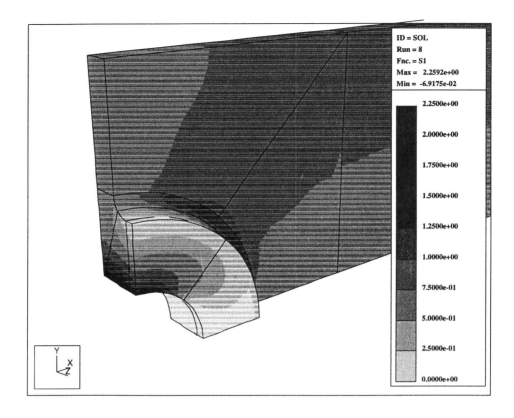

Figure 4   Three-dimensional model of the reinforced hole problem. Contours of the first principal stress.

rather as a numerical modeling exercise.

Using the mesh shown in Fig. 7 and p-extension in the range of $p = 1$ to $p = 8$ (trunk space), the maximal stress plotted against the number of degrees of freedom will appear to have converged to a finite value and the indicated location of the maximal stress will be at the largest circular cut-out in the web. The reason for this is that the stress intensity factor along the re-entrant edge is a small number and hence the effects of the singularity are not visible on this mesh within the range of p-values employed. This example demonstrates that it is not sufficient to look at the numerical results only. It is also necessary take into consideration a priori information concerning the nature of the solution, as noted in Section 1.2.2. In this case the stress report would be correctly phrased as follows: "Excluding the neighborhood of the re-entrant edge, the location and magnitude of the maximum principal stress are ..." Without such qualification, the report would be false.

Stresses are computed in order to determine a factor of safety or margin of safety with respect to failure. Therefore an infinite maximal stress predicted by a

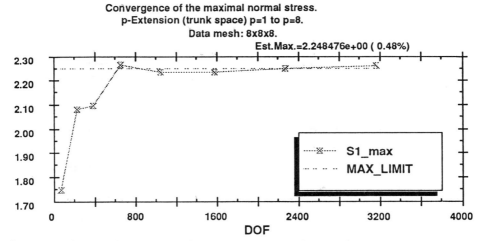

Figure 5 Three-dimensional model of the reinforced hole problem. Convergence of the maximum normal stress.

mathematical model based on the linear theory of elasticity has to be interpreted from the point of view of failure, such as crack initiation. In the case of the bracket shown in Fig. 7 the re-entrant edge may not be critical from the point of view of failure, nevertheless this cannot be decided without a properly formulated failure criterion.

1.2.4 THE PROBLEMS OF MESH DESIGN AND POLLUTION.

Mesh generation is typically the most laborious part of finite element analysis in three dimensions. Various sophisticated mesh generators are available. These mesh generators are connected with computer-aided design (CAD) software because usually the geometric representation is first constructed in a CAD environment. The quality of the mesh from the point of view of computational accuracy is not considered in general. In addition some details, such as as fillets (the rounding of corners and edges), are usually omitted.

A mesh, typical of meshes currently used in industrial applications for simple three-dimensional domains, is shown in Fig. 8. Note that some edges were filleted while others were not. Typically, finite element meshes are constructed for representation of the main topological features, rather than for ensuring the accuracy of the finite element solution. In this case the solution is singular along the edges which leads to a potentially large error in the finite element solution. The size of the error depends on the loading. In this regard we note that the errors in design that led to the sinking of the Sleipner A-1 platform were caused by improper meshing which described the geometry well but was grossly inadequate from the point of view of accuracy.

It is generally thought that one does not need to be concerned about the mesh in those areas which are far from the area of interest when a posteriori local extraction of the stresses with error estimation (recoveries principle) is used. This can lead to very unreliable results due to pollution caused by coarse meshes in the neighborhoods of

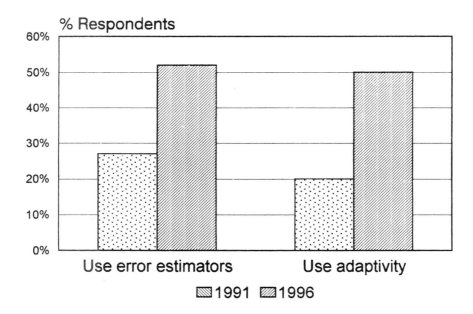

Figure 6   Incrèased use of error estimation and adaptivity in industry within the past five years.

the edges and corners. Capabilities for a posteriori estimation of pollution errors are needed. Unfortunately, there is a general lack of awareness of pollution errors in the engineering community.

Problems of a similar character but different nature occur in the modeling of structural connections. Typically, fasteners (rivets) are modeled by point constraints or point attachments. In such cases the results strongly depend on the mesh not only in the neighborhood of the fasteners but also far from the fasteners. For details we refer to [Sza90], [SB91], [BS92a], [BS92b].

1.2.5   THE P- AND HP-VERSIONS OF FEM.

There are three versions of FEM: In the classical h-version the errors of discretization are controlled by making the mesh sufficiently fine while keeping the degree p of the elements fixed, usually at $p = 1$ or $p = 2$. In the p-version the errors of discretization are reduced by keeping the mesh fixed, and increasing the polynomial degrees of elements (p). The hp-version combines both approaches. The p- and hp-versions are of substantial interest to software houses today because they provide effective means for controlling errors of discretization for a large and important class of problems.

The leading vendors of conventional finite element analysis software have implemented the p-version so that today the largest commercial finite element codes

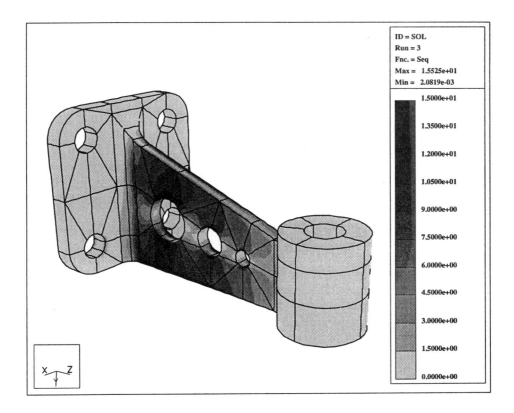

Figure 7 Model problem. Contours of the equivalent (von Mises) stress.

MSC/NASTRAN[4], ANSYS[5], I-DEAS[6] offer at least some p-version capabilities. There are several other codes which are based entirely on the p-version. These include Pro/MECHANICA; PolyFEM[7]; STRESS CHECK; PHLEX[8], and STRIPE[9]. The p-version has some important advantages:

1. Although the mesh design has to follow some rules, the meshing is much simpler than in the case of the h-version and there is much less sensitivity to the mesh than in the h-version.
2. The method is robust with respect to locking.
3. In practical cases a much faster rate of convergence can be achieved.
4. It is possible to obtain in practice a guarantee of the accuracy of *any* value of

[4] MSC/NASTRAN is a trademark of the MacNeal-Schwendler Corporation.
[5] ANSYS is a trademark of Ansys, Inc.
[6] I-DEAS is a trademark of Structural Dynamics Research Corporation.
[7] PolyFEM is a trademark of IBM Corporation.
[8] PHLEX is a trademark of Computational Mechanics Company.
[9] Flygtekniska Försöksanstalten, Bromma, Sweden.

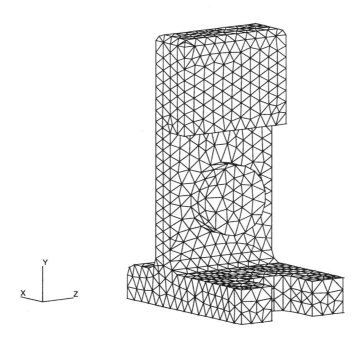

Figure 8 An example of meshes produced by automatic mesh generators.

   interest by comparing the values for increasing p.
5. The method is useful and effective for both design and certification computations.

   Mesh generators appropriate to the *p*-version have not yet been developed. An adaptive *hp*-version meshing is available in the program PHLEX.

1.2.6 THE PROBLEM OF MODELING

Proper selection and definition of the mathematical model is the most important prerequisite of reliability in numerical simulation. This involves (a) the formulation of the mathematical problem to be solved and (b) the selection of specific numerical data.

   By the term "formulation" we mean various simplifications, such as selection of plate or shell theories; the formulation of boundary conditions; modeling of fasteners; neglecting nonlinear effects; selecting constitutive laws; describing composite materials, etc. Identification of the goals of computation, such as, for example, determination of maximum equivalent stress; displacement; stress intensity factors; limit loads etc. is part of the formulation.

In many cases the data of interest are very sensitive to modeling decisions. An interesting example is the Girkmann-Timoshenko problem. An axially symmetric spherical shell with a stiffening ring shown in Fig. 9, was analysed by eminent engineers using the classical methods of structural analysis ([Gir56], [TWK59]) before computer-based methods were available to solve such problems. Fig. 9 is not to scale, the shell is quite thin, the radius to thickness ratio is 389.4. The material is isotropic, elastic, the modulus of elasticity is 3×10^6 lbf/in^2 and Poisson's ratio is zero. The shell is loaded by a uniformly distributed load of 0.28472 lbf/in^2 acting in the middle surface of the shell parallel to the axis of symmetry. This load is equilibrated by uniformly distributed normal traction acting on the base of the ring. The goals of the computation are to determine the radial force (F_r) and the bending moment (M) acting between the shell and the stiffening ring. These data are very sensitive to the stiffness of the ring and the boundary conditions used.

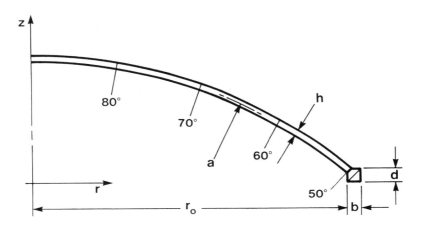

Figure 9 The Girkmann-Timoshenko problem: Generating section of a spherical shell with stiffening ring (not to scale). The dimensions are: $a = 919.20$ in, $h = 2.36$ in, $b = 23.64$ in, $d = 19.68$ in.

Analyses were performed using a fully axisymmetric model, i.e., the shell and stiffening ring were modeled as an axisymmetric elastic body, using three types of boundary conditions: (a) equilibrium loading (the stiffening ring was not constrained against rotation or radial displacement); (b) simple support (the stiffening ring was constrained against rotation but not against radial displacement); and (c) rigid support (the stiffening ring was constrained against both radial displacement and rotation). Using various meshes (one of which in shown in Fig. 9) and p-extension, the possibility of a significant discretization error was eliminated [Sza88] [SB91]. The results of the analysis are shown in Table 2.

It is seen that very substantial differences exist between the solutions obtained with

Table 2   The Girkmann-Timoshenko problem.

Description of the model	Force F_r (lbf/in)	Moment M (lbf in/in)
Fully axisymmetric model		
(a) Equilibrium loading	-105.2	-7.84
(b) Simple support	-93.0	-231.1
(c) Rigid support	-109.9	-43.8
The Girkmann-Timoshenko model	-8.95	-24.84

the Girkmann-Timoshenko model and the fully axisymmetric model. The reason for this is that the reactions F_r, M are very sensitive to the stiffness of the ring and the boundary conditions imposed on the ring. In the classical solution ([Gir56], [TWK59]) equilibrium loading was assumed. By the methods available at that time, the stiffness of the ring could not be estimated with sufficient accuracy.

The Girkmann-Timoshenko problem is an example that demonstrates the great importance of proper definition of the model and taking the sensitivity of the computed data to modeling decisions into consideration. Other examples are available in [SB91], [BS92a], [BS92b], [SO96].

Another source of uncertainties is the constitutive law, especially for nonlinear material behavior. There is a large literature about constitutive laws but there is insufficient information about the *values* of the parameters in the laws. In addition, there is substantial sensitivity to various factors. For example, it has been reported that the yield stress for the 5454-H32 aluminum alloy differs by 22% for the sheet (0.2 in. normal thickness) and plate (0.4 in. normal thickness) [WBSC87].

An extensive analysis of the properties of aluminum alloy 5454-H32 in the one dimensional setting and cyclic strain was performed [BJLS93]. Denoting $\sigma^{MEAS}(t)$ (resp. $\sigma^{PRED}(t)$) the measured stress (resp. the predicted stress) from a particular law, the accuracy measure Θ (in percent) was defined as

$$\Theta = 100 \frac{\max_{i=1 \to 20000} |\sigma^{MEAS}(t_i) - \sigma^{PRED}(t_i)|}{\max_{i=1 \to 20000} \frac{1}{2}|\sigma^{MEAS}(t_i) + \sigma^{PRED}(t_i)|}. \tag{1.1}$$

It was found that for the kinematic law and data taken from the material handbook leads to the mean value $\bar{\Theta} = 33\%$ and the standard deviation is 15% of the mean. Replacing σ^{MEAS} and σ^{PRED} in 1.1 by the two measured stresses for the same strain, a measure of the reproducibility is obtained. The mean was about 5% and the standard deviation was 20% of the mean. See [BJLS93] for additional data and assessment of other laws.

Obviously, the reliability of the computed data depends strongly on the reliability of the constitutive law, which is very low in the case of inelastic deformation.

1.2.7 SUMMARY OF TRENDS

1. A major emphasis in the last few years has been on the development of FEM for design purposes. No consensus has emerged concerning an optimal approach. Various alternatives are under consideration.
2. There is a decline in the development and use of in-house software. They are being replaced by commercial programs. Commercial programs have many proprietary details which complicate the assessment of their quality and the results are prone to misinterpretation.
3. There is a widespread and growing demand for a posteriori error estimation and adaptive approaches. The presently available error measures are often inadequate with respect to the goals of computation and are prone to severe misinterpretation. The programs do not indicate points where the stresses are infinite. Very likely, various expert systems will appear in the future.
4. The simplified formulation used in the design phase needs some approaches for the assessment of the reliability of the conclusions from these models. Also, there is a need for estimating the effects of uncertainties on the computed data.

1.3 RESEARCH TOPICS.

Some current research topics are discussed in the following which are likely to impact professional practice in the not-too-distant future.

1.3.1 THE P- AND HP-VERSIONS OF FEM

During the last few years significant progress in the theory of the p- and hp-versions of the finite element method has been made in the fields of elliptic equations (solids) and hyperbolic equations (fluids). The progress in industrial applications was discussed in Section 1.2.5. In this section some recent results and research directions for elliptic equations are discussed.

The essential feature of a very large class of problems in solid mechanics is that the input data (i.e., the boundary, the coefficients, the boundary conditions, etc.) are piecewise analytic. For this type of problems it is possible describe the regularity of the solution in the framework of countably normed spaces. See [BG88b], [GB93], [Guo94], [GB96]. Exact and accurate characterization of the regularity of the solution, which is as narrow as possible, but encompasses the majority of engineering problems, is essential for the analysis of the performance of the p- and hp-versions. Using these results, it was possible to prove that the hp-version converges exponentially for problems characterized by piecewise analytic data. More precisely, denoting the energy norm by $\| \cdot \|_E$, the exact (resp. finite element) solution by u (resp. u_{FE}), the number of degrees of freedom by N, then for the hp-version we have the estimate:

$$\|u - u_{FE}\|_E < Ce^{-\gamma N^{1/\alpha}} \tag{1.2}$$

where the coefficient α depends on the dimension of the problem. In two dimensions $\alpha = 3$ and in three dimensions $\alpha = 5$.

The coefficient $\gamma > 0$ depends on the strength of the singularity of the solution in the neighborhood of the boundary (edges, corners) and interfaces. There exist cases where γ is very small. The meshes must be refined in the neighborhood of the corners of the domain in two dimensions and vertices and edges in three dimensions. The elements in the neighborhood of the edges have very large aspect ratios ("needle elements") reflecting the fact that the solutions are smooth along the edges but singular in the direction perpendicular to the edge. For additional information see [BG86a], [BG86b], [BG88a], [Guo94], [BG92], [BG96].

In [GB86] a detailed analysis of a one-dimensional model problem was performed. It was shown that $\alpha = 2$ and it cannot be improved. More precisely, no sequence of meshes and elements of arbitrary degrees can lead to smaller α. We conjecture that $\alpha = 3$ (resp. $\alpha = 5$) in two (resp. three) dimensions are the optimal values, but this has not been proven. Further, in [GB86] we showed that the optimal value of γ depends on the strongest singularity of the solution. The first results about the hp-version were obtained in [BD81].

In [BSK81] the p-version was analysed and it was shown that the rate of convergence of the p-version is at least twice that of the h-version with uniform or quasi-uniform meshes for problems characterized by piecewise analytic data.

Due to difficulties in implementation in its asymptotically correct form, the hp-version is very often the p-version on judiciously constructed finite element meshes in practice. The principles of mesh construction are simple. When the correct mesh is used then, in the range of accuracy normally expected in engineering computations, the performance of the p-version is similar to that of the hp-version.

In [BS87] the results of [BSK81] were generalized and in [BS82] some aspects of p-version with emphasis on applications were discussed.

As noted above, the coefficient γ in equation (1.2) can be very small in some cases. In such cases neither the hp-version nor p- or h-versions can be effective. Nevertheless it was shown in [OB95] that the problem can be circumvented by using non-polynomial basis functions.

Although the basic properties of the hp-version are well understood with respect to the error in the energy norm, some important problems remain. One such problem is error estimation in L_∞ norm. For example, it is known that the derivatives (stresses) computed directly from a finite element solution obtained with the p-version have very low accuracy in those elements where the exact solution is singular, such as at re-entrant corners, vertices or edges. A detailed theoretical investigation of this subject is needed.

Typically, a-posteriori estimation of error is based on extrapolation which involves comparison of the solutions with different polynomial degrees. This approach usually gives good results not only for the energy norm but also for any other value of interest. Nevertheless, theoretical analysis is needed. See, for example, [BG86a], [BG86b], [Sza86], [YS94]. Also, various approaches described in the next section are applicable for the estimates in the energy norm.

In p-adaptive procedures polynomial degrees are adaptively assigned to finite elements. There can be large differences in polynomial degrees when elements of high degree are used. Experience has shown that the a posteriori error estimation of stresses based on comparing solutions with various degrees of freedom are less reliable when non-uniform p-distribution is used than in the case of uniform p-distributions.

In hp-adaptive procedures either the polynomial degree is increased or the mesh is refined. The only finite element code at present which has such a capability is PHLEX.

There are important aspects of implementation of the p- and hp-versions. One is to compute the stiffness matrix efficiently. The method used in the h-version for low p is very ineffective for the p-version but can be improved significantly [AFBP95].

The problem of solving large systems of linear equations generated by hp-extensions is complicated by the fact that the condition number is typically large. For various aspects of this problem we refer to [Man92], [BEM92], [OPF94], [GC96], [GM96].

The p-version has excellent robustness with respect to locking in shells and plates, see [HLP96]. For surveys of the hp-method we refer to [BG92], [BS94].

1.3.2   A-POSTERIORI ERROR ESTIMATION AND ADAPTIVE PROCEDURES FOR THE H-VERSION

In the last few years large emphasis has been placed on a-posteriori error estimation and adaptive procedures in both academic research and industrial applications. Error estimates in the energy norm measure are most typical although estimation of other data of interest have been introduced. A large number of papers have been written on this subject. Essentially there are two types of estimators in the h-version:

1. residual estimations;
2. estimations based on the smoothing (recovery).

Various estimators have been proposed and analysed either theoretically or computationally. A major problem in the selection of an estimator for practical use is the assessment of its quality and robustness.

Error indicators are defined locally and hence assessment of an indicator cannot depend on the pollution i.e., on the effects of the finite element error outside of the elements, because this effect is a global one while the estimation itself is local and hence it cannot "see" the pollution.

In [BSU94a], [BSU⁺94b], [BSU96] a theory was developed which allows the assessment of a particular error indicator. Given a translation-invariant mesh (e.g., Fig. 10), the asymptotic lower and upper bound C_L and C_U of the (elemental) effectivity index for particular estimators can be computed. The bound is over a set of solutions, (smooth ones or ones that satisfy, in addition, a differential equation or a set of equations e.g., characterized by the coefficients) and over a class of meshes. It was shown that in the case of smooth solutions it is possible to consider only the solutions which are locally polynomials of degree $p+1$, when p is the degree of the element used.

The robustness index R is introduced

$$R := \min\left(|1 - C_U| + |1 - C_L|, \ |1 - \frac{1}{C_U}| + |1 - \frac{1}{C_L}|\right). \tag{1.3}$$

Obviously, $R = 0$ is the optimal value of the robustness index. In [BSU94a] the indicators inside the domain were considered and in [BSU96] the elements at the boundary were investigated. In [BSU⁺94b] it was shown that the robustness index depends on the pattern but only weakly depends on the mesh in the neighborhood of the elements under consideration. The robustness index characterizes the quality of the

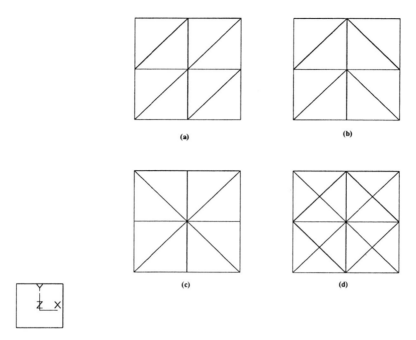

Figure 10   Types of mesh patterns considered: a) regular pattern, b) chevron pattern, c) union jack pattern, d) criss-cross pattern.

estimators well. Various estimators have large robustness index, especially for elements with large aspect ratios, anisotropic materials, etc. and therefore should not be used in practice. Based on the results of these investigations, it is possible to conclude the following:

1. Comparing the existing estimators for the error measured in energy norm based on the residual and smoothening (recovery) approach, for large classes of solutions, meshes and materials, the Zienkiewicz-Zhu (ZZ) estimator proposed in [ZZ92a], [ZZ92b] is the most robust one, although sufficiently rigorous mathematical theory that would explain this conclusion is not available at present.
2. Among the residual estimators the equilibrated one is in general the most robust one. The unequilibrated ones are not robust, in general. The theoretical understanding of this type of estimator is well advanced. The estimate (in the energy norm) could be a guaranteed upper bound on the entire domain although the robustness index is not necessarily small.

The above theory is asymptotic for the h-version (with respect to $h \to 0$), assuming that the exact solution is smooth. For elements of high degree the asymptotic theory can be misleading because the prescribed accuracy is achieved in practice in the preasymptotic range. Also, conclusions from the asymptotic theory cannot be used directly when the solution is not smooth, as in the neighborhoods of corners. The theory has to be adjusted and we expect that this will be done in the future.

Although a-posteriori error estimation is applied mostly in the energy norm, it is possible to derive the estimators for the other norms also because in the absence of pollution the recovered solution is closer to the exact solution than the finite element solution. We emphasize that this pollution control is an essential attribute of the estimator. The problem of pollution is discussed in the Section 1.3.4. The indicators (estimators) derived for linear equations can be used also for nonlinear ones because essentially we estimate the error for the linearized problem. For analyses of nonlinear equations we refer to [Ver94], [Ver95].

Error estimators used in conjunction with adaptive procedures should be related to the goals of analysis, not to the accuracy in the energy norm only. There are various adaptive procedures, but essentially all are aiming to create meshes in which all elements have approximately the same error, i.e., the mesh is equilibrated. In [BV84] the optimality of various adaptive procedures was analysed on the basis of a simple one-dimensional model problem. It was shown that the equilibration (equidistribution) principle does not lead to asymptotically optimal meshes when the solution is not "reasonable". The a-posteriori error estimators were first addressed in conjunction with elliptic equations (see [BR78], [LL83], [BW85], [BM87], [Bab86], [OD87], [EJ88], [Ewi90], [AC91], [DMR92], [AO92], [BDR92], [JPH92], [AO93], [Rod94], [Ain94]. For survey papers on the state of the art we refer to [OD87] and [Ver95].

The a-posteriori error estimators and adaptive techniques were developed also for other types of equations but we will not discuss them here. We refer, for example, to [OWA94].

Although significant progress has been made in the last 10 years, especially with respect to the error measured in the energy norm, many important questions still remain, because the aim is to have reliable and accurate estimates of *any data of interest* computed from a finite element solution. The estimators should be of high quality in the preasymptotic range and should not contain any constants which are not computable.

1.3.3 SUPERCONVERGENCE AND RECOVERY TECHNIQUES

Superconvergence means that some values can be locally extracted so that higher accuracy is obtained from a finite element solution than with direct computation. This possibility exists because the finite element solution oscillates about the exact solution. Typically, one is interested in the points where the accuracy in (for example) the first derivative of the finite element solution is better by an order than in a general point.

For a survey of the state of the art in the theory we refer to [Wah95]. In [BSUG96] the theory of the computer-based proof of superconvergence was developed. It was shown that the superconvergence point in an element (if such a point exists) can be determined by a finite number of operations. As in the case of a posteriori error estimation, it is essential that the pollution error be negligible and the exact solution must be smooth.

We considered the pattern of the meshes shown in Figure 10 and analysed the superconvergence points for the derivatives of the Laplace and elasticity equations.

We analysed the classical superconvergence when the set of admissible solutions is the entire set of smooth solutions. We also analysed superconvergence for the set of the solutions which satisfy the homogenous differential equation under consideration. This is obviously a very practical case because in most applications the "volume forces" are zero. Such sets are said to be "harmonic".

Let us consider for example the Poisson problem $-\Delta u = f$ for elements of degree 3 and let us be interested in the superconvergence point for the value $\partial u / \partial x$ and the regular pattern shown in Fig. 10. For general f there is only one superconvergence point while for the set of "harmonic" functions we have exactly six superconvergence points, see [BSUG96].

In the general case, especially for the elasticity equations, no superconvergence point has to exist even for "harmonic" solutions. Consider, for example, the problem of the superconvergence for the stress component σ_{12} for Poisson's ratio $\nu = 0.3$ and the chevron pattern shown in Fig. 10. In this case no superconvergence point exists. We define η % superconvergence set of points in the element, see [BSUG96], [BSGU96], [BSU95]. The points $x \in \tau$ (τ is the element) belong to the η % superconvergence set of σ_{12} if

$$100 \frac{|\sigma_{12}(x)|}{\max |\sigma_{12}(x)|_{x \in \tau}} \leq \eta(x)\% \tag{1.4}$$

for all admissible solutions whether general or "harmonic". Further, we define

$$\eta_{\min}\% = \min \eta(x)\%, \qquad x \in \tau. \tag{1.5}$$

Obviously a superconvergence point exists only if $\eta_{\min} = 0$. In Fig. 11 we show the $\eta\%$ superconvergence set for "harmonic" solutions (for σ_{12}, $\nu = 0.3$) and element degree $p = 2$. In this case $\eta_{\min}\% = 11.79$. Because $\eta_{min}\% > 0$, there is no superconvergence point.

We have to distinguish between superconvergence in the interior elements or the elements at the boundary or on the mesh with refinement pattern, see [BSGU94]. It is also possible to recover the derivatives of the solutions or stresses by various approaches suggested in the literature. Some of those were analysed in [BSGU95].

This discussion was restricted to pointwise superconvergence. Note that L_2-type superconvergence has been analysed also, see e.g., [Zlá77] [LZ79]. It is also possible to obtain the data of interest by special techniques, called *extraction methods*. These techniques yield more accurate results than direct computations do and do not require special meshes. Nevertheless, this extraction technique is more complicated than the simple pointwise superconvergence or the recovery technique. For more see [BM84], [Sza90], [SY96a], [YS95], [SY96b].

For a survey of the results up to 1987 we refer to [KN87], [ZL89]. For some additional results we refer to [GW89], [SSW96].

1.3.4   THE POLLUTION PROBLEM

A posteriori error estimators and superconvergence points in the elements are defined locally utilizing only the element under consideration and possibly some of its neighbors. The quality of the estimator or $\eta\%$ superconvergence was defined

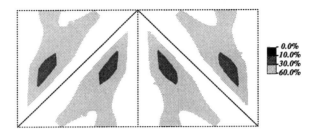

Figure 11   The $\eta\%$ superconvergence regions for the elasticity problem ($\nu = 0.3$) for σ_{12}

elementwise. This definition has meaning only if the pollution in the given element is negligible. Pollution describes the effects of errors originating outside of the element under consideration on the error within the element. More precisely, denoting the error in the element τ by V, we can write

$$V = V_1 + V_2 \tag{1.6}$$

where V_1 is the error governed by the finite element solution in the element under consideration and its small neighborhood and V_2 is the effect of the errors outside of the element.

For example, consider the problem of solving the Laplace or the elasticity equations on $\Omega \in R^2$, where Ω has a boundary with reentrant corners. Assuming that the mesh is quasiuniform, the pollution error is dominant [BSMU94], [BSUG95]. Even in the presence of reentrant corners elements of high polynomial degree will be subject to pollution because Green's function is not smooth at the reentrant corners of Ω.

The pollution is said to be negligible if $\|V_2\|/\|V_1\| << 1$. If the pollution error is dominant then the a-posteriori error estimation, recovered values, superconvergence, etc., are not reliable because $\|V_1\| << \|V_2\|$. For example, the (elemental) effectivity index of the error indicator (estimator) could be of order 0.1 or even 0.01 in practical cases.

It is virtually impossible to estimate the size of the pollution error a priori. Therefore an a posteriori estimator of pollution error has to be used. The estimator is necessarily global, nevertheless it has to be computable efficiently. Such an estimator was proposed and analysed in [BSUG95], [BSGU96], [BSG96]. Such an estimate can then be used in an adaptive procedure which leads to prescribed accuracy in the data of interest (for example, the stress in the single element) even if the error in the energy norm is large.

Consider, for example, an L-shaped domain with a crack. In Fig. 12 we show a

sequence of meshes adaptively constructed to increase the accuracy only in the single
(shaded) element in a given patch of elements. Observe that the refinement changes
substantially in the neighborhoods of the corner and tip of the crack as the accuracy
is increased. For details we refer to [BSG96].

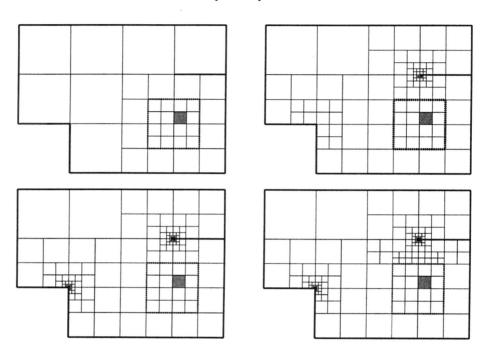

Figure 12   Meshes constructed adaptively with the objective to increase the accuracy in
the single (shaded) element.

Understanding pollution errors, and having a capability for assessing their
magnitude, are of major practical importance because in practice refinements far from
the area of interest are generally neglected. The so-called "global-local approach" is of
this type when the refined mesh is used only in the area of interest. See also [Ode95].

Although the basic principles of pollution estimates and control are available, further
theoretical and practical investigations are needed.

1.3.5   THE PARTITION OF UNITY METHOD

A frequently asked question is: "why are polynomial basis functions used in the h-, p-
and hp-versions of the finite element method?". There are a few reasons for this:

a) polynomials approximate *smooth* functions well;

b) monomials are homogenous with respect to scaling;

c) polynomials have certain advantages from the point of view of implementation.

Nevertheless, if the exact solution is highly irregular then the polynomials *do not* approximate it well. For example, consider the problem

$$-\frac{d}{dx}\left(a(x)\frac{du}{dx}\right) = f(x) \qquad 0 < x < 1, \qquad u(0) = u(1) = 0 \qquad (1.7)$$

with $0 < \alpha_1 \leq a(x) \leq \alpha_2$, but only measurable. Then in [BO] we have shown that the finite element method using polynomial functions converges arbitrarily slowly. More precisely, given $\psi(N) \to 0$, $N = 1, 2, \ldots$, it is possible to find $a(x)$, $0 < \alpha_1 \leq a(x) \leq \alpha_2$ corresponding to ψ such that

$$\lim_{N\to\infty} \sup \|u - u_{FE}\|_{H^\ell}/\psi(N) \geq C > 0, \qquad \ell = 0, 1. \qquad (1.8)$$

Here u_{FE} is the finite element solution and N is number of degrees of freedom. Furthermore, adaptive selection of the mesh affects only the constant C in (1.8).

In higher dimensions the finite element method based on polynomial basis functions can also perform poorly. The reason is that coefficients can be selected so that the solution is very irregular, more precisely, the solution is not in any $H^\alpha(\Omega^\star)$, $\alpha > 1$ in any $\Omega^\star \subset \Omega$. Analogous difficulties occur when solving the Helmholtz equation with high wave number because the solution is highly oscillatory.

Consider solving the Laplace equation $\Delta w = 0$. In this case the harmonic polynomials have the same approximation properties as the set of all polynomials. Use of the set of all polynomials is necessary in the standard FEM because otherwise the requirement of continuity could not be satisfied. Therefore two separate questions have to be addressed in selecting the basis functions:

a) What set of basis functions approximate the solution well?

b) How to enforce continuity?

In [BCO94], [Mel95], [BM96], it was shown that by the method of partition of unity conforming shape functions can be constructed from local nonconforming ones. In addition, if the solution can be approximated well by the local (nonconforming) shape functions then the conforming shape functions constructed by the partition of unity method approximate the solution with the same accuracy. Therefore the method gives freedom in selecting the local shape functions. In [BCO94] it was shown that in two dimensions the rate of convergence $0(N^{-1/2})$ can be achieved even if the coefficients are arbitrarily rough, provided that the proper shape functions are used.

In [Mel95], [BM96] it was shown that when solving the Helmholtz equation

$$\Delta u + k^2 u = 0 \qquad (1.9)$$

with a high wave number, it is very advantageous to use as the (local) basis functions the generalized harmonic functions of Bergman and Vekua, or a set of plane wave solutions of (1.9). Note that these shape functions satisfy the differential equation (1.9). If the right hand side of eq. 1.9 is not zero, additional shape functions are necessary. In [Mel95], [BM96] the a priori estimates were given and numerical examples showing effectiveness of this method were presented.

Because of its flexibility, this method has a high potential. The partition of unity method has some commonality with *the mesh-free finite element method* [BKO+96] and the "cloud method" described in [OD95].

1.3.6   THE PROBLEM OF THIN DOMAINS

Models of homogeneous or laminated plates and shells are understood as approximate solutions of three-dimensional problems, which have to be developed adaptively based on an a-posteriori error estimation. Each model is, of course, solved numerically and hence there is an interplay between the error in the model and the error of the approximate solution of the model.

The adaptive selection of the model must account for singularities and boundary layer effects associated with the model.

This approach is different from both the classical modeling based on physical derivation or asymptotic approaches. See [GR91], [Amb91], [MNP91], [Cia92]. Adaptive selection of models is based on the dimensional reduction method (see, for example, [VB80], [SS88], [Sza89], [Sch89], [SA92], [BSA92], [BL92], [BLS94], [BS96]) which establishes a theoretical basis and practical means for the construction of hierarchic models.

The hierarchic dimensional reduction method for homogeneous plates and shells is closely related to the hp-version of FEM where the model is characterized by the use of the polynomial shape functions through the thickness. The p-version is well suited for the treatment of plates and shells because it is very robust with respect to locking. For more information we refer to [ZD94], [HLP96], [OC96], [CO96].

Hierarchic modeling, as defined above, is a methodology for obtaining approximate solutions for three-dimensional problems. It is assumed that this three-dimensional problem is exactly formulated and the aim is to find its solution accurately. A major difficulty is that large uncertainties may have to be incorporated in the formulation. Consider, for example, zero displacement prescribed on the boundary of an elastic body. This is a crude idealization, especially with respect to any data of interest in the neighborhood of the boundaries. Fixed (built-in) boundaries should be modeled more precisely, nevertheless some uncertainty is usually unavoidable. Therefore quantitative assessment of the effects of uncertainties is an important aspect of computation. Theoretically and practically justified approaches are needed. We also refer to the Girkmann-Timoshenko problem discussed in Section 1.2.6 which exemplifies the high degree of sensitivity of plate and shell-like bodies to various boundary conditions.

1.3.7   SUMMARY

We have discussed a few directions of theoretical research on the finite element method, which, in our opinion, are important for the practical uses of FEM and therefore new results are desirable. The common focus of the research topics discussed herein is reliability of the computed data. An essential prerequisite of reliability in engineering computations is that the mathematical problem must be properly formulated.

1.4   CONCLUSIONS

The authors' objective in writing this paper was to identify some trends in engineering computation and their relation to research in FEM. It is obvious that successful use of FEM depends on the reliability of the formulation; the reliability of the parameters used; the reliability of the numerical solution, and proper interpretation of the results. Failure to ensure reliability in either one of these areas can have disastrous consequences. The Sleipner disaster was mentioned as an example. In many cases the consequences are less obvious: Lack of confidence in the reliability of computed information causes engineering managers to undertake expensive and time-consuming tests, or bypass finite element analysis entirely.

Only a few aspects of the finite element method could be discussed within the framework of this paper. Many relevant references were cited but certainly not all. Additional details can be found in these references.

ACKNOWLEDGEMENTS

The writers wish to thank Dr. R. L. Actis for providing Figures 2, 3, 4, 5, 7; Dr. M. Halpern for providing Fig. 6 and other information concerning his survey, and Dr. J. Z. Zhu for providing Fig. 8. The authors also thank their sponsoring agencies for their support: The work of Ivo Babuška is partially supported by the Office of Naval Research under Contract N 000 14-90-J-1030 and the National Science Foundation under Grant DMS-95-01841. The work of Barna Szabó is partially supported by the Air Force Office of Scientific Research under Grant No. F49620-95-1-0196.

References

[AC91]   Ainsworth M. and Craig A. W. (1991) A-posteriori error estimators in the finite element method. *Num. Math.* 60: 429–463.

[AFBP95]   Andersson B., Falk V., Babuška I., and Petersdorff T. (1995) Reliable stress and fracture mechanics analyses of complex components using the h-p version of FEM. *Int. J. Num. Meth. Engrg.* 38: 2135–2163.

[Ain94]   Ainsworth M. (1994) The performance of Bank-Weiser's error estimator for quadrilaterial finite elements. *Numer. Meth. for PDE's* 10: 609–623.

[Amb91]   Ambartsumjan S. A. (1991) *The Theory of Anistrotropic Plates.* Hemisphere, New York.

[AO92]   Ainsworth M. and Oden J. T. (1992) A procedure for a-posteriori estimation for h-p finite element methods. *Comp. Meth. Appl. Mech. Engrg.* 101: 73–96.

[AO93]   Ainsworth M. and Oden J. T. (1993) A unified approach to a posteriori error estimation based on element residual methods. *Num. Math.* 65: 23–50.

[Bab86]   Babuška I. (1986) Accuracy estimates and adaptive refinements in finite element computations. In Babuška I., Zienkiewicz O. C., Gago J., and de A. Oliveira E. R. (eds) *Accuracy Estimates and Adaptive Refinement in Finite Element Computations*, pages 3–23. John Wiley & Sons, Inc., New York.

[BCO94]   Babuška I., Caloz G., and Osborn J. (1994) Special finite element methods for a class of second order elliptic problems with rough coefficients. *SIAM J. Numer. Anal.* 31: 945–981.

[BD81]   Babuška I. and Dorr M. R. (1981) Error estimates for the combined h- and p-versions of the finite element method. *Num. Math.* 37: 257–277.

[BDR92]   Babuška I., Duran R., and Rodriguez R. (1992) Analysis of the efficiency of an a posteriori error estimator for linear triangular elements. *SIAM J. Numer. Anal.* 29: 947–964.

[BEM92]   Babuška I., Elman H. C., and Markley K. (1992) Parallel implementation of the h-p version of the finite element method for a shared memory architecture. *SIAM J. Sci. Stat. Comp.* 13: 1433–1459.

[BG86a]   Babuška I. and Guo B. (1986) The h-p version of the finite element method. Part I the basic approximation results. *Comp. Mech.* 1: 21–41.

[BG86b]   Babuška I. and Guo B. (1986) The h-p version of the finite element method. Part II general results and applications. *Comp. Mech* 1: 203–220.

[BG88a]   Babuška I. and Guo B. Q. (1988) The h-p version of the finite element method for domains with curved boundaries. *SIAM J. Numer. Anal.* 25: 837–861.

[BG88b]   Babuška I. and Guo B. Q. (1988) Regularily of the solutions of elliptic problems with piecewise analytic data. Part I boundary value problems for linear elliptic equations of second order. *SIAM J. Math. Anal.* 19: 172–203.

[BG92]   Babuška I. and Guo B. Q. (1992) The h, p and h-p version of the finite element method; basic theory and applications. *Advances in Engineering Software* 15: 159–174.

[BG96]   Babuška I. and Guo B. (1996) Approximation properties of the h-p version of the finite element method. *To appear in Comp. Meth. Appl. Mech. Engrg.* .

[BJLS93] Babuška I., Jerina K., Li Y., and Smith P. (1993) Quantitative assessment of the accuracy of constitutive laws for plasticity with an emphasis on cyclic deformation. In Bertram L. A., Brown S. B., and Freed A. D. (eds) *Material Parameter Estimation for Modern Constitutive Equations*, volume MD-Vol. 43, AMD-Vol. 168, pages 113–171. American Society of Mechanical Engineers.

[BKO+96] Belytschko T., Krongauz Y., Organ D., Fleming M., and Kryse P. (1996) Meshless methods: An overview and recent developments. *To appear in Comp. Meth. Appl. Mech. Eng.* .

[BL92]   Babuška I. and Li L. (1992) The problem of plate modeling. Theoretical and computational results. *Comp. Meth. Appl. Mech. Engrg.* 100: 249–273.

[BLS94]   Babuška I., Lee I., and Schwab C. (1994) On the a posteriori estimation of the modelling error for the heat conduction in plates and its use for adaptive hierarchic modelling. *Appl. Numer. Math.* 14: 5–21.

[BM84]   Babuška I. and Miller A. (1984) The post-processing approach in the finite element method. Part I calculation of displacements, stresses and other higher derivatives of displacements. Part II the calculation of stress intensity factors. *Int. J. Num. Meth. Engrg.* 20: 1085–1110, 1111–1129.

[BM87]   Babuška I. and Miller A. (1987) A feedback finite element method with a-posteriori error estimation. Part I: The finite element method and some basic properties of the a-posteriori error estimator. *Comp. Meth. Appl. Mech. Engrg.* 61: 1–40.

[BM96]   Babuška I. and Melenk J. M. (1996) The partition of unity method. *To appear in Int. J. Num. Meth. Engrg.* .

[BO]   Babuška I. and Osborn J.Finite element method for the one dimensional problem with rough coefficients. *To appear* .

[BR78]   Babuška I. and Rheinboldt W. C. (1978) Error estimates for adaptive finite element computations. *SIAM J. Numer. Anal.* 15: 736–754.

[BS82]   Babuška I. and Szabó B. A. (1982) On the rates of convergence of the finite element method. *Int. J. Numer. Meth. Engrg.* 18: 323–341.

[BS87]   Babuška I. and Suri M. (1987) The optimal convergence rate of the p-version of the finite element method. *SIAM J. Numer. Anal.* 24: 750–776.

[BS92a]   Bortman J. and Szabó B. (1992) Analysis of fastened structural connections. *AIAA Journal* 30: 2758–2764.

[BS92b]   Bortman J. and Szabó B. (1992) Nonlinear models for fastened structural connections. *Computers and Structures* 43: 909–923.

[BS94]   Babuška I. and Suri M. (1994) The p and h-p versions of the finite element method, basic principles and properties. *SIAM Review* 36: 578–632.

[BS96]   Babuška I. and Schwab C. (1996) A-posteriori error estimation for hierarchic models of elliptic boundary value problems on thin domains. *SIAM J. Numer. Anal.* 33: 221–246.

[BSA92]   Babuška I., Szabó B. A., and Actis R. L. (1992) Hierarchic models for laminated composites. *Int. J. Numer. Meth. Engrg.* 33: 506–533.

[BSG96]   Babuška I., Strouboulis T., and Gangaraj S. K. (1996) Practical aspects of a posteriori estimation and adaptive control of the pollution error for reliable finite element analysis. Technical Report TICAM 96-0, The University of Texas at Austin, Austin, TX 78712.

[BSGU94] Babuška I., Strouboulis T., Gangaraj S. K., and Upadhyay C. S. (1994) $\eta\%$ - superconvergence of finite element approximation in the interior of locally refined meshes of quadrilaterals. *J. Appl. Anal.* 16: 3–49.

[BSGU95] Babuška I., Strouboulis T., Gangaraj S. K., and Upadhyay C. S. (1995) Validation of recepies for the recovery of stresses and derivatives by a computer based approach. *Math. Comp. Model* 20 45: 45–89.

[BSGU96] Babuška I., Strouboulis T., Gangaraj S. K., and Upadhyay C. S. (1996) Pollution error in the h-version of the finite element method and local quality of recovered derivatives. *To appear in: Comp. Meth. Appl. Mech. Engrg.* .

[BSK81] Babuška I., Szabó B. A., and Katz I. N. (1981) The p-version of the finite element method. *SIAM J. Numer. Anal.* 18: 515–545.

[BSMU94] Babuška I., Strouboulis T., Mathur A., and Upadhyay C. S. (1994) Pollution error in the h-version of the finite-element method and local quality of a-posteriori error estimators. *Finite Elem. Anal. Design* 17: 273–305.

[BSU94a] Babuška I., Strouboulis T., and Upadhyay C. S. (1994) A model study of the quality of a-posteriori error estimators for linear elliptic problems. error estimation in the interior of patchwise uniform grids of triangles. *Comp. Mech. Appl. Mech. Engrg.* 114: 307–378.

[BSU$^+$94b] Babuška I., Strouboulis T., Upadhyay C. S., Gangaraj S. K., and Capps K. (1994) Validations of a posteriori error estimators by numerical approach. *Int. J. Num. Meth. Engrg.* 37: 1073–1121.

[BSU95] Babuška I., Strouboulis T., and Upadhyay C. S. (1995) $\eta\%$ superconvergence of finite elements in the interior of general meshes of triangles. *Comp. Meth. Appl. Mech. Engrg.* 122: 273–305.

[BSU96] Babuška I., Strouboulis T., and Upadhyay C. S. (1996) A model sudy of the quality of a-posteriori error estimators for linear elliptic problems Part II, error estimation at the boundary of patchwise uniform grids of triangles. Technical Report TICAM 96-20, The University of Texas at Austin, Austin, TX 78712.

[BSUG95] Babuška I., Strouboulis T., Upadhyay C. S., and Gangaraj S. K. (1995) A-posteriori estimation and adaptive control of the pollution error in the h-version of the finite element method. *Int. J. Numer. Meth. Engrg.* 38: 4207–4235.

[BSUG96] Babuška I., Strouboulis T., Upadhyay C. S., and Gangaraj S. K. (1996) Computer-based proof of the existence of superconvergence points in the finite element method; superconvergence of the derivatives in finite element solutions of Laplace's, Poisson's and the elasticity equations. *Numer. Meth. Part. Diff. Eq.* 12: 347–392.

[BV84]   Babuška I. and Vogelius M. (1984) Feedback and adaptive finite element
            solution of one dimensional boundary value problems. *Num. Math.* 44:
            75–102.

[BW85]   Bank R. E. and Weiser A. (1985) Some a posteriori error estimators for
            elliptic partial differential equations. *Math. Comp.* 44: 283–301.

[Cia92]   Ciarlet P. G. (1992) *Plates and Functions in Elastic Multistructures: An
            Asymptotic Analysis.* Masson, Paris.

[CO96]   Cho J. R. and Oden J. T. (1996) A-priori error estimation of the h-p
            finite element approximations for hierarchical models of plate and shell-
            like structures. Technical Report TICAM 95-15, The University of Texas
            at Austin, Austin, TX 78712.

[Cur96]   Curry T. (March, 1996) From the President. *MSC/WORLD Magazine*
            VI, No. 1: 3.

[DMR92]   Duran R., Muschietti T. A., and Rodriguez R. (1992) On the asymototic
            exactness of error estimators for linear triangular elements. *SIAM J. Num.
            Anal.* 29: 78–88.

[EJ88]   Eriksson K. and Johnson C. (1988) An adaptive finite element method for
            linear elliptic problems. *Math. Comp* 50: 361–383

[Ewi90]   Ewing R. E. (1990) A-posteriori error estimation. *Comp. Meth. Appl.
            Mech. Engrg.* 82: 59–72.

[GB86]   Gui W. and Babuška I. (1986) The h, p and h-p versions of the finite
            element method in 1 dimension, Part II the error analysis of the h and
            h-p versions. *Num. Math.* 49: 613–657.

[GB93]   Guo B. Q. and Babuška I. (1993) On the regularity of elasticity problems
            with piecewise analytic data. *Advances in Appl. Mathematics* 14: 307–347.

[GB96]   Guo B. Q. and Babuška I. (1996) Regularity of the solutions for elliptic
            problems on nonsmooth domains in R^3. Part I, Countably normed spaces
            on polyhedral domains. Part II, Regularity in the neighborhood of edges.
            To appear in Proc. Royal Society, Edinborough .

[GC96]   Guo B. Q. and Cao W. (1996) A preconditioner of the h-p version of the
            finite element method in two dimensions. *To appear in Num. Math.* .

[Gir56]   Girkmann K. (1956) *Flächentragwerke.* Springer-Verlag, Vienna, 4 edition.

[GM96]   Guo X. Z. and Myers K. W. (1996) A parallel solver for the hp-version
            of finite element methods. *To appear in Comput. Methods Appl. Mech.
            Engrg* .

[GR91]   Gileweski W. and Radwanska M. (1991) A survey of finite element models
            for the analyses of moderately thick shells finite elements. *Finite Elem.
            Anal. Design* 9: 1–21.

[Guo94]   Guo B. Q. (1994) The h-p version of the finite element method for solving
            boundary value problems in polyhedral domains. In Costabel M., Dauge
            M., and Nicoise S. (eds) *Boundary Value Problems and Integral Equations*,
            pages 101–120. Marcel Dekker.

[GW89]   Goodsell G. and Whiteman R. R. (1989) Pointwise superconvergence of
            recovered gradients for piecewise linear finite element approximation. *Int.
            J. Numer. Meth. Engrg.* 27: 469–481.

[Hal96]   Halpern M. R. (1996) 1996 Simulation survey. Technical report, D. H.
            Brown Associates, Port Chester NY.

[HBD91]   Halpern M. R., Brown D., and Desai T. (1991) 1991 FEA survey. Technical
            report, D. H. Brown Associates, Port Chester NY.

[HLP96]   Hakula H., Leino Y., and Pitkaranta J. (1996) Scale resolution, locking
            and high order finite element modeling of shells. *To appear in Comp.
            Meth. Appl. Mech. Engrg.* .

[Hol94]   Holand I. (1994) The Sleipner accident. In Bell K. (ed) *From Finite
            Elements to the Troll Platform*, pages 157–168. Department of Structural
            Engineering, The Norvegian Institute of Technology, Trondheim, Norway.

[JPH92]   Johnson C. and P. Hansbo P. (1992) Adaptive finite element methods in
            computational mechanics. *Comp. Mech. Appl. Mech. Engrg.* 101: 143–181.

[JR94]   Jacobsen B. and Rosendahl F. (March, 1994) The Sleipner platform
            accident. *Structural Engineering International* pages 190–193.

[KN87]   Křižek M. and Neittaanmäki P. (1987) On superconvergence techniques.
            Acta. Applic. Math. 9: 175–187.

[Kur95]   Kurowski P. M. (1995) When good engineers deliver bad FEA. *Machine
            Design, November 9* .

[LL83]   Ladeveze P. and Legullion D. (1983) Error estimation procedure in the
            finite element method and application. *SIAM J. Numer. Anal.* 20, 48,5:
            485–509.

[LZ79]   Lesaint P. and Zlámal M. (1979) Superconvergence of the gradient of finite
            element solutions. *RAIRO Anal. Numer.* 13: 139–166.

[Mac86]   Mackerle J. (1986) MAKABASE, an information system on structural
            mechanics software and applications. *Adv. Engrg. Software* 8: 81–87.

[Man92]   Mandel J. (1992) Adaptive estimate solvers in finite elements solving large
            scale problems in mechanics. In Papadrakis M. (ed) *Development and
            Application of Computational Solution Methods*. John Wiley, Chichester.

[Mel95]   Melenk J. M. (1995) *On Generalized Finite Element Methods, Ph. D.
            Thesis*. PhD thesis, University of Maryland.

[MNP91] Mazja V. G., Nazarov S. A., and Plamenevskii B. A. (1991) *Asymptotic Theory of Elliptic Boundary Value Problems in Singularly Perturbed Domains, 2.* Akademik Verlag, Berlin.

[OB95] Oh H.-S. and Babuška I. (1995) The method of auxiliary mapping for the finite element solutions of elasticity problems containing singularities. *J. Comp. Phys.* 121: 193–212.

[OC96] Oden J. T. and Cho T. R. (1996) Adaptive hpq finite element methods of hierarchical models of plate and shell-like structures. Technical Report TICAM 95-16, The University of Texas at Austin, Austin, TX 78712.

[OD87] Oden J. T. and Demkowicz L. (1987) Advances in adaptive improvements. a survey of adaptive finite element methods in computational mechanics, accuracy estimates and adaptive refinements in finite element computations. *ASME* pages 1–43.

[OD95] Oden J. T. and Duarte A. (1995) hp Clouds: a meshless method to solve boundary value problems. Technical Report TICAM 95-05, The University of Texas at Austin, Austin, TX 78712.

[Ode95] Oden J. T. (1995) Local and pollution error estimation for finite element approximations of elliptic boundary value problems. Technical Report TICAM 95-03, The University of Texas at Austin, Austin, TX 78712.

[OPF94] Oden J. T., Patra A., and Feng X. (1994) Parallel domain decomposition solver for adaptive h-p finite element methods. Technical Report TICAM 94-11, The University of Texas at Austin, Texas, 78712.

[OWA94] Oden J. T., Wu W., and Ainsworth M. (1994) A-posteriori error estimators for the Navier-Stokes problem. *Comp. Meth. Mech. Engrg.* 11: 185–202.

[Pet95] Petroski H. (1995) The Boeing 777. *American Scientist* 83: 519–522.

[RGA93] Rettedal W. K., Gudmestad O. T., and Aarum T. (1993) Design of concrete platforms after Sleipner A-1 sinking. In Chakrabarti S. K., Agee C., Maeda H., Williams A. N., and Morrison D. (eds) *Offshore Technology. Proc. 12th Int. Conf. on Offshore Mechanics and Arctic Engineering (OMAE)*, volume 1, pages 309–319. ASME, New York.

[Rod94] Rodriguez R. (1994) Some remarks on the Zienkiewicz-Zhu estimator. *Numer. Meth. for PDE's* 10: 625–635.

[SA92] Szabó B. A. and Actis R. L. (1992) Hierarchic models for laminated plates. In Noor A. K. (ed) *Symposium on Adaptive, Multilevel and Hierarchical Computational Strategies*, volume AMD-Vol. 157, pages 69–94. American Society of Mechanical Engineers, New York.

[SB91] Szabó B. and Babuška I. (1991) *Finite Element Analysis.* John Wiley & Sons, Inc., New York.

[Sch89]   Schwab C. (1989) *Dimensional Reduction for Elliptic Boundary Value Problems*. PhD thesis, University of Maryland, College Park, MD 20742.

[SO96]   Stein E. and Ohnimus S. (1996) Dimensional adaptivity in linear elasticity with hierarchical test-spaces for h- and p-refinement processes. *Engineering with Computers* 12: 107–119.

[SS88]   Szabó B. and Sahrmann G. (1988) Hierarchic plate and shell models based on p-extension. *Int. J. Num. Meth. Engrg.* 26: 1855–1881.

[SSW96]   Schatz A. H., Sloan I. H., and Wahlbin L. B. (1996) Superconvergence in finite element methods and meshes which are locally symmetric with respect to a point. *SIAM J. Numer. Anal.* 33: 505–521.

[SY96a]   Szabó B. and Yosibash Z. (1996) Numerical analysis of singularities in two dimensions. Part 2: Computation of generalized flux/stress intensity factors. *Int. J. Num. Meth. Engrg.* 39: 409–434.

[SY96b]   Szabó B. and Yosibash Z. (1996) Superconvergent extraction of flux intensity factors and first derivatives from finite element solutions. *Comput. Meth. Appl. Mech. Engrg.* 129: 349–370.

[Sza86]   Szabó B. A. (1986) Estimation and control of error based on p-convergence. In Babuška I., Zienkiewicz O. C., Gago J., and de A. Oliveira E. R. (eds) *Accuracy Estimates and Adaptive Refinement in Finite Element Computations*, pages 61–70. John Wiley & Sons, Inc., New York.

[Sza88]   Szabó B. (1988) Geometric idealizations in finite element computations. *Comm. Num. Meth. Engrg.* 4: 393–400.

[Sza89]   Szabó B. A. (1989) Hierarchic plate and shell models based on p-extension. In Noor A. K., Belytschko T., and Simo J. C. (eds) *Analytical and Computational Models for Shells*, volume CED-Vol. 3, pages 317–331. American Society of Mechanical Engineers, New York.

[Sza90]   Szabó B. (1990) Superconvergent procedures for the computation of engineering data from finite element solutions. *Computers & Structures* 35: 441–444.

[TWK59]   Timoshenko S. and Woinowsky-Krieger S. (1959) *Theory of Plates and Shells*. McGraw-Hill, New York, 2 edition.

[VB80]   Vogelius M. and Babuška I. (1980) On a dimensional reduction method. *Math. Comp.* 37: 31–46.

[Ver94]   Verfürth R. (1994) A-posteriori error estimation and adaptive mesh-refinement techniques. *J. Copm. and Appl. Math.* 50: 67–83.

[Ver95]   Verfürth R. (1995) A review of a posteriori error estimation and adaptive mesh-refinement techniques. *Preprint* .

[Wah95]   Wahlbin L. B. (1995) *Superconvergence in Galerkin Finite Element Methods. Lecture Notes in Mathematics 1605*. Springer-Verlag, New York.

[WBSC87]   Wong W. A., Bucci R. J., Stentz R. H., and Conway J. B. (1987) Tensile and strain controlled fatigue data for certain aluminum alloys for application in the transport industry. *SAE Technical Paper Series, No. 870094, International Congress and Exposition, Detroit* .

[YS94]   Yosibash Z. and Szabó B. (1994) Convergence of stress maxima in finite element computations. *Comm. Num. Meth. Engrg.* 10: 683–697.

[YS95]   Yosibash Z. and Szabó B. (1995) Numerical analysis of singularities in two dimensions. Part 1: Computation of eigenpairs. *Int. J. Num. Meth. Engrg.* 38: 2055–2082.

[ZD94]   Zboinski G. and Demkowicz L. (1994) Application of the 3d hpq adaptive finite element method for plate and shell analyses. Technical Report TICAM 94-13, The University of Texas at Austin, Austin, TX 78712.

[ZL89]   Zhu O. D. and Lin Q. (1989) *Superconvergence Theory of FEM*. Hunan Science Press.

[Zlá77]   Zlámal M. (1977) *Some Superconvergence Results in the Finite Element Method. Lecture Notes in Mathematics 660*. Springer-Verlag, New York.

[ZZ92a]   Zienkiewicz O. C. and Zhu J. Z. (1992) The superconvergent patch recovery and a posteriori error estimation. Part I the recovery technique. *Int. J. Numer. Meth. Engrg.* 33: 1365–1382.

[ZZ92b]   Zienkiewicz O. C. and Zhu J. Z. (1992) The superconvergent patch-recovery and a posteriori error estimates. Part II, error estimates and adaptivity. *Int. J. Num. Meth. Engrg.* 33: 1365–1382.

2

Solution of Singular Problems Using *hp* Clouds

J. Tinsley Oden and C. Armando Duarte

Texas Institute for Computational and Applied Mathematics
The University of Texas at Austin
Taylor Hall 2.400
Austin, Texas, 78712, U.S.A.

2.1   INTRODUCTION

The *hp* cloud technique [DO, DO95a] is a generalization of a family of so-called meshless methods (see, e.g., [BKOF96, LCJ+, Dua95] for an overview) proposed in recent months that provide both *h* and *p* (spectral) type approximations of boundary-value problems while freeing the analyst from traditional difficulties due to mesh connectivities. In these methods, the bounded domain of the solution of an elliptic boundary-value problem is covered by the union of a collection of open sets (the *clouds*) over which spectral-type approximation can be constructed. The mathematical foundation of these techniques applied to linear elliptic problems is discussed in [DO]. A-posteriori estimates and adaptive *hp* clouds are developed in [DO95a].

   In the present investigation, the generalization of *hp* clouds to problems with singularities is presented. The theory of *hp* clouds is reviewed following this introduction and adaptive *hp* cloud techniques are outlined as well. Then the application of *hp* clouds to problems with singularities is developed, particular attention being given to the calculation of stress intensity factors in linear fracture mechanics. A new scheme is presented for computing very accurate approximations of stress intensity factors. Several numerical examples are presented to support the theoretical developments.

2.2   *H-P* CLOUD APPROXIMATIONS

The fundamental idea in the *hp* cloud method is to use a partition of unity to construct a hierarchy of functions that can represent polynomials of any degree. This family

The Mathematics of Finite Elements and Applications
Edited by J. R. Whiteman © 1997 John Wiley & Sons Ltd.

of functions, called $\mathcal{F}_N^p$, has many interesting properties like compact support and the ability to reproduce, through linear combinations, polynomials of any degree. In addition, the functions $\mathcal{F}_N^p$ can be built with any degree of regularity. Another remarkable feature of the functions $\mathcal{F}_N^p$ is that there is no need to partition the domain into smaller subdomains, e.g., finite elements, to construct these functions—all that is required is an arbitrarily placed set of nodes in the domain Ω. The construction of the functions $\mathcal{F}_N^p$, also known as hp cloud functions, is described in the next section.

2.2.1 CONSTRUCTION OF A PARTITION OF UNITY

Let Ω be an open bounded domain in $\mathbb{R}^n$, $n = 1, 2$ or 3 and $\boldsymbol{Q}_N$ denote an arbitrarily chosen set of N points $\boldsymbol{x}_\alpha \in \bar{\Omega}$ referred to as *nodes*:

$$\boldsymbol{Q}_N = \{\boldsymbol{x}_1, \boldsymbol{x}_2, \ldots, \boldsymbol{x}_N\}, \ \boldsymbol{x}_\alpha \in \bar{\Omega}$$

Let $\mathcal{T}_N := \{\omega_\alpha\}_{\alpha=1}^N$ denote a finite open covering of Ω consisting of N *clouds* ω_α with centers at $\boldsymbol{x}_\alpha$, $\alpha = 1, \ldots, N$ and having the following property

$$\bar{\Omega} \subset \bigcup_{\alpha=1}^N \bar{\omega}_\alpha \tag{2.1}$$

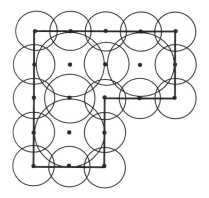

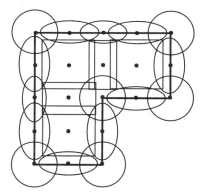

(a) Open covering build using circles.

(b) Open covering build using circles, ellipses and rectangles.

Figure 1   Examples of open coverings.

A cloud ω_α can be almost anything. In two dimensions, for example, it can be a rectangle, an ellipse or a circle. Figure 1 shows examples of valid open coverings and associated clouds. In this paper we shall restrict ourselves to the case where a cloud ω_α is defined by

$$\omega_\alpha := \{\boldsymbol{y} \in \mathbb{R}^n | \ \|\boldsymbol{x}_\alpha - \boldsymbol{y}\|_{\mathbb{R}^n} < h_\alpha\} \tag{2.2}$$

This corresponds to the case shown in Fig. 1(a).

A class of functions $\mathcal{S}_N := \{\varphi_\alpha\}_{\alpha=1}^N$ is called a *partition of unity* subordinated to the open covering $\mathcal{T}_N$ if it possesses the following properties [OR76]:

1) $\varphi_\alpha \in C_0^\infty(\omega_\alpha)$, $1 \le \alpha \le N$

2) $\sum_{\alpha=1}^N \varphi_\alpha(x) = 1$, $\forall\, x \in \Omega$

Note that $\varphi_\alpha(x)$ may be negative.

There is no unique way to build functions φ_α satisfying the above requirements. Each approach has its own merits and embedded costs. The choice of a particular partition of unity should be based on

- the class of problems to be solved, e.g. linear or non-linear problems,
- the complexity of the geometry of the domain,
- the regularity required from the approximation, e.g. C^0, C^1, or higher,
- the importance of the meshless character of the approximation, etc.

The following algorithm is currently used in the *hp* cloud method to build partitions of unity.

Let $\mathcal{W}_\alpha : \mathbb{R}^n \to \mathbb{R}$ denote a *weighting function* with compact support ω_α that belongs to the space $C_0^s(\omega_\alpha)$, $s \ge 0$ and suppose that

$$\mathcal{W}_\alpha(x) \ge 0 \qquad \forall\, x \in \Omega$$

In the case of the clouds defined in (2.2), the weighting functions $\mathcal{W}_\alpha$ can be implemented with any degree of regularity using "ridge" functions. More specifically, the weighting functions $\mathcal{W}_\alpha$ can be implemented through the composition

$$\mathcal{W}_\alpha(x) := g(r_\alpha)$$

where g is, e.g., a B-spline with compact support $[-1, 1]$ and r_α is the functional

$$r_\alpha := \frac{\|x - x_\alpha\|_{\mathbb{R}^n}}{h_\alpha}$$

In the computations presented in Section 2.3, g is a quartic $C^3(\Omega)$ B-spline. Details on the construction of the B-splines can be found in, e.g., [deB78].

The partition of unity functions φ_α can then be defined by

$$\varphi_\alpha(x) = \frac{\mathcal{W}_\alpha(x)}{\sum_\beta \mathcal{W}_\beta(x)} \tag{2.3}$$

which are known as Shepard functions [She68]. The main advantages of this particular partition of unity are

- low computational cost and simplicity of computation,
- it is meshless—there is no need to partition the domain to build this partition of unity,
- it can easily be implemented in any dimension,
- it can be constructed with any degree of regularity
- it allows easy implementation of h adaptivity, as is demonstrated in Section 2.3.1.

2.2.2   THE FAMILY $\mathcal{F}_N^p$

The construction of the hp cloud functions is very straightforward after a partition of unity is derived, such as the one described above, for the domain Ω.

Let $\widehat{\mathcal{L}}_p := \left\{\widehat{L}_i\right\}_{i \in \mathcal{I}}$, denote a set of *basis functions* $\widehat{L}_i$ defined on the unit cloud

$$\widehat{\omega} := \{\boldsymbol{\xi} \in \mathbb{R}^n | \ \|\boldsymbol{\xi}\|_{\mathbb{R}^n} < 1\}$$

satisfying

$$\begin{align}
\widehat{L}_1(\boldsymbol{\xi}) &\equiv 1 \tag{2.4}\\
\mathcal{P}_p(\widehat{\omega}) &\subset \text{span}(\widehat{\mathcal{L}}_p)
\end{align}$$

In the above, $\mathcal{I}$ denotes an index set and $\mathcal{P}_p$ denotes the space of polynomials of degree less or equal to p.

The family of functions $\mathcal{F}_N^p$ is defined by

$$\mathcal{F}_N^p := \left\{\varphi_\alpha(\boldsymbol{x})\left(\widehat{L}_i \circ \mathbf{F}_\alpha^{-1}(\boldsymbol{x})\right) \mid \alpha = 1,\ldots,N; \ i \in \mathcal{I}\right\} \tag{2.5}$$

where N is the number of nodes in the domain and

$$\mathbf{F}_\alpha^{-1} : \omega_\alpha \to \widehat{\omega} \tag{2.6}$$

$$\mathbf{F}_\alpha^{-1}(\boldsymbol{x}) := \frac{\boldsymbol{x} - \boldsymbol{x}_\alpha}{h_\alpha}$$

According to the above definition, $\mathcal{F}_N^p$ is constructed by multiplying each partition of unity function $\varphi_\alpha \in \mathcal{S}_N$ by the elements from the set $\widehat{\mathcal{L}}_p$. One element from the space of hp cloud functions can therefore be written as

$$u^{hp}(\boldsymbol{x}) = \sum_{\alpha=1}^N \sum_{i \in \mathcal{I}} \left[a_{\alpha i}\varphi_\alpha(\boldsymbol{x})(\widehat{L}_i \circ \mathbf{F}_\alpha^{-1}(\boldsymbol{x}))\right]$$

The following theorems are proved in [DO].

Theorem 2.1 *Let $\mathcal{L}_p := \{L_i\}_{i \in \mathcal{I}}$ and L_i be the same functions $\widehat{L}_i$ but defined on Ω. Then $\mathcal{L}_p \subset \text{span}\{\mathcal{F}_N^p\}$,*

Therefore the elements from the set $\mathcal{L}_p$ can be recovered through linear combinations of the hp clouds functions. This is one of the most fundamental properties of the hp cloud functions.

Theorem 2.2 *Let $\widehat{L}_i$, $i \in \mathcal{I}$, be the basis functions from the set $\widehat{\mathcal{L}}_p$ and $\mathcal{W}_\alpha$, $\alpha = 1,\ldots,N$ be the weighting functions used to construct the partition of unity functions φ_α defined in (2.3). Suppose that $\widehat{L}_i$, $i \in \mathcal{I} \in C^l(\widehat{\omega})$ and $\mathcal{W}_\alpha$, $\alpha = 1,\ldots,N \in C_0^q(\omega_\alpha)$. Then the hp cloud functions $\mathcal{F}_N^p$ defined in (2.5) belong to the space $C_0^{min(l,q)}(\Omega)$.*

Figure 2(a) shows the partition of unity function φ_α associated with a node $\boldsymbol{x}_\alpha$ at the origin. A uniform 5×5 node arrangement and quartic splines are used to build it. Figure 2(b) shows the function $xy\varphi_\alpha$ from the families $\mathcal{F}_{N=25}^{p \geq 2}$.

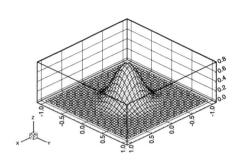 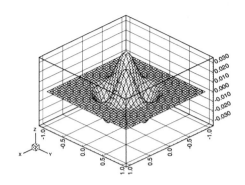

(a) 2-D partition of unity function φ_α.

(b) 2-D *hp* cloud function $xy\varphi_\alpha$ from the families $\mathcal{F}^{p\geq 2}_{N=25}$.

Figure 2   2-D *hp* cloud functions.

Remark 2.1 *There are many cases in which there is some knowledge about the function being approximated and the requirement that $\mathcal{P}_p(\widehat{\omega}) \subset \text{span}(\widehat{\mathcal{L}}_p)$ can be weakened without deteriorating the approximating properties of the set $\widehat{\mathcal{L}}_p$ [BM95]. There are also situations in which the inclusion in the set $\widehat{\mathcal{L}}_p$ of specially tailored functions to model, e.g., boundary layers, shocks, singularities, etc., can be very advantageous. This is the case in the stress analysis of cracks where the quantities of interest are the stress intensity factors. This case is discussed in detail in Section 2.3.2.*

An a-priori estimate

In this section we summarize some results presented in [DO].

Let $\boldsymbol{X}^{hp}_\alpha$ be the restriction to $(\Omega \cap \omega_\alpha)$ of the elements from the two parameters (p and $h(N)$) family of spaces

$$\widetilde{\boldsymbol{X}}^{hp}_\alpha(\omega_\alpha) := \text{span}\left\{\widehat{L}_i \circ \mathbf{F}^{-1}_\alpha\right\}_{i \in \mathcal{I}}$$

where the mapping $\mathbf{F}^{-1}_\alpha$ is defined in (2.6) and $\widehat{L}_i \in \widehat{\mathcal{L}}_p$.

Theorem 2.3 *Let $u \in H^{r+1}(\Omega)$, $r \geq 0$. Suppose that the following hold for all local spaces $\boldsymbol{X}^{hp}_\alpha$*

$$\mathcal{P}_r(\Omega \cap \omega_\alpha) \subset \boldsymbol{X}^{hp}_\alpha \subset H^1(\Omega \cap \omega_\alpha)$$

Then for fixed h and p there is $u_{hp} \in \boldsymbol{X}^{hp} := \text{span}\{\mathcal{F}^p_N\}$ such that

$$\|u - u_{hp}\|_{L^2(\Omega)} \leq C_1 h^{r+1} \|u\|_{H^{r+1}(\Omega)}, \tag{2.7}$$

$$\|u - u_{hp}\|_{H^1(\Omega)} \leq C_2 h^r \|u\|_{H^{r+1}(\Omega)} \tag{2.8}$$

where the constants C_1 and C_2 do not depend on h or u but depend on p, and $h = \max_{\alpha=1,\ldots,N(h)} h_\alpha$.

An a-posteriori error estimate

Let $\Omega \subset \mathbb{R}^2$ be a bounded domain with Lipschitz boundary $\partial\Omega$. Consider the model elliptic boundary-value problem of finding the solution u of

$$-\Delta u + cu = f \qquad in \ \Omega$$

subject to the boundary conditions

$$\frac{\partial u}{\partial n} = g \qquad on \ \Gamma_N$$
$$u = 0 \qquad on \ \Gamma_D$$

The variational form of this problem is to find $u \in V_D$ such that

$$B(u,v) = L(v) \qquad \forall \ v \in V_D$$

where V_D is the space

$$V_D = \{v \in H^1(\Omega) : v = 0 \ on \ \Gamma_D\}$$

and where

$$B(u,v) = \int_\Omega (\nabla u \cdot \nabla v + cuv)d\boldsymbol{x}$$
$$L(v) = \int_\Omega fv d\boldsymbol{x} + \int_{\Gamma_N} gv d\boldsymbol{x}$$

Suppose that $\boldsymbol{X}^{hp} \subset V_D$ is a subspace built using hp clouds. Then the hp cloud-Galerkin approximation of this problem is to find $u_{hp} \in \boldsymbol{X}^{hp}$ such that

$$B(u_{hp}, v_{hp}) = L(v_{hp}) \qquad \forall \ v_{hp} \in \boldsymbol{X}^{hp}$$

The following is proved in [DO95a].

Theorem 2.4 *Suppose that $\boldsymbol{X}^{hp} \subset (C^2(\Omega) \cap V_D)$. Let r denote the interior residual*

$$r = f + \Delta u_{hp} - cu_{hp} \qquad in \ \Omega$$

and R denote the boundary residual

$$R = g - \frac{\partial u_{hp}}{\partial n} \qquad on \ \Gamma_N$$

Then the energy norm of the discretization error $e = u - u_{hp}$ satisfies

$$\|e\|_{E,\Omega} \leq \rho^{1/2} \bar{C} \left(\sum_\alpha \eta_\alpha^2\right)^{1/2}$$

where the contributions η_α from the clouds ω_α are error indicators and are given by

$$\eta_\alpha^2 = \frac{h_\alpha^2}{p_\alpha^2} \|r\|_{L^2(\omega_\alpha \cap \Omega)}^2 + \frac{h_\alpha}{p_\alpha} \|R\|_{L^2(\partial(\omega_\alpha \cap \Omega) \cap \Gamma_N)}^2 \tag{2.9}$$

and $\rho \in I\!N$ satisfies

$$card\{\alpha | x \in \omega_\alpha\} \leq \rho \qquad \forall\, x \in \Omega$$

The error indicators η_α are used in Section 2.3.1 to implement h, p and hp adaptivity in the hp cloud method.

2.3 NUMERICAL EXAMPLES

In this section, the techniques described in Section 2.2 are used to construct appropriate finite dimensional subspaces of functions used in the Galerkin method. The resulting approach is denoted as the hp *cloud method*. Two problems are solved in this section using the hp cloud method. The first problem demonstrates how h, p and hp adaptivity can be implemented in the hp cloud method and how the method can handle the presence of a singularity. The second problem deals with the stress analysis of cracks in linear elasticity and the computation of stress intensity factors. We explore the fact that the hp cloud spaces are able include almost any kind of function and develop a novel and very efficient approach to extract stress intensity factors. It is demonstrated numerically that the computed stress intensity factors converge at the same rate as the error in energy and, therefore, at twice the rate of the error in the energy norm.

In all problems solved the following is adopted:

- The essential boundary conditions are imposed using the method of Lagrange multipliers.
- The domain integrations are performed using a background cell structure that exactly covers the domains. Nonetheless, there is no relationship between the background cell structures and the nodes x_α used in the discretizations. It should be noted that the generation of a background cell structure is much easier than the generation of a traditional finite element mesh.
- The algorithm presented in [DO95a] is used to handle re-entrant corners.
- In all problems analyzed, the size of the supports of the hp cloud functions are automatically set using the algorithm presented in [DO95b, DO].

2.3.1 POTENTIAL FLOW IN A L-SHAPED DOMAIN

In this section, a potential flow problem in an L-shaped domain is solved using the hp cloud method. The problem statement is given below. The domain Ω and the boundary segments Γ_1 and Γ_2 are depicted in Fig. 3. The value of u is set to zero at $(1, 1)$ in order to make the solution unique. This simple boundary-value problem is used to demonstrate how h, p and hp adaptivity can be implemented in the hp cloud method without using the concept of a mesh and how adaptivity can be used in the hp cloud method to handle the presence of singularities.

Find u such that

$$-\Delta u \;=\; 0 \qquad in\ \Omega$$
$$-\frac{\partial u}{\partial n} \;=\; 5 \qquad on\ \Gamma_1 \qquad (2.10)$$
$$-\frac{\partial u}{\partial n} \;=\; -5 \qquad on\ \Gamma_2$$
$$-\frac{\partial u}{\partial n} \;=\; 0 \qquad on\ \partial\Omega\backslash(\Gamma_1\cup\Gamma_2)$$

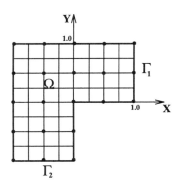

Figure 3   Boundary conditions, initial nodal arrangement and background cell structure used for quadrature.

For this problem, the set of functions $\widehat{\mathcal{L}}_p(\widehat{\omega})$ defined in (2.4) is composed of the following monomials:

$$\widehat{\mathcal{L}}_p(\widehat{\omega}) = \{\xi^i\eta^j\,|\,0 \le i + j \le p\}$$

That is, the family of hp cloud functions, $\mathcal{F}_N^p$, is constructed by multiplying the partition of unity functions φ_α, $\alpha = 1,\ldots,N$, by the smallest set of complete polynomials of degree less or equal to p.

The background cell structure used to perform the numerical quadrature is represented in Fig. 3. The nodal arrangement used in the first step of h, p and hp adaptation is also shown in the figure.

h adaptivity

Theorem 2.3 demonstrates that the discretization error of the hp cloud solution can be controlled by decreasing the radius h_α of the clouds. This, of course, has to be followed by the addition of more nodes to the discretization in order to guarantee that a valid covering for the domain still exists. It should be noted, however, that there is no constraint on how these nodes are placed in the domain—they can simply be inserted into the regions of interest. This strategy is denoted the h *version* of the hp cloud method. This approach is demonstrated in the solution of problem (2.10). Details of the algorithm used in the h version of the hp cloud method can be found [DO95a]. The nodal arrangement represented in Fig. 3 and clouds with polynomial degree $p = 1$ are used to obtain the first approximate solution. The error indicators given by (2.9) are then computed for each cloud ω_α. Clouds with errors above a preset value are selected to be refined while the polynomial order is kept fixed. The refinement of clouds involves the addition of news nodes around each of the refined clouds. The size of the supports ω_α are then automatically reset to take into account the new nodes added to the discretization (details of the algorithm can be found in [DO]). A new solution is then computed using the new discretization and the process is repeated until the quality of the solution is deemed acceptable.

Figure 4 shows the nodal arrangement and the covering obtained after seven steps of h adaptation. A high concentration of nodes near the corner at $(0,0)$ is observed.

Figure 5 depicts the fluxes computed using the discretization of Fig. 4. The flux distribution indicates the presence of a singularity at the re-entrant corner at $(0, 0)$.

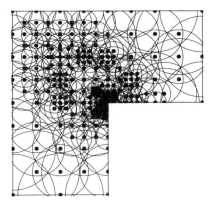

Figure 4   Discretization obtained using h adaptivity.

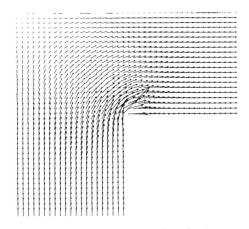

Figure 5   Flux distribution.

p adaptivity

In the p version of the hp cloud method, the solution space is enriched by keeping fixed the size h_α of the clouds and increasing the polynomial order p associated with each cloud. It is noted that each cloud ω_α can have a different polynomial order associated with it, *independently* of the polynomial order associated with neighboring clouds.

The polynomial order associated with each cloud after six steps of p adaptation is indicated in Fig. 6(a). The polynomial orders assigned with each cloud is automatically chosen using the error indicators given by (2.9). The polynomial orders range from $p = 1$ to $p = 7$. Figure 6(b) represents the potential u_p computed using the discretization of Fig. 6(a).

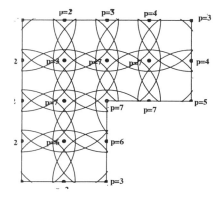

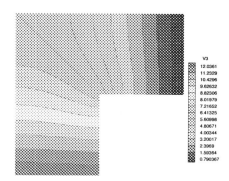

(a) Discretization obtained using p adaptivity.   (b) Solution obtained using p adaptivity.

Figure 6   Results from the p version of the hp cloud method

hp adaptivity

The hp version of the hp cloud method is implemented by combining the h and p algorithms described previously. Two or more steps of h adaptation are initially performed in order to isolate any singularity present in the solution. This is followed by a number of p steps until the estimated error is below a preset value. In the case of problem (2.10), two h steps are initially performed, starting from the nodal arrangement of Fig. 3. After that, the error is controlled through p enrichment of the clouds. The final discretization is presented in Fig. 7. Figure 8(a) shows the computed fluxes and potential. The three dimensional plot of Fig. 8(b) presents the computed flux in the y direction using the discretization of Fig. 7. The presence of a singularity at $(0,0)$ is evident.

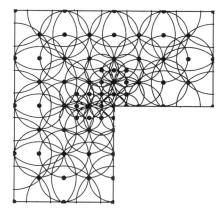

Figure 7   Discretization obtained using hp adaptivity.

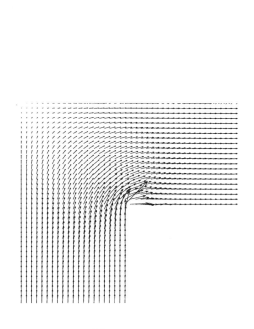

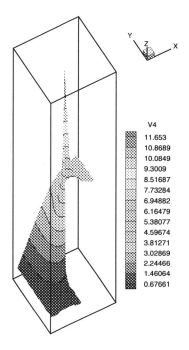

(a) Flux distribution.

(b) Flux in the y direction obtained using *hp* adaptivity.

Figure 8 Results obtained using *hp* adaptivity.

2.3.2 *MODELING OF CORNER SINGULARITIES AND COMPUTATION OF STRESS INTENSITY FACTORS IN LINEAR FRACTURE MECHANICS*

In this section, we demonstrate how corner singularities in plane elasticity problems can be efficiently modeled using the framework of *hp* clouds. These singularities occur when the solution domain has corners, abrupt changes in boundary data or consists of two or more materials [OB95]. The *hp* cloud results are compared with those of the *p* version of the finite element method. We also present a new approach for the extraction of the amplitude of the singular terms associated with corner singularities. The accuracy and simplicity of this novel approach is demonstrated numerically.

Near crack tip expansion

In the neighborhood of a crack, assuming traction-free crack surfaces and in the absence of body forces, each component of the displacement vector $\boldsymbol{u} = \{u_x, u_y\}^T$

can be written as [SB91, SB88]

$$u_x(r,\theta) \;=\; \sum_{j=1}^{M}\left(A_j^{(1)}u_{xj}^{(1)}(r,\theta)+A_j^{(2)}u_{xj}^{(2)}(r,\theta)\right)+\tilde{u}_x(r,\theta) \qquad (2.11)$$

$$u_y(r,\theta) \;=\; \sum_{j=1}^{M}\left(A_j^{(1)}u_{yj}^{(1)}(r,\theta)+A_j^{(2)}u_{yj}^{(2)}(r,\theta)\right)+\tilde{u}_y(r,\theta) \qquad (2.12)$$

where (r,θ) are the polar coordinates associated with the crack tip (cf. Fig. 9), $\tilde{u}_x(r,\theta)$ and $\tilde{u}_y(r,\theta)$ are functions smoother than any term in the sum and the eigenfunctions are given by

$$u_{xj}^{(1)}(r,\theta) \;=\; \frac{r^{\lambda_j}}{2G}\left\{[\kappa-Q_j^{(1)}(\lambda_j+1)]\cos\lambda_j\theta-\lambda_j\cos(\lambda_j-2)\theta\right\}$$

$$u_{xj}^{(2)}(r,\theta) \;=\; \frac{r^{\lambda_j}}{2G}\left\{[\kappa-Q_j^{(2)}(\lambda_j+1)]\sin\lambda_j\theta-\lambda_j\sin(\lambda_j-2)\theta\right\}$$

$$u_{yj}^{(1)}(r,\theta) \;=\; \frac{r^{\lambda_j}}{2G}\left\{[\kappa+Q_j^{(1)}(\lambda_j+1)]\sin\lambda_j\theta+\lambda_j\sin(\lambda_j-2)\theta\right\}$$

$$u_{yj}^{(2)}(r,\theta) \;=\; -\frac{r^{\lambda_j}}{2G}\left\{[\kappa+Q_j^{(2)}(\lambda_j+1)]\cos\lambda_j\theta+\lambda_j\cos(\lambda_j-2)\theta\right\}$$

where the eigenvalues λ_j are

$$\lambda_1=\frac{1}{2},\qquad \lambda_j=\frac{j+1}{2}\qquad j\ge 2$$

and

$$Q_j^{(1)}=\left\{\begin{array}{ll}-1 & j=3,5,7,\ldots\\ -\Lambda_j & j=1,2,4,6,\ldots\end{array}\right. \qquad Q_j^{(2)}=\left\{\begin{array}{ll}-1 & j=1,2,4,6,\ldots\\ -\Lambda_j & j=3,5,7,\ldots\end{array}\right.$$

where

$$\Lambda_j=\frac{\lambda_j-1}{\lambda_j+1}$$

and the material constant κ and G are

$$\kappa=\left\{\begin{array}{ll}3-4\nu & \text{plane strain}\\ \frac{3-\nu}{1+\nu} & \text{plane stress}\end{array}\right. \qquad G=\frac{E}{2(1+\nu)}$$

where E is the Young's modulus and ν is the Poisson's ratio.

The coefficients $A_1^{(1)}$ and $A_1^{(2)}$ in (2.11) and (2.12) are related to the Mode I and Mode II stress intensity factors of linear elasticity fracture mechanics, usually denoted by K_I and K_{II}, as follows:

$$A_1^{(1)}=\frac{K_I}{\sqrt{2\pi}}\qquad A_1^{(2)}=\frac{K_{II}}{\sqrt{2\pi}}$$

Enrichment of the *hp* cloud spaces and extraction of stress intensity factors

Suppose that the eigenfunctions $u_{xj}^{(1)}$, $u_{xj}^{(2)}$, $u_{yj}^{(1)}$, $u_{yj}^{(2)}$, $j = 1, \ldots, M$ are added to the set $\hat{\mathcal{L}}_p$ defined in Section 2.2.2 giving

$$\hat{\mathcal{L}}_p^* := \left\{ \hat{L}_i \right\}_{i \in \mathcal{I}} \cup \left\{ u_{xj}^{(1)}, \ u_{xj}^{(2)}, \ u_{yj}^{(1)}, \ u_{yj}^{(2)} \right\}_{j=1,\ldots,M} \tag{2.13}$$

Then, the enriched *hp* cloud functions are defined as follows

$$\mathcal{F}_N^{p*} := \left\{ \varphi_\alpha(x) \left[\hat{L}_i \circ \mathbf{F}_\alpha^{-1}(x) \right] \cup \varphi_\alpha(x) \left[\bar{u}_{xj}^{(1)}(x), \bar{u}_{xj}^{(2)}(x), \bar{u}_{yj}^{(1)}(x), \bar{u}_{yj}^{(2)}(x) \right] \right.$$

$$\left. \mid \alpha = 1, \ldots, N; \ i \in \mathcal{I}; \ j = 1, \ldots, M \right\} \tag{2.14}$$

where

$$\bar{u}_{xj}^{(1)}(x) := u_{xj}^{(1)} \circ \mathbf{T}^{-1}(x)$$

and $\mathbf{T}^{-1}$ is the transformation from rectangular to polar coordinates. The other bar quantities are computed similarly and the transformation $\mathbf{F}^{-1}$ is defined in (2.6). Note that all elements from $\mathcal{F}_N^{p*}$ have compact support.

The *hp* cloud approximation to the displacement field (2.11),(2.12) can be written as

$$u_x^{hp}(x) = \sum_{\alpha=1}^{N} \sum_{i \in \mathcal{I}} \left[a_x^{\alpha i} \varphi_\alpha(x)(\hat{L}_i \circ \mathbf{F}_\alpha^{-1}(x)) \right]$$

$$+ \sum_{\alpha=1}^{N} \sum_{j=1}^{M} \left[b_{\alpha j}^{(1)} \varphi_\alpha(x) \bar{u}_{xj}^{(1)}(x) + b_{\alpha j}^{(2)} \varphi_\alpha(x) \bar{u}_{xj}^{(2)}(x) \right]$$

$$u_y^{hp}(x) = \sum_{\alpha=1}^{N} \sum_{i \in \mathcal{I}} \left[a_y^{\alpha i} \varphi_\alpha(x)(\hat{L}_i \circ \mathbf{F}_\alpha^{-1}(x)) \right]$$

$$+ \sum_{\alpha=1}^{N} \sum_{j=1}^{M} \left[b_{\alpha j}^{(1)} \varphi_\alpha(x) \bar{u}_{yj}^{(1)}(x) + b_{\alpha j}^{(2)} \varphi_\alpha(x) \bar{u}_{yj}^{(2)}(x) \right]$$

In the above expansion, for a fixed $x \in \bar{\Omega}$, only a few terms are nonzero since all the functions have compact support. The expansion for, e.g., u_x^{hp} can be rewritten as

$$u_x^{hp}(x) = \underbrace{\sum_{\alpha=1}^{N} \sum_{i \in \mathcal{I}} \left[a_x^{\alpha i} \varphi_\alpha(x)(\hat{L}_i \circ \mathbf{F}_\alpha^{-1}(x)) \right]}_{\text{polynomial reproducing part}} + \underbrace{\sum_{j=1}^{M} [\bar{u}_{xj}^{(1)}(x) \overbrace{\sum_{\alpha=1}^{N} b_{\alpha j}^{(1)} \varphi_\alpha(x)}^{A_j^{(1)}}]}_{\text{singular part}}$$

$$\overbrace{+ \sum_{j=1}^{M} [\bar{u}_{xj}^{(2)}(\boldsymbol{x}) \underbrace{\sum_{\alpha=1}^{N} b_{\alpha j}^{(2)} \varphi_\alpha(\boldsymbol{x})]}_{\text{singular part}}}^{A_j^{(2)}} \tag{2.15}$$

Comparing (2.15) with (2.11) leads to the following conclusion: the term $\tilde{u}_x$ can be approximated very accurately by the polynomial part of u_x^{hp} since it is a smooth function. Therefore the difference between $\tilde{u}_x$ and the polynomial part of u_x^{hp} can be made as small as we want (possibly exponentially) by increasing the polynomial order associated with the clouds; consequently we must expect that

$$\sum_{\alpha=1}^{N} b_{\alpha j}^{(1)} \varphi_\alpha \overset{p \to \infty}{\longrightarrow} A_j^{(1)} \qquad \text{and} \qquad \sum_{\alpha=1}^{N} b_{\alpha j}^{(2)} \varphi_\alpha \overset{p \to \infty}{\longrightarrow} A_j^{(2)}$$

Therefore the amplitude of *any number* of generalized stress intensity factors can be obtained directly from the coefficient of the hp cloud functions without any additional work. The first and second stress intensity factors from linear fracture mechanics are therefore given by

$$K_I = \sqrt{2\pi} A_1^{(1)} = \sqrt{2\pi} \sum_{\alpha=1}^{N} b_{\alpha 1}^{(1)} \varphi_\alpha$$

$$K_{II} = \sqrt{2\pi} A_1^{(2)} = \sqrt{2\pi} \sum_{\alpha=1}^{N} b_{\alpha 1}^{(2)} \varphi_\alpha$$

Remark 2.2 *The case of more complicated boundary conditions at the crack surfaces can be treated exactly as above [OB95]. In the case of anisotropic or nonhomogenous materials, the numerical approach proposed by [PB95] can be used to compute the eigenvalues and eigenfunctions of the asymptotic expansion near the crack tip. The approach described above can then be used without any modifications.*

Cracked panel

In this section, the edge-cracked panel shown in Fig. 9 is analyzed using the enriched hp cloud spaces $\mathcal{F}_N^{p*}$ defined in (2.14). The hp cloud results are compared with those presented by Szabo [Sza86] and Duarte and Barcellos [DdB91] for the p version of the finite element method using strongly graded meshes.

A state of plane-strain, Poisson's ratio of 0.3 and unity thickness are assumed. The tractions corresponding to the first symmetric term of the asymptotic expansion (2.11), (2.12) are applied on $\partial\Omega$. The two components of the displacement at $(0,0)$ and the vertical component at $(d,0)$ are set to zero to make the solution unique. The components of the stress tensor associated with the first term of the asymptotic expansion are given by [SB91]

$$\sigma_{ij}^{(1)} = \frac{K_I}{\sqrt{2\pi r}} f_{ij}^{(1)}(\theta) \tag{2.16}$$

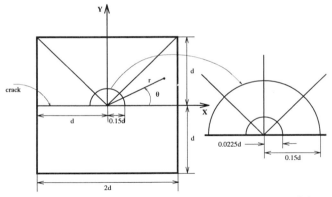

Figure 9   Cracked panel and geometric mesh with three layers of finite elements.

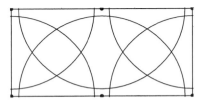

Figure 10   *hp* cloud discretization for the cracked panel.

$$f_{11}^{(1)}(\theta) = \cos\left(\frac{\theta}{2}\right)\left[1 - \sin\left(\frac{\theta}{2}\right)\sin\left(\frac{3\theta}{2}\right)\right] \tag{2.17}$$

$$f_{22}^{(1)}(\theta) = \cos\left(\frac{\theta}{2}\right)\left[1 + \sin\left(\frac{\theta}{2}\right)\sin\left(\frac{3\theta}{2}\right)\right] \tag{2.18}$$

$$f_{12}^{(1)}(\theta) = \cos\left(\frac{\theta}{2}\right)\sin\left(\frac{\theta}{2}\right)\cos\left(\frac{3\theta}{2}\right) \tag{2.19}$$

where (r, θ) is the polar coordinate system shown in Fig. 9.

Due to symmetry, only half of the panel is modeled. The nodal arrangement and associated covering used in the *hp* cloud discretization are depicted in Fig. 10. No special nodal arrangement is used near the singularity at $(0,0)$. Figure 9 shows the finite element mesh used by Szabo [Sza86]. The dimensions of the finite elements decreases in geometric progression towards the singularity. Duarte and Barcellos [DdB91] used a mesh with the same geometric progression but with five layers of elements instead of the three layers as shown in Fig. 9.

The exact strain energy for half of the panel is [Sza86]

$$U(u_{x1}^{(1)}, u_{y1}^{(1)}) = 0.23706469 K_I^2 \frac{d}{E}$$

The values $E = K_I = d = 1$ are adopted in the calculations.

In the following, an *hp* cloud node is said to be *enriched* if the set $\widehat{\mathcal{L}}_p^*$ defined in (2.13) is used to build the *hp* cloud functions associated with that node. For this problem the set $\widehat{\mathcal{L}}_p^*$ has always the following elements

$$\widehat{\mathcal{L}}_p^* := \{1, \xi, \eta\} \cup \left\{ u_{xj}^{(1)}, \ u_{xj}^{(2)}, \ u_{yj}^{(1)}, \ u_{yj}^{(2)} \right\}_{j=1,\dots,p-1}$$

That is, the set $\widehat{\mathcal{L}}_p^*$ has linear polynomials and $p-1$ symmetric and antisymmetric modes from the near crack tip expansion.

Figure 11 shows the convergence in the energy norm of the hp cloud and finite element solutions. There are two curves for the hp cloud method. The solid curve corresponds to the case in which all six nodes shown in Fig. 10 are enriched nodes. The error associated with this discretization is zero for $p \geq 2$ because $\widehat{\mathcal{L}}_{p \geq 2}^*$ contain the exact solution (cf. Theorem 2.1). The error (of $\mathcal{O}(10^{-6})$ in strain energy) shown in Fig. 11 for $p = 2$ is due to integration errors.

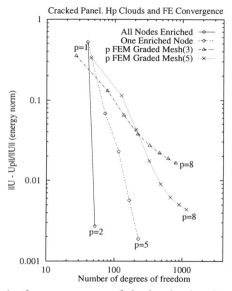

Figure 11   Convergence in the energy norm of the hp cloud and p finite element solutions.

This problem was constructed so that the exact solution is equal to the first symmetric mode all over the domain. In practical problems no such restriction applies and the expansion (2.11) (2.12) is valid only in a neighborhood $r < r_0$, $r_0 > 0$ [SB88]. However, far from the crack tip the solution is smooth and can be approximated well by polynomials. Therefore in the general case enriched nodes should be used only in the vicinity of the crack tip. The second curve for the hp cloud method shown in Fig. 11 corresponds to the use of this strategy. More precisely, the curve corresponds to the use of the discretization shown in Fig. 10 with only one enriched node–the one at the crack tip. It can be observed that the hp cloud solution converges at a very high rate, as if the solution was smooth. The convergence of the p finite element solution using graded meshes with three (cf. Fig. 9) and five layers of elements is also shown in Fig. 11. It can be observed that the hp cloud discretization requires much fewer degree of freedoms than the finite element counterpart. An additional advantage of using special functions, made possible by the partition of unity methods framework, is that the stress intensity factors can be obtained, without using elaborate extraction processes, via the approach described in Section 2.3.2.

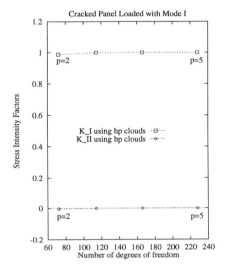

(a) Convergence of the Mode 1 and Mode 2 stress-intensity factors extracted from *hp* cloud solutions.

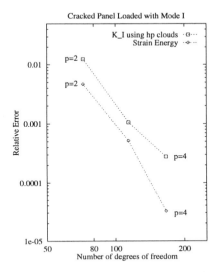

(b) Convergence of the strain energy and the Mode 1 stress-intensity factor extracted from *hp* cloud solutions.

Figure 12   Convergence of stress intensity factors.

Figure 12(a) shows the values of K_I and K_{II} obtained using the *hp* cloud discretization with only one enriched node and the technique described in Section 2.3.2. The relative errors in strain energy (not energy norm) and the absolute value of the relative error in K_I, computed from the coefficients of the *hp* cloud solution, are plotted against the number of degrees of freedom in Fig. 12(b). The stress intensity factor converges at about the same rate as the strain energy and therefore much faster than the error in the energy norm.

Figure 13 shows the computed stress component σ_{22}. The *hp* cloud solution corresponds to the discretization with only one enriched node (at $(0,0)$) and $p = 5$. The plot of Fig. 13(a) is in the (θ, r), $0 \leq \theta \leq \pi$, $0.00001 \leq r \leq 1$, plane; θ corresponds to the x direction and r corresponds to the y direction shown in the picture. The effectiveness of the *hp* cloud approximation to capture the singularity at $(0,0)$ is evident. Indeed

$$\frac{\text{Max}|\sigma_{22} - \sigma_{22}^{hp}|}{\text{Max}|\sigma_{22}|} = 0.000144$$

where the maximum is taken over all points used to plot Fig. 13(a). Figure 13(b) shows the same results of Fig. 13(a) but this time the plot is done in the (x, y) plane.

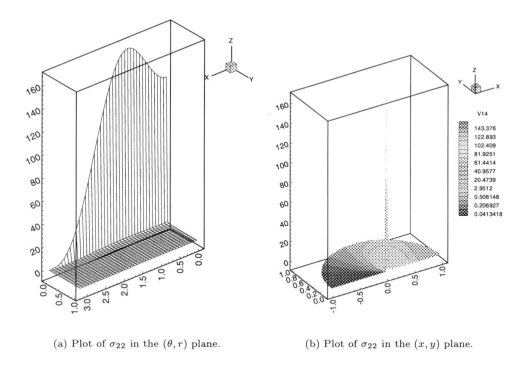

(a) Plot of σ_{22} in the (θ, r) plane.   (b) Plot of σ_{22} in the (x, y) plane.

Figure 13   Stress component σ_{22} computed using hp clouds.

2.4   CONCLUSIONS

The hp cloud method is shown to be an extremely effective tool for handling singularities and is particularly effective in computing stress intensity factors in linear elastic fracture mechanics. Experiments suggests the method is exponentially convergent and superior in performance to comparable hp finite element methods.

Acknowledgment: Preliminary work on this meshless technology was performed under the support of the Army Research Office (J. T. Oden) under contract DAAH04-96-0062. The support of the present work by the National Science Foundation under contract DMI-9561365 and of the CNPq Graduate Fellowship Program (C. A. Duarte) grant # 200498/92-4 is gratefully acknowledged. The authors thank Professor Ivo Babuška from the University of Texas at Austin for fruitful discussion during the course of this investigation.

References

[BKOF96] Belytschko T., Krongauz Y., Organ D., and Fleming M. (1996) Meshless methods: An overview and recent developments. To appear in a Special Issue of *Computer Methods in Applied Mechanics and Engineering* on meshless methods.

[BM95] Babuska I. and Melenk J. M. (July 1995) The partition of unity finite element method. Technical Report BN-1185, Inst. for Phys. Sc. and Tech., University of Maryland.

[DdB91] Duarte C. A. M. and de Barcellos C. S. (December 1991) The element residual method for elasticity and potential problems. In de Las Casas E. B. and de Paula F. A. (eds) *Symposium on Computational Mechanics*, pages 328–335. Belo Horizonte, MG, Brazil. In Portuguese.

[deB78] deBoor C. (1978) *A Practical Guide to Splines*. Springer-Verlag, New York.

[DO] Duarte C. A. M. and Oden J. T.*Hp* clouds—an *hp* meshless method. *Numerical Methods for Partial Differential Equations* (to appear).

[DO95a] Duarte C. A. M. and Oden J. T. (1995) An *hp* adaptive method using clouds. Technical report, TICAM, The University of Texas at Austin. To appear in *Computer Methods in Applied Mechanics and Engineering*.

[DO95b] Duarte C. A. M. and Oden J. T. (1995) Hp clouds–a meshless method to solve boundary-value problems. Technical Report 95-05, TICAM, The University of Texas at Austin.

[Dua95] Duarte C. A. M. (1995) A review of some meshless methods to solve partial differential equations. Technical Report 95-06, TICAM, The University of Texas at Austin.

[LCJ⁺] Liu W. K., Chen Y., Jun S., Belytschko T., Pan C., Uras R. A., and Chang C. T.Overview and applications of the reproducing kernel particle methods. *Archives of Computational Methods in Engineering. State of Art Reviews* (to appear).

[OB95] Oh H.-S. and Babuska I. (1995) The method of auxiliary mapping for the finite element solutions of elasticity problems containing singularities. *Journal of Computational Physics* 121: 193–212.

[OR76] Oden J. T. and Reddy J. N. (1976) *An Introduction to the Mathematical Theory of Finite Elements*. John Wiley and Sons, New York.

[PB95] Papadakis P. J. and Babuska I. (1995) A numerical procedure for the determination of certain quantities related to the stress intensity factors in two-dimensional elasticity. *Computer Methods in Applied Mechanics and Engineering* 122: 69–92.

[SB88]   Szabo B. A. and Babuska I. (1988) Computation of the amplitude of
            stress singular terms for cracks and reentrant corners. In Cruse T. A.
            (ed) *Fracture Mechanics: Nineteenth Symposium, ASTM STP 969*, pages
            101–124.

[SB91]   Szabo B. and Babuska I. (1991) *Finite Element Analysis*. John Wiley and
            Sons, New York.

[She68]   Shepard D. (1968) A two-dimensional function for irregularly spaced data.
            In *ACM National Conference*, pages 517–524.

[Sza86]   Szabo B. A. (1986) Estimation and control of error based on p convergence.
            In Babuska I., Zienkiewicz O. C., Gago J., and Oliveira E. R. d. A.
            (eds) *Accuracy Estimates and Adaptive Refinements in Finite Elements
            Computations*. John Wiley and Sons, New York.

3

Towards robust adaptive finite element methods for partial differential Volterra equation problems arising in viscoelasticity theory

Simon Shaw and J. R. Whiteman

BICOM, Institute of Computational Mathematics,
Brunel University,
Uxbridge UB8 3PH,
U.K.

3.1 INTRODUCTION

Physical problems involving viscoelastic materials give rise to mathematical models wherein the canonical elliptic, parabolic and wave equations are each augmented with a memory term. The resulting partial differential Volterra (PDV) equations will then, in general, have to be numerically approximated in order to obtain useful solutions. This approximation procedure should be adaptive in that it should reliably and accurately control the approximation error on a user-specified tolerance level or, minimally, report an accurate estimate of this error back to the user.

   Such a solver would need to be built on *a posteriori* error estimates but, although the literature is rich with *a priori* analysis of such PDV equations, there is—as far as we are aware—no corresponding body of knowledge on *a posteriori* error estimation. Accordingly, motivated by problems of viscoelasticity, in this article we want to consider the possibility of providing computable error estimates for space-time finite element discretization of partial differential equations with memory. In particular we shall examine the potential for a specific example of the generic *elliptic-Volterra* problem: find u such that

$$Au(t) = f(t) + \int_0^t B(t,s)u(s)\,ds \quad \text{in} \quad Q := \Omega \times \mathcal{J}, \tag{3.1}$$

The Mathematics of Finite Elements and Applications
Edited by J. R. Whiteman © 1997 John Wiley & Sons Ltd.

with appropriate boundary data specified. Here: for a positive real number T, $\mathcal{J} :=$ $[0, T]$ is a time interval; for $n \in \{1, 2, 3\}$, Ω an open bounded (spatial) domain in $\mathbb{R}^n$; for $t \in \mathcal{J}$, $x \in \Omega$, $f(t) = f(x, t)$ is a given load function; and, A and $B(t, s)$ are (at most for B) second-order partial differential operators, acting in Ω, with A assumed self-adjoint and strongly elliptic and $B(t, s)$ "similar" to A in a way that will be made precise below. To ease the notation we shall often suppress the dependence of these and other quantities on the space variable x. A concrete example of (3.1) may be obtained by setting $A := -\nabla^2$ and $B(t, s) := -\nabla \cdot \beta(t, s)\nabla$, where ∇^2, $\nabla \cdot$ and ∇ are the usual Laplacian, divergence and gradient operators in Ω. However, there is a much more practically important form of (3.1) that arises in connection with the stress analysis of polymers, this *quasistatic* problem is outlined later in Section 2.

A problem related to (3.1) occurs when attempting to model diffusion processes in viscoelastic media, here one seeks a solution to the *parabolic-Volterra* problem: find u such that

$$\dot{u}(t) + Au(t) = f(t) + \int_0^t B(t, s)u(s)\,ds, \qquad (3.2)$$

plus initial and boundary data, and where the overdot indicates time differentiation. The theory of hereditary heat conduction is discussed, for example, by Nunziato [Nun71]. Also, in the *viscodynamic* problem Newton's second law gives that $\ddot{u}(t)$ should be added to the left of (3.1) (this term is assumed negligible in the quasistatic theory) thus altering (3.1) to the *hyperbolic-Volterra* problem: find u such that

$$\ddot{u}(t) + Au(t) = f(t) + \int_0^t B(t, s)u(s)\,ds, \qquad (3.3)$$

plus initial and boundary data. See Shaw, Warby & Whiteman [Whi94, Chap. 3]. Thus the canonical second order partial differential equation problems each have a physically meaningful Volterra analogue.

In terms of error analysis the problem (3.1), discretized in space using finite elements and in time by a quadrature rule, has been examined by Shaw, Warby, Whiteman, Dawson & Wheeler in [SWW+94], under fairly general assumptions, and the result improved upon in the viscoelasticity context (in terms of the constant in the error bound) by Shaw, Warby & Whiteman in [SWW96]. The parabolic-Volterra problem (3.2) has received the most attention in the literature with the works of Sloan & Thomée [ST86], Greenwell-Yanik & Fairweather [GYF88], Cannon & Lin [CL90], Thomée & Wahlbin [TW94] and Pani & Peterson [PP96] being representative of the current state-of-the-art. A reformulation of the so-called non-ageing viscodynamic problem, (3.3), yields a form like (3.2) but with $A \equiv 0$ (see [Whi94, Chap. 3], equation (3.63) for example). Such a problem has been considered very recently by Lubich, Sloan & Thomée in [LST96]. All of these methods employ the finite element method in space and a combination of finite difference and quadrature replacements in time in order to arrive at a numerical scheme. Similar discretization strategies have also been analyzed for the hyperbolic-Volterra problem (3.3), see Greenwell-Yanik & Fairweather [GYF88] again, Shaw, Warby & Whiteman [SWW95], and also Pani, Thomée & Wahlbin [PTW92]. The earliest numerical analysis that we know of for problems of type (3.2), (3.3) is that of Douglas Jr. & Jones Jr. [DJ62], who use finite differences in space as well as time.

Numerical schemes for Volterra problems are inherently "expensive". To see this, admit that any discretization of the Volterra term will (in general) involve some type of quadrature replacement. Thus, at each discrete time level t_i, about $O(i)$ operations are required to evaluate the contribution of the Volterra term, and so to solve to the final time $t_N = T$ will require about $1 + 2 + \cdots + N = O(N^2)$ operations. We term this the "N^2 problem". Also, it is evident that all previous solutions need to be referenced at each time level and this will demand around $O(N)$ storage locations in computer memory.

The error analyses referred to above are all *a priori* and, on the whole, are relevant only to quasi-uniform space and time discretizations. This means that all the time steps, $k_1, k_2, \ldots, k_N$, are about the same size, and also that at each time level the spatial mesh size, h say, is roughly constant over Ω. Moreover, it as always assumed that the space discretization does not change with time. So, for a solution that is only locally non-smooth in Q we are forced, if we want "accuracy", to take h and k "small" everywhere in Q and, in view of the operation count and storage estimates given above, we could easily overwhelm all but the most powerful computational resource. In this respect the sparse quadrature rules developed by Sloan & Thomée in [ST86], see also [TW94] and [PTW92], yield dramatic speed-up when the Volterra kernels are smooth enough but have so far been applied only to schemes with constant time step.

Now, although the error analyses could (at least for (3.1) and (3.2)) probably be enhanced to allow for variable time steps, and possibly also for nested refinements in the space mesh, the given error bounds are still not computable because they involve unknown bounds on the derivatives of u. In most cases useful such bounds are not available. Thus we obtain an algorithm that has the potential for adaptively selecting the discretization throughout Q, for optimal performance, but no criteria by which this selection can be made. It is our view that the provision of adaptive schemes for (3.1), (3.2) and (3.3) is essential to problems of viscoelasticity in order to combat the N^2 problem and hence arrive at practical and reliable solvers for physically interesting problems. This implies that we must select discretization methods to which reliable *a posteriori* error estimates can be attached, and hence computed in terms of known quantities, in order to dynamically and adaptively determine $h(x, t)$ and $k(x, t)$ during the computation.

In this article we begin to set out a plan as to how this might be achieved by deriving in Section 3 a residual-based *a posteriori* error estimator for a discontinuous-in-time, continuous-in-space Galerkin finite element discretization of an example of (3.1). Our approach is based on the pioneering work of Eriksson, Johnson *et al.*, see for example Johnson in [Whi94, Chap. 6], or Eriksson & Johnson [EJ91], who have devised an "algorithm" for error estimation based on Galerkin orthogonality and the stability properties of associated dual problems. The methodology may be regarded as a far reaching extension of the Aubin-Nitsche duality technique ("Nitsche's trick", see e.g. Oden & Reddy [OR76]). This approach has already been applied by Johnson *et al.* to a number of the classical differential equation problems of mathematical physics, see the content and references of Johnson in [Whi94, Chap. 6] and Eriksson *et al.* [EEHJ95], and a start has been made for pure-time dependent Volterra problems by Shaw & Whiteman in [SW96b] and [SW96a]. We leave *a priori* error analysis for another time since our first goal is to find an effective adaptive solution technique.

Throughout we restrict ourselves to a model problem with $n = 1$ (so that $\Omega \subset \mathbb{R}$)

in order to keep the exposition simple. Also, although the notation we use is fairly standard, and will be defined at its point of introduction, it is useful to establish some conventions here at the outset.

If $(\mathcal{B}, \| \cdot \|_{\mathcal{B}})$ is a Banach space then for real $p \in [1, \infty]$ we denote by $L_p(\mathcal{R}; \mathcal{B})$ the usual Banach space of "functions", defined on a domain $\mathcal{R}$ and valued in $\mathcal{B}$, the pth powers of which are absolutely Lebesgue integrable over $\mathcal{R}$. For $p \in [1, \infty)$ the norm for this space is

$$\| \cdot \|_{L_p(\mathcal{R};\mathcal{B})} := \left(\int_{\mathcal{R}} \| \cdot \|_{\mathcal{B}}^p \, d\mathcal{R} \right)^{1/p}, \quad \text{while} \quad \| \cdot \|_{L_\infty(\mathcal{R};\mathcal{B})} := \underset{y \in \mathcal{R}}{\text{ess sup}} \{ \| \cdot \|_{\mathcal{B}}(y) \}. \quad (3.4)$$

In addition, if $(\mathcal{H}, (\cdot, \cdot)_{\mathcal{H}})$ is a Hilbert space then $L_2(\mathcal{R}; \mathcal{H})$ is also a Hilbert space with inner product

$$(u, v)_{L_2(\mathcal{R};\mathcal{H})} := \int_{\mathcal{R}} (u, v)_{\mathcal{H}} \, d\mathcal{R}, \quad (3.5)$$

which induces the norm $\| \cdot \|_{L_2(\mathcal{R};\mathcal{H})} := \sqrt{(\cdot, \cdot)_{L_2(\mathcal{R};\mathcal{H})}}$. Below we will subscript all required norms and inner products in this way to avoid confusion, the exception to this will be for the space $L_2(\Omega)$, the inner product and induced norm of which we denote simply by $(\cdot, \cdot)$ and $\| \cdot \|$ respectively.

For an integer $m \geq 0$ and real $p \in [1, \infty]$ we write $W_p^m(\mathcal{R}; \mathcal{B})$ for the usual Sobolev space of "functions" which, along with all of their distributional derivatives up to and including those of order m, belong to $L_p(\mathcal{R}; \mathcal{B})$. In the usual multi-index notation (e.g. Oden & Reddy [OR76]) the norm is given by

$$\| \cdot \|_{W_p^m(\mathcal{R};\mathcal{B})} := \left(\sum_{|\alpha| \leq m} \| D^\alpha \cdot \|_{L_p(\mathcal{R};\mathcal{B})}^p \right)^{1/p} \quad \text{for } p \in [1, \infty), \quad (3.6)$$

with the "ess sup$_y$" and "max$_\alpha$" modifications when $p = \infty$. The space $W_2^m(\mathcal{R}; \mathcal{H})$ is a Hilbert space and, following convention, we write $H^m(\mathcal{R}; \mathcal{H}) := W_2^m(\mathcal{R}; \mathcal{H})$. Finally, if either $\mathcal{B} = \mathbb{R}$ or $\mathcal{H} = \mathbb{R}$ in any of these then we shorten the notation by writing $W_p^m(\mathcal{R})$ instead of $W_p^m(\mathcal{R}; \mathcal{B})$ etc.

The material in this paper improves on that in [SW96c] in that we suggest a way of dealing with inhomogeneous Neumann data, and we provide sharp data stability estimates for the viscoelasticity problem.

3.2   THE MODEL PROBLEM

The quasistatic viscoelasticity model that we adopt below is discussed at some length by Shaw, Warby & Whiteman in the proceedings of the last MAFELAP conference, so here it is appropriate to include only a brief summary of the governing equations and refer to [Whi94, Chap. 3] for the details. In this regard see also [SWW+94] and [SWW96].

For every time $t \in \mathcal{J}$ the interior of a linearly viscoelastic body is presumed to occupy Ω and be subject to a system of body forces, $f = (f_i)_{i=1}^n$, acting throughout its volume, and a system of surface tractions, $f_\Gamma = (f_{\Gamma i})_{i=1}^n$, acting on a part of the boundary $\Gamma_N \subset\subset \partial\Omega$. The remainder of the boundary, $\Gamma_D := \partial\Omega\setminus\overline{\Gamma_N} \neq \emptyset$, is assumed to be rigidly fixed in space. Assuming "small" deformations the state of stress and strain in the body are described respectively by the stress tensor, $(\sigma_{ij})_{i,j=1}^n$, and the strain tensor, $(\varepsilon_{ij})_{i,j=1}^n$. Let $u = (u_i)_{i=1}^n$ denote the deformation of the body from its undeformed configuration, then the components of ε are given by

$$\varepsilon_{ij}(u) := \frac{1}{2}\left(\frac{\partial u_i}{\partial x_j} + \frac{\partial u_j}{\partial x_i}\right) \qquad \text{for } i,j \in \mathcal{N}(1,n) := \{1,2,\ldots,n\}. \tag{3.7}$$

Also, in the quasistatic theory $\partial^2 u/\partial t^2$ is deemed negligible and so, from Newton's second law, we find that the stresses are related to the applied forces by the equilibrium statements (summation implied), for $t \in \mathcal{J}$:

$$\sigma_{ij,j}(x,t) = f_i(x,t) \text{ for } x \in \Omega,$$
$$u_i(x,t) = 0 \text{ for } x \in \Gamma_D, \qquad \text{and} \qquad \sigma_{ij}(x,t)\widehat{n}_j(x) = f_{\Gamma i}(x,t) \text{ for } x \in \Gamma_N, \tag{3.8}$$

where $i,j \in \mathcal{N}(1,n)$ and $\widehat{n} = (\widehat{n}_i)_{i=1}^n$ is a.e. the unit outward normal to Γ_N. To close this problem we need a constitutive relation between σ and ε which, from linear viscoelasticity theory, we take as

$$\sigma(t) = D(t,t)\varepsilon\big(u(t)\big) - \int_0^t D_s(t,s)\varepsilon\big(u(s)\big)\,ds, \tag{3.9}$$

where: subscript s indicates s-differentiation; for $0 \le s \le t \in \mathcal{J}$ D is a symmetric positive-definite matrix of material functions; we have hidden the potential dependence on $x \in \overline{\Omega}$ of these quantities; and, the body is assumed quiescent prior to $t = 0$—that is $u(t) = 0$ for all $t < 0$.

Often in linear viscoelasticity theory it is assumed that the constitutive matrix D is of convolution form—in which case we replace $D(t,t)$ by $D(0)$ and $D(t,s)$ by $D(t-s)$— and also that $D(t) = \varphi(t)D$, where D is a constant matrix and $\varphi: \mathcal{J} \to [0,\infty)$ is a so-called $stress\ relaxation\ function$. A typical form for φ is

$$\varphi(t) = \varphi_0 + \sum_{i=1}^{N_\varphi} \varphi_i e^{-\alpha_i t}, \tag{3.10}$$

with $\alpha_1,\ldots,\alpha_{N_\varphi} > 0$ and $\varphi_0,\ldots,\varphi_{N_\varphi} > 0$ such that $\varphi(0) = 1$. We note that the normalization $\varphi(0) = 1$ is always possible by appropriately scaling the matrix D.

For more on the continuum mechanics concepts alluded to above we may refer, for example, to Hunter [Hun83], and for more on this viscoelasticity problem see Shaw, Warby, Whiteman, Dawson & Wheeler [SWW+94], Shaw, Warby & Whiteman [Whi94, Chap. 3], Shaw, Warby & Whiteman [SWW96] as well as the texts by Golden & Graham [GG88] and Ferry [Fer70].

As mentioned above, a priori error analyses for various discretizations of this and related problems have been given elsewhere. It is our intention here to provide an a $posteriori$ error estimate which will allow the construction of software with rational adaptivity and useful error feedback to the user. Toward this end, and in order to focus more on the steps in the analysis as opposed to the complexity of the problem, we collapse this model to one space dimension and consider the specific example of (3.1): find u such that

$$Au(t) - \Lambda u(t) = f(t) \quad \text{in } Q := \Omega \times \mathcal{J}. \tag{3.11}$$

In this problem: $f \in L_\infty(\mathcal{J}; L_2(\Omega))$ is a given load; $\Omega := (0, 1)$; $\mathcal{J} := [0, T]$, for given $T \in (0, \infty)$;

$$\begin{aligned}
Au(t) &:= -\left(P(x)u'(x, t)\right)' + \sigma(x)u(x, t); \\
\Lambda u(t) &:= \int_0^t B(t, s)u(x, s)\, ds; \\
B(t, s)u(s) &:= -\left(P_B(t, s)u'(x, s)\right)' + \sigma_B(t, s)u(x, s);
\end{aligned} \tag{3.12}$$

the prime denotes x-differentiation; and, $P > 0$, P', $\sigma \geq 0$, P_B, σ_B are assumed to be smooth enough functions of their respective arguments. In the sequel we shall often refrain from explicitly displaying the x dependence of these and other quantities. Now defining the *Neumann-Volterra trace operator*:

$$Ku(x, t) := P(x)u'(x, t) - \int_0^t P_B(t, s)u'(x, s)\, ds, \tag{3.13}$$

and adopting the convention $Ku(t) := Ku(1, t)$, we supply this differential-Volterra problem with the mixed boundary conditions:

$$u(0, t) = 0 \quad \text{and} \quad Ku(t) = f_\Gamma(t) \quad \forall t \in \mathcal{J}, \tag{3.14}$$

for $f_\Gamma \in L_\infty(\mathcal{J})$ a given traction.

Multiplying (3.11) by a smooth function $v(x, t)$ such that $v(0, t) = 0$ for all $t \in \mathcal{J}$, and then integrating by parts over Ω, we can weaken (3.11) by introducing the test space $\mathcal{H} := \{v \in H^1(\Omega) : v(0) = 0\}$, and defining the (semi- or spatial-) weak formulation as the problem: find $u \in L_\infty(\mathcal{J}; \mathcal{H})$ such that, a.e. in $\mathcal{J}$,

$$A(u(t), v) = (f(t), v) + v(1)f_\Gamma(t) + \int_0^t B(t, s; u(s), v)\, ds \quad \forall v \in \mathcal{H}. \tag{3.15}$$

Here the bilinear forms, $A(\cdot, \cdot)$ and $B(t, s; \cdot, \cdot)$, from $\mathcal{H} \times \mathcal{H}$ into $\mathbb{R}$ are given for all v, $w \in \mathcal{H}$ by

$$\begin{aligned}
A(v, w) &:= \int_0^1 Pv'w' + \sigma vw\, dx, \\
B(t, s; v, w) &:= \int_0^1 P_B(t, s)v'w' + \sigma_B(t, s)vw\, dx \quad \text{for } 0 \leq s \leq t \in \mathcal{J}.
\end{aligned} \tag{3.16}$$

In fact (see Theorem 3.1 later) $A(\cdot, \cdot)$ is a continuous and $\mathcal{H}$-coercive bilinear form and thus $(\mathcal{H}, A(\cdot, \cdot))$ is a Hilbert space with the natural or "energy" norm $\|\cdot\|_{\mathcal{H}} := \sqrt{A(\cdot, \cdot)}$. This energy norm is equivalent to $\|\cdot\|_{H^1(\Omega)}$.

Seeking a solution in a space such as $L_\infty(\mathcal{J}; \mathcal{H})$ seems natural for this problem because (3.11) does not involve time derivatives. However, such functions do not, in general, meaningfully possess pointwise values on $\mathcal{J}$ and so we define the "fully weak" problem as: find $u \in L_\infty(\mathcal{J}; \mathcal{H})$ such that

$$a(u, v) = L(v) \qquad \forall v \in L_1(\mathcal{J}; \mathcal{H}), \tag{3.17}$$

where:

$$\begin{aligned}
a(u, v) &:= \int_0^T A\big(u(t), v(t)\big)\, dt - \int_0^T \int_0^t B\big(t, s; u(s), v(t)\big)\, ds\, dt, \\
L(v) &:= \int_0^T \big(f(t), v(t)\big)\, dt + \int_0^T f_\Gamma(t) v(1, t)\, dt.
\end{aligned} \tag{3.18}$$

We assume that a unique solution exists solving this problem, and we note that $w \in \mathcal{H} \Rightarrow w \in C(\overline{\Omega})$, this is a consequence of the Sobolev embedding theorem (see e.g. Adams [Ada75]).

Central to the methodology of Eriksson, Johnson *et al.* is the introduction of a related dual problem. It is the data-stability properties of this problem, coupled with Galerkin-orthogonality (see (3.22) later), that provides the link between the Galerkin-error, which is not computable, and the Galerkin-residual, which is. So, as a device used in deriving the *a posteriori* error estimate in the next section we define the continous dual backward problem: find $\chi \in L_1(\mathcal{J}; \mathcal{H})$ such that

$$a^*(\chi, v) = L^*(v) \qquad \forall v \in L_\infty(\mathcal{J}; \mathcal{H}). \tag{3.19}$$

Here (compare (3.18)):

$$\begin{aligned}
a^*(\chi, v) &:= \int_0^T A\big(v(t), \chi(t)\big)\, dt - \int_0^T \int_t^T B\big(s, t; v(t), \chi(s)\big)\, ds\, dt; \\
L^*(v) &:= \int_0^T \big(v(t), g(t)\big)\, dt + \int_0^T v(1, t) g_\Gamma(t)\, dt;
\end{aligned} \tag{3.20}$$

and judicious choices for $g \in L_1\big(\mathcal{J}; L_2(\Omega)\big)$ and $g_\Gamma \in L_1(\mathcal{J})$ will be made later. This definition is motivated by the following lemma.

Lemma 3.1 $a^*(w, v) = a(v, w)$ $\forall v \in L_\infty(\mathcal{J}; \mathcal{H})$ *and* $\forall w \in L_1(\mathcal{J}; \mathcal{H})$.

Proof. Interchange the order of time integration in the Volterra term.
This lemma, along with the stability estimates to follow, will be exploited in the next section. Now, motivated by the physically reasonable form of the stress relaxation function (3.10), we introduce a useful nomenclature.

Definition 3.1 *A function* $\phi \in C\big(\mathcal{J}; [0, \infty)\big)$ *will be called a* function with fading memory *if the quantity* $\Phi(t) := 1/\big(1 - \|\phi\|_{L_1(0,t)}\big)$ *is defined in* $\mathcal{J}$ *and is such that* $\Phi(t) \in (0, \infty)$ *for all* $t \in \mathcal{J}$. *We will call the nondecreasing function* Φ *the* long term effect *of* ϕ.

Equipped with this important notion of fading memory we can now state a data-stability estimate for (3.11). This theorem requires a rather detailed proof which space prevents us from including, full details can be found in the technical report: Shaw & Whiteman [SW96d] (the reasoning is similar to that used in [SWW96]).

Theorem 3.1 *Consider (3.11) in the two cases $f \in L_p(\mathcal{J}; L_2(\Omega))$, $f_\Gamma \in L_p(\mathcal{J})$ for $p = 1$ and $p = \infty$, and assume that in each case a unique solution $u \in L_p(\mathcal{J}; \{v \in \mathcal{H} : Av \in L_2(\Omega)\})$ exists. Also, assume that $B(t, s)$ is similar to A in the following two senses. For all $v, w \in \mathcal{H}$ and a.e. in $\{0 \le s \le t \in \mathcal{J}\}$:*

(a) $\|B(t, s)v\| \le \phi(t - s)\|Av\|$,

(b) $|B(t, s; w, v)| \le \phi(t - s)\|w\|_\mathcal{H}\|v\|_\mathcal{H}$,

where ϕ is a function with fading memory with long term effect Φ. Then, using (a) the elliptic stability estimates hold:

$$\|Au\|_{L_p(0,t;L_2(\Omega))} \le C_1(p, t)\|f\|_{L_p(0,t;L_2(\Omega))} \qquad (A)$$

for $p = 1$ or $p = \infty$ where $C_1(1, t) := S_1(t)$ and $C_1(\infty, t) := \Phi(t)$. Also, defining: $\hat{P}(t) := \|P(t)\|_{L_\infty(\Omega)}$, $\hat{\sigma}(t) := \|\sigma(t)\|_{L_\infty(\Omega)}$, $\check{P}(t) := \text{ess inf}\{P(t) : x \in \Omega\}$ and $\check{\sigma}(t) := \text{ess inf}\{\sigma(t) : x \in \Omega\}$, we have for a.e. $t \in \mathcal{J}$ the equivalence of norms,

$$C_2(t)\|u\|_{H^1(\Omega)} \le \|u\|_\mathcal{H} \le C_3(t)\|u\|_{H^1(\Omega)}, \qquad (B)$$

where: if $\check{\sigma} > 0$,

$$C_2(t) := \sqrt{\min\{\check{P}(t), \check{\sigma}(t)\}} \qquad and \qquad C_3(t) := \sqrt{\max\{\hat{P}(t), \hat{\sigma}(t)\}},$$

otherwise if $\check{\sigma} \ge 0$, the Friedrichs inequality

$$\|u\| \le \frac{1}{\sqrt{2}}\|u'\| \le \frac{1}{\sqrt{2}}\|u\|_{H^1(\Omega)} \qquad (C)$$

gives $C_2(t) := \sqrt{2\check{P}(t)/3}$ instead, with $C_3(t)$ unchanged. Incorporating assumption (b), on $|B(t, s; \cdot, \cdot)|$, the following energy- and L_2-stability estimates hold:

$$\|u\|_{L_p(0,t;\mathcal{H})} \le C_4(p, t)\left(\|f\|_{L_p(0,t;L_2(\Omega))} + \sqrt{2}\|f_\Gamma\|_{L_p(0,t)}\right),$$
$$\|u\|_{L_p(0,t;L_2(\Omega))} \le C_6(p, t)\left(\|f\|_{L_p(0,t;L_2(\Omega))} + \sqrt{2}\|f_\Gamma\|_{L_p(0,t)}\right), \qquad (D)$$

for $p = 1$ or $p = \infty$ where $C_6(p, t) := 2^{-1/2}\|C_2^{-1}\|_{L_\infty(0,t)}C_4(p, t)$,

$$C_4(1, t) := 2^{-1/2}\|C_2^{-1}\|_{L_\infty(0,t)}S_1(t) \qquad and \qquad C_4(\infty, t) := 2^{-1/2}\|C_2^{-1}\|_{L_\infty(0,t)}\Phi(t).$$

Finally, if $P'(x, t) \in L_\infty(\Omega)$ for all $t \in \mathcal{J}$, the following strong stability estimate holds

$$\|u''\|_{L_p(0,t;L_2(\Omega))} \le C_7(p, t)\left(\|f\|_{L_p(0,t;L_2(\Omega))} + \sqrt{2}\|f_\Gamma\|_{L_p(0,t)}\right), \qquad (E)$$

for $p = 1$ or $p = \infty$ where

$$C_7(p, T) := \max\left\{C_4(p, t)\left\|\frac{C_5(t)}{\check{P}(t)C_2(t)}\right\|_{L_\infty(0,t)}, C_1(p, t)\left\|\frac{\sqrt{2}}{\check{P}(t)}\right\|_{L_\infty(0,t)}\right\},$$

and $C_5(t) := \|P'(t)\|_{L_\infty(\Omega)} + \hat{\sigma}(t)$.

Note that the role of ϕ in this theorem qualifies our remark in the introduction that $B(t, s)$ is "similar" to A. Note also that the backward problem (3.19) is no more than a forward problem in reversed time. Thus we also have dual stability estimates. Again the proof can be found in [SW96d].

Theorem 3.2 *Consider the dual problem (3.19) in the two cases* $g \in L_p(\mathcal{J}; L_2(\Omega))$, $g_\Gamma \in L_p(\mathcal{J})$ *for* $p = 1$ *and* $p = \infty$, *and assume that in each case a unique solution* $\chi \in L_p(\mathcal{J}; \{v \in \mathcal{H} : Av \in L_2(\Omega)\})$ *exists. Then the dual problem may be expressed in classical forward form as the problem: find* $\widehat{\chi}$ *such that*

$$A\widehat{\chi}(t) = \widehat{g}(t) + \int_0^t \widehat{B}(t, s)\widehat{\chi}(s) \, ds \quad \text{in } Q, \tag{A}$$

subject to $\widehat{\chi}(0, t) = 0$ *and* $\widehat{K}\widehat{\chi}(t) = \widehat{g}_\Gamma(t)$ *at* $x = 0$ *and* $x = 1$ *respectively. This problem is related to (3.19) by:*

$$\widehat{\chi}(t) \equiv \chi(T - t), \quad \widehat{g}(t) \equiv g(T - t), \quad \widehat{g}_\Gamma(t) \equiv g_\Gamma(T - t), \quad \widehat{B}(t, s) \equiv B(T - s, T - t),$$

and $\widehat{K}$ *is the Neumann-Volterra trace operator, associated with* A *and* $\widehat{B}(t, s)$, *analogous to (3.13). Moreover, the similarity assumptions (a) and (b) in Theorem 3.1 imply that*

$$\|\widehat{B}(t, s)v\| \leq \phi(t - s)\|Av\| \quad \text{and} \quad |\widehat{B}(t, s; w, v)| \leq \phi(t - s)\|w\|_\mathcal{H}\|v\|_\mathcal{H}, \tag{B}$$

for all $v, w \in \mathcal{H}$ *and a.e. in* $\{0 \leq s \leq t \in \mathcal{J}\}$. *The bilinear form associated with* $\widehat{B}(t, s)$ *is defined in the obvious way. These imply that a direct analogue of Theorem 3.1 applies also to (A), and hence also to (3.19). In particular, from Theorem 3.1 parts (D) and (E) we have:*

$$\|\chi\|_{L_p(T-t, T; \mathcal{H})} \leq C_4(p, t) \left(\|g\|_{L_p(T-t, T; L_2(\Omega))} + \sqrt{2}\|g_\Gamma\|_{L_p(T-t, T)} \right), \tag{C}$$

$$\|\chi\|_{L_p(T-t, T; L_2(\Omega))} \leq C_6(p, t) \left(\|g\|_{L_p(T-t, T; L_2(\Omega))} + \sqrt{2}\|g_\Gamma\|_{L_p(T-t, T)} \right), \tag{D}$$

$$\|\chi''\|_{L_p(T-t, T; L_2(\Omega))} \leq C_7(p, t) \left(\|g\|_{L_p(T-t, T; L_2(\Omega))} + \sqrt{2}\|g_\Gamma\|_{L_p(T-t, T)} \right), \tag{E}$$

for $p = 1$ *or* $p = \infty$, *all* $t \in \mathcal{J}$ *and where* C_4, C_6 *and* C_7 *are as given in Theorem 3.1.*

Remark 3.1 *We remark that in more general instances of (3.11), where the fading memory assumption does not apply, the form of these stability estimates still hold but the constant will usually be exponentially large in* t. *This is due to the Gronwall lemma.*

We now move on to give a residual-based *a posteriori* error estimate.

3.3 DISCRETIZATION AND AN *A POSTERIORI* GALERKIN-ERROR ESTIMATE

To effect a finite element discretization of (3.17) we firstly partition $\mathcal{J}$ into N time intervals $\{\mathcal{J}_i := (t_{i-1}, t_i)\}_{i=1}^N$, each of length $k_i := t_i - t_{i-1} > 0$, and such that

$$\mathcal{J} = \overline{\mathcal{J}_1 \cup \mathcal{J}_2 \cup \cdots \cup \mathcal{J}_N},$$

with $t_0 = 0$ and $t_N = T$. For each of these $\mathcal{J}_i$ we also partition $\Omega = \Omega^i$ into M^i intervals $\{\Omega^i_j := (x_{j-1}, x_j)\}_{j=1}^{M^i}$, each of length $h_{ij} := x_j - x_{j-1} > 0$, and such that

$$\overline{\Omega} = \overline{\Omega^i_1 \cup \Omega^i_2 \cup \cdots \cup \Omega^i_{M^i}} \qquad \forall i \in \mathcal{N}(1, N),$$

with $x_0 = 0$ and $x_{M^i} = 1$. By $h_i = h_i(x)$ we denote the piecewise constant mesh function on the time interval $\mathcal{J}_i$. We need to assume that this discretization of Q always exists because in an adaptive scheme the h_{ij} and k_i are of course not known *a priori*.

Thus we have broken the prismatic domain Q into a sequence of consecutive "slabs" indexed by the time levels $i \in \mathcal{N}(1, N)$. For each Ω^i we define the semidiscrete (spatial) finite element spaces

$$\mathcal{H}_i := \{v \in \mathcal{H} : v \text{ is linear on } \Omega^i_j \text{ for each } j \in \mathcal{N}(1, M^i)\} \quad \forall i \in \mathcal{N}(1, N),$$

and then define the space-time finite element space as

$$V_r := \{v \in L_\infty(\mathcal{J}; \mathcal{H}) : v|_{\mathcal{J}_i} \in \mathbb{P}_r(\mathcal{J}_i; \mathcal{H}_i) \quad \forall i \in \mathcal{N}(1, N)\},$$

where $\mathbb{P}_r(\mathcal{J}_i; \mathcal{H}_i)$ is the space of polynomials of degree r, defined on $\mathcal{J}_i$, and with coefficients in $\mathcal{H}_i$. We also define the sets of internal nodes of Ω^i for each time level:

$$\mathcal{N}^i := \{x_j, \ j \in \mathcal{N}(1, M^i - 1), \text{ is an internal node in } \Omega^i\}.$$

Below we work exclusively with V_0 so that we have an approximating space consisting of continuous piecewise linears in space with discontinuous piecewise constant coefficients in time. The scheme is then an application of the standard continuous Galerkin method in space coupled to a discontinuous Galerkin method in time and, including the polynomial degrees, may be abbreviated to the dG(0)cG(1) method. With the space V_0 we form the Galerkin finite element approximation to (3.17) as: find $U \in V_0$ such that

$$a(U, v) = L(v) \qquad \forall v \in V_0, \tag{3.21}$$

and subtracting this from (3.17) we immediately obtain the Galerkin orthogonality relationship

$$a(u - U, v) = 0 \qquad \forall v \in V_0, \tag{3.22}$$

which plays a crucial role in finite element error analysis.

A fundamentally important role in our *a posteriori* error estimator is played by the *Volterra projection*, to which we now turn.

Definition 3.2 *For any* $w \in L_\infty(\mathcal{J}; \mathcal{H})$ *we define the Volterra projection,* Y, *such that* $Y: w \to Yw$ *almost everywhere in* $\mathcal{J}$ *in the following piecewise manner.* $Y_i: L_\infty(\mathcal{J}_i; \mathcal{H}) \to L_\infty(\mathcal{J}_i; \mathcal{H}_i)$ *for all* $i \in \mathcal{N}(1, N)$, *where* Y_i *is defined by the variational equation:* $(Y_i w(t), v) = \Psi_w(t; v)$ *for all* $v \in \mathcal{H}_i$, *a.e. in* $\mathcal{J}_i$, *and where*

$$\Psi_w(t; v) := -A(w(t), v) + \int_0^t B(t, s; w(s), v)\, ds + v(1) f_\Gamma(t).$$

Treating t *as a parameter and using a trace theorem to see that* $|v(1)| \leq C\|v\|_{\mathcal{H}}$, *it is clear that* $\Psi_w(t; v)$ *is, for fixed* w, *a continuous linear form on* $\mathcal{H}_i$ *and thus* $Y_i w$ *exists and is unique (by the Riesz theorem).*

This projection will allow us to separate completely the space and time errors when deriving the error estimate below, to see this it is useful to think in terms of a semidiscrete approximation to (3.11). Based on (3.15) we will call the function $\widetilde{U}$, which is such that $\widetilde{U}|_{\mathcal{J}_i} \in L_\infty(\mathcal{J}_i; \mathcal{H}_i)$, a semidiscrete approximation to u if

$$A(\widetilde{U}(t), v) = (f(t), v) + v(1)f_\Gamma(t) + \int_0^t B(t, s; \widetilde{U}(s), v)\, ds \qquad \forall v \in \mathcal{H}_i, \qquad (3.23)$$

almost everywhere in $\mathcal{J}_i$, for all $i \in \mathcal{N}(1, N)$. In terms of a notional semidiscrete dual problem another useful notation is

$$\mathcal{H}_{\mathcal{J}} := \{v \in L_1(\mathcal{J}; \mathcal{H}) : v|_{\mathcal{J}_i} \in L_1(\mathcal{J}_i; \mathcal{H}_i) \quad \forall i \in \mathcal{N}(1, N)\},$$

which we might think of as a space of "semidiscrete dual functions". The key point is this.

Proposition 3.1 *Let $\mathcal{P}_i: L_2(\Omega) \to \mathcal{H}_i$ denote the $L_2(\Omega)$ projection on $\mathcal{H}_i$ for each $i \in \mathcal{N}(1, N)$. Then, with regard to the semidiscrete problem (3.23), we have $-Y_i\widetilde{U}(t) = \mathcal{P}_i f(t)$ a.e. in $\mathcal{J}_i$ for each i. Moreover,*

$$(f + Y_iU, v) = (Y_i(U - \widetilde{U}), v) = (Y_iU + \mathcal{P}_i f, v) \qquad \forall v \in \mathcal{H}_i, \text{ a.e. in } \mathcal{J}_i,$$

where $\mathcal{P}_i f$ is given by $(\mathcal{P}_i f, v) = (f, v)$ for all $v \in \mathcal{H}_i$.

Proof. Firstly, working from (3.23) it is clear from Definition 3.2 that $(f(t), v) = (-Y_i\widetilde{U}, v)$ for all $v \in \mathcal{H}_i$. Now, using this we get:

$$(f + Y_iU, v) = (f + Y_i\widetilde{U} - Y_i\widetilde{U} + Y_iU, v) = (Y_i(U - \widetilde{U}), v) = (Y_iU + \mathcal{P}_i f, v),$$

as required.

And so the Volterra projection will allow us to study the time discretization error $U - \widetilde{U}$ in terms of projections of the known quantities U and f. Our next result is an error representation formula in terms of the data to the dual problem, the L_2-projection referred to below is given in more detail later in Lemma 3.5.

Lemma 3.2 *Let $\{P_i\}_{i=1}^N$ be a collection of maps $P_i: C(\overline{\Omega}) \to \mathcal{H}_i$ and let $L: \mathcal{H}_{\mathcal{J}} \to V_0$ be the L_2 projection on V_0. If P and P_B are continuous on $\overline{\Omega}$, and defining θ and ρ such that $\theta|_{\mathcal{J}_i} = \theta_i := \chi - P_i\chi \in L_1(\mathcal{J}_i; \mathcal{H})$ and $\rho|_{\mathcal{J}_i} = \rho_i := P_i\chi - LP_i\chi \in L_1(\mathcal{J}_i; \mathcal{H}_i)$, we have the following error representation formula:*

$$L^*(e) = \int_0^T \left(f - AU + \int_0^t B(t, s)U(s)\, ds, \theta \right) dt + \int_0^T \theta(1, t)\Big(f_\Gamma(t) - KU(t)\Big) dt$$

$$- \sum_{i=1}^N \int_{t_{i-1}}^{t_i} \left[\sum_{k=1}^i \int_{t_{k-1}}^{\min\{t_k, t\}} \sum_{j \in \mathcal{N}^k} \theta_i(x_j, t)P_B(t, s, x_j)[U_j'(s)]\, ds \right.$$

$$\left. - \sum_{j \in \mathcal{N}^i} \theta_i(x_j, t)P(x_j)[U_j'(t)] \right] dt + \sum_{i=1}^N \int_{t_{i-1}}^{t_i} (\mathcal{P}_i f + Y_iU, \rho_i)\, dt.$$

Here: $[U_j'(t)] := U'(x_j^+, t) - U'(x_j^-, t)$ is the jump in first derivative at x_j, where x_j a node on any space mesh; and $AU(t)$, $B(t, s)U(s)$ must be interpreted in a piecewise manner on each space mesh.

Proof. In the dual problem (3.19) take $v = e \in L_\infty(\mathcal{J}; \mathcal{H})$ and use Lemma 3.1 with the Galerkin orthogonality (3.22) to write

$$\begin{aligned} L^*(e) &= a^*(\chi, e) = a(e, \chi) = a(u, \chi - \pi\chi) - a(U, \chi - \pi\chi) \\ &= L(\chi - \pi\chi) - a(U, \chi - \pi\chi) \end{aligned}$$

for all $\pi\chi \in V_0$. Choosing $\pi\chi|_{\mathcal{J}_i} = LP_i\chi$ for all $i \in \mathcal{N}(1, N)$ we obtain

$$L^*(e) = \Big[L(\theta) - a(U, \theta)\Big] + \Big[L(\rho) - a(U, \rho)\Big] = I + II. \tag{i}$$

For the term I we integrate by parts over each spatial element, on each time level, and recall that $w(0) = 0$ and $w \in C(\overline{\Omega})$ for $w \in \mathcal{H}$. From (3.16) with $t \in \mathcal{J}_i$ and $s \in \mathcal{J}_k$, for $k \le i$, we therefore get:

$$\begin{aligned} A(U, \theta_i) &= \sum_{j=1}^{M^i} \int_{x_{j-1}}^{x_j} PU'\theta_i' + \sigma U\theta_i \, dx, \\ &= \sum_{j=1}^{M^i} \int_{x_{j-1}}^{x_j} \theta_i AU \, dx - \sum_{j \in \mathcal{N}_i} \theta_i(x_j, t) P(x_j)[U_j'(t)] + \theta_i(1, t) P(1) U'(1, t), \end{aligned}$$

and

$$\begin{aligned} \int_0^t B(t, s; U(s), \theta) \, ds &= \sum_{k=1}^i \int_{t_{k-1}}^{\min\{t_k, t\}} \sum_{j=1}^{M^k} \int_{x_{j-1}}^{x_j} P_B U'(s)\theta_i' + \sigma_B U(s)\theta_i \, dx \, ds \\ &= \sum_{k=1}^i \int_{t_{k-1}}^{\min\{t_k, t\}} \Bigg[\sum_{j=1}^{M^k} \int_{x_{j-1}}^{x_j} \theta_i(t) B(t, s) U(s) \, dx \\ &\quad - \sum_{j \in \mathcal{N}^k} \theta_i(x_j, t) P_B(t, s, x_j)[U_j'(s)] + \theta_i(1, t) P_B(1) U'(1, s)\Bigg] \, ds. \end{aligned}$$

Putting these together, and being careful to interpret $AU(t)$ and $B(t, s)U(s)$ in a piecewise manner on each space mesh, we get

$$\begin{aligned} I &= \int_0^T \Big(f(t) - AU(t) + \int_0^t B(t, s)U(s) \, ds, \theta(t)\Big) \, dt + \int_0^T \theta(1, t)\Big(f_\Gamma(t) - KU(t)\Big) \, dt \\ &\quad + \sum_{i=1}^N \int_{t_{i-1}}^{t_i} \Bigg[\sum_{j \in \mathcal{N}_i} \theta_i(x_j, t) P(x_j)[U_j'(t)] \\ &\quad - \sum_{k=1}^i \int_{t_{k-1}}^{\min\{t_k, t\}} \sum_{j \in \mathcal{N}^k} \theta_i(x_j, t) P_B(t, s, x_j)[U_j'(s)] \, ds\Bigg] \, dt. \end{aligned}$$

For the term II we use the fact that $\rho \in \mathcal{H}_\mathcal{J}$, then using the Volterra projection and Proposition 3.1 we get

$$II = \sum_{i=1}^N \int_{t_{i-1}}^{t_i} \Big[(f, \rho_i) + \rho_i(1, t) f_\Gamma(t) - A(U, \rho_i) + \int_0^t B(t, s; U(s), \rho_i) \, ds\Big] \, dt$$

$$= \sum_{i=1}^{N} \int_{t_{i-1}}^{t_i} (f + Y_i U, \rho_i) \, dt = \sum_{i=1}^{N} \int_{t_{i-1}}^{t_i} (\mathcal{P}_i f + Y_i U, \rho_i) \, dt.$$

Putting these representations of I and II together, and recalling (i), completes the proof.

To obtain an error bound we need to bound $|L^*(e)|$ from below by the error itself and then make an appropriate choice for the $\{P_i\}$ to bound $|L^*(e)|$ from above by computable quantities. In the first instance we have the following lower bound.

Lemma 3.3 *Let* $z \in L_1(\mathcal{J})$ *be arbitrary, then if in the dual problem (3.19) we set* $g = ze \in L_1(\mathcal{J}; \mathcal{H})$ *and* $g_\Gamma = 0$, *it holds that*

$$\sup \{ |L^*(e)| : \|z\|_{L_1(\mathcal{J})} = 1 \} = \|e\|^2_{L_\infty(\mathcal{J}; L_2(\Omega))},$$

where $e = u - U$ *is the Galerkin error.*

Proof. Simply: $\sup\{|L^*(e)| : \|z\|_{L_1(\mathcal{J})} = 1\} = \big\| \|e\|^2 \big\|_{L_\infty(\mathcal{J})}$, since $L_\infty(\mathcal{J})$ is dual to $L_1(\mathcal{J})$.

Before continuing we require the following interpolation error estimates.

Lemma 3.4 *For an integer* $m \geq 1$ *suppose that* $w: [a, b] \to \mathbb{R}$ *has pointwise values for real numbers* a *and* b *such that* $h := b - a > 0$. *Let* $w_I \in \mathbb{P}_1(a, b)$ *be the linear interpolate to* w *given by*

$$w_I(x) := \frac{x}{h} w(b) + \left(1 - \frac{x}{h}\right) w(a),$$

then for sufficiently regular w,

$$\|w - w_I\|_{L_2(a,b)} \leq \min \left\{ \Pi_1 \|hw'\|_{L_2(a,b)}, \Pi_2 \|h^2 w''\|_{L_2(a,b)} \right\},$$

where: $\Pi_1 = \sqrt{8/27}$; $\Pi_2 = \sqrt{72/3125}$.

Proof. Note that w has pointwise values so this interpolate is well defined. For the first derivative result expand w in Taylor series about $w(a)$ and $w(b)$, then substitute in w_I. The second derivative result is classical.

Lemma 3.5 *Define* $L: \mathcal{H}_\mathcal{J} \to V_0$ *as the* L_2 *projection in time acting on semidiscrete functions* w. *That is,* Lw *is obtained from*

$$\langle Lw, v \rangle = \langle w, v \rangle \qquad \forall v \in V_0 \qquad \text{where} \qquad \langle \cdot, \cdot \rangle := \int_0^T (\cdot, \cdot) \, dt,$$

then: $\|Lw\|_{L_1(\mathcal{J}_i; L_2(\Omega))} \leq \|w\|_{L_1(\mathcal{J}_i; L_2(\Omega))}$ *for all* $i \in \mathcal{N}(1, N)$.

Proof. Choose an $i \in \mathcal{N}(1, N)$ and take $v = 0$ everywhere except in $\mathcal{J}_i$ where we set $v = Lw$, hence $v \in V_0$ and

$$k_i \|Lw\|^2 = \int_{t_{i-1}}^{t_i} (w, Lw) \, dt \leq \|Lw\| \, \|w\|_{L_1(\mathcal{J}_i; L_2(\Omega))}.$$

Noting that $\|Lw\|_{L_1(\mathcal{J}_i; L_2(\Omega))} = k_i \|Lw\|$, since Lw is constant in time on $\mathcal{J}_i$, the desired result is immediate.

With these preliminaries out of the way we can state our *a posteriori* Galerkin-error estimate.

Theorem 3.3 *Let the solutions, u and χ, to (3.11) and (3.19) exist uniquely in $L_p(\mathcal{J}; \{v \in \mathcal{H} : Av \in L_2(\Omega)\})$ for $p = 1$ and $p = \infty$ respectively. Also let Theorems 3.1 and 3.2 hold in these specific cases. Impose the mesh condition $\mathcal{H}_{i-1} \subseteq \mathcal{H}_i$ for all $i \in \mathcal{N}(2, N)$, then the Galerkin solution satisfies the a posteriori error bound:*

$$\|u - U\|_{L_\infty(0,t_m;L_2(\Omega))} \leq \max_{1 \leq i \leq m} \left\{ \mathcal{C}_h(t_m)\mathcal{E}_i^h + \mathcal{C}_k(t_m, \hat{h}_i)\mathcal{E}_i^k \right\} \qquad \forall m \in \mathcal{N}(1, N),$$

where,

$$(a) \quad \mathcal{E}_i^h \quad := \quad \left\| h_i^2(f - AU + \Lambda U) \right\|_{L_\infty(\mathcal{J}_i;L_2(\Omega))},$$

$$(b) \quad \mathcal{E}_i^k \quad := \quad \left\| \mathcal{P}_i f + Y_i U \right\|_{L_\infty(\mathcal{J}_i;L_2(\Omega))},$$

and $f - AU + \Lambda U$ is interpreted piecewise on each space mesh. In addition:

$$\mathcal{C}_h(t_m) \quad := \quad \Pi_2 C_7(1, t_m),$$

$$\mathcal{C}_k(t_m, \hat{h}_m) \quad := \quad 2 C_6(1, t_m)$$

$$+ 2 \min \left\{ \Pi_1 \hat{h}_m C_4(1, t_m) \left\| C_2^{-1}(t) \right\|_{L_\infty(0,t_m)}, \Pi_2 \hat{h}_m^2 C_7(1, t_m) \right\},$$

$$\hat{h}_m \quad := \quad \max \left\{ h_{ij} : j \in \mathcal{N}(1, M^i) \text{ and } i \in \mathcal{N}(1, m) \right\},$$

and all other notation is as defined previously in Theorem 3.1 and Lemma 3.4.

Proof. We demonstrate the proof for $m = N$ only. Set $g_\Gamma = 0$ in the dual problem and let $P_i\chi \in L_1(\mathcal{J}_i; \mathcal{H}_i)$ denote the nodal interpolant of $\chi|_{\mathcal{J}_i} \in L_1(\mathcal{J}_i; \mathcal{H})$ for each $i \in \mathcal{N}(1, N)$. This implies that $P_i\chi - \chi = 0$ for all nodes $x_j \in \Omega^i$, for a.e. $t \in \mathcal{J}_i$ and for all $i \in \mathcal{N}(1, N)$. In this case, and because of the nested mesh condition, Lemma 3.2 simplifies to give

$$L^*(e) \quad = \quad \sum_{i=1}^N \int_{t_{i-1}}^{t_i} \left(f - AU + \int_0^t B(t,s)U(s)\, ds, (\chi - P_i\chi)(t) \right)$$

$$+ \left(\mathcal{P}_i f + Y_i U, (P_i\chi - L P_i\chi)(t) \right) dt,$$

$$= \quad \mathrm{I} + \mathrm{II},$$

where we recall that $f - AU + \Lambda U$ is interpreted in a piecewise manner on each space mesh. Taking this expression term by term we get firstly that

$$|\mathrm{I}| \quad \leq \quad \sum_{i=1}^N \int_{t_{i-1}}^{t_i} \|h_i^2(f - AU + \Lambda U)\| \, \|h_i^{-2}(\chi - P_i\chi)(t)\| \, dt,$$

$$\leq \quad \Pi_2 \sum_{i=1}^N \|h_i^2(f - AU + \Lambda U)\|_{L_\infty(\mathcal{J}_i;L_2(\Omega))} \int_{t_{i-1}}^{t_i} \|\chi''\| \, dt,$$

$$\leq \quad \Pi_2 \max_{1 \leq i \leq N} \left\{ \|h_i^2(f - AU + \Lambda U)\|_{L_\infty(\mathcal{J}_i;L_2(\Omega))} \right\} \|\chi''\|_{L_1(\mathcal{J};L_2(\Omega))},$$

after using the interpolation estimate from Lemma 3.4. We now use the projection estimate, $\|(I - L)P_i\chi\|_{L_1(\mathcal{J}_i;L_2(\Omega))} \leq 2\|P_i\chi\|_{L_1(\mathcal{J}_i;L_2(\Omega))}$ from Lemma 3.5, and from

Lemma 3.4 and Theorem 3.2 we get

$$\|P_i\chi\|_{L_1(\mathcal{J}_i;L_2(\Omega))} \quad \leq \quad \|\chi\|_{L_1(\mathcal{J}_i;L_2(\Omega))}$$
$$+ \min\left\{\Pi_1\hat{h}\|\chi'\|_{L_1(\mathcal{J}_i;L_2(\Omega))}, \Pi_2\hat{h}^2\|\chi''\|_{L_1(\mathcal{J}_i;L_2(\Omega))}\right\}$$
$$\implies \sum_{i=1}^{N}\int_{t_{i-1}}^{t_i}\|(I-L)P_i\chi\|\,dt \leq C_k(T,\hat{h}_N)\|g\|_{L_1(\mathcal{J};L_2(\Omega))}.$$

Using these, and arguing similarly as for I, we now find that

$$|\text{II}| \leq C_k(T,\hat{h}_N)\max_{1\leq i\leq N}\left\{\|\mathcal{P}_if + Y_iU\|_{L_\infty(\mathcal{J}_i;L_2(\Omega))}\right\}\|g\|_{L_1(\mathcal{J};L_2(\Omega))}.$$

Using now the stability estimate $\|\chi''\|_{L_1(\mathcal{J};L_2(\Omega))} \leq C_7(1,T)\|g\|_{L_1(\mathcal{J};L_2(\Omega))}$, from Theorem 3.2, we then get

$$|L^*(e)| \quad \leq \quad C_h(T)\max_{1\leq i\leq N}\left\{\|h_i^2(f - AU + \Lambda U)\|_{L_\infty(\mathcal{J}_i;L_2(\Omega))}\right\}\|g\|_{L_1(\mathcal{J};L_2(\Omega))}$$
$$+C_k(T,\hat{h}_N)\max_{1\leq i\leq N}\left\{\|\mathcal{P}_if + Y_iU\|_{L_\infty(\mathcal{J}_i;L_2(\Omega))}\right\}\|g\|_{L_1(\mathcal{J};L_2(\Omega))}.$$

In the dual problem we now set $g = z(t)e(t)$, for arbitrary $z \in L_1(\mathcal{J})$, and note that

$$\|g\|_{L_1(\mathcal{J};L_2(\Omega))} = \int_0^T |z(t)|\,\|e(t)\|\,dt \leq \|z\|_{L_1(\mathcal{J})}\|\|e(t)\|\|_{L_\infty(\mathcal{J})}.$$

So, with this and Lemma 3.3 we obtain

$$\|e\|_{L_\infty(\mathcal{J};L_2(\Omega))}^2 \quad = \quad \sup\left\{|L^*(e)| : \|z\|_{L_1(\mathcal{J})} = 1\right\}$$
$$\leq \quad \max_{1\leq i\leq N}\left\{C_h(T)\mathcal{E}_i^h + C_k(T,\hat{h}_N)\mathcal{E}_i^k\right\}\|e\|_{L_\infty(\mathcal{J};L_2(\Omega))}.$$

An obvious cancellation of the error term completes the proof.

3.4 IMPLEMENTATION AND NUMERICAL EXPERIMENTS

So far we have produced an *a posteriori* Galerkin-error bound for the fully discrete finite element approximation (3.21) to the classical problem (3.11). Unfortunately the scheme (3.21) is not fully practical because the integrals associated with the Volterra operator and inner products need to be evaluated. To arrive at a fully practical discrete scheme we will have to replace these integrals with numerical quadrature rules and this will introduce an error into the computed solution that is *not bounded by Theorem 3.3*. This situation is unfortunate but not uncommon in finite element methodology.

In this section we will use a simple template to generate exact solutions to (3.11), and then use these solutions to demonstrate Theorem 3.3. Our approach to the quadrature problem is, in this article, somewhat heuristic. The Galerkin scheme (3.21) can be written in the form

$$\int_{t_{i-1}}^{t_i}\left(AU(t) - \int_{t_{i-1}}^{t}B(t,s)U(s)ds\right)dt = \int_{t_{i-1}}^{t_i}F(t)\,dt + \int_{t_{i-1}}^{t_i}\sum_{j=1}^{i-1}\int_{t_{j-1}}^{t_j}B(t,s)U(s)\,ds\,dt,$$

for all $i \in \mathcal{N}(1, N)$. Here $\boldsymbol{F}$ is the load vector and $\boldsymbol{A}$, $\boldsymbol{B}(t, s)$ are the stiffness and stiffness-history matrices generated by the bilinear forms $A(\cdot, \cdot)$ and $B(t, s; \cdot, \cdot)$. In the test problems below the datum $P_B(t, s)$ is simple enough for the Volterra integrals of $\boldsymbol{B}(t, s)$ to be performed exactly, which we do, and all other (space and time) integrals are replaced by the Gauss rule:

$$\int_{-1}^{1} w(y)\, dy \approx w(-1/\sqrt{3}) + w(1/\sqrt{3}), \tag{3.24}$$

which is exact for cubics. Therefore this quadrature rule has a truncation error much smaller than the expected Galerkin error since, with piecewise linears in space and piecewise constants in time, we expect the latter error in $L_\infty(\mathcal{J}; L_2(\Omega))$ to be of second- and first-order respectively.

Also, in the results below $\|\cdot\|_{L_2^h(\Omega)}$, or $\|\cdot\|_h$ for short, is the "norm" resulting from applying (3.24) to $\|\cdot\|$, and in Theorem 3.3 we replace the L_∞ norm in $\mathcal{E}_i^h$ by the maximum of the $L_2^h(\Omega)$ norm sampled at the ends and midpoint of $\mathcal{J}_i$, and in $\mathcal{E}_i^k$ we replace $\|\mathcal{P}_i f + Y_i U\|_{L_\infty(\mathcal{J}_i; L_2(\Omega))}$ by $\|(\mathcal{P}_i f + Y_i U)(t_i)\|_{L_2^h(\Omega)}$.

To illustrate the performance of this scheme we construct a template for generating exact solutions to (3.11) in the following way. The solution is required to be of the separable form $u(x, t) = X(x)Z(t)$, then

$$\begin{aligned}
Au(t) &= -Z(t)\big(P(x, t)X''(x) + P'(x, t)X'(x) - \sigma(x, t)X(x)\big), \\
\Lambda u(t) &= -X''(x)V_1(x, t) - X'(x)V_2(x, t) + X(x)V_3(x, t),
\end{aligned}$$

where:

$$V_1(x, t) := \int_0^t P_B(x, t, s)Z(s)\, ds, \qquad V_2(x, t) := \int_0^t P_B'(x, t, s)Z(s)\, ds,$$

$$V_3(x, t) := \int_0^t \sigma_B(x, t, s)Z(s)\, ds.$$

Hence, for the load we have

$$\begin{aligned}
f(t) = Au(t) - \Lambda u(t) = \; & X''(x)\big(V_1(x, t) - P(x, t)Z(t)\big) \\
& + X'(x)\big(V_2(x, t) - P'(x, t)Z(t)\big) \\
& + X(x)\big(\sigma(x, t)Z(t) - V_3(x, t)\big).
\end{aligned}$$

Also:

$$f_\Gamma(t) = X'(1)\big(P(1, t)Z(t) - V_4(t)\big) \qquad \text{where} \qquad V_4(t) := \int_0^t P_B(1, t, s)Z(s)\, ds.$$

Equipped with these formulae we now outline the following test problem which is supposed to represent a "real" viscoelasticity problem.

Test Problem. Set $T = 1$, $X(x) = a_n x^n + \cdots + a_0$, $Z(t) = b_m t^m + \cdots + b_0$, $\sigma = \sigma_B = 0$ and $P_B(t, s) = P\varphi_s(t - s)$, where $P > 0$ and $\varphi(t)$ is the normalized stress relaxation function given by (3.10). Noting the reduction formula:

$$E_m(\alpha, t) := \int_0^t e^{-\alpha(t-s)} s^m\, ds = \frac{1}{\alpha}\left\{\begin{array}{ll} t^m - mE_{m-1}(\alpha, t), & \text{for } m \geq 1, \\ 1 - e^{-\alpha t}, & \text{for } m = 0, \end{array}\right.$$

Table 1   Result of solving the Test Problem with constant time step k and two spatial elements with $X(x)$ and $Z(t)$ given by (3.25). Here $e = u - U$ is the error and $\hat{t}_i := t_{i-1/2}$ are the midpoints of the time intervals.

k	$\|e\|_{L^h_\infty(\mathcal{J};L^h_2(\Omega))}$	$\max_i\{\mathcal{C}_h(t_i)\mathcal{E}^h_i\}$	$\max_i\{\mathcal{C}_k(t_i)\mathcal{E}^k_i\}$	$\max_i\|e(\hat{t}_i)\|_h$
1/2	5.076(−1)	0.000	1.362(+1)	1.780(−1)
1/4	3.653(−1)	0.000	9.368(+0)	6.765(−2)
1/8	2.129(−1)	0.000	5.507(+0)	3.021(−2)
1/16	1.121(−1)	0.000	3.139(+0)	1.230(−2)
1/32	5.500(−2)	0.000	1.840(+0)	3.215(−3)
1/64	2.708(−2)	0.000	1.060(+0)	1.397(−3)
1/128	1.351(−2)	0.000	5.786(−1)	5.029(−4)
1/256	6.759(−3)	0.000	3.027(−1)	1.527(−4)
1/512	3.381(−3)	0.000	1.548(−1)	4.208(−5)
1/1024	1.691(−3)	0.000	7.828(−2)	1.104(−5)
1/2048	8.456(−4)	0.000	3.936(−2)	2.826(−6)
1/4096	4.228(−4)	0.000	1.973(−2)	7.149(−7)

we find that $V_2 = V_3 = 0$ and

$$V_1(t) = V_4(t) = P\sum_{i=1}^{N_\varphi}\alpha_i\varphi_i\sum_{j=0}^{m}b_j E_j(\alpha_i, t).$$

In terms of Theorem 3.1 and 3.2 we see that $\|B(t,s)v\| \le \varphi_s(t-s)\|Av\|$ and so we take $\phi(s) = -\varphi'(s)$, this implies that $S_1(t) = \Phi(t) = 1/\varphi(t)$ since $\varphi(0) = 1$. Now, in order to determine $\mathcal{C}_h$ and $\mathcal{C}_k$ for Theorem 3.3 we return to the constants in Theorems 3.1 and 3.2:

$$C_1(1,t) = S_1(t), \qquad C_2(t) = \sqrt{2P/3}, \qquad C_4(1,t) = \sqrt{3/4P}S_1(t),$$
$$C_5 = 0, \qquad C_6(1,t) = 3S_1(t)/4P, \qquad C_7(1,t) = \sqrt{2}S_1(t)/P.$$

Inserting these into $\mathcal{C}_h$ and $\mathcal{C}_k$ and recalling the constants in Lemma 3.4 now gives

$$\mathcal{C}_h(t_m) = 0.215S_1(t_m)/P \qquad \text{and} \qquad \mathcal{C}_k(t_m, \hat{h}_m) \le 1.93S_1(t_m)/P,$$

since $\hat{h}_m \le \text{meas}(\Omega) = 1$.

For the numerical experiments we use the following data for this test problem:

$$\varphi_0 = 0.183\,43, \qquad \varphi_1 = 0.430\,765\,9, \qquad \varphi_2 = 0.385\,804\,1,$$

$$P = 402\,635.4, \qquad \alpha_1 = 4.7756, \qquad \alpha_2 = 161.3532,$$

which corresponds to a stress relaxation function suitable for a certain nylon 66 compound, see [SWW+94]. One unit of time here corresponds to three years, which illustrates the typical time scale of viscoelastic effects and why it is important to have stability factors that are consistent with long time integration.

The first two sets of results are based, respectively, on constant time steps with zero spatial errors, and uniform mesh refinement with zero temporal errors. For the results

Table 2   Result of solving the Test Problem with a single time step $k = 1.0$ and M equi-spaced spatial elements, with $X(x)$ and $Z(t)$ given by (3.26).

M	$\|e\|_{L^h_\infty(\mathcal{J};L^h_2(\Omega))}$	$\max_i\{\mathcal{C}_h(t_i)\mathcal{E}^h_i\}$	$\max_i\{\mathcal{C}_k(t_i)\mathcal{E}^k_i\}$
2	4.192(+1)	5.362(+1)	4.673(−13)
4	2.205(+0)	5.784(+0)	6.662(−14)
8	1.758(−1)	3.467(+0)	8.033(−14)
16	5.878(−2)	9.698(−1)	1.566(−13)
32	1.691(−2)	2.445(−1)	6.430(−13)
64	4.382(−3)	6.115(−2)	1.382(−12)
128	1.105(−3)	1.529(−2)	4.589(−12)
256	2.769(−4)	3.822(−3)	1.543(−11)
512	6.927(−5)	9.555(−4)	7.291(−11)
1024	1.732(−5)	2.389(−4)	2.720(−10)
2048	4.330(−6)	5.972(−5)	1.057(−9)
4096	1.083(−6)	1.493(−5)	4.451(−9)
8192	2.706(−7)	3.733(−6)	1.717(−8)
16384	6.766(−8)	9.331(−7)	7.012(−8)
32768	1.692(−8)	2.333(−7)	2.788(−7)

in Table 1 we set

$$X(x) := 0.01x \qquad \text{and} \qquad Z(t) = 500t^2 - 400t, \qquad (3.25)$$

and we see that $\mathcal{E}^h_i \equiv 0$ as expected. The first order temporal convergence of the scheme in $L_\infty(\mathcal{J};L_2(\Omega))$ is clear, as is the fact that the *a posteriori* estimate provides a true reflection of the error up to a multiplicative factor lying between 25 and 46. In [SW96b] it was noticed that the dG(0) scheme is apparently superconvergent at the midpoints of the $\mathcal{J}_i$, and the maximum $L_2(\Omega)$ error is shown at these midpoints, $\{\hat{t}_1, \ldots, \hat{t}_N\}$, in the fifth column of Table 1. The second-order superconvergence is evident for the smaller time steps but not for the larger ones and this is apparently due to the computer round-off error that results from the exponential function acting on the large α_2 value. (The second-order accuracy is recovered if one reduces α_2.)

The results in Table 2 correspond to approximating the solution given by:

$$\begin{aligned} X(x) &= -0.216x + 5.832x^2 - 56.016x^3 + 240.4x^4 \\ &\quad - 535x^5 + 645x^6 - 400x^7 + 100x^8, \qquad (3.26) \\ Z(t) &= -100. \end{aligned}$$

The true and estimated errors again behave as expected and are of the order $O(M^{-2})$, and this time the estimated error is a factor of only about 14 too large. For this problem we should have $\mathcal{E}^h_i \equiv 0$, which is effectively the case for small M, the drift away from zero as M increases seems again to be due to round-off.

We now want to use Theorem 3.3 to control that part of the error, on the given tolerance level TOL_Ω, that is bounded by the term containing $\mathcal{E}^h_i$. That is, we seek to choose the mesh functions $\{h_i\}^N_{i=1}$ such that

$$\mathcal{C}_h(t_i)\mathcal{E}^h_i = \text{TOL}_\Omega \quad \forall i \in \mathcal{N}(1, N).$$

Table 3 Result of solving the Test Problem with a single time step $k = 1.0$ and M equi-spaced spatial elements, with $X(x)$ and $Z(t)$ given by (3.27).

M	$\|e\|_{L^h_\infty(J;L^h_2(\Omega))}$	$\max_i\{C_h(t_i)\mathcal{E}_i^h\}$	$\max_i\{C_k(t_i)\mathcal{E}_i^k\}$
8	1.105(−1)	1.090(+0)	6.096(−14)
16	2.155(−2)	2.798(−1)	2.561(−14)
32	5.143(−3)	7.011(−2)	1.269(−13)
64	1.274(−3)	1.753(−2)	2.308(−13)
128	3.180(−4)	4.382(−3)	1.037(−12)
256	7.945(−5)	1.096(−3)	3.000(−12)
512	1.986(−5)	2.739(−4)	1.009(−11)

Setting $\mathcal{R}_h := f - AU + \Lambda U$, equidistribution of the error and some simplification of the norms leads, for each $\Omega_j^i \subset \Omega^i$, to the formulae

$$h_{ij}^4 \int_{\Omega_j^i} \|\mathcal{R}_h\|^2_{L^h_\infty(\mathcal{J}_i)} \, dx = \frac{\text{TOL}^2_\Omega}{C^2_h(t_i)M^i} \quad \Rightarrow \quad h_{ij} = \sqrt{\frac{\text{TOL}_\Omega}{C_h(t_i)\sqrt{M^i \int_{\Omega_j^i} \|\mathcal{R}_h\|^2_{L^h_\infty(\mathcal{J}_i)} \, dx}}},$$

and this provides a rule for selecting the new mesh sizes. Strictly this relation is nonlinear because M^i is not independent of $h_i(x)$, however, in practice we use the existing value for M^i. The signal to move forward to the next time level is when $h_{ij}^{\text{new}} \geq h_{ij}$ for all $j \in \mathcal{N}(1, M^i)$.

Due to the nested mesh assumption in Theorem 3.3 we permit only nested refinement of the space mesh so that h_{ij} may only change by a power of 2, then we set

$$h_{ij}^{\text{new}} \longleftarrow 2^p h_{ij} \qquad \text{where} \qquad p := \lfloor \log_2(h_{ij}^{\text{new}}/h_{ij}) \rfloor,$$

and the "floor" function, $\lfloor \cdot \rfloor$, denotes greatest lower integer bound.

To illustrate this spatial adaptivity we pick an extreme example:

$$X(x) = 0.01x^{20} \qquad \text{and} \qquad Z(t) = -100, \tag{3.27}$$

and show in Tables 3 and 4 the results for uniform and adaptive mesh refinement for one time step $k = 1.0$. The second column in Table 4 gives $\widehat{M}$, which is the number of elements generated by the adaptivity, and if we compare the actual errors and the corresponding number of elements between Tables 3 and 4 we notice that the adaptivity has little effect on the true error for a fixed M. However, more importantly, we see that if we judge the accuracy of the computation on the estimated errors, then far fewer elements are required to obtain a given accuracy. For example, consider the estimated errors for the pairs $(M, \widehat{M}) = (32, 11)$, $(128, 41)$ and $(512, 158)$.

So, the spatial adaptivity appears to work well, the problem that we have is in finding a means of interpreting Theorem 3.3 in terms of selecting appropriate time steps. Unlike $\mathcal{E}_i^h$, the error term $\mathcal{E}_i^k$ does not explicitly contain the current time step and so we cannot determine these in the same way as we found a new space mesh. In fact it is not even true in general that $\mathcal{E}_i^k$ may be made arbitrarily small just by an appropriate choice for k_i. The error at the current time level is influenced by all

Table 4 Result of solving the Test Problem adaptively with a single time step $k = 1.0$, and with $X(x)$ and $Z(t)$ given by (3.27). Here $\widehat{M}$ gives the number of spatial elements generated by the adaptivity.

TOL_Ω	$\widehat{M}$	$\|e\|_{L^h_\infty(\mathcal{J};L^h_2(\Omega))}$	$\max_i\{\mathcal{C}_h(t_i)\mathcal{E}^h_i\}$	$\max_i\{\mathcal{C}_k(t_i)\mathcal{E}^k_i\}$
$2.500(-1)$	11	$5.728(-3)$	$7.374(-2)$	$2.005(-13)$
$6.250(-2)$	20	$1.540(-3)$	$2.069(-2)$	$3.436(-13)$
$1.563(-2)$	41	$3.233(-4)$	$4.336(-3)$	$1.355(-12)$
$3.906(-3)$	78	$9.807(-5)$	$1.251(-3)$	$3.219(-11)$
$9.766(-4)$	158	$3.003(-5)$	$2.911(-4)$	$1.823(-10)$
$2.441(-4)$	312	$1.674(-5)$	$7.548(-5)$	$3.717(-10)$
$6.104(-5)$	629	$1.393(-6)$	$1.854(-5)$	$1.955(-9)$

Table 5 Result of solving the Test Problem with the time step control (3.28) and 2 equi-spaced spatial elements, with $X(x)$ and $Z(t)$ given by (3.25).

$\text{TOL}_\mathcal{J}$	$\|e\|_{L^h_\infty(\mathcal{J};L^h_2(\Omega))}$	$\max_i\{\mathcal{C}_h(t_i)\mathcal{E}^h_i\}$	$\max_i\{\mathcal{C}_k(t_i)\mathcal{E}^k_i\}$
0.4	$5.663(-1)$	$0.000(+0)$	$1.447(+1)$
0.2	$5.692(-1)$	$0.000(+0)$	$1.451(+1)$
0.1	$6.133(-2)$	$0.000(+0)$	$5.706(-1)$
0.05	$6.133(-2)$	$0.000(+0)$	$5.706(-1)$
0.025	$6.133(-2)$	$0.000(+0)$	$9.989(-1)$

previous errors which have been "memorized" by the Volterra term. This problem is not simply a feature of this particular numerical scheme, it has also been reported by: Blom & Brunner [BB87], for a collocation scheme; by Jones & McKee [JM85], for a linear multistep method; and, by Shaw, Warby & Whiteman [SWW96], for a trapezoidal algorithm. This feature seems to imply that a "one pass" solver algorithm with guaranteed error control is simply not (yet?) possible for second-kind Volterra problems.

Multi-pass algorithms on the other hand are of course possible, here the error information from one pass is used to determine a sequence of time steps for the next pass, and so on. However, this is potentially expensive (recall the N^2 problem) and so it is also of interest to construct a heuristic time step control so as to obtain a one pass solver. The a posteriori error bound can always be used to check the integrity of the discrete solution.

For the scalar pure-time analogue of (3.11) described by Shaw & Whiteman in [SW96b], the time step control was arrived at by using the discrete solution to emulate the truncation error associated with piecewise constant interpolation, here we look at three other time step controllers based more directly on the a posteriori error estimate.

Looking back at Table 1 we see that $\mathcal{E}^k_i$ converges to zero at the same rate as the temporal error and this suggests the simple time step modification:

$$k_i^{\text{new}} = \gamma k_i \frac{\text{TOL}_\mathcal{J}}{\mathcal{C}_k(t_i)\mathcal{E}^k_i}, \qquad (3.28)$$

where $\text{TOL}_\mathcal{J} > 0$ is a user-specified tolerance. So, at each time level a new time

Table 6 Result of solving the Test Problem with the time step control (3.29) and 2 equi-spaced spatial elements, with $X(x)$ and $Z(t)$ given by (3.25).

TOL$_{\mathcal{J}}$	$\|e\|_{L_\infty^h(\mathcal{J};L_2^h(\Omega))}$	$\max_i\{\mathcal{C}_h(t_i)\mathcal{E}_i^h\}$	$\max_i\{\mathcal{C}_k(t_i)\mathcal{E}_i^k\}$
0.4	$1.102(-1)$	$0.000(+0)$	$3.423(-1)$
0.2	$1.067(-1)$	$0.000(+0)$	$1.987(-1)$
0.1	$6.898(-3)$	$0.000(+0)$	$9.702(-2)$
0.05	$\cdots$	$\cdots$	$\cdots$

Table 7 Result of solving the Test Problem with the time step control (3.28) modified by (3.30) with $\gamma = 1.0$, 2 equi-spaced spatial elements and with $X(x)$ and $Z(t)$ given by (3.25).

TOL$_{\mathcal{J}}$	$\|e\|_{L_\infty^h(\mathcal{J};L_2^h(\Omega))}$	$\max_i\{\mathcal{C}_k(t_i)\mathcal{E}_i^k\}$	time (hours)
0.4	$1.065(-2)$	$3.921(-1)$	0.11
0.2	$5.090(-3)$	$1.961(-1)$	0.34
0.1	$2.390(-3)$	$9.805(-2)$	1.2
0.05	$1.142(-3)$	$4.903(-2)$	3.8
0.025	$5.503(-4)$	$2.451(-2)$	13.2
0.0125	$2.695(-4)$	$1.226(-2)$	48.2

Table 8 Result of solving the Test Problem with the time step control (3.28) modified by (3.30) with $\gamma = 0.9$, 2 equi-spaced spatial elements and with $X(x)$ and $Z(t)$ given by (3.25).

TOL$_{\mathcal{J}}$	$\|e\|_{L_\infty^h(\mathcal{J};L_2^h(\Omega))}$	$\max_i\{\mathcal{C}_k(t_i)\mathcal{E}_i^k\}$	time (hours)
0.4	$9.987(-3)$	$3.558(-1)$	0.012
0.2	$4.724(-3)$	$1.772(-1)$	0.052
0.1	$2.236(-3)$	$8.842(-2)$	0.212
0.05	$1.090(-3)$	$4.417(-2)$	0.86
0.025	$5.405(-4)$	$2.207(-2)$	3.22
0.0125	$2.669(-4)$	$1.103(-2)$	12.85

step is obtained from a simple scaling dependent upon the estimated error and the requested tolerance, the constant $\gamma \in (0,1]$ is introduced (quite arbitrarily) as a "safety factor", we use $\gamma = 1/2$. If $\mathcal{C}(t_i)\mathcal{E}_i^k > \text{TOL}_{\mathcal{J}}$ then the solution is recomputed on $\mathcal{J}_i^{\text{new}}$, otherwise we set $k_{i+1} := k_i^{\text{new}}$ and march forward in time. At the first time level (i.e. when $i = 1$) this iteration can be repeated as often as necessary and we will always obtain $\mathcal{C}_k(k_1)\mathcal{E}_1^k \leq \text{TOL}_{\mathcal{J}}$, but at subsequent time levels this iteration is allowed only once for each i because it is not guaranteed to "converge". Some results for this scheme are shown in Table 5. The fact that identical results are obtained for $\text{TOL}_{\mathcal{J}} = 0.1$ and $\text{TOL}_{\mathcal{J}} = 0.05$ seems to indicate that allowing only one iteration of (3.28) per value of i is not adequate. The problem with increasing this allowance is that we have no guide as to what the maximum number of iterations should be.

The second method that we test consists of replacing $\mathcal{C}_k(t_i)\mathcal{E}_i^k$ with a "discrete derivative" which we hope will be independent of the time step. We use (3.28) for the first time level and thereafter take

$$k_i^{\text{new}} = \gamma\text{TOL}_{\mathcal{J}} \bigg/ \left| \frac{\mathcal{C}_k(t_i)\mathcal{E}_i^k - \mathcal{C}_k(t_{i-1})\mathcal{E}_{i-1}^k}{\frac{1}{2}(k_i + k_{i-1})} \right|. \tag{3.29}$$

Some results for this, again with $\gamma = 1/2$, are shown in Table 6. We see that $\max_i\{\mathcal{C}_k(t_i)\mathcal{E}_i^k\} \leq \text{TOL}_{\mathcal{J}}$ is more or less achieved by the scheme in the instances shown, but this should not be taken as an indication of success. In fact (3.29) typically always forced at least two solves for each i because $k_{i+1} := k_i^{\text{new}}$ was always predicted to be much too large. Also, for $\text{TOL}_{\mathcal{J}} = 0.05$ the scheme "hung up" at $t_i \approx 0.33$, and could not choose k_i^{new} small enough to satisfy $\mathcal{C}_k(t_i)\mathcal{E}_i^k \leq \text{TOL}_{\mathcal{J}}$. This suggests that (3.29) should only be applied for a fixed maximum number of times for each i as well.

We have to be unsatisfied with the time step controllers (3.28) and (3.29) for obvious reasons. However, a simple modification to (3.28) produces much better results. In (3.28) we replace $\text{TOL}_{\mathcal{J}}$ with

$$\text{tol}_{\mathcal{J}} := \text{TOL}_{\mathcal{J}}\Phi(t)/\Phi(\infty). \tag{3.30}$$

Since $\Phi(\infty) \approx \Phi(T)$ for large T this allows a progressive increase in the tolerable error through time in accordance with the stability of the dual problem, at an intuitive level this seems a very reasonable thing to do. Some results for this adaptive scheme are shown in Table 7 for $\gamma = 1$, and in Table 8 for $\gamma = 0.9$. The choice $\gamma = 0.9$ is to be preferred since less iterations are required at each time level and the execution times, shown in the fourth columns, are correspondingly lower. For each of these calculations the maximum number of iterations at each time level is unrestricted, unlike for the results in Table 5.

3.5   DISCUSSION

There are several important issues to discuss in connection with the foregoing. We take them one at a time.

• **Variational crimes.** Theorem 3.3 bounds only the Galerkin error. In any practical scheme not only the inner products but also the Volterra operator will be replaced by some form of quadrature. This will introduce another source of error into the computed solution and any robust *a posteriori* error estimate should contain terms to account for this. A full *a posteriori* analysis of quadrature error has been given by Shaw & Whiteman in [SW96a] for a prototype pure-time analogue of (3.2). A similar analysis could be repeated for the present context. In particular it would be necessary to consider "mass lumping" in order to calculate $\mathcal{E}_i^k$, and keep track of the incurred quadrature error.

• **Mesh de-refinement.** In the proof of Theorem 3.3 (Lemma 3.2) we chose $P_i\chi$ to be the nodal interpolate of χ during each time level and this had the effect of making the first derivative jumps at the element edges disappear. For higher space dimensional problems this will not happen and so, in such cases, we might make a different choice for the $\{P_i\}$. If, following Eriksson & Johnson [EJ91], we choose $P_i\chi$ to be the L_2 projection of χ on $\mathcal{H}_i$ the resulting estimate becomes more complicated because it contains the history of the jumps. However, we should then be able to avoid the mesh constraint $\mathcal{H}_{i-1} \subseteq \mathcal{H}_i$, and remove the dependence of C_k oh $\hat{h}$. We leave these considerations for another time.

• $n > 1$ **problems.** Energy norm estimates are appropriate for higher space dimensional problems, since we do not then necessarily have the required amount of elliptic regularity. In [Hor95] some *a priori* estimates on the constants in Korn's inequality are available, and thus the argument for the average- and maximum-energy stability estimates can be repeated for the viscoelasticity problem.

• **Increased accuracy: dG(1)cG(1).** If the exact solution to (3.11) is smooth enough we can achieve higher accuracy in time by using $V_{r>0}$ as the trial and test space. For example, to formulate the dG(1)cG(1) method we begin by defining basis functions for the time intervals $\mathcal{J}_i$. Set $\psi_{i-1}^i(t) := (t_i - t)/k_i$ and $\psi_i^i(t) := (t - t_{i-1})/k_i$ and write the fully discrete solution on $\mathcal{J}_i$ as $U(t)|_{\mathcal{J}_i} = U_0^i\psi_{i-1}^i(t) + U_1^i\psi_i^i(t)$. Note that $\psi_{i-1}^i \neq \psi_i^{i-1}$ because the approximation is discontinuous. Then, for $v = \psi_{i-1}^i$ or $v = \psi_i^i$ we have (from (3.23) for example) by integrating over $\mathcal{J}_i$ that

$$A \int_{t_{i-1}}^{t_i} [U_0^i\psi_{i-1}^i + U_1^i\psi_i^i] v - v \int_{t_{i-1}}^t \varphi_s(t - s) [U_0^i\psi_{i-1}^i(s) + U_1^i\psi_i^i(s)] \, ds \, dt = G_i(v),$$

where the $G_i(v)$ are column arrays that depend upon the loading and known history of the displacement, and we have assumed that $B(t, s) = \varphi_s(t - s)A$, in accordance with the viscoelasticity model described earlier. Now, using

$$\psi_{i-1}^i(t) - \int_{t_{i-1}}^t \varphi_s(t - s)\psi_{i-1}^i(s) \, ds = \varphi(t - t_{i-1}) - \frac{1}{k_i} \int_{t_{i-1}}^t \varphi(t - s) \, ds,$$

$$\psi_i^i(t) - \int_{t_{i-1}}^t \varphi_s(t - s)\psi_i^i(s) \, ds = \frac{1}{k_i} \int_{t_{i-1}}^t \varphi(t - s) \, ds,$$

the above can be written equivalently as

$$A \int_{t_{i-1}}^{t_i} U_0^i \left[\varphi(t - t_{i-1}) - \frac{1}{k_i} \int_{t_{i-1}}^t \varphi(t - s) \, ds \right] v$$

$$+ U_1^i \left[\frac{1}{k_i} \int_{t_{i-1}}^t \varphi(t - s) \, ds \right] v \, dt = G_i(v).$$

Taking $v = \psi_{i-1}^i$ and $v = \psi_i^i$ in turn produces a two-by-two block system of equations which may be solved for the array unknowns $\boldsymbol{U}_0^i$ and $\boldsymbol{U}_1^i$. All of the analysis presented in this paper extends to this method with the exception of Lemma 3.5

- **N^2 problem.** For Volterra kernels of the form (3.10) the N^2 problem disappears due to a simple recurrence formula based on the properties of the exponential, see for example equation (5.3) in [SWW$^+$94]. To implement this recurrence in the (spatially) adaptive context is straightforward if we retain the finest space $\mathcal{H}_i$ so far encountered in the computation, and represent all solution vectors (exactly due to $\mathcal{H}_{i-1} \subseteq \mathcal{H}_i$) in this space. However, another popular choice for φ is the *power law*: $\varphi(t) = \varphi_0 t^{-\alpha}$, where $\varphi_0 > 0$ and $\alpha \in (0,1)$. For this no simple recurrence formula exists and we have no choice but to confront the N^2 problem head-on. The timings shown in Tables 7 and 8 demonstrate that we *cannot afford to ignore this issue*. Firstly: some crude numerical experiments have suggested to us that the sparse quadrature rules of Sloan & Thomée in [ST86] can be profitably applied to this scheme—which is surprising since the integrand is not smooth. Secondly: we could exploit the fading memory once again and replace $\varphi(t)$ with a "chopped" function $\varphi(\tau;t)$ where $\varphi'(\tau;t) = \varphi'(t)$ for $t \in [0,\tau]$ and $\varphi'(\tau;t) = 0$ otherwise. Here $\tau \in \mathcal{J}$ is adaptively defined and the prime denotes t-differentiation. An *a posteriori* data-error estimate appropriate to such a scheme has been given by Shaw, Warby & Whiteman in [SWW96].
- **A priori error analysis.** Although we have focused on *a posteriori* error analysis it is also important to make an *a priori* analysis of the error since this may reveal extra constraints on the scheme. For example, in [EJ91] the gradients of the mesh functions $h_i(x)$ are required to be small enough everywhere in Ω.
- **And finally....** This work, along with [SW96c], appears to represent the first time that rigorous *a posteriori* error analysis has been attempted for PDV equations, and although much remains to be done, we believe there is scope for treating the general problems (3.1), (3.2) and (3.3).

References

[Ada75]   Adams R. (1975) *Sobolev spaces.* Academic Press.

[BB87]   Blom J. G. and Brunner H. (1987) The numerical solution of nonlinear Volterra integral equations of the second kind by collocation and iterated collocation methods. *SIAM J. Sci. Stat. Comput.* 8: 806—830.

[CL90]   Cannon J. R. and Lin Y. (1990) *A priori* L^2 error estimates for finite-element methods for nonlinear diffusion equations with memory. *SIAM J. Numer. Anal.* 27: 595—607.

[DJ62]   Douglas J. and Jones B. F. (1962) Numerical methods for integro-differential equations of parabolic and hyperbolic types. *Numer. Math.* 4: 96—102.

[EEHJ95]   Eriksson K., Estep D., Hansbo P., and Johnson C. (1995) Introduction to adaptive methods for differential equations. *Acta Numerica* pages 105—158.

[EJ91]   Eriksson K. and Johnson C. (1991) Adaptive finite element methods for parabolic problems. I: a linear model problem. *SIAM J. Numer. Anal.* 28: 43—77.

[Fer70]   Ferry J. D. (1970) *Viscoelastic properties of polymers.* John Wiley and Sons Inc.

[GG88]   Golden J. M. and Graham G. A. C. (1988) *Boundary value problems in linear viscoelasticity.* Springer-Verlag.

[GYF88]   Greenwell-Yanik E. and Fairweather G. (1988) Finite element methods for parabolic and hyperbolic partial integro-differential equations. *Nonlinear Analysis, Theory, Methods and Applications* 12: 785—809.

[Hor95]   Horgan C. O. (1995) Korn's inequalities and their applications in continuum mechanics. *SIAM Review* 37: 491—511.

[Hun83]   Hunter S. C. (1983) *Mechanics of continuous media (second edition).* Mathematics and its Applications, Ellis Horwood, England.

[JM85]   Jones H. M. and McKee S. (1985) Variable step size predictor-corrector schemes for second kind Volterra integral equations. *Math. Comp.* 44: 391—404.

[LST96]   Lubich C., Sloan I. H., and Thomée V. (1996) Nonsmooth data error estimates for approximations of an evolution equation with a positive-type memory term. *Math. Comp.* 65: 1—17.

[Nun71]   Nunziato J. W. (1971) On heat conduction in materials with memory. *Quart. Appl. Maths.* 29: 187—204.

[OR76]   Oden J. T. and Reddy J. N. (1976) *An introduction to the mathematical theory of finite elements.* A Wiley Interscience Publication, John Wiley and Sons.

[PP96]   Pani A. K. and Peterson T. E. (1996) Finite element methods with numerical quadrature for parabolic integrodifferential equations. *SIAM J. Numer. Anal.* 33: 1084—1105.

[PTW92]   Pani A. K., Thomée V., and Wahlbin L. B. (1992) Numerical methods for hyperbolic and parabolic integro-differential equations. *J. Integral Equations Appl.* 4: 533—584.

[ST86]   Sloan I. H. and Thomée V. (1986) Time discretization of an integro-differential equation of parabolic type. *SIAM J. Numer. Anal.* 23: 1052—1061.

[SW96a]   Shaw S. and Whiteman J. R. (1996) Backward Euler and Crank-Nicolson finite element variants with rational adaptivity and *a posteriori* error estimates for an integrodifferential equation. Submitted to Math. Comp.

[SW96b]   Shaw S. and Whiteman J. R. (1996) Discontinuous Galerkin method with *a posteriori* $L_p(0, t_i)$ error estimate for second-kind Volterra problems. To appear in Numer. Math.

[SW96c]   Shaw S. and Whiteman J. R. (1996) Towards adaptive finite element schemes for partial differential Volterra equation solvers. Submitted to Advances in Computational Mathematics.

[SW96d]   Shaw S. and Whiteman J. R. (1996) Toward robust adaptive finite element methods for partial differential volterra equation problems arising in viscoelasticity theory. Technical report, TR96/2, BICOM, Brunel University, Uxbridge, U.K.

[SWW+94] Shaw S., Warby M. K., Whiteman J. R., Dawson C., and Wheeler M. F. (1994) Numerical techniques for the treatment of quasistatic viscoelastic stress problems in linear isotropic solids. *Comput. Methods Appl. Mech. Engrg.* 118: 211—237.

[SWW95]   Shaw S., Warby M. K., and Whiteman J. R. (1995) An implicit history discretization of a hereditary wave equation. To appear in Seminario de Analisis Numerico, ed. P. Michavila, Universidad Politecnica de Madrid.

[SWW96]   Shaw S., Warby M. K., and Whiteman J. R. (1996) Error estimates with sharp constants for a fading memory Volterra problem in linear solid viscoelasticity. To appear in SIAM J. Numer. Anal. (expected SINUM 34-3, summer 1997).

[TW94]   Thomée V. and Wahlbin L. B. (1994) Long-time numerical solution of a parabolic equation with memory. *Math. Comp.* 62: 477—496.

[Whi94]   Whiteman J. R. (ed) (1994) *The mathematics of finite elements and applications, highlights 1993.* John Wiley and Sons Ltd., Chichester.

4

Preconditioners for the adaptive *hp* version finite element method

Mark Ainsworth, Bill Senior[1] and Derek Andrews

Department of Mathematics and Computer Science,
Leicester University,
Leicester LE1 7RH,
U.K.

4.1 INTRODUCTION

4.1.1 SUMMARY

A domain decomposition preconditioner suitable for *hp* finite element approximation on adaptively refined meshes with non-uniform polynomial degree is described. The preconditioner is highly suited for parallel computation and generalizes methods proposed by Smith [Smi91] for the *h*-version finite element method with piecewise linear basis functions, and by Mandel [Man90] for the *p*-version finite element method. The preconditioner was recently analysed by Ainsworth [Ain96a, Ain96b] where it was shown that the condition number of the preconditioned system grows at most logarithmically in the degree p and mesh size h. This result generalizes the known sharp estimates in each of the cases mentioned above. The preconditioner is used to solve the linear systems on highly non-uniform adaptive *hp* meshes arising from the adaptive solution of a problem with a crack singularity.

4.1.2 MODEL PROBLEM

Let $\Omega \in \mathbb{R}^2$ be a domain with polygonal boundary $\partial\Omega$. Consider the problem:

$$-\Delta U = f \text{ in } \Omega \tag{4.1}$$

---

[1] The support of the Engineering and Physical Sciences Research Council through a research studentship is gratefully acknowledged.

The Mathematics of Finite Elements and Applications
Edited by J. R. Whiteman © 1997 John Wiley & Sons Ltd.

subject to $U = 0$ on the boundary $\partial\Omega$. The bilinear form $B : H^1(\Omega) \times H^1(\Omega) \mapsto \mathbb{R}$ is given by

$$B(U, V) = \int_\Omega \nabla U \cdot \nabla V \, \mathrm{d}\boldsymbol{x} \qquad (4.2)$$

where $H^1(\Omega)$ is the usual Sobolev space of distributions with square integrable derivatives. The subspace $H_0^1(\Omega) \subset H^1(\Omega)$ is the completion of smooth functions with support in Ω, with respect to the $H^1(\Omega)$ norm. The variational formulation of the problem defined by (4.1) is

Find $U \in H_0^1(\Omega)$ such that $B(U, V) = (f, V)$ for all $V \in H_0^1(\Omega)$. $\qquad (4.3)$

4.1.3 COARSE GRID

The finite element analysis of problem (4.3) begins by first subdividing the domain Ω into an initial 'coarse' quasi-uniform partitioning $\mathcal{P}_H$ consisting of non-overlapping quadrilateral elements $\{\Omega_H^K\}$ of size $O(H)$. The space X_H consists of piecewise bilinear functions defined on the partitioning $\mathcal{P}_H$. An approximation of problem (4.1) can then be obtained by solving the discrete problem

Find $u_H \in X_H$ such that $B(u_H, v_H) = (f, v_H)$ for all $v_H \in X_H$. $\qquad (4.4)$

4.1.4 ADAPTIVE REFINEMENT

Typically, owing to the geometry of the domain (e.g. re-entrant corners) the true solution U of problem (4.1) will be non-smooth in certain regions. It is often necessary to enhance the approximation space X_H in the neighbourhood of those areas where the true solution is less regular.

Various strategies are possible for improving the accuracy. The simplest techniques are to either subdivide the mesh uniformly throughout the domain (the h-version finite element method) or to enrich the polynomial degree of the functions on each of subdomains (the p-version finite element method [BKS81]). Alternatively, one can use a *posteriori* error estimates to identify specific regions in which the approximation is poor and perform a selective enrichment by locally subdividing the elements and increasing the polynomial degree non-uniformly. The latter approach is in the spirit of the hp version of the finite element method [BG88, BS87].

The chief advantage of the uniform refinement strategies is that the data structure is kept simple, meaning that the algorithms can often be coded extremely efficiently. However, the disadvantages are that degrees of freedom are introduced in regions where the solution can be adequately represented using the initial coarse approximation. Conversely, the hp version can lead to approximations in which the required accuracy is obtained using a minimal number of degrees of freedom. However, at present, there is no universally agreed general strategy for adaptive hp refinement.

Suppose that by following a particular adaptive strategy, a refined partition $\mathcal{P}_h$ has been generated starting from an initial coarse partitioning. The final partition then consists of elements from the original coarse partitioning $\mathcal{P}_H$ along with elements obtained by successive local subdivision of elements from the coarse partitioning. In addition, the polynomial degree in the elements may have been increased resulting in quite complicated final meshes. For instance, a typical sequence of adaptive refinements is shown in Figures 3-4. The final mesh (Mesh 8 in Figure 4) contains elements of degree ranging from $p = 1$ to $p = 8$.

4.1.5 PRECONDITIONING

Suppose that appropriate mesh refinements and polynomial enrichments have been performed and let the associated finite element subspace be denoted by X. The finite element approximation on the refined subspace X is defined by

$$\text{Find } u \in X \text{ such that } B(u,v) = (f,v) \text{ for all } v \in X. \tag{4.5}$$

The discrete form of this problem is then

$$B\boldsymbol{x} = \boldsymbol{f} \tag{4.6}$$

where B is a symmetric positive definite matrix. The basic approach for solving the matrix equation (4.6) will be the conjugate gradient method. The condition number κ governs the performance of the conjugate gradient solution routine: each iteration reducing the error by at least a factor $(\sqrt{\kappa}-1)/(\sqrt{\kappa}+1)$. Unfortunately, it is generally found that the condition number grows rapidly as the mesh is refined or the polynomial degree increased [Ain96a]. One possibility is to apply a preconditioner to the linear system. A preconditioning form $C(\cdot,\cdot)$ is constructed for which there exist μ, Υ such that

$$\mu C(v,v) \le B(v,v) \le \Upsilon C(v,v) \text{ for all } v \in X \tag{4.7}$$

where μ and Υ depend on the mesh and the polynomial degree, but are independent of the function v. The preconditioning form should be chosen with two properties in mind. First, the problem

$$\text{Find } w \in X \text{ such that } C(w,v) = g(v) \text{ for all } v \in X. \tag{4.8}$$

where $g(\cdot)$ is an appropriate linear functional on X, must be capable of being solved efficiently. In particular, the work required in applying the preconditioner should be modest in comparison with simply solving the problem (4.5) directly. Secondly, since the rate of convergence for the preconditioned algorithm is controlled by Υ/μ, this ratio must be controlled as the mesh is refined and the polynomial degree increased.

The construction of suitable preconditioners will be discussed in the next section. The aim is to produce a flexible preconditioning algorithm that effectively controls the growth of the condition number *regardless of the particular adaptive refinement strategy employed*. Equally importantly, the given preconditioner will be found to be computationally attractive.

4.2 PRECONDITIONING ALGORITHM

4.2.1 THE BASIC PROCEDURE

The elements Ω_K constructed during the initial coarse partitioning $\mathcal{P}_H$ will be referred to as *subdomains*. As remarked earlier, the unrefined subdomains exist as elements in the final partitioning $\mathcal{P}_h$. The refined subdomains serve to group the elements in the final partitioning into local sets of elements. The preconditioner will be based on this natural domain decomposition. The overall data structure can also exploit this situation:

Step 1 Coarse Grid

- Construct the initial coarse partitioning $\mathcal{P}_H$.
- Assemble and solve the finite element problem (4.4) on the coarse mesh.
- Use the coarse grid solution to design the partitioning $\mathcal{P}_h$ and the space X.

Step 2 Assembly of Schur Complement

- Assemble the contributions to the global stiffness matrix and global load vector from the subdomain Ω_K.
- Apply static condensation to the subdomain stiffness matrix and subdomain load vector to eliminate the internal degrees of freedom on the subdomain.
- Assemble the reduced matrix and vector into a global matrix and vector thereby obtaining a system of the form $S\boldsymbol{x} = \boldsymbol{b}$. The Schur complement is assembled by the standard finite element sub-assembly (of the local Schur complements).

Step 3 Solve Schur complement $S\boldsymbol{x} = \boldsymbol{b}$

- Apply preconditioned conjugate gradient with preconditioner described below.

Step 4 Internal Degrees of Freedom

- Apply back-substitution to obtain the values of the interiors of the subdomains.

One advantage of this approach is that the internal unknowns are eliminated at a local level thereby simplifying the overall data structure since the number of degrees of freedom 'visible' at a global level is drastically reduced. A second is that the problems associated with the interiors are independent and only require information available locally. These features can be effectively exploited in both a shared memory and distributed memory parallel environment.

4.2.2   THE PRECONDITIONER

The preconditioner is based on a decomposition of the degrees of freedom (dofs) on the edges of the subdomains into three sets:

$$
\begin{aligned}
\mathcal{I}_H &= \{\text{DOFs associated with the vertices of the coarse grid } \mathcal{P}_H\} \\
\mathcal{I}_h &= \{\text{DOFs associated with linear functions introduced by refinement of } \mathcal{P}_H\} \\
\mathcal{I}_p &= \{\text{DOFs associated with higher order functions}\}
\end{aligned}
$$

For ease of exposition, suppose that the degrees of freedom have been ordered so that the set $\mathcal{I}_H$ is numbered first, followed by $\mathcal{I}_h$ and finally $\mathcal{I}_p$. The Schur complement and reduced load vector may be partitioned into blocks corresponding to this ordering:

$$
S = \begin{bmatrix} S_{HH} & S_{Hh} & S_{Hp} \\ S_{Hh}^t & S_{hh} & S_{hp} \\ S_{Hp}^t & S_{hp}^t & S_{pp} \end{bmatrix} ; \quad b = \begin{bmatrix} b_H \\ b_h \\ b_p \end{bmatrix} \tag{4.9}
$$

The solution of the preconditioning problem

$$
Cx = b \tag{4.10}
$$

consists of three main steps:

Linear DOFs   The degrees of freedom within the set $\mathcal{I}_h$ are partitioned further into subsets corresponding to the separate edges $E_1, \ldots E_n$ of the coarse mesh $\mathcal{P}_H$. The submatrix S_{hh} then has the block structure

$$
S_{hh} = \begin{bmatrix} S_{hh}^{(1,1)} & \cdots & S_{hh}^{(1,n)} \\ \vdots & \ddots & \vdots \\ S_{hh}^{(1,n)\,t} & \cdots & S_{hh}^{(n,n)} \end{bmatrix} \tag{4.11}
$$

Let

$$
D_{hh} = \mathrm{diag}(S_{hh}^{(1,1)}, \ldots, S_{hh}^{(n,n)}) \tag{4.12}
$$

then inverting D_{hh} to obtain

$$
x_h = D_{hh}^{-1} b_h \tag{4.13}
$$

corresponds to solving independent problems over each edge of the coarse mesh $\mathcal{P}_H$.

Higher Order DOFs   Similarly, dividing the higher order degrees of freedom in the set $\mathcal{I}_p$ into sets corresponding to each of the separate edges in the fine mesh $\mathcal{P}_h$ gives the matrix S_{pp} a block structure. Inverting the block diagonal D_{pp} of the matrix S_{pp} corresponds to solving independent problems over each edge of the fine mesh to compute

$$
x_p = D_{pp}^{-1} b_p \tag{4.14}
$$

Coarse Grid Correction   The problem associated with the coarse grid involves inverting the same stiffness matrix B_{HH} arising from the original bilinear discretization on the coarse grid $\mathcal{P}_H$. Firstly it is necessary to construct a *restriction* of the data b_H and b_h to the space X_H. This transformation is effected by writing the bilinear functions on the original coarse mesh as linear combinations of the linear functions on the edges of the refined mesh. That is, on the edges of $\mathcal{P}_H$ one has for a suitable constant matrix R

$$
\psi_H = \phi_H + R\phi_h \tag{4.15}
$$

where ψ_H are the bilinear functions on the coarse grid $\mathcal{P}_H$, and ϕ_H and ϕ_h are the functions corresponding to the degrees of freedom in the sets $\mathcal{I}_H$ and $\mathcal{I}_h$ respectively.

The matrix R can be given explicitly if we let x_j denote the node on the edge of $\mathcal{P}_h$ at which the j-th linear degree of freedom $\phi_{h,j} \in \mathcal{I}_h$ is based, then

$$R = [\psi_H(x_j)]_{j \in \mathcal{I}_h} \tag{4.16}$$

In practice, it is unnecessary to assemble the matrix explicitly since the data structure provides for an efficient and simple computation of the inter-level transfer operations. The matrix R is unchanged even if higher order approximation is used provided that the functions $\phi_{h,j}$ are bilinear (such is the case when using hierarchical element basis functions).

The steps to construct the coarse grid correction may be summarized as:

- Form the *restriction* of the data

$$\hat{b}_H = b_H + Rb_h.$$

- Solve the coarse grid problem

$$B_{HH}\hat{x}_H = \hat{b}_H$$

where B_{HH} is the stiffness matrix for the discretization (4.4).
- *Prolongate* the solution $\hat{x}_H$ to the space spanned by elements of $\mathcal{I}_H$ and $\mathcal{I}_h$ according to

$$x_H^c = \hat{x}_H; \quad x_h^c = R^t \hat{x}_H.$$

Having completed these steps, the solution of the preconditioning problem (4.10) is then given by

$$x = \begin{bmatrix} x_H^c \\ x_h + x_h^c \\ x_p \end{bmatrix} \tag{4.17}$$

The action of the inverse of the preconditioner C therefore corresponds to solving independent problems for the linear degrees of freedom over each of the edges of $\mathcal{P}_H$ separately; independent solves for the higher order degrees of freedom over each of the edges of $\mathcal{P}_h$ separately; and a global solve over the degrees of freedom in the coarse grid discretization. Moreover, each of these overall steps may be performed concurrently. In particular, the computations are ideally suited to a parallel programming environment.

4.2.3 GENERALIZATION TO NON-UNIFORMLY REFINED MESHES

Suppose now that the fine grid $\mathcal{P}_h$ is obtained by *selectively* refining elements from the coarse grid. The set $\partial \mathcal{P}_H$ of edges in the coarse partition is defined as before. However, the definition of the set $\partial \mathcal{P}_h$ must be generalized to encompass constrained degrees of freedom between the subdomains. Therefore, the set $\partial \mathcal{P}_h$ consists of the edges joining unconstrained vertices lying on the edges of the subdomains. These edges need not coincide with the actual interelement edges as, for example, is the case in the example shown in Figure 1.

As remarked earlier, the performance of the preconditioner is governed by the quantities μ and Υ in the equivalence:

$$\mu C(v,v) \leq B(v,v) \leq \Upsilon C(v,v) \text{ for all } v \in X \tag{4.18}$$

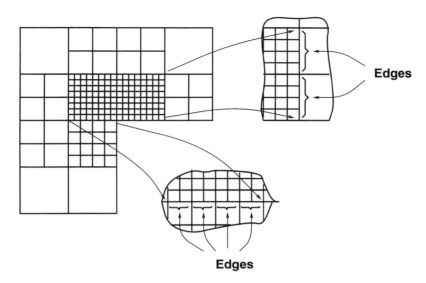

Figure 1   Illustration of the edges contained in the set $\partial\mathcal{P}_h$ for an adaptively refined mesh.

The algorithm presented above generalizes both an algorithm proposed by Smith [Smi91] for the h-version finite element method and an algorithm proposed by Mandel [Man90] for the p-version. For the case of first order $(p = 1)$ approximation, Smith [Smi91] has shown that the ratio grows at most as

$$\Upsilon/\mu \le C(1 + \log^2(H/h)) \tag{4.19}$$

where the constant C is independent of H and h. Conversely, for the p-version finite element method $(H/h = 1)$, it has been shown [BCMP91] that the algorithm given by Mandel [Man90] yields

$$\Upsilon/\mu \le C(1 + \log^2 p) \tag{4.20}$$

where the constant C is independent of p and the number of elements.

The preconditioner described above retains the theoretical properties of each of these schemes as the following result (obtained by simple modifications of the arguments used by Ainsworth [Ain96a]) shows:

Theorem 4.1 *Suppose that on each of the refined subdomains Ω_K, the mesh is quasi-uniform of size h_K and the maximum polynomial degree is p_K. Suppose that the polynomial degree on the subdomain is such that the degree on the boundary does not exceed the polynomial degree on the interior. Then the condition number of the preconditioned system $\kappa(C^{-1}S)$ is bounded by*

$$\kappa(C^{-1}S) \le K \min\left(H/\underline{h}, 1 + \log^2 \bar{p}\right) \cdot (1 + \log(H\bar{r}))^2 \tag{4.21}$$

where $\bar{p} = \max_K p_K$, $\underline{h} = \min_K h_K$ and $\bar{r} = \max_K p_K/h_K$. The constant K is independent of the parameters H, $\underline{h}$ and $\bar{p}$.

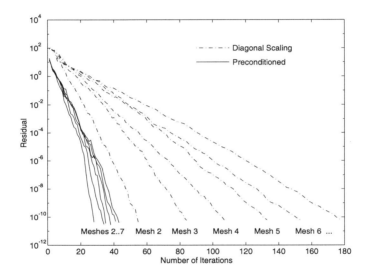

Figure 2   Comparison of preconditioner versus diagonal scaling for adaptively refined meshes.

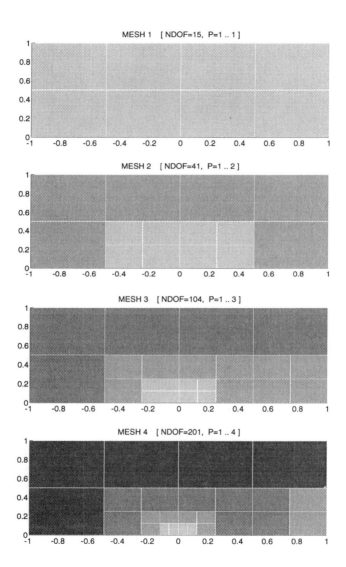

Figure 3 Adaptively designed meshes 1-4 for numerical example. The shading indicates the polynomial order of the element interior (lower order elements have lighter shading).

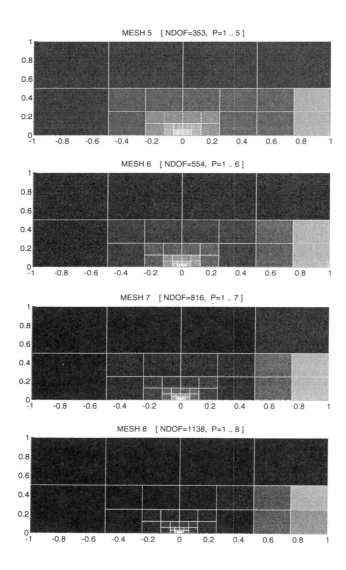

Figure 4 Adaptively designed meshes 5-8 for numerical example. The shading indicates the polynomial order of the element interior (lower order elements have lighter shading).

4.3 NUMERICAL EXAMPLE

The performance of the algorithm will be illustrated by using it to solve the linear systems designed during an adaptive *hp* version analysis of a Poisson equation on a slit domain. Thanks to the symmetry, only half of the physical domain is used in the computations. The meshes obtained are shown in Figures 3-4.

The systems were solved using the preconditioner described above in conjunction with the conjugate gradient method. For comparison, the systems were also solved using simple diagonal scaling as a preconditioner. The behaviour of the residuals during the solutions of the linear systems are shown in Figure 2. It will be observed that the number of iterations required for the solver based on simple diagonal scaling to reduce the residual to machine accuracy increases significantly as the mesh is progressively refined. However, the number of iterations required for the new preconditioner hardly increases despite the highly refined meshes with non-uniform polynomial degree ranging from 1 to 8. This is precisely what is predicted by Theorem 4.1.

References

[Ain96a]   Ainsworth M. (1996) A hierarchical domain decomposition preconditioner for *h-p* finite element approximation on locally refined meshes. *SIAM J. Sci. Comput. (To appear)* .

[Ain96b]   Ainsworth M. (1996) A preconditioner based on domain decomposition for *h-p* finite element approximation on quasi-uniform meshes. *SIAM J. Numer. Anal. (To appear)* .

[BCMP91] Babuska I., Craig A., Mandel J., and Pitkaranta J. (1991) Efficient preconditioning for the *p*-version finite element method in two dimensions. *SIAM J. Numer. Anal.* 28: 624–661.

[BG88]   Babuska I. and Guo B. Q. (1988) The h–p version of the finite element method for domains with curved boundaries. *SIAM J. Num. Anal.* 25: 837–861.

[BKS81]   Babuska I., Katz I., and Szabo B. (1981) The *p* version of the finite element method. *SIAM J. Numer. Anal.* 18: 515–545.

[BS87]   Babuska I. and Suri M. (1987) The *h-p* version of the finite element method with quasi-uniform meshes. *RAIRO M^2AN* 21: 199–238.

[Man90]   Mandel J. (1990) Iterative solvers by substructuring for the *p*-version finite element method. *Comp. Meth. Appl. Mech. Eng.* 80: 117–128.

[Smi91]   Smith B. (1991) Domain decomposition algorithms for the partial differential equations of linear elasticity. *PhD Thesis* .

5

A Modified FE Assembling Procedure with Applications to Electromagnetics, Acoustics and *hp*-Adaptivity

L. Demkowicz

Texas Institute for Computational and Applied Mathematics
The University of Texas at Austin
Taylor Hall 2.400
Austin, Texas 78712, USA

5.1 INTRODUCTION

A modified FE assembling procedure is discussed. Restriction of a global basis function to a contributing element is assumed in the form of a linear combination of the element shape functions. This is a generalization of the classical approach where the linear combination reduces to just a single shape function. This generalized approach allows naturally the enforcement of various continuity requirements, with applications to acoustics, electromagnetics ($\boldsymbol{H}(\Omega, div)$ and $\boldsymbol{H}(\Omega, \mathbf{curl})$ spaces), adaptivity (constrained approximations), implementation of essential boundary conditions, all in context of arbitrary *hp*-approximations. The technique allows for an optimal choice of element shape functions to optimize local element calculations and simultaneously satisfy the necessary continuity requiremenets.

5.2 THE CONCEPT

In the classical Finite Element (FE) discretization, the global *basis functions* are constructed as unions of the element *shape functions*. The construction is illustrated in Fig. 1. If e_j denotes a basis function corresponding to the j-th node, and K is one of the elements adjacent to the node, the restriction of function e_j to element K, $e_j|_K$,

The Mathematics of Finite Elements and Applications
Edited by J. R. Whiteman © 1997 John Wiley & Sons Ltd.

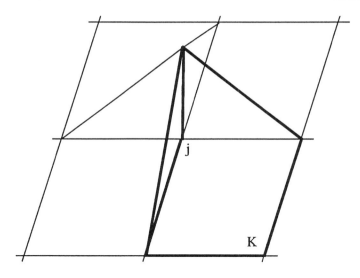

Figure 1   Construction of a FE basis function on a regular mesh

coincides with an l-th shape function for element K,

$$e_j|_K = \chi_{l,K} \qquad (5.1)$$

The basis function number j corresponding to the element K shape function number l, is called the *connectivity*, and typically it is stored in a data structure array [Bec81]

$$j = NODES(l, K) \qquad (5.2)$$

The situation changes if one has to construct a basis function on an *irregular*, locally refined mesh, as illustrated in Fig. 2. The restriction of a basis function to a *small* element K, adjacent to node j, no longer coincides with a single shape function. Rather, it is represented as a linear combination of the element shape functions.

This motivates introducing the following generalizations of the construction of the global FE basis functions.

Scalar case

$$e_j|_K = \sum_l c_{jl,K} \chi_{l,K} \qquad (5.3)$$

Vector case

$$e_j|_K = \sum_{J=1}^{2(3)} \sum_l c_{jl,K}^J \chi_{l,K} i_J \qquad (5.4)$$

where i_J denote the canonical basis in $\mathbb{R}^2$ or $\mathbb{R}^3$.

Obviously, the coefficients $c_{jl,K}^J$ or $c_{jl,K}$ must be known.

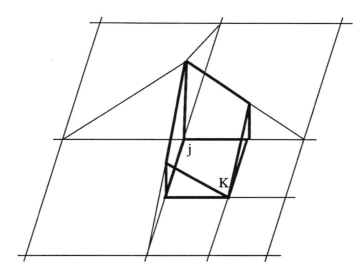

Figure 2 Construction of a FE basis function on an irregular mesh

A MODIFIED FE ASSEMBLING PROCEDURE

We begin with the usual abstract variational formulation,

$$
\begin{cases}
u \in V \\
b(u, v) = l(v) \quad \forall v \in V
\end{cases}
\tag{5.5}
$$

where V is a Hilbert space, b is a bilinear (sesquilinear) form defined on $V \times V$, and l is a linear (antilinear) form on V.

Introducing a set of linearly independent basis functions $e_j \in V, j = 1, \ldots, N$, we construct the usual Galerkin approximation,

$$
\begin{cases}
u = \displaystyle\sum_{j=1}^{N} u_j e_j \\
\displaystyle\sum_{j} b(e_j, e_i) u_j = l(e_i), \quad i = 1, \ldots, N
\end{cases}
\tag{5.6}
$$

With b_K and l_K denoting contributions of element K to the global bilinear and linear forms, the classical assembling procedure is:

$$
l(e_j) = l(e_j) + l_K(\chi_{l,K})
$$
$$
b(e_i, e_j) = b(e_i, e_j) + b_K(\chi_{k,K}; \chi_{l,K})
\tag{5.7}
$$

The *modified assembling procedure* will involve an extra summation:

$$
l(e_j) = l(e_j) + \sum_{l} c_{jl,K} l_K(\chi_{l,K})
$$
$$
b(e_i, e_j) = b(e_i, e_j) + \sum_{k}\sum_{l} c_{ik,K} c_{jl,K} b_K(\chi_{k,K}, \chi_{l,K})
\tag{5.8}
$$

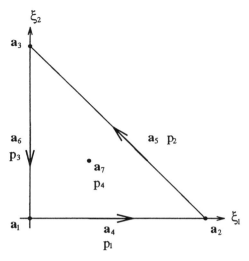

Figure 3 A triangular element of variable order of approximation

For a vector-valued problem, basis functions e_j are vector-valued, and the assembling will involve extra summations with respect to the global components,

$$l(e_j) = l(e_j) + \sum_J \sum_l c^J_{jl,K} l_K(\chi_{l,K} i_J)$$

$$b(e_i, e_j) = b(e_i, e_j) + \sum_I \sum_J \sum_k \sum_l c^I_{ik,K} c^J_{jl,K} b_K(\chi_{k,K} i_I, \chi_{l,K} i_J)$$

(5.9)

5.3 AN HP ELEMENT

The new approach towards the construction of the FE approximation allows the definition of the local element shape functions to be independent of the final form of the global basis functions. From the algebraic point of view, the shape functions are now used only for setting up the local approximation, i.e. the *spaces of element shape functions* $X_h(K)$, see [Cia78]. In particular, their choice may be optimized to minimize the number of operations involved in the element calculations.

As an example, we present the definition of a master triangular element of the variable order introduced in [Dem91] which allows for a contruction of hp-adaptive FE approximations. The element is illustrated in Fig. 3. It consists of seven nodes: three vertex nodes a_1, a_2, a_3, three mid-side nodes $a_4, a_5, a_6,$, and a middle node a_7. Each of the three mid-side nodes and the middle node may have a different, corresponding order of approximation p_1, p_2, p_3, p_4, respectively. Introducing the area coordinates $\lambda_1 = 1 - \xi_1 - \xi_2, \lambda_2 = \xi_1, \lambda_3 = \xi_2$, we define the corresponding *scalar-valued* shape functions as follows:

- vertex node shape functions,

$$\hat{\chi}_i = \lambda_i, \quad i = 1, 2, 3$$

(5.10)

- mid-side node shape functions,

$$\hat{\chi}_{1,i} = \frac{\displaystyle\prod_{j=0,j\neq i}^{p_1-1} (\lambda_2 - \frac{j}{p_1})\lambda_1}{\displaystyle\prod_{j=0,j\neq i}^{p_1-1} (\frac{i}{p_1} - \frac{j}{p_1})(1 - \frac{i}{p_1})}, \quad i = 1,\ldots,p_1-1 \tag{5.11}$$

with formulas for $\hat{\chi}_{2,i}$ and $\hat{\chi}_{3,i}$ obtained by permutating indices,
- middle node shape functions,

$$\hat{\chi}_{4,i,j,k} = \frac{\displaystyle\prod_{m=0}^{i-1}(\lambda_1 - \frac{m}{p_4})\prod_{n=0}^{j-1}(\lambda_2 - \frac{n}{p_4})\prod_{l=0}^{k-1}(\lambda_3 - \frac{m}{p_4})}{\displaystyle\prod_{m=0}^{i-1}(\frac{i}{p_4} - \frac{m}{p_4})\prod_{n=0}^{j-1}(\frac{j}{p_4} - \frac{n}{p_4})\prod_{l=0}^{k-1}(\frac{k}{p_4} - \frac{l}{p_4})} \tag{5.12}$$

with $1 \leq i, j, k \leq p_4 - 1, i + j + k = p_4$.

It can be proved that the shape functions satisfy the *approximability condition*. Namely, if p denotes the minimum order of approximation for the element,

$$p = \min\{1, p_1, p_2, p_3, \max\{p_4, 2\}\} \tag{5.13}$$

then the element space of shape functions contain complete polynomials of order p,

$$\mathcal{P}^p \subset X_h(\hat{K}) \tag{5.14}$$

where $X_h(\hat{K})$ is the space of master element shape functions,

$$X_h(\hat{K}) = \text{span}\{ \quad \chi_i, i = 1,2,3, \ \chi_{i,j}, j = 1,\ldots,p_i - 1, i = 1,2,3,$$
$$\chi_{4,ijk}, 1 \leq i, j, k \leq p_4 - 1, i + j + k = p_4\} \tag{5.15}$$

The definition can be easily extended to curved elements using either CAD based transformations (*exact* elements), or the usual parametric maps [Dem91].
  We shall show in the following examples how the new assembling procedure combined with the element defined above can be used to construct global *hp* approximations corresponding to several non-classical continuity requirements.

5.4   EXAMPLES

APPLICATION 1: IMPLEMENTATION OF ESSENTIAL BOUNDARY CONDITIONS

One possible essential boundary condition in 2-D elasticity is that which involves only the *normal component* of the displacement vector,

$$u_n = \mathbf{u} \circ \mathbf{n} = 0 \tag{5.16}$$

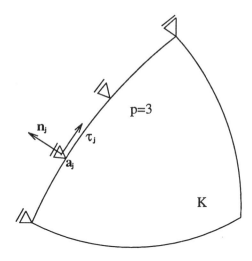

Figure 4   Essential boundary conditions in elasticity

Here u denotes the displacement vector, and n is the outward normal unit vector to the boundary, see Fig. 4. We define formally the *finite element space* X_h as:

$$X_h = \{u_h = (u_{h1}, u_{h2}) \ : \ u_{hi}|_K \in X(K), \text{ and } u_{hi} \text{ is continuous}, i = 1, 2\} \quad (5.17)$$

Let p denote the order of approximation for the edge illustrated in Fig. 4. We introduce $p+1$ Lagrange-like nodes a_j, located along the edge [1], and consider the corresponding unit tangential and normal vectors τ_j and n_j, respectively.

The global degrees of freedom (d.o.f.) are now constructed as follows:

$$\left. \begin{array}{l} u \to u(a_j) \circ n_j \\[2mm] u \to u(a_j) \circ \tau_j \end{array} \right\} \quad j = 0, \dots, p \qquad (5.18)$$

The corresponding global basis functions form a dual basis in the finite element space X_h:

$$e_j^J \in X_{hp} \subset H^1(\Omega) \quad j = 0, \dots, p, \ J = 1, 2 \qquad (5.19)$$

With the basis functions constructed in such a way, the boundary condition (5.16) is implemented simply by dropping in calculations the basis functions corresponding to the normal displacement.

EVALUATION OF COEFFICIENTS $C^J_{JL,K}$

It remains to show how to evaluate coefficients $c^J_{jl,K}$ involved in the transformation from the local to the global basis. A natural way to go is to consider a dual basis ϕ_l

---

[1] More precisely, the nodes are images of the corresponding Lagrange nodes for the master element under the particular map defining the curved element.

to the element shape functions χ_k, i.e. an arbitrary set of functionals defined on the space of element shape functions $X(K)$, such that

$$\phi_l(\chi_k) = \delta_{lk} \tag{5.20}$$

For examples of such *element degrees of freedom* for the presented triangular element, see [Dem91]. Multiplying

$$e_j^J|_K = \sum_{J=1}^{2} \sum_{l} c_{jl,K}^J \chi_{l,K} i_J \tag{5.21}$$

first by i_J, and then applying functional ϕ_l to both sides, we simply get:

$$c_{jl,K}^J = \phi_l(e_j^J|_K \circ i_J) \tag{5.22}$$

This procedure requires, however, that

- the global basis functions must be known explicitly,
- the element d.o.f. should be easy to calculate.

None of the two conditions may be satisfied in practice.

An alternative, more practical approach, is to use the *original*, global d.o.f. ψ_i^J. Applying ψ_i^J to both sides of (5.21) , we end up with a system of linear equations for the unknown coefficients $c_{jl,K}^J$,

$$\sum_{J=1}^{2} \sum_{l} c_{jl,K}^J \psi_i^I(\chi_{l,K} i_J) = \delta_{ij} \delta^{IJ}, \quad i = 0, \ldots, p, \ I = 1, 2 \tag{5.23}$$

In practice, it is sufficient to formulate and solve such systems only for the Lagrange nodes sitting on one element side, i.e. the d.o.f. corresponding to the vertex nodes and the mid-side node.

APPLICATION 2: ELECTROMAGNETICS

A natural functional space for electromagnetic computations is (see e.g. [Mon93]),

$$H(\Omega, \text{curl}) = \{E \in L^2(\Omega) \ : \ \nabla \times E \in L^2(\Omega)\} \tag{5.24}$$

Let E be now a FE approximation of an exact solution from the space $H(\Omega, \text{curl})$. Let $E_i \in X(K_i)$ denote the restriction of E to an element K_i. One can prove that the global function E belongs to the space $H(\Omega, \text{curl})$ *if and only if*

$$n \times [E_i - E_j] = 0 \text{ on } \Gamma_{ij} \tag{5.25}$$

where Γ_{ij} denotes the interface between elements K_i and K_j, see Fig. 5. We now proceed in a way similar to the previous example. With p denoting the order of approximation along the common side, we introduce the $p + 1$ Lagrange-like nodes, $p + 1$ tangential unit vectors τ_j but only $p - 1$ normal unit vectors n_j. The global d.o.f. are now defined as follows:

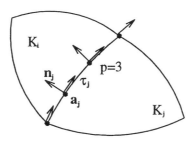

Figure 5   Tangential continuity in electromagnetics

- d.o.f shared by adjacent elements:

$$\boldsymbol{E} \to \boldsymbol{E}(\boldsymbol{a}_j) \circ \boldsymbol{\tau}_j, \quad j = 0, \dots, p \tag{5.26}$$

- d.o.f. internal to the element:

$$\boldsymbol{E} \to \boldsymbol{E}(\boldsymbol{a}_j) \circ \boldsymbol{n}_j, \quad j = 1, \dots, p-1 \tag{5.27}$$

Calculation of the coefficients $c_{lj,K}^J$ follows the same lines as in the previous example.

APPLICATION 3: ACOUSTICS

The situation is analogous to that in electromagnetics. A natural space for the solution is

$$\boldsymbol{H}(\Omega, \mathrm{div}) = \{ v \in \boldsymbol{L}^2(\Omega) \; : \; \boldsymbol{\nabla} \circ v \in \boldsymbol{L}^2(\Omega) \} \tag{5.28}$$

Again, a FE solution v sits in the space $\boldsymbol{H}(\Omega, \mathrm{div})$ if an only if

$$\boldsymbol{n} \circ [v_i - v_j] = 0 \text{ on } \Gamma_{ij}$$

The continuity of the normal component can be enforced by introducing $p+1$ normal unit vectors $\boldsymbol{n}_j$ and only $p-1$ tangential vectors $\boldsymbol{\tau}_j$, and defining the global d.o.f. as follows:

- d.o.f. shared by adjacent elements:

$$v \to v(\boldsymbol{a}_j) \circ \boldsymbol{n}_j, \quad j = 0, \dots, p \tag{5.29}$$

- d.o.f.internal to the element;

$$v \to v(\boldsymbol{a}_j) \circ \boldsymbol{\tau}_j, \quad j = 1, \dots, p-1 \tag{5.30}$$

APPLICATION 4: CONSTRAINED APPROXIMATION

In the final example, we return to the notion of the constrained approximation discussed in the beginning of this note to make one extra remark in context of the hp-adaptive approximations.

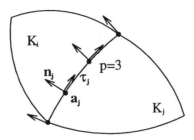

Figure 6   Normal continuity in acoustics

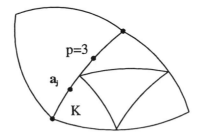

Figure 7   Constrained approximation

The task is now to enforce the continuity of the approximation along a *big* element side, with two adjacent *small* elements, see Fig. 7. First of all, the order of approximation along the side for all three elements must be the same. In [Dem89] we used a procedure based on the degrees of freedom corresponding to the big element. As those d.o.f for the hp-elements are quite cumbersome, the implementation, and in particular the evaluation of coefficients $c^J_{lj,K}$ may be difficult to program [2]. However, if one is also willing to transform the basis for the big elements, the procedure may be significantly simplified by introducing standard, Lagrange global d.o.f. defined as:

$$u \rightarrow u(a_j), \quad j = 0, \ldots, p \qquad (5.31)$$

where $a_j, j = 0, \ldots, p$ are the standard Lagrange nodes distributed along the big element side. The global basis functions,

$$e_j \in X_{hp} \subset H^1(\Omega) \qquad (5.32)$$

will again form a dual basis to the global d.o.f. in the FE space X_h. The evaluation

of coefficients $c_{jl,K}$ is almost trivial. All we need to do, is evaluate both sides of

$$e_j|_K = \sum_{l=0}^{p} c_{jl,K} \chi_{l,K} \qquad (5.33)$$

---

[2] The gain in using the big element d.o.f is that the assembling procedure for the big element reduces to the standard one.

at the nodes a_j, to get a system of equations:

$$\sum_{l=0}^{p} c_{jl,K} \chi_{l,K}(a_i) = \delta_{ij}, \quad i = 0, \ldots, p \tag{5.34}$$

5.5   CONCLUSIONS

In the note a new assembling procedure for Finite Element calculations has been presented. The procedure *decouples* the definition of the element shape functions from the construction of the global basis functions in the sense that the global basis functions are not longer unions of the contributing elements' shape functions. Instead, the restriction of a basis function to an element coincides with a linear combination of the element shape functions. The main advantages of the new approach are:

- The choice of the element shape functions can be optimized with respect to the element calculations (use of recursive formulas, inexpensive integration),
- The construction of the global basis functions, following the choice of the global d.o.f. may reflect global continuity requirements and/or essential boundary conditions.

References

[Bec81]   Becker, E. B. and Carey, G. F., and Oden, J. T. (1981) *Finite Elements. An Introduction*. Prentice-Hall, New Jersey.

[Cia78]   Ciarlet, P. G. (1978) *The Finite Element Method for Elliptic Problems*. North Holland, New York.

[Dem89]   Demkowicz, L. and Oden, J. T. and Rachowicz, W. and Hardy, O. (1989) Toward a Universal *hp* Adaptive Finite Element Strategy. Part1: Constrained Approximation and Data Structure. *Computer Methods in Applied Mechanics and Engineering* 77(1-2): 79–112.

[Dem91]   Demkowicz, L., Karafiat, A. and Oden, J. T. (October 1991) Solution ofElastic Scattering Problems in Linear Acoustics Using *hp* Boundary ElementMethod. In Demkowicz L., Oden J. T., and Babuska I. (eds) *Proceedings of Second Workshop on Reliability and Adaptive Methods in Computational Mechanics*, volume 101, pages 251–282. Computer Methods in Applied Mechanics and Engineering. Cracow.

[Mon93]   Monk, P. (1993) An Analysis of Nédélec's Method for the Spatial Discretization of Maxwell's Equations. *Journal of Computational and Applied Mathematics* 47: 101–121.

6

Adaptive Domain Decomposition Methods in FEM and BEM

M. Kuhn and U. Langer

Institute of Mathematics,
Johannes Kepler University,
A-4040 Linz,
Austria.

6.1 INTRODUCTION

The Domain Decomposition (DD) approach offers many opportunities for marrying the advantages of the Finite Element Method (FEM) with those of the Boundary Element Method (BEM) in many practical applications. For instance, in magnetic field computations for electric motors, we can use the BEM in the air subdomains, including the exterior of the motor and the air gap, more successfully than the FEM which is preferred in ferromagnetic materials where nonlinearities can occur in the partial differential equation (PDE), or in subdomains where the right–hand side does not vanish [HK96, KMZ90]. The same is true for many problems in solid mechanics [Hol92] and in other areas of research. A very straightforward and promising technique for the coupling of FEM and BEM was proposed by Costabel [Cos87] and others [HW91, CW96]. In the different subdomains of a non–overlapping domain decomposition, we use either the standard finite element (FE) Galerkin method or a mixed–type boundary element (BE) Galerkin method and couple these weakly over the coupling boundaries (interfaces) Γ_C. The mixed BE Galerkin method makes use of the full Cauchy data representation on the BE subdomain boundaries via the Calderón projector.

The main aim of the paper is the design, analysis and implementation of fast and well adapted parallel solvers for large-scale coupled FE/BE–equations approximating plane, linear and nonlinear magnetic field problems including technical magnetic field problems (e.g. electric motors). To be specific, we consider a characteristic cross-section, which lies in the (x, y)-plane of the $\mathbb{R}^3$ of the original electromagnetic device that is to be modelled. Let us assume that $\Omega_0 \subset \mathbb{R}^2$ is a bounded simply connected domain and that homogeneous Dirichlet boundary conditions are given on $\Gamma_D = \partial \Omega_0$.

The Mathematics of Finite Elements and Applications
Edited by J. R. Whiteman © 1997 John Wiley & Sons Ltd.

Formally the nonlinear magnetic field problem can be written as follows [Hei94]

$$-\text{div}\ (\nu(x,|\nabla u(x)|)\nabla u(x)) \quad = \quad S(x) + \frac{\partial H_{0y}(x)}{\partial x} - \frac{\partial H_{0x}(x)}{\partial y}, \quad x \in \Omega, \quad (6.1)$$

$$u(x) \quad = \quad 0, \quad x \in \Gamma_D, \quad\quad\quad\quad\quad\quad (6.2)$$

$$|u(x)| \quad \to \quad 0 \quad \text{for } |x| \to \infty, \quad\quad\quad\quad (6.3)$$

with $\Omega := R^2 \setminus \bar{\Omega}_0$. The solution u is the z–component A_z of the vector potential $\vec{A} = (A_x, A_y, A_z)^T$ introduced in the Maxwell equation. The component of the current density, which acts orthogonally to the cross-section being considered, is represented by $S(x)$, whereas H_{0x} and H_{0y} stand for sources associated with permanent magnets that may occur, and $\nu(.)$ denotes a coefficient depending on the material and on the gradient $|\nabla u(x)|$ (induction). We now introduce the exterior domain Ω_+ by defining a, so called, coupling boundary $\Gamma_+ := \partial\Omega_+$. The definition of Γ_+ is restricted by the conditions

$$\nu(x) = \nu_p \ \forall x \in \Omega_+, \ (\text{supp }S \cup \text{supp }H_0) \subset \bar{\Omega}_-, \ \text{diam}(\bar{\Omega}_0 \cup \bar{\Omega}_-) < 1, \quad (6.4)$$

where $\Omega_- := R^2 \setminus (\bar{\Omega}_+ \cup \bar{\Omega}_0)$. Note that the condition $\text{diam}(\bar{\Omega}_0 \cup \bar{\Omega}_-) < 1$ is only technical and can be fulfilled by scaling the problem appropriately. Besides the decomposition $\bar{\Omega} = \bar{\Omega}_- \cup \bar{\Omega}_+$, we allow the inner domain Ω_- to be decomposed further following the natural decomposition of Ω_- according to the change of data:

$$\bar{\Omega}_- = \bigcup_{j=1}^{N_M} \bar{\Omega}_j, \quad \text{with} \quad \hat{\Omega}_i \cap \hat{\Omega}_j = \emptyset \quad \forall i \neq j. \quad (6.5)$$

In Section 6.2, we present an automatic and adaptive domain decomposition procedure which produces a decomposition of Ω into p subdomains (p = number of processors to be used) and such controlling data for the distributed mesh generator [Glo95] that we can expect a well load-balanced performance of our solver.

In Section 6.3, we consider linear plane magnetic field problems for which a domain decomposition according to Section 6.2 is available, and introduce the mixed variational DD FE/BE discretization.

Section 6.4 is devoted to algorithms for solving the resulting linear, symmetric, but indefinite system coupled FE/BE equations. First of all, the system of coupled FE/BE equations can be reformulated as a symmetric and positive definite (spd) system on the basis of Bramble/Pasciak's transformation [BP88]. We provide a preconditioning and a parallelization of the Conjugate Gradient (CG) method applied to this spd system. Using a special DD data distribution, we parallelize the preconditioning equation and the remaining algorithm in such a way that the same amount of communication is needed as in the earlier introduced and well studied parallel PCG method for solving symmetric and positive definite FE equations [HLM90, HLM91].

Section 6.5 deals with the adaption of the components of the preconditioner to the specific problem under consideration. These components can be chosen such that the resulting algorithm is, at least, almost asymptotically optimal with respect to the operation count and quite robust with respect to complicated geometries, jumping coefficients and mesh grading near singularities (see numerical results given in Section 6.6 and in [HK96]).

In Section 6.6, we present some numerical results obtained for three test problems. The first of these is an academic one (Subsection 6.6.1). We study the influence of graded meshes, used to provide a good approximation to the singularity situated at the end of a slit, on the parallel solver. In Subsection 6.6.2, we apply our linear DD–solver to a plane linear elasticity test problem modelled by Lamé's system of PDEs. An appropriate adaptation of the components of the DD–preconditioner results in a parallel solver the efficiency of which is comparable to that of the solver for the potential equation. Finally, we solve a technical nonlinear magnetic field problem (Subsection 6.6.3). The nonlinear magnetic field problem is solved by the Full-DD-Newton–Method developed in [Hei93, HK96]. In every nested Newton step we use basically the linear DD–solver described in this paper. All numerical results presented in this paper were obtained by the use of the package FEM∞BEM [HHJ+96]. The code runs on various parallel computers and programming platforms including PVM (see, e.g., [HJ95]).

6.2 ADAPTIVE DOMAIN DECOMPOSITION PREPROCESSING

6.2.1 THE DD-DATA PARTITIONING

In this section, we focus our interest on how a decomposition of the domain Ω into a given number of subdomains can be obtained by exploiting the natural decomposition into domains based on the different materials (6.5). We are interested in well load-balanced decompositions especially in the case of discretizations which are adapted to singularities.

We assume that a triangle-based description of the geometry of the problem under consideration is given. Besides the geometrical data each triangle is characterized by a parameter pointing to that of the N_M material-regions to which the triangle belongs. Note that interfaces between different materials, i.e. the boundaries of the $\hat{\Omega}_j$'s (cf. (6.5)), are represented by edges of the triangulation. The aim of the automatic preprocessing is a decomposition into $p \geq N_M$ subdomains

$$\bar{\Omega} = \bigcup_{i \in \mathcal{I}} \bar{\Omega}_i, \quad \text{where} \quad \Omega_p := \Omega_+ \quad \text{and} \quad \hat{\bar{\Omega}}_j = \bigcup_{i \in \mathcal{I}_j} \bar{\Omega}_i \quad \forall j = 1, \ldots, N_M \qquad (6.6)$$

where the sets of indices are given by $\mathcal{I} := \{1, \ldots, p\}$ and

$$\mathcal{I}_j \subset \mathcal{I}^\star := \{1, \ldots, p-1\}, \quad \bigcup_{j=1}^{N_M} \mathcal{I}_j = \mathcal{I}^\star, \quad \mathcal{I}_j \cap \mathcal{I}_k = \emptyset \quad \forall j \neq k,$$

i.e., the subdomains $\hat{\Omega}_j$ determined by the materials may be decomposed further (see, e.g., [HLM92]). We assume that there exist open balls B_{r_i} and $B_{\bar{r}_i}$ $(i \in \mathcal{I}^\star)$ with positive radii $\underline{r}_i$ and $\bar{r}_i$, such that $B_{r_i} \subset \Omega_i \subset B_{\bar{r}_i}$ and $0 < \underline{c} \leq \bar{r}_i/\underline{r}_i \leq \bar{c}$ $\forall i \in \mathcal{I}^\star$ with fixed (i-independent) constants $\underline{c}$ and $\bar{c}$. Note, in the case of Ω being bounded we would have $\mathcal{I}^\star := \{1, \ldots, p\}$ and in the following all terms induced by p, which then stands for the exterior domain, would vanish.

Besides the natural triangular-based description of the geometry we introduce a special DD-data structure described below which is well suited as input for DD-based algorithms running on massively parallel computers. Thus, starting off with a given triangle-based geometrical description (*.tri–file) we wish to end up with a well-balanced decomposition of our problem which is described by using some DD-data format (*.dd–file). Fig. 1 shows the interactions between the preprocessing components *Decomp, Tri2DD* and *AdapMesh* and the data- or file-types *.tri, *.dd, and *.fb being involved. In the simplest case the process starts, on the left in the diagram, with applying *Decomp* to a *.tri–file which results in a decomposition as defined in (6.6), i.e. each triangle is assigned to one of the Ω_i's, $i \in \mathcal{I}^\star$. The output of this process is also a *.tri–file which is then converted into a *.dd–file by applying *Tri2DD*. In

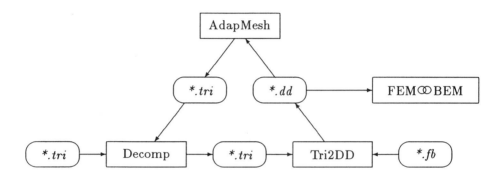

Figure 1   The components of the preprocessing.

our case, such a *.dd–file is the input for the parallel code FEM⊗BEM [HHJ+96]. Because of the simple structure of the DD-data format being used it is easy to change the refinement information given in the *.dd–file, e.g., in order to adapt the mesh to singularities. In the latter case the new mesh created from the updated *.dd–file may differ significantly from the original mesh described by the initial *.tri–file. As a consequence we have to expect a bad load balance. We can overcome this difficulty by, first, applying *AdapMesh* which simulates the mesh generation as it is realized in the parallel program and creates finally a new *.tri–file and, second, restarting the cycle with *Decomp*. Note, in the optimal case with respect to the load balance the mesh used for computations (i.e. the one created from a *.dd–file within FEM⊗BEM) would coincide with the mesh being used for the decomposition, see also the example in Subsection 6.6.1.

6.2.2   A SHORT DESCRIPTION OF THE PREPROCESSING CODES

At this point, we describe the codes and the main ideas on which they are based. More information and technical details can be found in the documentation [GHHK96].

Decomp decomposes single-material domains using the spectral bisection method (sbm) [Sim91]. For each of the N_M material domains the number of elements of $\mathcal{I}_j$ is determined such that the maximum number of triangles per subdomain becomes minimal. As a result, each triangle is assigned to one of the subdomains.

Tri2DD converts triangular-based data into the DD-data format. The algorithm is based on the definition of edges which then define the faces. Note, interfaces between different materials will be maintained as they were given in the original *.tri*-file. On the other hand the artificially created boundaries within one material are smoothed.

AdapMesh creates a mesh based on DD-data using, optionally, adaptivity information. The resulting *.tri*-file can be used as input for *Decomp.*

During the preprocessing we are concerned with two types of describing data. The first (*.tri*-files) is based on nodes (characterized by their coordinates) which define the edges (straight lines or arcs of circles are allowed) and, finally, triangles defined by their edges. Each of the triangles is characterized by two additional parameters, where only one of these pointing to the material to which the triangle belongs has to be initialized from the very beginning. The second type describes the mapping of the triangles on the subdomains and it is defined as a result of decomposing Ω.

The second type of data (*.dd*-files) follows the DD-data format described in [HHJ+96]. It is based on defining cross-points numbered globally which then define edges. Main objects are faces described by edges. The faces, or a union of these, are mapped onto the array of processors. The *.fb*-files are auxiliary files (optional) and contain controlling data or fixed cross-points.

6.2.3 EXAMPLE: THE SLIT PROBLEM

Starting with any of the data formats the preprocessor has to maintain the original information throughout all stages ending up with a dd-file which describes finally the decomposition of our domain into subdomains according to the a-priori given refinement information.

In Fig. 2 we start with a manual decomposition of the problem which is not well-balanced and apply, from left to right, the components *AdapMesh, Decomp,* and *Tri2DD.* The preprocessor performs a quasi-static decomposition. It is possible to

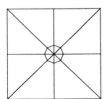 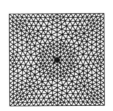  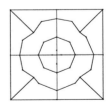

Figure 2 The four stages of preprocessing starting with an initial decomposition (D1) on the left and ending up with a load-balanced decomposition (D2) on the right.

restart the preprocessing procedure optionally. This may be necessary if the refinement information (grading or density of the mesh) has been modified, e.g., according to the behaviour of the solution obtained from a coarse grid calculation. The *a-posteriori* change of the refinement information, i.e., of the discretization, in general destroys the load-balance, and one has to restart the decomposition procedure. The numerical results obtained on the parallel computer GC–PowerPlus are presented in Subsection 6.6.1.

6.3 COUPLED FE/BE - DD - DISCRETIZATION

6.3.1 A MIXED VARIATIONAL FORMULATION

Let us first consider a linear ($\nu = \nu(x)$) magnetic field problem of the form (6.1) - (6.3) for which a domain decomposition according to Section 6.2 is available. In particular, we assume that the index set $\mathcal{I} = \mathcal{I}_F \cup \mathcal{I}_B$ can be decomposed into two disjoint sets of indices $\mathcal{I}_F$ and $\mathcal{I}_B$ such that

$$p \in \mathcal{I}_B, \tag{6.7}$$

$$(\operatorname{supp} S(.) \cup \operatorname{supp} H_0(.)) \cap \Omega_i = \emptyset \quad \forall i \in \mathcal{I}_B, \tag{6.8}$$

$$\nu(x) = \nu_i = const \quad \forall i \in \mathcal{I}_B. \tag{6.9}$$

For each Ω_i ($i \in \mathcal{I}$) the index i belongs to one of the two index sets $\mathcal{I}_B$ and $\mathcal{I}_F$ according to the discretization method applied to Ω_i, where $\mathcal{I}_B$ and $\mathcal{I}_F$ stand for BEM and FEM, respectively.

Following Costabel [Cos87], Hsiao and Wendland [HW91] and others, we can rewrite the weak formulation of the boundary value problem (6.1) - (6.3) by means of integration by parts in the boundary element subdomains Ω_i, $i \in \mathcal{I}_B$ and by the use of Calderón's representation of the full Cauchy data as a mixed DD coupled domain and boundary integral variational problem: Find $(\lambda, u) \in \mathbf{V} := \Lambda \times \mathbf{U}_0$ such that

$$a(\lambda, u; \eta, v) = \langle F, v \rangle \quad \forall (\eta, v) \in \mathbf{V}, \tag{6.10}$$

where

$$
\begin{aligned}
a(\lambda, u; \eta, v) &:= a_B(\lambda, u; \eta, v) + a_F(u, v), \\
a_B(\lambda, u; \eta, v) &:= \sum_{i \in \mathcal{I}_B \setminus \{p\}} \nu_i \Big\{ \langle \mathcal{D}_i u_i, v_i \rangle_{\Gamma_i} + \frac{1}{2} \langle \lambda_i, v_i \rangle_{\Gamma_i} + \langle \lambda_i, \mathcal{K}_i v_i \rangle_{\Gamma_i} + \\
&\qquad \langle \eta_i, \mathcal{V}_i \lambda_i \rangle_{\Gamma_i} - \langle \eta_i, \mathcal{K}_i u_i \rangle_{\Gamma_i} - \frac{1}{2} \langle \eta_i, u_i \rangle_{\Gamma_i} \Big\} \\
&\qquad + \nu_p \Big\{ \langle \mathcal{D}_p u_p, v_p \rangle_{\Gamma_+} - \frac{1}{2} \langle \lambda_p, v_p \rangle_{\Gamma_+} + \langle \lambda_p, \mathcal{K}_p v_p \rangle_{\Gamma_+} + \\
&\qquad \langle \eta_p, \mathcal{V}_p \lambda_p \rangle_{\Gamma_+} - \langle \eta_p, \mathcal{K}_p u_p \rangle_{\Gamma_+} + \frac{1}{2} \langle \eta_p, u_p \rangle_{\Gamma_+} \Big\}, \\
a_F(u, v) &:= \sum_{i \in \mathcal{I}_F} \int_{\Omega_i} \nu(x) \nabla^T u(x) \nabla v(x) \, dx, \\
\langle F, v \rangle &:= \sum_{i \in \mathcal{I}_F} \int_{\Omega_i} \left[S(x) v(x) - H_{0y}(x) \frac{\partial v(x)}{\partial x} + H_{0x}(x) \frac{\partial v(x)}{\partial y} \right] dx, \\
\langle \lambda_i, v_i \rangle_{\Gamma_i} &:= \int_{\Gamma_i} \lambda_i v_i \, ds \text{ and } v_i = v|_{\partial \Omega_i}, \; u_i = u|_{\partial \Omega_i}, \; \Gamma_i := \partial \Omega_i,
\end{aligned}
$$

with the well-known boundary integral operators $\mathcal{V}_i, \mathcal{K}_i, \mathcal{D}_i$ defined by the relation

$$
\begin{aligned}
\mathcal{V}_i \lambda_i(x) &:= \int_{\Gamma_i} \mathcal{E}(x,y)\lambda_i(y)\, ds_y, \\
\mathcal{K}_i v_i(x) &:= \int_{\Gamma_i} \partial_y \mathcal{E}(x,y)v_i(y)\, ds_y, \\
\mathcal{D}_i u_i(x) &:= -\partial_x \int_{\Gamma_i} \partial_y \mathcal{E}(x,y)u_i(y)\, ds_y
\end{aligned}
\tag{6.11}
$$

and with the fundamental solution

$$
\mathcal{E}(x,y) = -\frac{1}{2\pi}\log|x-y|
\tag{6.12}
$$

of the Laplacian. The mapping properties of the boundary integral operators (6.11) on Sobolev spaces are now well known [Cos88]. The spaces $\mathbf{U}_0$ and $\boldsymbol{\Lambda}$ are defined by the relations

$$
\begin{aligned}
\mathbf{U}_0 &:= \{u \in H^1(\Omega_-) : \ u|_{\Gamma_{BE}} \in H^{1/2}(\Gamma_{BE}), u|_{\partial\Omega_0} = 0\}, \\
\boldsymbol{\Lambda} &:= \{\lambda = (\lambda_i)_{i\in\mathcal{I}_B} : \ \lambda_i \in H^{-1/2}(\Gamma_i),\ i \in \mathcal{I}_B\} = \prod_{i\in\mathcal{I}_B}\Lambda_i,
\end{aligned}
\tag{6.13}
$$

with $\Lambda_i = H^{-1/2}(\Gamma_i),\ i \in \mathcal{I}_B$. Further we use the notation $\Gamma_{BE} := \bigcup_{i\in\mathcal{I}_B}\partial\Omega_i \setminus \Gamma_D, \Gamma_{FE} := \bigcup_{i\in\mathcal{I}_F}\partial\Omega_i \setminus \Gamma_D, \Gamma_C := \Gamma_{BE}\cup\Gamma_{FE}$ and $\Omega_F := \bigcup_{i\in\mathcal{I}_F}\Omega_i$. Introducing in $\mathbf{V} := \boldsymbol{\Lambda} \times \mathbf{U}_0$ the norm

$$
\|(\lambda,u)\|_{\mathbf{V}} := \left(\|\lambda\|_{\boldsymbol{\Lambda}}^2 + \|u|_{\Gamma_{BE}}\|_{H^{1/2}(\Gamma_{BE})}^2 + \|u\|_{H^1(\Omega_F)}^2\right)^{1/2}
\tag{6.14}
$$

with

$$
\|\lambda\|_{\boldsymbol{\Lambda}}^2 = \sum_{i\in\mathcal{I}_B}\|\lambda_i\|_{H^{-1/2}(\Gamma_i)}^2 \quad\text{and}\quad \|u|_{\Gamma_{BE}}\|_{H^{1/2}(\Gamma_{BE})}^2 = \sum_{i\in\mathcal{I}_B}\|u_i\|_{H^{1/2}(\Gamma_i)}^2,
$$

one can prove that the bilinear form $a(.,.)$ is $\mathbf{V}$–elliptic and $\mathbf{V}$–bounded provided that the domain decomposition satisfies the conditions imposed as above (see also [HW91]). Therefore, the existence and uniqueness of the solution are a direct consequence of the Lax–Milgram theorem.

6.3.2 THE COUPLED BE/FE DISCRETIZATION

Now, we can define the nodal FE/BE basis of piecewise linear trial functions based upon a regular triangulation of the subdomains $\Omega_i, i \in \mathcal{I}_F$ and the corresponding discretization of the boundary pieces $\Gamma_{ij} = \bar{\Omega}_i \cap \bar{\Omega}_j,\ i,j \in \mathcal{I}$:

$$
\Phi = [\phi_1, \ldots, \phi_{N_\Lambda}, \phi_{N_\Lambda+1}, \ldots, \phi_{N_\Lambda+N_C}, \phi_{N_\Lambda+N_C+1}, \ldots, \phi_N],
$$

where $N = N_\Lambda + N_C + N_I$ and $N_I = \sum_{i\in\mathcal{I}_F} N_{I,i}$, $N_\Lambda = \sum_{i\in\mathcal{I}_B} N_{\Lambda,i}$. Here, $\phi_1, \ldots, \phi_{N_\Lambda}$ are the basis functions for approximating λ on $\Gamma_i,\ i \in \mathcal{I}_B$, $\phi_{N_\Lambda+1}, \ldots, \phi_{N_\Lambda+N_C}$ represent u on Γ_C and $\phi_{N_\Lambda+N_C+1}, \ldots, \phi_{N_\Lambda+N_C+N_I}$ approximate u in $\Omega_i,\ i \in \mathcal{I}_F$. The definition of the finite dimensional subspaces of $\boldsymbol{\Lambda}, \mathbf{U}_0$ and $\mathbf{V}$

$$
\begin{aligned}
\boldsymbol{\Lambda}_h &:= \text{span } [\phi_1, \phi_2, \ldots, \phi_{N_\Lambda}], \\
\mathbf{U}_h &:= \text{span } [\phi_{N_\Lambda+1}, \ldots, \phi_{N_\Lambda+N_C}, \phi_{N_\Lambda+N_C+1}, \ldots, \phi_N], \\
\mathbf{V}_h &:= \boldsymbol{\Lambda}_h \times \mathbf{U}_h
\end{aligned}
$$

allows us to formulate the discrete problem as follows: Find $u_h \in \mathbf{V}_h$ such that

$$a(u_h, v_h) = \langle F, v_h \rangle \quad \forall v_h \in \mathbf{V}_h. \tag{6.15}$$

The isomorphism $\Phi : \mathbb{R}^N \to \mathbf{V}_h$ leads to the linear system:

$$\begin{pmatrix} K_\Lambda & -K_{\Lambda C} & 0 \\ K_{C\Lambda} & K_C & K_{CI} \\ 0 & K_{IC} & K_I \end{pmatrix} \begin{pmatrix} \mathbf{u}_\Lambda \\ \mathbf{u}_C \\ \mathbf{u}_I \end{pmatrix} = \begin{pmatrix} \mathbf{f}_\Lambda \\ \mathbf{f}_C \\ \mathbf{f}_I \end{pmatrix}, \tag{6.16}$$

where the block entries are defined by

$$(K_\Lambda \mathbf{u}_\Lambda, \mathbf{v}_\Lambda) = \sum_{i \in \mathcal{I}_B} \nu_i \langle \eta_i, \mathcal{V}_i \lambda_i \rangle_{\Gamma_i} \quad \text{with } \lambda_i = \Phi_{\Lambda_i} \mathbf{u}_{\Lambda_i}, \eta_i = \Phi_{\Lambda_i} \mathbf{v}_{\Lambda_i},$$

$$(K_{C\Lambda} \mathbf{u}_\Lambda, \mathbf{v}_C) = \sum_{i \in \mathcal{I}_B \backslash \{p\}} \nu_i \{ \langle \lambda_i, \mathcal{K}_i v_i \rangle_{\Gamma_i} + \frac{1}{2} \langle \lambda_i, v_i \rangle_{\Gamma_i} \}$$

$$+ \nu_p \{ \langle \lambda_p, \mathcal{K}_p v_p \rangle_{\Gamma_p} - \frac{1}{2} \langle \lambda_p, v_p \rangle_{\Gamma_p} \},$$

$$K_{\Lambda C} = K_{C\Lambda}^T,$$

$$K_C = K_{CB} + K_{CF}, \quad \text{with}$$

$$(K_{CB} \mathbf{u}_C, \mathbf{v}_C) = \sum_{i \in \mathcal{I}_B} \nu_i \langle \mathcal{D}_i u_i, v_i \rangle_{\Gamma_i}, \ u_i = \Phi_{C_i} \mathbf{u}_{C_i}, \ v_i = \Phi_{C_i} \mathbf{v}_{C_i} \text{ and}$$

$$\left(\begin{pmatrix} K_{CF} & K_{CI} \\ K_{IC} & K_I \end{pmatrix} \begin{pmatrix} \mathbf{u}_C \\ \mathbf{u}_I \end{pmatrix}, \begin{pmatrix} \mathbf{v}_C \\ \mathbf{v}_I \end{pmatrix} \right) = \sum_{i \in \mathcal{I}_F} \int_{\Omega_i} \nu(x) \nabla^T u \nabla v \, dx,$$

where $u|_{\Omega_F} = \Phi_F \mathbf{u}_F$, $v|_{\Omega_F} = \Phi_F \mathbf{v}_F$. Here, Φ_{Λ_i} $(i \in \mathcal{I}_B)$ and Φ_{C_i} $(i \in \mathcal{I})$ contain the basis functions for approximating λ and u on $\partial \Omega_i$, respectively. The basis functions in Φ_F are used to approximate u in Ω_i $(i \in \mathcal{I}_F)$. The FE entries, especially K_I, are sparse matrices, whereas the BE blocks are fully populated.

6.4 PARALLEL SOLVER

6.4.1 BRAMBLE-PASCIAK TRANSFORMATION AND SPECTRAL EQUIVALENCE RESULTS

The nonsymmetric, positive definite system (6.16) can be solved approximately by the Bramble/Pasciak CG method [BP88]. The method requires a preconditioner C_Λ which can be inverted easily and which fulfills the spectral equivalence inequalities

$$\underline{\gamma}_\Lambda C_\Lambda \leq K_\Lambda \leq \overline{\gamma}_\Lambda C_\Lambda, \quad \text{with} \quad \underline{\gamma}_\Lambda > 1. \tag{6.17}$$

With the definitions

$$K_1 = K_\Lambda, \quad \mathbf{f}_1 = \mathbf{f}_\Lambda, \quad K_{12} = K_{21}^T = (-K_{\Lambda C} \quad 0),$$

$$K_2 = \begin{pmatrix} K_C & K_{CI} \\ K_{IC} & K_I \end{pmatrix}, \quad \mathbf{f}_2 = \begin{pmatrix} -\mathbf{f}_C \\ -\mathbf{f}_I \end{pmatrix}.$$

we can reformulate (6.16) as a symmetric but indefinite system:

$$\begin{pmatrix} K_1 & K_{12} \\ K_{21} & -K_2 \end{pmatrix} \begin{pmatrix} \mathbf{u}_1 \\ \mathbf{u}_2 \end{pmatrix} = \begin{pmatrix} \mathbf{f}_1 \\ \mathbf{f}_2 \end{pmatrix}. \tag{6.18}$$

Following Bramble and Pasciak [BP88] this system can be transformed into

$$G\mathbf{u} = \mathbf{p}, \quad \text{where} \tag{6.19}$$

$$G := \begin{pmatrix} C_\Lambda^{-1} K_1 & C_\Lambda^{-1} K_{12} \\ K_{21} C_\Lambda^{-1}(K_1 - C_\Lambda) & K_2 + K_{21} C_\Lambda^{-1} K_{12} \end{pmatrix}, \quad \mathbf{p} = \begin{pmatrix} C_\Lambda^{-1} \mathbf{f}_1 \\ K_{21} C_\Lambda^{-1} \mathbf{f}_1 - \mathbf{f}_2 \end{pmatrix}.$$

Then, the matrix G is self-adjoint and positive definite with respect to the scalar product $[.,.]$ which is defined by

$$[\mathbf{w}, \mathbf{v}] := ((K_1 - C_\Lambda) \mathbf{w}_1, \mathbf{v}_1) + (\mathbf{w}_2, \mathbf{v}_2). \tag{6.20}$$

Moreover, G is spectrally equivalent to the regulisor R, where

$$R := \begin{pmatrix} I & 0 \\ 0 & K_2 + K_{21} K_1^{-1} K_{12} \end{pmatrix}.$$

Bramble and Pasciak [BP88] proved the spectral equivalence inequalities

$$\underline{\lambda} [R\mathbf{v}, \mathbf{v}] \le [G\mathbf{v}, \mathbf{v}] \le \overline{\lambda} [R\mathbf{v}, \mathbf{v}] \quad \forall \mathbf{v} \in \mathbb{R}^N, \tag{6.21}$$

where

$$\underline{\lambda} = \left(1 + \frac{\alpha}{2} + \sqrt{\alpha + \frac{\alpha^2}{4}}\right)^{-1} \quad \text{and} \quad \overline{\lambda} = \frac{1 + \sqrt{\alpha}}{1 - \alpha} \tag{6.22}$$

with $\alpha = 1 - (1/\overline{\gamma}_\Lambda)$. Thus, we have to find a preconditioner C_2 for the matrix

$$K_2 + K_{21} K_1^{-1} K_{12} = \begin{pmatrix} K_C + K_{C\Lambda} K_\Lambda^{-1} K_{\Lambda C} & K_{CI} \\ K_{IC} & K_I \end{pmatrix}. \tag{6.23}$$

The DD preconditioner defined by

$$C_2 = \begin{pmatrix} I_C & K_{CI} B_I^{-T} \\ 0 & I_I \end{pmatrix} \begin{pmatrix} C_C & 0 \\ 0 & C_I \end{pmatrix} \begin{pmatrix} I_C & 0 \\ B_I^{-1} K_{IC} & I_I \end{pmatrix} \tag{6.24}$$

is spectrally equivalent to $K_2 + K_{21} K_1^{-1} K_{12}$ if we have preconditioners C_I and C_C fulfilling the inequalities

$$\underline{\gamma}_C C_C \le \tilde{S}_C + K_{C\Lambda} K_\Lambda^{-1} K_{\Lambda C} \le \overline{\gamma}_C C_C, \tag{6.25}$$

$$\underline{\gamma}_I C_I \le K_I \le \overline{\gamma}_I C_I, \tag{6.26}$$

where $\tilde{S}_C = K_C - K_{CI} K_I^{-1} K_{IC} + K_{CI}(K_I^{-1} - B_I^{-T})K_I(K_I^{-1} - B_I^{-1})K_{IC}$, and B_I is an appropriately chosen non-singular matrix [Lan94].

Lemma 6.1 *If the symmetric and positive definite block preconditioners $C_I = diag(C_{I,i})_{i \in \mathcal{I}_F}$ and C_C satisfy the spectral equivalence inequalities (6.25) and (6.26) with positive constants $\underline{\gamma}_C, \overline{\gamma}_C, \underline{\gamma}_I, \overline{\gamma}_I$, then the spectral equivalence inequalities*

$$\underline{\gamma}_2 C_2 \leq K_2 + K_{21} K_1^{-1} K_{12} \leq \overline{\gamma}_2 C_2 \tag{6.27}$$

hold for the preconditioner C_2 defined in (6.24) with the constants

$$\underline{\gamma}_2 = \min\{\underline{\gamma}_C, \underline{\gamma}_I\} \left(1 - \sqrt{\tfrac{\mu}{1+\mu}}\right), \quad \overline{\gamma}_2 = \max\{\overline{\gamma}_C, \overline{\gamma}_I\} \left(1 + \sqrt{\tfrac{\mu}{1+\mu}}\right). \tag{6.28}$$

Here $\mu = \rho(S_C^{-1} T_C)$ denotes the spectral radius of $S_C^{-1} T_C$, with the FE Schur complement S_C and the operator T_C being defined by

$$S_C = \overset{\circ}{K}_C - \overset{\circ}{K}_{CI} K_I^{-1} \overset{\circ}{K}_{IC} \text{ and } T_C = \overset{\circ}{K}_{CI}(K_I^{-1} - B_I^{-T})K_I(K_I^{-1} - B_I^{-1}) \overset{\circ}{K}_{IC},$$

respectively. $\overset{\circ}{K}_C$, $\overset{\circ}{K}_{CI}$ and $\overset{\circ}{K}_{IC}$ denote the non-zero FE blocks of K_C, K_{CI} and K_{IC}, respectively.

The proof, given in [Lan94], applies the classical FE DD spectral equivalence result proved in [HLM90, HLM91]. With (6.22), we conclude the following theorem.

THEOREM 6.1 *If the conditions imposed on C_A, C_C, C_I, and B_I, especially (6.17), (6.25) and (6.26) are satisfied, then the FE/BE DD preconditioner*

$$C = diag(I_1, C_2) \tag{6.29}$$

is self-adjoint and positive definite with respect to the inner product $[.,.]$ and satisfies the spectral equivalence inequalities

$$\underline{\gamma}[C\mathbf{v}, \mathbf{v}] \leq [G\mathbf{v}, \mathbf{v}] \leq \overline{\gamma}[C\mathbf{v}, \mathbf{v}] \quad \forall \mathbf{v} \in \mathbb{R}^N, \tag{6.30}$$

with the constants

$$\underline{\gamma} = \underline{\lambda} \min\left\{1, \underline{\gamma}_2\right\} \quad and \quad \overline{\gamma} = \overline{\lambda} \max\left\{1, \overline{\gamma}_2\right\},$$

where $\underline{\lambda}, \overline{\lambda}, \underline{\gamma}_2, \overline{\gamma}_2$ are given in (6.22) and (6.28), respectively.

6.4.2   THE PARALLEL PCG ALGORITHM

For the vectors belonging to the inner coupling boundary Γ_C we define two types of distribution called overlapping (type 1) and adding (type 2):

<u>type 1:</u> $\mathbf{u}_C, \mathbf{w}_C, \mathbf{s}_C$ are stored in P_i as $\mathbf{u}_{C,i} = A_{C,i}\mathbf{u}_C$ (analogous $\mathbf{w}_{C,i}, \mathbf{s}_{C,i}$)

<u>type 2:</u> $\mathbf{r}_C, \mathbf{v}_C, \mathbf{f}_C$ are stored in P_i as $\mathbf{r}_{C,i}, \mathbf{v}_{C,i}, \mathbf{f}_{C,i}$ such that
    $\mathbf{r}_C = \sum_{i=1}^p A_{C,i}^T \mathbf{r}_{C,i}$ (analogous $\mathbf{v}_C, \mathbf{f}_C$),

where the matrices $A_{C,i}$ are the "C-block" of the Boolean subdomain connectivity matrix A_i which maps some overall vector of nodal parameters into the superelement vector of parameters associated with the subdomain $\bar{\Omega}_i$ only. P_i denotes the ith processor.

Using this notation and the operators introduced in the previous section we can formulate an improved version of the PCG-algorithm presented in [Lan94] with a given accuracy ε as stopping criterion. This parallel PCG-algorithm is given in [HK96].

Note the vectors $\mathbf{z}_i = (\mathbf{z}_{\Lambda,i}, \mathbf{z}_{C,i}, \mathbf{z}_{\Lambda C,i})^T$ and $\mathbf{h}_i = (\mathbf{h}_{\Lambda,i}, \mathbf{h}_{C,i}, \mathbf{h}_{\Lambda C,i})^T$ which have been inserted additionally in order to achieve a synchronization between the FEM and BEM processors especially in step 1 (matrix-times-vector operation). Without this synchronization one has to expect a computation time per iteration which is, depending on the problem, up to 30 per cent higher. The definition of the vector $\mathbf{p}_i$ avoids the computation of $C_{\Lambda,i}\mathbf{r}_{\Lambda,i}$ (occurred originally in step 4, cf. [HK96]) which is not necessarily available ($C_{\Lambda,i}$ is defined such that the inverse operation $C_{\Lambda,i}^{-1}\mathbf{w}_{\Lambda,i}$ can be performed easily).

6.5   COMPONENTS OF THE PRECONDITIONER

The performance of our algorithm depends heavily on the right choice of the components C_Λ, C_C, C_I and B_I defining the preconditioner C (see Theorem 6.1). C_Λ, C_I and B_I are block-diagonal matrices with the blocks $C_{\Lambda,i}$, $C_{I,i}$ and $B_{I,i}$, respectively. In our experiments, the following components have turned out to be the most efficient ones:

$C_{I,i}$: (**Vmn**) Multigrid V-cycle with m pre- and n post-smoothing steps in the Multiplicative Schwarz Method [HLM91, HL92].

$C_{\Lambda,i}$: (**Circ**) Scaled single layer potential BE matrix for a uniformly discretized circle. This matrix is circulant and easily invertible [Rja90]. (**mgV**) $\delta_i \cdot K_{\Lambda,i}(I_{\Lambda,i} - M_{\Lambda,i})^{-1}$. $M_{\Lambda,i}$ is the multigrid iteration operator satisfying the conditions formulated in [JLM$^+$89].

$B_{I,i}$: (**HExt**) Implicitly defined by hierarchical extension (formally $E_{IC,i} = -B_{I,i}^{-1}K_{IC,i}$) [HLMN94].

C_C: (**S-BPX**) Bramble/Pasciak/Xu type preconditioner [TCK91]. (**BPS-D**) Bramble/Pasciak/Schatz type preconditioner [BPS89, Dry82]. (**mgD**) $\tilde{K}_C(I_C - M_C)^{-1}$, $(\tilde{K}_C\mathbf{u}_C, \mathbf{v}_C) := \sum_{i \in \mathcal{I}} \nu_i \langle \mathcal{D}_i u_i, v_i \rangle$, as described in [CKL96].

These preconditioners C_I, C_C, C_Λ, and the basis transformation B_I satisfy the conditions stated in Theorem 6.1. In particular, inequalities (6.26) for C_I are fulfilled with constants $\underline{\gamma}_I$, $\overline{\gamma}_I$ independent of the discretization parameter h [HLM91, HL92, HLMN94]. The preconditioner C_Λ is scaled such that $\underline{\gamma}_\Lambda > 1$, and $\overline{\gamma}_\Lambda$ in (6.17) remains independent of h for both, (Circ) and (mgV).

With respect to B_I, the constant μ in (6.28) can be estimated by

$$\mu \leq \eta^{2k}(1 + c_1 l)^2 \leq \eta^{2k}(1 + c_2(\ln h^{-1}))^2,$$

cf. [HLMN94], with k being the number of local multigrid iterations, l being the number of grids, the h-independent multigrid rate $\eta < 1$, and the h-independent constants c_i. Thus, μ is independent of h if $k = \mathcal{O}(\ln \ln h^{-1})$.

In the (S-BPX) case, the inequalities in (6.25) hold with an h-independent constant $\underline{\gamma}_C$, and $\overline{\gamma}_C \leq c_3(1 + \mu)$ [TCK91, HLMN94]. Therefore, we can prove for (S-BPX) that $\overline{\gamma}/\underline{\gamma} \leq c_4(1 + \mu)(\sqrt{\mu} + \sqrt{1 + \mu})^2 = \mathcal{O}(1)$ if $k = \mathcal{O}(\ln \ln h^{-1})$. However, in the range of practical applications, this means $k = 1$! For (BPS-D), the estimate $\overline{\gamma}_C/\underline{\gamma}_C \leq c_5(1 + (\ln h^{-1})^2)(1 + \mu)$ has been proved [BPS89]. In the case (mgD), C_C arises from the hypersingular operator and C_C^{-1} is realized via a standard multi-grid procedure applied to the global operator (assembled over the subdomains) $\tilde{K}_C$ which is the discretization of a pseudo-differential operator of order one [vPS90]. $\tilde{K}_C$ becomes positive definite after implementing the Dirichlet boundary conditions. We can get an estimate of the same type as for (S-BPX). Note that the FE/BE–Schur-complement energy is equivalent to the $\|.\|^2_{H^{1/2}(\Gamma_C)}$–norm (see, e.g., [Dry82, BPS89, CKL96]).

Consequently, we can estimate the numerical effort Q required to obtain a relative accuracy ε by $Q = \mathcal{O}(h^{-2} \ln h^{-1} \ln \ln h^{-1} \ln \varepsilon^{-1})$ for the (BPS-D) case, and by $Q = \mathcal{O}(h^{-2} \ln \ln h^{-1} \ln \varepsilon^{-1})$ in the (S-BPX) case, i.e. almost optimal. If a BPX-type extension [Nep95] is applied instead of (HExt) in a nested iteration approach [Hac85], we can prove that $Q = \mathcal{O}(h^{-2})$, i.e., we obtain an optimal method.

Preconditioners $C_{\Lambda,i}$ for $K_{\Lambda,i}$ and C_C for the FE/BE Schur complement derived on the basis of boundary element techniques can also be found in [Rja90, KW92, SW95]. The construction of efficient FE Schur complement preconditioners was one of the main topics in the research on FE-DD-methods (see Proceedings of the DD-conferences held annually since 1987).

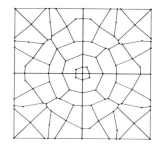 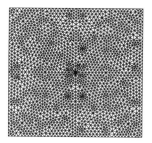

Figure 3 The decomposition (D3) into 64 subdomains (left) and the resulting FEM mesh of the first grid (right).

6.6 NUMERICAL EXPERIMENTS ON PARALLEL COMPUTERS

6.6.1 ACADEMIC PROBLEM

As a first test problem we consider the Dirichlet boundary value problem

$$\Delta u(x) = 0 \quad \text{in } \Omega \quad \text{and} \quad u(x) = r^{1/2} \sin \phi/2 \quad \text{on } \partial\Omega \tag{6.31}$$

with Ω being the square $(-1,1) \times (-1,1)$ with a slit $([0,1],0)$ and (r,ϕ) being the polar coordinates with respect to the origin.

	BEM		FEM	
	min	max	min	max
(D2)	25	26	68 (63)	68 (78)
(D3)	20	25	65 (41)	66 (89)

Table 1   The direct results of the preprocessor (number of BE/FE–elements, respectively) and the number of FE–elements (in parentheses) after applying the mesh–generator.

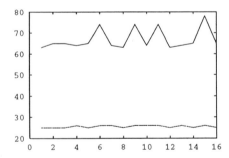 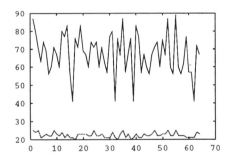

Figure 4   The local number of finite elements (upper curves) and boundary elements (lower curves) for 16 subdomains (D2) (left) and 64 subdomains (D3) (right) after the discretization.

The results of the preprocessing for (D2) and (D3) are given in Table 1; Fig. 4 shows the resulting local number of elements after remeshing the subdomains as realized in the parallel code.

The components of the preconditioner have been chosen as follows:

$C_{\Lambda,i}$: mgV (multigrid) C_C : mgD (multigrid)
$C_{I,i}$: V11 $B_{I,i}$: HExt.

The BE–matrices were computed fully analytically using piecewise linear functions for the displacements and piecewise constant functions for the tractions [Ste96].

(D2) 16 BEM		(D3) 64 BEM			(D2) 16 FEM		(D3) 64 FEM	
$I(\epsilon)$	CPU	$I(\epsilon)$	CPU	l	$I(\epsilon)$	CPU	$I(\epsilon)$	CPU
18	2.9	18	4.7	1	12	2.0	13	3.7
19	4.2	19	6.5	2	15	2.8	16	5.0
19	5.0	20	7.9	3	16	4.2	17	6.5
20	9.9	20	16.3	4	17	6.6	18	10.2
20	25.1	20	26.9	5	17	15.2	18	21.0
8897		32699		N(5)	136081		550795	
$1.00 \rightarrow 0.86$				eff.	$1.00 \rightarrow 0.79$			

Table 2 The slit problem. Iteration count ($I(\epsilon)$, $\epsilon = 10^{-6}$), CPU time (system generation and solution) in seconds for BEM and FEM discretizations and scaled efficiency according to the CPU time and the number of unknowns (N(5)) for the 5th level. The experiments were carried out on a GC–PowerPlus using one processor per subdomain.

We present the results for two different decompositions: an automatic decomposition into 16 subdomains (D2) (see Fig. 2d) and an automatic decomposition into 64 subdomains (D3) (see Fig. 3), both resulting from a manual decomposition into 16 subdomains (D1) (see Fig. 2a) with mesh-grading towards the origin with $h_{max}/h_{min} > 100$. The underlying discretization for (D3) is roughly double as fine as for (D2), such that the local problem size should be constant, i.e., independent of p.

Looking at the results for (D2) and (D3) in Table 2 we observe constant iteration numbers with respect to both, l and p. Even the BEM and FEM results are similar. The values obtained for the scaled efficiency are quite satisfying taking into account that the process of decomposition has been performed fully automatically for a non-standard graded mesh.

6.6.2 ELASTICITY PROBLEM

We now want to extend the ideas discussed above to problems of plane linear elasticity in which the displacement $u(x) = (u_1(x), u_2(x))^T$ satisfies formally the system of Lamé equations

$$-\mu(x)\Delta u(x) - (\lambda(x) + \mu(x))\mathrm{grad}\,\mathrm{div}\,u(x) = 0 \quad \text{in } \Omega,$$
$$u(x) = 0 \text{ on } \Gamma_D, \quad \sum_{l=1}^{2} \sigma_{kl}(u(x))n_l = g_k(x) \text{ on } \Gamma_N, \ (k = 1, 2),$$
(6.32)

where Ω is a bounded Lipschitz domain, $\sigma_{kl}(u)$ are the components of the stress tensor $\sigma(u)$ and $n(x) = (n_1(x), n_2(x))^T$ is the outward normal vector to $\Gamma_D \cup \Gamma_N = \Gamma := \partial\Omega$ ($\Gamma_D \neq \emptyset$) and λ and μ, $\lambda, \mu > 0$, are the Lamé coefficients of the elastic material. In (6.32), $g = (g_1, g_2)^T$ is the vector of boundary tractions. The extension of the

(D4) 16 BEM		(D5) 64 BEM			(D4) 16 FEM		(D5) 64 FEM	
$I(\epsilon)$	CPU	$I(\epsilon)$	CPU	l	$I(\epsilon)$	CPU	$I(\epsilon)$	CPU
47	3.9	53	24.1	1	32	3.7	30	19.1
51	5.1	56	27.6	2	42	4.8	37	21.5
53	8.8	57	33.8	3	49	6.6	41	24.6
54	24.3	58	45.1	4	53	14.2	44	32.1
54	83.7	60	92.7	5	55	40.8	47	58.2
11647		49146		N(5)	160287		641594	
$1.00 \rightarrow 0.95$				eff.	$1.00 \rightarrow 0.70$			

Table 3 The dam problem. Iteration count ($I(\epsilon)$, $\epsilon = 10^{-6}$), CPU time (system generation and solution) in seconds for BEM and FEM discretizations and scaled efficiency according to the CPU time and the number of unknowns (N(5)) for the 5th level. The experiments were carried out on a GC–PowerPlus using one processor per subdomain.

theory presented above for potential problems to linear elasticity is straightforward. The equations and definitions can be found, e.g., in [HHKL96].

As a test problem we consider a dam filled with water as sketched in Figure 5. As indicated there, Dirichlet boundary conditions (b.c.) are given on Γ_D (zero displacement) and Neumann b.c. on Γ_N (the tractions are equal to zero, or they are chosen according to the water pressure). The Lamé constants are given for rock by $\mu_r = 7.26e4MPa$, $\lambda_r = 3.74e4MPa$ and for concrete by $\mu_c = 9.2e6MPa$, $\lambda_c = 9.2e6MPa$.

We consider two automatic decompositions: one into 16 subdomains (D4) (as drawn in the middle in Fig. 5) and one into 64 subdomains (D5) (as drawn on the right in Fig. 5), where for the latter the discretization on Γ is approximately double as fine as for (D4). Both decompositions are the result of the automatic preprocessing started with an a–priori graded grid towards the regions where high stresses are expected.

For the results presented in Table 3, the components of the preconditioner have

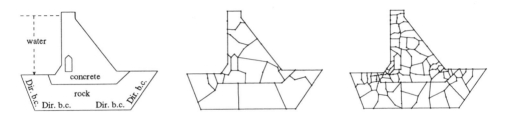

Figure 5 The dam (left) and its decomposition into 16 subdomains (D4) and 64 subdomains (D5).

been chosen as follows:

$C_{\Lambda,i}$: mgV (multigrid) C_C : S–BPX
$C_{I,i}$: V01 (as symmetric Multiplicative Schwarz method) $B_{I,i}$: HExt.

The BE–matrices were computed fully analytically using piecewise linear functions for the displacements and piecewise constant functions for the tractions [Ste96]. Looking at the results for (D4) and (D5) we observe almost constant iteration numbers with respect to l and p. The lower efficiency in the FEM–case is also due to the higher local problem size for (D5) compared to (D4).

6.6.3 TECHNICAL MAGNETIC FIELD PROBLEM

The third example, a direct current motor (dc motor), should demonstrate that the algorithm also works for real-life, nonlinear magnetic field problems (see [HHKL96] for more details). Starting with a user mesh, we apply an automatic domain decomposition procedure (see Figure 6) and a parallel mesh generator, the basic ideas of which are presented in [Glo95], to obtain the initial mesh ($q = 1$), which is to be refined four times to get the final mesh for our computations ($q = l = 5$).

Example	Dirichlet b. c.	Dirichlet b. c.	radiation condition	radiation condition
Choice for C_C	BPS-D	S-BPX	BPS-D	S-BPX
subdomains	32 FEM	32 FEM	31 FEM 1 BEM	31 FEM 1 BEM
No. of unknowns	417 328	417 328	414 568	414 568
Newton it. 1st grid CG iter. 1st grid	7 16,12,13, 13,13,17,24	6 10,12,12 12,12,11	7 19,14,16, 16,12,19,28	7 10,14,14 14,14,13,17
Newton it. 2nd grid CG iter. 2nd grid	2 16,23	2 13,16	2 18,25	2 15,20
Newton it. 3rd grid CG iter. 3rd grid	2 17,27	2 15,17	2 18,31	2 19,24
Newton it. 4th grid CG iter. 4th grid	2 17,35	2 16,18	2 18,34	2 19,29
Newton it. 5th grid CG iter. 5th grid	4 18,38,22,30	4 16,21,18,20	4 18,42,22,33	4 20,32,27,33
Time (generation)	25.2	25.2	33.6	33.6
Time (linear solver)	168.9	137.8	178.8	198.5
Total time	194.1	163.0	212.4	232.1

Table 4   Performance for the dc motor. CPU time in seconds on a GC-Power Plus using 32 processors; relative accuracy $\varepsilon = 10^{-6}$.

We present the numerical results in Table 4 and the level lines of the solution in Figure 6. Again, we have done calculations for both the machine with homogeneous

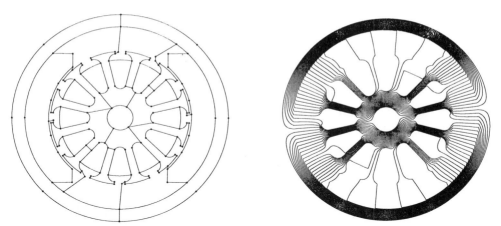

Figure 6 Domain decomposition and level lines for the direct current machine.

Dirichlet conditions on its boundary and the infinite domain with the radiation condition. The best results with respect to the total computing time have been achieved with $\varepsilon_{\text{lin}} = 0.01$. The components of the PCG solver have been chosen as follows:

$$C_{\Lambda,i} \quad : \quad \text{Circ} \qquad\qquad C_C \quad : \quad \text{see Table}$$
$$C_{I,i} \quad : \quad \text{V11} \qquad\qquad B_{I,i} \quad : \quad \text{HExt} \ .$$

A comparison of different parallel solvers for nonlinear magnetic field problems as well as a detailed description of the Newton algorithm being applied can be found in [HJ95].

References

[BP88]   Bramble J. H. and Pasciak J. E. (1988) A preconditioning technique for indefinite systems resulting from mixed approximations of elliptic problems. *Mathematics of Computation* 50(181): 1–17.

[BPS89] Bramble J. H., Pasciak J. E., and Schatz A. H. (1986, 1987, 1988, 1989) The construction of preconditioners for elliptic problems by substructuring I – IV. *Mathematics of Computation* 47, 103–134, 49, 1–16, 51, 415–430, 53, 1–24.

[CKL96] Carstensen C., Kuhn M., and Langer U. (1996) Fast parallel solvers for symmetric boundary element domain decomposition equations. Report 500, Institute of Mathematics, Johannes Kepler University Linz.

[Cos87] Costabel M. (1987) Symmetric methods for the coupling of finite elements and boundary elements. In Brebbia C. A., Wendland W. L., and Kuhn G. (eds) *Boundary Elements IX*, pages 411–420. Springer–Verlag.

[Cos88]   Costabel M. (1988) Boundary integral operators on Lipschitz domains: Elementary results. *SIAM J. Math. Anal.* 19: 613–626.

[CW96]   Carstensen C. and Wriggers P. (1996) On the symmetric boundary element method and the symmetric coupling of boundary elements and finite elements. *IMA Journal of Numerical Analysis* Submitted.

[Dry82]   Dryja M. (1982) A capacitance matrix method for Dirichlet problems on polygonal regions. *Numerische Mathematik* 39(1): 51–64.

[GHHK96]   Goppold M., Haase G., Heise B., and Kuhn M. (1996) Preprocessing in BE/FE domain decomposition methods. Technical Report 96-2, Institute of Mathematics, Johannes Kepler University Linz.

[Glo95]   Globisch G. (1995) PARMESH - a parallel mesh generator. *Parallel Computing* 21(3): 509–524.

[Hac85]   Hackbusch W. (1985) *Multi–Grid methods and applications*, volume 4 of *Springer Series in Computational Mathematics*. Springer–Verlag, Berlin.

[Hei93]   Heise B. (1993) Nonlinear field calculations with multigrid-Newton methods. *IMPACT of Computing in Science and Engineering* 5: 75–110.

[Hei94]   Heise B. (1994) Analysis of a fully discrete finite element method for a nonlinear magnetic field problem. *SIAM J. Numer. Anal.* 31(3): 745–759.

[HHJ^{+}96] Haase G., Heise B., Jung M., Kuhn M., and Steinbach O. (1996) FEM∞BEM - a parallel solver for linear and nonlinear coupled FE/BE-equations. DFG-Schwerpunkt "Randelementmethoden", Report 94-16, University of Stuttgart.

[HHKL96]   Haase G., Heise B., Kuhn M., and Langer U. (1996) Adaptive domain decomposition methods for finite and boundary element equations. In *Reports from the Final Conference of the Priority Research Programme Boundary Element Methods 1989–1995, (W. Wendland ed.), Stuttgart, October 1995*. Springer Verlag, Berlin. Also Technical Report 95-2, Institute of Mathematics, Johannes Kepler University Linz.

[HJ95]   Heise B. and Jung M. (1995) Comparison of parallel solvers for nonlinear elliptic problems based on domain decomposition ideas. Report 494, Institute of Mathematics, Johannes Kepler University Linz.

[HK96]   Heise B. and Kuhn M. (1996) Parallel solvers for linear and nonlinear exterior magnetic field problems based upon coupled FE/BE formulations. *Computing* 56(3): 237–258.

[HL92]   Haase G. and Langer U. (1992) The non-overlapping domain decomposition multiplicative Schwarz method. *International Journal of Computer Mathematics* 44: 223–242.

[HLM90]   Haase G., Langer U., and Meyer A. (1990) A new approach to the Dirichlet domain decomposition method. In Hengst S. (ed) *Fifth Multigrid Seminar, Eberswalde 1990*, pages 1–59. Karl–Weierstrass–Institut, Berlin. Report R–MATH–09/90.

[HLM91]   Haase G., Langer U., and Meyer A. (1991) The approximate dirichlet domain decomposition method. Part I: An algebraic approach. Part II: Applications to 2nd-order elliptic boundary value problems. *Computing* 47: 137–151 (Part I), 153–167 (Part II).

[HLM92]   Haase G., Langer U., and Meyer A. (1992) Domain decomposition preconditioners with inexact subdomain solvers. *J. of Num. Lin. Alg. with Appl.* 1: 27–42.

[HLMN94]   Haase G., Langer U., Meyer A., and Nepomnyaschikh S. V. (1994) Hierarchical extension operators and local multigrid methods in domain decomposition preconditioners. *East-West J. Numer. Math.* 2(3): 173–193.

[Hol92]   Holzer S. (1992) On the engineering analysis of 2D problems by the symmetric Galerkin boundary element method and coupled BEM/FEM. In Kane J. H., Maier G., Tosaka N., and Atluri S. N. (eds) *Advances in boundary element techniques.* Springer, Berlin.

[HW91]   Hsiao G. C. and Wendland W. L. (1991) Domain decomposition in boundary element methods. In *Proc. of IV Int. Symposium on Domain Decomposition Methods, (R. Glowinski, Y. A. Kuznetsov, G. Meurant, J. Périaux, O. B. Widlund eds.), Moscow, May 1990*, pages 41–49. SIAM Publ., Philadelphia.

[JLM+89]   Jung M., Langer U., Meyer A., Queck W., and Schneider M. (1989) Multigrid preconditioners and their applications. In Telschow G. (ed) *Third Multigrid Seminar, Biesenthal 1988*, pages 11–52. Karl–Weierstrass–Institut, Berlin. Report R–MATH–03/89.

[KMZ90]   Khoromskij B. N., Mazurkevich G. E., and Zhidkov E. P. (1990) Domain decomposition method for magnetostatics nonlinear problems in combined formulation. *Sov. J. Numer. Anal. Math. Modelling* 5(2): 111–136.

[KW92]   Khoromskij B. N. and Wendland W. L. (1992) Spectrally equivalent preconditioners for boundary equations in substructuring techniques. *East-West Journal of Numerical Mathematics* 1(1): 1–26.

[Lan94]   Langer U. (1994) Parallel iterative solution of symmetric coupled FE/BE-equations via domain decomposition. *Contemporary Mathematics* 157: 335–344.

[Nep95]   Nepomnyaschikh S. V. (1995) Optimal multilevel extension operators. Preprint SPC 95–3, Faculty of Mathematics, Technical University Chemnitz–Zwickau.

[Rja90]   Rjasanow S. (1990) Vorkonditionierte iterative Auflösung von Randele-
           mentgleichungen für die Dirichlet–Aufgabe. Wissenschaftliche Schriften-
           reihe 7/1990, Technical University Chemnitz.

[Sim91]   Simon H. D. (1991) Partitioning of unstructured problems for parallel
           processing. *Comput. System in Eng.* 2: 135–148.

[Ste96]   Steinbach O. (1996) Galerkin– und Kollokationsdiskretisierungen für
           Randintegralgleichungen in 2D. Report 96–5, University of Stuttgart.

[SW95]   Steinbach O. and Wendland W. L. (1995) Efficient preconditioners for
           boundary element methods and their use in domain decomposition.
           DFG-Schwerpunkt "Randelementmethoden", Report 95-19, University of
           Stuttgart.

[TCK91]   Tong C. H., Chan T. F., and Kuo C. J. (1991) A domain decomposition
           preconditioner based on a change to a multilevel nodal basis. *SIAM J.
           Sci. Stat. Comput.* 12(6): 1486–1495.

[vPS90]   von Petersdorf T. and Stephan E. P. (1990) On the convergence of the
           multigrid method for a hypersingular integral equation of the first kind.
           Numerische Mathematik 57: 379–391.

7

Additive Schwarz Methods for Integral Equations of the First Kind

Ernst P. Stephan

Institut für Angewandte Mathematik,
Universität Hannover,
Welfengarten 1,
30167 Hannover, Germany

7.1 INTRODUCTION

This paper gives a survey on additive Schwarz methods applied to the h and p versions of the boundary element method. We consider weakly singular and hypersingular integral equations, namely the single layer potential operator and the normal derivative of the double layer potential operator. That means we have to consider weak formulations of the form

$$a(u,v) := \langle Au, v \rangle = \langle g, v \rangle \quad \text{for all } v \in \tilde{H}^{\alpha/2}(\Gamma), \tag{7.1}$$

where $A : \tilde{H}^{\alpha/2}(\Gamma) \to H^{-\alpha/2}(\Gamma)$ and $a(\cdot, \cdot)$ is symmetric and positive definite. Here, $\langle \cdot, \cdot \rangle$ denotes the $L^2(\Gamma)$ inner product. An approximation u_M of u is obtained by solving the Galerkin scheme: find $u_M \in S_M \subset \tilde{H}^{\alpha/2}(\Gamma)$ such that

$$\langle Au_M, v \rangle = \langle g, v \rangle \quad \text{for all } v \in S_M. \tag{7.2}$$

where S_M is a finite-dimensional subspace. The condition number of this linear system grows at least like $h^{-1}p^2$ if the h-p version is used for quasi-uniform meshes, and it grows exponentially for geometric meshes. Here h is the mesh size and p is the degree of the splines used in the Galerkin scheme. Therefore in order to use the conjugate gradient algorithm to solve the system efficiently we need a good preconditioner.

The hypersingular and weakly singular integral equations we consider are

$$Wu(x) := -\frac{1}{\pi} f.p. \int_\Gamma \frac{u(y)}{|x-y|^2} \, ds_y = g_1(x), \qquad x \in \Gamma, \tag{7.3}$$

The Mathematics of Finite Elements and Applications
Edited by J. R. Whiteman © 1997 John Wiley & Sons Ltd.

and

$$Vu(x) := -\frac{1}{\pi} \int_\Gamma u(y) \log|x - y| \, ds_y = g_2(x), \qquad x \in \Gamma, \tag{7.4}$$

where f.p. denotes a finite part integral in the sense of Hadamard and $g_1 \in H^{-1/2}(\Gamma)$, $g_2 \in H^{1/2}(\Gamma)$. Note that the order α of the pseudodifferential operator W is 1 whereas V has order -1.

Different choices of the trial space S_M in (7.2) lead to different types of the Galerkin scheme, namely to the h, p, and h-p versions defined as follows. For simplicity we consider only $\Gamma = (-1, 1)$ and a regular mesh of size h on Γ defined as

$$x_j = -1 + jh, \qquad h = \frac{2}{N}, \qquad j = 0, \dots, N. \tag{7.5}$$

In *the h version* we take on this mesh the space S^h of continuous piecewise-linear functions for the hypersingular operator W while using piecewise-constant functions for the weakly singular operator V. For W in order that S^h may be a subspace of $\tilde{H}^{1/2}(\Gamma)$ it is necessary that the functions vanish at the endpoints ± 1. For the h version we solve (7.2) in S^h and increase the accuracy by reducing h.

In *the p version* the space S^p on this mesh is defined as the space of functions on Γ whose restrictions to $\Gamma_j := (x_{j-1}, x_j)$, $j = 1, \dots, N$, are polynomials of degree at most p. For W it is required that these functions are continuous and vanish at the endpoints of Γ. For the p version we solve (7.2) with functions in S^p and increase the accuracy by increasing p.

The Galerkin discretizations of boundary integral equations lead to very large systems of linear equations $Au = f$ with full $M \times M$ matrices. The direct solution of these linear systems can be very expensive in terms of storage and computational work, i.e. a Gauss solver requires $\mathcal{O}(M^3)$ operations. So we need a good iterative method to approximate the Galerkin solution, keeping the error in the energy norm at the order of the Galerkin error. For a positive definite matrix A the conjugate gradient method needs $\mathcal{O}(M^{1/2} \log M)$ iterations and thus $\mathcal{O}(M^{5/2} \log M)$ operations to achieve this order of accuracy. A multigrid method for the hypersingular integral equation has been analyzed in [vPS90] and has been used in [vPS92] as a preconditioner for the conjugate gradient method to reduce the number of operations to $\mathcal{O}(M^2)$.

Additive Schwarz methods were originally designed for finite element discretisations of differential equations (see e.g. [BCMP91, BLP94, BP93, BPX90, DW91, Lio88, Pav94, Wid89, Zha92]). Our analysis for the h version Galerkin boundary element method for the hypersingular integral equation extends the approach by Bramble et.al. [BLP94, BP93, BPX90] and Xu [Xu92] whose applications were on finite element methods. Since the boundary integral operators are non-local the finite element theory does not carry over directly to the boundary element method. The latter needs a new analysis which for the hypersingular integral equation is based on the fractional Sobolev space $H_\infty^{1/2}(\Gamma)$.

The applicability of Schwarz iterations with overlapping domains to first kind integral equations was recently investigated in [HS96] for the h version of the boundary element method. Multilevel methods for the h version were considered in [TS96] for the standard nodal basis, in [TSM97] for the hierarchical basis and in [STZ97] for a wavelet basis. The p version for weakly singular and hypersingular integral operators was analyzed in [ST95]. The results in [ST95, TS96, TSM97, STZ97] can be summarised as follows. For the h version the 2-level and multilevel additive Schwarz methods yield preconditioned systems which have bounded condition numbers when the standard nodal basis or a wavelet basis are used, whereas a hierarchical basis preconditioner gives logarithmically growing condition numbers. For the p version considered in [ST95] the condition number is proved to behave like $1 + \log^3 p$. In [HTS97] we design a multilevel method for the p version, which is simply preconditioning by the inverse of the diagonal matrix and which results in a preconditioned system having condition numbers behaving like $p(1 + \log^3 p)$. In [HTS97] we also prove the efficiency of the 2-level and multilevel methods for the h-p version, with both quasi-uniform and geometric meshes. For the p-version we note that even though the condition number of the multilevel method increases faster than that of the 2-level method, which may lead to a bigger number of iterations to solve the linear system, the implementation of the multilevel method is recommended since for each iteration it is actually the diagonal preconditioner, and is therefore cheaper for each iteration.

The additive multilevel methods usually need slightly more iteration steps than the multigrid methods but the higher flexibility of these algorithms simplifies the use of parallel computing (the simple subspace corrections are not applied in a sequential order but in parallel). Another advantage is that it allows a simpler, more natural data structure, and is therefore much better for non-uniformly refined grids. Consequently it is highly efficient to combine additive multilevel methods with adaptive methods (cf. Section 7.5).

The paper is organized as follows. In Section 7.2 we report on the additive Schwarz methods applied to the h and p-versions for positive definite symmetric problems resulting from the first kind integral equations (7.3) and (7.4). In Section 7.3 we consider indefinite problems resulting from integral equation formulations for the Helmholtz equation. In Section 7.4 we report on 3D problems, in Section 7.5 on adaptive methods and in Section 7.6 we present numerical experiments.

7.2   ADDITIVE SCHWARZ METHODS FOR POSITIVE DEFINITE PROBLEMS

Let
$$S_M = S_0 + S_1 + \cdots + S_N \tag{7.6}$$
denote a decomposition of S_M into subspaces S_j. Then the additive Schwarz method (ASM) requires the solving, by an iterative method, of the equation
$$Pu_M := (P_0 + P_1 + \cdots + P_N)u_M = g_M, \tag{7.7}$$
where the projections $P_j : S_M \to S_j$, $j = 0, \ldots, N$, are defined for any $v_M \in S_M$ by
$$a(P_j v_M, \phi_j) = a(v_M, \phi_j) \quad \text{for any } \phi_j \in S_j. \tag{7.8}$$
Note the right hand side of (7.7), $g_M = \sum_{j=0}^N g_{M,j}$, can be computed without knowing the solution u_M of (7.2) by
$$a(g_{M,j}, \phi_j) = \langle g, \phi_j \rangle \quad \text{for any } \phi_j \in S_j, \ j = 0, \ldots, N. \tag{7.9}$$

Note that equations (7.2) and (7.7) have the same solutions and an explicit form for P is not necessary in solving (7.7).

In fact, if we use the Richardson method to solve (7.7), then given an iterate u_M^n we compute u_M^{n+1} by

$$u_M^{n+1} = u_M^n - \tau(Pu_M^n - g_M),$$

where the residual $r^n := (Pu_M^n - g_M)$ is computed as $r^n = \sum_{j=0}^N r_j^n$ with $r_j^n :=$ $P_j u_M^n - g_{M,j}$ being solutions of

$$a(r_j^n, w_j) = a(u_M^n, w_j) - \langle g, w_j \rangle \quad \text{for any } w_j \in S_j. \tag{7.10}$$

Here τ is a damping parameter whose optimal value is given by

$$\tau_{opt} = \frac{2}{\lambda_{\max}(P) + \lambda_{\min}(P)}.$$

In the following we comment on the performance of the additive Schwarz operator as preconditioner for the CG or GMRES method, for the use of the additive Schwarz method as a linear solver, see [HS96].

Let A_M be the discrete form of the pseudodifferential operator A, i.e. $A_M : S_M \to S_M$, and be defined for any $v_M \in S_M$ as

$$\langle A_M v_M, w \rangle = a(v_M, w) \quad \forall w \in S_M. \tag{7.11}$$

Then the additive Schwarz operator P can be written as $P = BA_M$, where the preconditioner B is given by

$$B = \sum_{j=0}^M A_j^{-1} Q_j.$$

Here Q_j is the L^2-projection from S_M onto S_j, and A_j is the restriction of A_M onto S_j, i.e.

$$\langle Q_j v_M, w_j \rangle = \langle v_M, w_j \rangle \quad \forall v_M \in S_M \text{ and } w_j \in S_j,$$

and

$$\langle A_j v_j, w_j \rangle = a(v_j, w_j) \quad \text{for any } v_j, w_j \in S_j.$$

This representation of $P = BA_M$ can be seen from the fact that $P_j = A_j^{-1} Q_j A_M$. In the implementation of the preconditioned conjugate gradient method, an explicit form for B is not important. Indeed, we only need to know the action of B on any $v_M \in S_M$.

It is known that the rates of convergence of the above-mentioned methods to solve (7.7) depend on the condition number $\kappa(P)$ of P. Moreover, if there exist positive constants c_0 and c_1 such that

$$c_0 a(v_M, v_M) \le a(Pv_M, v_M) \le c_1 a(v_M, v_M) \quad \forall v_M \in S_M,$$

then $\kappa(P) \le c_1/c_0$.

The following lemma is standard in proving bounds for the maximum and minimum eigenvalues of the additive Schwarz operator P.

Lemma 7.1 *Let* $v_M = \sum_{j=0}^{N} v_{M,j}$, *where* $v_{M,j} \in S_j$, *be a representation of* $v_M \in$ $S_M = S_0 + \cdots + S_N$.

(i) *If a representation can be chosen such that, for some* $C_1 > 0$,

$$\sum_{j=0}^{N} a(v_{M,j}, v_{M,j}) \leq C_1^{-1} a(v_M, v_M), \tag{7.12}$$

then $\lambda_{\min}(P) \geq C_1$.

(ii) *If there exists* $C_2 > 0$ *such that for any representation of* v_M

$$a(v_M, v_M) \leq C_2 \sum_{j=0}^{N} a(v_{M,j}, v_{M,j}) \tag{7.13}$$

then $\lambda_{\max}(P) \leq C_2$.

7.2.1 HYPERSINGULAR INTEGRAL EQUATIONS.

In this subsection we consider different additive Schwarz methods for the hypersingular integral operator. We recall the results for the 2-level and multilevel methods for the h-version from [TS96], for the p version from [ST95] and for the h-p version from [HTS97]. Since the additive Schwarz method is generally defined by (7.6)–(7.9) it suffices to give for each version the decomposition of the ansatz space corresponding to (7.6).

h version

We decompose S_h, the space of continuous piecewise-linear functions, as

$$S_h = S_H + S_{h,1} + \cdots + S_{h,N_h-1}, \tag{7.14}$$

where S_H is the space of continuous piecewise-linear functions on the coarse mesh with size $H = 2h$ and $S_{h,j} = \text{span}\{\phi_{h,j}\}$. Here $\phi_{h,j}$ is the hat function which takes the value 1 at the mesh point x_j and 0 at the other mesh points. In order to show that the additive Schwarz operator for (7.14) has bounded condition number one proceeds as follows (see [TS96]).

In view of Lemma 7.1 one verifies (7.12) in order to get a good bound for $\lambda_{\min}(P)$. It is necessary to define an appropriate decomposition for any $v_h \in S_h$. This is done in [TS96] via a partition of unity consisting of suitable piecewise-linear functions $\{\Theta_j\}_{j=1,\ldots,N_h-1}$ with supp $\Theta_j = \bar{\Gamma}'_j := [x_{j-1}, x_j]$ and $\left|\frac{d\Theta_j}{dx}\right| \leq \frac{c}{h}$ with a constant $c > 0$. Then

$$v_h = v_H + v_{h,1} + \cdots + v_{h,N_h-1} \tag{7.15}$$

where $v_H := \tilde{P}_{S_H} v_h$ and $v_{h,j} := \Pi_h(\Theta_j w_h)$ with $w_h := v_h - w_H$. Here $\tilde{P}_{S_H}$ is the Galerkin projection from $H_\infty^{1/2}(\Gamma)$ onto S_H, and Π_h is the interpolation operator from $C(\Gamma)$ onto S_h. Then

$$\|v_{h,j}\|^2_{H_\infty^{1/2}(\Gamma'_j)} \leq C \|\Theta_j w_h\|^2_{H_\infty^{1/2}(\Gamma'_j)} \leq C \left(h^{-1} \|w_h\|^2_{L^2(\Gamma'_j)} + h \|w_h\|^2_{H^1(\Gamma'_j)} \right). \tag{7.16}$$

On the other hand (see e.g. [SW90])

$$\|w_h\|_{L^2(\Gamma)} \le c\, H^{1/2}\|v_h\|_{H^{1/2}_\infty(\Gamma)}, \qquad \|w_h\|_{H^{1/2}_\infty(\Gamma)} \le c\,\|v_h\|_{H^{1/2}_\infty(\Gamma)}. \tag{7.17}$$

Hence summing over j yields with the inverse property

$$\sum_{j=1}^{N_h-1} \|v_{h,j}\|_{H^{1/2}_\infty(\Gamma'_j)} \le c\left(h^{-1}\|w_h\|^2_{L^2(\Gamma)} + h\|w_h\|^2_{\overset{\circ}{H}{}^1(\Gamma)}\right) \le c\|v_h\|^2_{H^{1/2}_\infty(\Gamma)} \tag{7.18}$$

implying (7.12) since $\|v_H\|_{H^{1/2}_\infty(\Gamma)} \le c\|v_h\|_{H^{1/2}_\infty(\Gamma)}$.

In order to show the boundedness of $\lambda_{\max}(P)$ one checks on (7.13). In order to do so one splits $P = P_0 + T$ with $T = \sum_{j=1}^{N_h-1} P_j$ and since P_0 is bounded in $H^{1/2}_\infty(\Gamma)$ one has left to consider

$$a(Tv_h, v_h) = \sum_{i=1}^{N_h-1} \frac{a(v_h, \phi_{hi})^2}{a(\phi_{hi}, \phi_{hi})} \le c\, a(v_h, v_h). \tag{7.19}$$

Let W_h be the discrete form of the hypersingular operator W, i.e. $W_h : S_h \to S_h$ defined $\forall v_h \in S_h$ as

$$\langle W_h v_h, w_h\rangle = a(v_h, w_h) \quad \forall w_h \in S_h. \tag{7.20}$$

Then with supp $\phi_{h,i} = \bar{\Gamma}'_i = [x_{i-1}, x_{i+1}]$

$$a(v_h, \phi_{hi}) \le \|W_h v_h\|_{L^2(\Gamma'_i)}\|\phi_{hi}\|_{L^2(\Gamma'_i)} \le h^{1/2}\|W_h v_h\|_{L^2(\Gamma'_i)}\|\phi_{hi}\|_{H^{1/2}_\infty(\Gamma'_i)} \tag{7.21}$$

Hence

$$a(Tv_h, v_h) \le c\, h\|W_h v_h\|^2_{L^2(\Gamma)}. \tag{7.22}$$

But $\|W_h v_h\|^2_{L^2(\Gamma)} \le c\, h^{-1}\|v_h\|^2_{H^{1/2}_\infty(\Gamma)}$ yielding (7.19).

If one continues the 2-level method for the global problem on S_H

$$S_h = S_{h_1} + \sum_{l=2}^{L}\left(S_{h_l,1} + \cdots + S_{h_l,N_{h_l}-1}\right) \tag{7.23}$$

where $h_{l-1} = 2h_l$ then one has the multilevel method.

For brevity we only comment on the boundedness of $\lambda_{\max}(P)$, for the lower bound for $\lambda_{\min}(P)$ see [TS96]. The multilevel additive Schwarz operator is now defined as

$$P = \sum_{l=1}^{L}\sum_{i=1}^{N_{h_l}-1} P_{S_{h_l,i}} \tag{7.24}$$

where $P_{S_{h_l,i}} : S_h \to S_{h_l,i}$ is defined for any $v \in S_h$ by

$$a(P_{S_{h_l,i}}v, w) = a(v, w) \quad \forall w \in S_{h_l,i}. \tag{7.25}$$

For $T_l := \sum_{i=1}^{N_{h_l}-1} P_{S_{h_l,i}}$ the strengthend Cauchy-Schwarz inequality holds (see [TS96]): There exists a constant $c > 0$ and $\gamma \in (0,1)$ such that for any $v \in S^k = \sum_{i=1}^{N_{h_k}-1} S_{h_k,i}$ where $1 \le k \le l \le L$ there holds

$$a(T_l v, v) \le c\,\gamma^{2(l-k)} a(v, v). \tag{7.26}$$

Then application of a general argument by Bramble and Pasciak [BP93, Theorem 3.1] yields for $P = \sum_{l=1}^{L} T_l$ (cf. Lemma 2.8 in [TS96]) with a constant c_1

$$a(Pv, v) \leq c_1 a(v, v) \text{ for any } v \in S_h. \tag{7.27}$$

Altogether we have the boundedness of the condition number of the 2-level additive and the multilevel additive Schwarz operators.

Theorem 7.1 [TS96, STZ97] *The additive Schwarz operator corresponding to (7.14) or (7.23) has bounded condition number independent of h, the mesh size of the finest level, and the number of the levels L.*

p version

In the 2 level method the ansatz space S^p is decomposed as follows. We denote by S^1 the space of continuous piecewise-linear functions which vanish at the endpoints ± 1. This space serves the same purpose as the coarse grid space in the h version. To each subinterval $\Gamma_j = (x_{j-1}, x_j)$, $j = 1, \ldots, N_0$, we associate the space S_j^p which is the affine image on Γ_j of the space span $\{\mathcal{L}^2, \ldots, \mathcal{L}^p\}$, where $\mathcal{L}^k(x) = \int_{-1}^{x} L^{k-1}(y)\, dy$ with the Legendre polynomial L^{k-1} of degree $k-1$. Note that $\mathcal{L}^k$ vanishes at ± 1 and therefore functions in S_j^p vanish at the endpoints of Γ_j. We can decompose S^p as a direct sum

$$S^p = S^1 \oplus S_1^p \oplus \cdots \oplus S_{N_0}^p. \tag{7.28}$$

Theorem 7.2 *[ST95] For the additive Schwarz operator P according to (7.28) there holds*

$$\lambda_{\min}(P) \geq C_1 \left(1 + \log^3 p\right)^{-1} \quad and \quad \lambda_{\max}(P) \leq C_2,$$

with positive constants C_1 and C_2 independent of p and N_0 and hence $\kappa(P) \leq C_2 C_1^{-1}(1 + \log^3 p)$.

Proof. In order to prove the estimate for $\lambda_{\min}(P)$ it suffices in view of Lemma 7.1 to prove that for any $u_p = u_1 + \sum_{j=1}^{N_0} u_{p,j}$ where $u_1 \in S^1$, and $u_{p,j} \in S_j^p$ there holds

$$\|u_1\|_{\tilde{H}_{\infty}^{1/2}(\Gamma)}^2 + \sum_{j=1}^{N_0} \|u_{p,j}\|_{\tilde{H}_{\infty}^{1/2}(\Gamma_j)}^2 \leq C_1^{-1} \left(1 + \log^3 p\right) \|u_p\|_{\tilde{H}_{\infty}^{1/2}(\Gamma)}^2. \tag{7.29}$$

First we note that since $u_{p,j}$ vanishes at the mesh points, its linear interpolant is identically zero, and therefore u_1 is the linear interpolant of u_p. Hence there exists a positive constant c independent of p and N_0 such that

$$\|u_1\|_{\tilde{H}_{\infty}^{1/2}(\Gamma)}^2 \leq c\,(1 + \log p)\|u_p\|_{\tilde{H}_{\infty}^{1/2}(\Gamma)}^2. \tag{7.30}$$

Next let $w_p := u_p - u_1 = \sum_{j=1}^{N_0} u_{p,j}$. Then for any $j = 1, \ldots, N_0$ since $w_p|_{\Gamma_j} = u_{p,j}$ we have

$$\sum_{j=1}^{N_0} \|u_{p,j}\|_{H^{1/2}(\Gamma_j)}^2 = \sum_{j=1}^{N_0} \|w_p\|_{H^{1/2}(\Gamma_j)}^2 \leq \|w_p\|_{H^{1/2}(\Gamma)}^2 \leq \|w_p\|_{\tilde{H}_{\infty}^{1/2}(\Gamma)}^2, \tag{7.31}$$

where the first inequality characterizing the property of the $H^{1/2}$-norm was proved in [vP89]. On the other hand, [BCMP91, Theorem 6.5] gives

$$\|\psi\|_{\tilde{H}_\infty^{1/2}(\Gamma_j)} \le c\,(1+\log p)\,\|\psi\|_{H_\infty^{1/2}(\Gamma_j)} \quad \text{for any } \psi \in S_j^p,\, j=1,\dots,N_0, \qquad (7.32)$$

where c is a constant independent of p and ψ. Hence (7.30)–(7.32) yield (7.29). $\square$

We get a multilevel method when we further decompose S_j^p by $S_j^p = \oplus_{k=2}^p \tilde{S}_j^k$, where $\tilde{S}_j^k = \operatorname{span}\{\mathcal{L}_j^k\}$ and $\mathcal{L}_j^k$ is the affine image of $\mathcal{L}^k$ onto Γ_j. Hence

$$S^p = S^1 \oplus \left(\oplus_{k=2}^p \oplus_{j=1}^{N_0} \tilde{S}_j^k\right). \qquad (7.33)$$

Therefore a function u_p in S^p can be represented as

$$u_p = u_1 + \sum_{k=2}^p \sum_{j=1}^{N_0} u_{k,j}, \qquad (7.34)$$

where $u_1 \in S^1$ and $u_{k,j} \in \tilde{S}_j^k$ for $k=2,\dots,p$, $j=1,\dots,N_0$.

With the following result one obtains a bound for the additive Schwarz operator of the multilevel p-version

Lemma 7.2 *[HTS97] Let $u_k = c_k \mathcal{L}_k^*$ where $\mathcal{L}_k^* = \mathcal{L}_k/\|\mathcal{L}_k\|_{L^2(I)}$ with $I = (-1,1)$, and $u = \sum_{k=2}^p u_k$. Then there exist positive constants C_1 and C_2 independent of the polynomial degrees such that*

$$C_1\|u\|_{H_\infty^{1/2}(I)}^2 \le \sum_{k=2}^p \|u_k\|_{H_\infty^{1/2}(I)}^2 \le C_2 p\,\|u\|_{H_\infty^{1/2}(I)}^2.$$

In view of Lemma 7.1 one shows

$$\|u_1\|_{H_\infty^{1/2}(\Gamma)}^2 + \sum_{j=1}^{N_0}\sum_{k=2}^p \|u_{k,j}\|_{H_\infty^{1/2}(\Gamma_j)}^2 \le C_1^{-1} p(1+\log^3 p)\|u_p\|_{H_\infty^{1/2}(\Gamma)}^2 \qquad (7.35)$$

which yields a lower bound for $\lambda_{\min}(P)$. Due to Lemma 2.4 in [ST95] we have

$$\|u_1\|_{H_\infty^{1/2}(\Gamma)}^2 \le C(1+\log p)\|u_p\|_{H_\infty^{1/2}(\Gamma)}^2 \qquad (7.36)$$

and with Lemma 7.2 we obtain

$$\sum_{k=2}^p \|u_{k,j}\|_{H_\infty^{1/2}(\Gamma_j)}^2 \le cp\|\sum_{k=2}^p u_{k,j}\|_{H_\infty^{1/2}(\Gamma_j)}^2. \qquad (7.37)$$

Summing over j and using (7.36) one obtains (7.35). Since (7.33) is a direct sum decomposition the representation (7.34) is unique and application of Lemma 7.1 yields the boundedness of $\lambda_{\max}(P)$.

Theorem 7.3 *[HTS97] There exist positive constants C_1, C_2 independent of p and N_0 such that $\lambda_{\min}(P) \ge C_1 p^{-1}(1+\log^3 p)^{-1}$ and $\lambda_{\max}(P) \le C_2$ for the multilevel additive Schwarz operator defined via (7.33). Hence $\kappa(P) \le \frac{C_2}{C_1} p(1+\log^3 p)$.*

h-p version

In [TS97] we consider a preconditioner based on a two-level mesh for the h-p version of the Galerkin boundary element method for the hypersingular equation. We show that the condition number of the additive Schwarz operator is of order $\max_i \left(1 + \log \frac{H_i p_i}{h_i}\right)(1 + \log p_i)$ where p_i is the maximal polynomial degree used in the i-th subdomain Γ_i. Here the possibly irregular coarse mesh is obtained by dividing Γ into disjoint subdomains Γ_i of length H_i whereas the quasi-uniform fine mesh is obtained by further dividing Γ_i into subintervals Γ_{ij}, the maximal length of the subintervals Γ_{ij} being denoted by h_i.

For preconditioning on a geometric mesh we take for simplicity $\Gamma = (0,1)$ and assume that the solution of (7.3) has a singularity only at the origin and belongs to the countably normed space $B_\beta^2(\Gamma)$, $0 < \beta < 1$ (see [Heu92]). The geometric mesh Γ_σ^n contains layers Γ_i, $1 \le i \le n$, according to the distance to the origin, and $H_i = h_i = \operatorname{diam}(\Gamma_i) \sim \sigma^{n-i+1}$, $1 \le i \le n$. In this case the 2-level h-p version becomes just the 2-level p-version on the geometric mesh and the corresponding additive Schwarz operator has condition number of order $(1 + \log n)^2$ with $p_i \sim i$ [HTS97].

7.2.2   WEAKLY SINGULAR INTEGRAL EQUATIONS

In this subsection we consider different additive Schwarz methods for the weakly singular integral operator recalling results from [ST95, TS96, HTS97].

h-version

In the 2-level method the ansatz space S_M in (7.2) is now the space $\bar{S}_h$ of piecewise-constant functions on Γ discretised by a mesh of width h. We decompose

$$\bar{S}_h = \bar{S}_H + \bar{S}_{h,1} + \cdots + \bar{S}_{h,N_h-1}, \tag{7.38}$$

where $H = 2h$ and $S_{h,j} = \operatorname{span}\{\chi_{h,j}\}$. Here the Haar function $\chi_{h,j}$ is the derivative of the hat function $\phi_{h,j}$ and has therefore support on $\bar{\Gamma}_j \cup \bar{\Gamma}_{j+1}$.

If we continue the 2-level method for the global problem on $\bar{S}_H$, we then have the multilevel method which was analysed in [TS96]. The decomposition of the ansatz space can then be described by

$$\bar{S}_h = \bar{S}_{h_1} + \sum_{l=2}^{L} \left(\bar{S}_{h_l,1} + \cdots + \bar{S}_{h_l,N_{h_l}-1}\right) \tag{7.39}$$

where $\bar{S}_{h_1}$ is the space with the coarsest mesh including the functions being constant on Γ and $h_{l-1} = 2h_l$.

Theorem 7.4 [TS96, STZ97] *There exist constants C_1 and C_2 independent of h (and the number of levels L) such that $\lambda_{\min}(P) \ge C_1$ and $\lambda_{\max}(P) \le C_2$ and $\kappa(P) \le (C_2/C_1)$ with P defined via (7.38) or (7.39).*

The proof of Theorem 7.4 is similar to the proof of Theorem 7.1. It uses i) that the antiderivative operator $J_{1/2}$ defined on $\tilde{H}_0^{-1/2}(I) = \{v \in \tilde{H}^{-1/2}(I) : \int_I v(x)\, dx = 0\}$

satisfies (cf. [HS96]) for $u \in \tilde{H}_0^{-1/2}(\Gamma)$

$$\|u\|_{\tilde{H}^{-1/2}(I)} \simeq \|J_{1/2} u\|_{\tilde{H}^{1/2}(I)} \tag{7.40}$$

and ii) that the Haar function $\chi_{h,j}$ has zero integral mean.

p-version

We define on the mesh (7.5) the space $\bar{S}^0$ of piecewise-constant functions on Γ. To each subinterval Γ_j, we associate the space $\bar{S}_j^p$ of polynomials of degree $p > 0$ with support in $\overline{\Gamma}_j$. The space $\bar{S}_j^p$ is the affine image onto Γ_j of the space spanned by the Legendre polynomials of degree p. We have

$$\bar{S}^p = \bar{S}^0 \oplus \bar{S}_1^p \oplus \cdots \oplus \bar{S}_{N_0}^p. \tag{7.41}$$

Then Theorem 7.2 holds with P defined via (7.41) instead of (7.28) (see [ST95]).

For a multilevel method we introduce on each subinterval Γ_j the space $\bar{S}_j^k = \mathrm{span}\,\{L_j^k\}$ where L_j^k is the affine image onto Γ_j of the Legendre polynomial L^k of degree k. Then

$$\bar{S}^p = \bar{S}^0 \oplus \left(\oplus_{k=1}^{p} \oplus_{j=1}^{N_0} \bar{S}_j^k \right). \tag{7.42}$$

and Theorem 7.3 holds with P defined via (7.42) instead of (7.33). Note that the Legendre polynomials L^k have zero integral mean and therefore the norm equivalence (7.40) holds for $\bar{S}_j^k$.

Hierarchical bases

As an alternative to the subspace decomposition (7.23) we consider now the following hierarchical basis decomposition of S_h:

$$S_h = S_{h_1} \oplus \bigoplus_{l=2}^{L} \bigoplus_{j=1}^{N_l-1} {}' S_{h_l,j}$$

where $'$ denotes the sum over odd values of j only. The corresponding multilevel additive Schwarz operator is defined as

$$P_{HB} = \sum_{l=1}^{L} \sum_{j=1}^{N_l-1} {}' P_{S_{h_l},j}.$$

In [TSM97] it will be shown that the largest eigenvalue of P_{HB} is bounded independently of L and the smallest eigenvalue behaves like L^{-3}. This correponds in some way to the hierarchical basis result in [Yse88] for the 2D-FEM. By using a different basis (wavelets) it will be shown in [STZ97] that the corresponding multilevel additive Schwarz operator has bounded condition number (independently of L). Corresponding results will also be obtained for the weakly singular case [TSM97, STZ97]. It can be shown by a simple example that the hierarchical basis preconditioner fails for the three-dimensional hypersingular case. However, the situation for the weakly singular case in 3D is currently under research and the numerical results indicate that in this case a hierarchical basis preconditioner reduces the condition number of the linear systems considerably.

7.3   ADDITIVE SCHWARZ METHODS FOR INDEFINITE PROBLEMS

Now let $B = A + K$ be a pseudodifferential operator of order $\alpha = 1$ or $\alpha = -1$ where A is positive definite and K is a compact operator from $\tilde{H}^{\alpha/2}(\Gamma)$ into $H^{-\alpha/2}(\Gamma)$. Then B satisfies a Gårding inequality, i.e., there exist $\gamma > 0$ and $\eta > 0$ such that the real part of $\langle Bu, u \rangle$ satisfies $\forall u \in \tilde{H}^{\alpha/2}(\Gamma)$,

$$\Re(B(u,u)) \geq \gamma \|u\|^2_{\tilde{H}^{\alpha/2}(\Gamma)} - \eta \|u\|^2_{\tilde{H}^{\alpha/2-1/2}(\Gamma)}. \tag{7.43}$$

Again with $f \in H^{-\alpha/2}(\Gamma)$ we solve $Bu = f$ by Galerkin's method, i.e. find $u \in S \subset \tilde{H}^{\alpha/2}(\Gamma)$ such that

$$\langle Bu, v \rangle = \langle f, v \rangle \quad \forall v \in S. \tag{7.44}$$

The following operators map the boundary element space S onto the subspaces S_i, $i = 0, \ldots, N$, and are defined in terms of the bilinear forms $b(u,v) := \langle Bu, v \rangle$ and $a(u,v) := \langle Au, v \rangle$.

Definition 7.1 *For any* $w \in V$, $Q_0 w \in S_0$ *is the solution of* $b(Q_0 w, v_0) = b(w, v_0)$ $\forall v_0 \in S_0$. *For* $i = 1, \ldots, N$ *and for any* $w \in V$, $P_i w \in S_i$ *is the solution of* $a(P_i w, v_i) = b(w, v_i)$ $\forall v_i \in S_i$.

The additive Schwarz operator is now defined as $Q = Q_0 + P_1 + \cdots + P_N$ and the additive Schwarz method consists of solving

$$Q u_M = b_M \tag{7.45}$$

with RHS $b_M = \sum_{j=0}^{N} b_j$ where

$$b(b_0, v_0) = \langle f, v_0 \rangle \quad \forall v_0 \in S_0, \quad a(b_j, v_j) = \langle f, v_j \rangle \quad \forall v_j \in S_j, j = 1, \ldots, N. \tag{7.46}$$

It is shown in [ST97a, ST97b] that this additive Schwarz algorithm when used with the GMRES method gives an efficient solver for the Galerkin scheme (7.44). The rate of convergence is bounded for the h-version and logarithmically growing in p for the p-version, if the mesh size of the coarse space is sufficiently small.

As proved in [EES83] the rate of convergence of the GMRES method when used to solve (7.45) is given as $1 - \frac{C_0^2}{C_1^2}$, where

$$C_0 := \inf_{v \in S} \frac{a(v, Qv)}{a(v, v)} \quad \text{and} \quad C_1 := \sup_{v \in S} \frac{a(Qv, Qv)}{a(v, v)}. \tag{7.47}$$

With this result we show in [ST97a, ST97b] that C_0 is independent of h in the h-version and behaves like $1 + \log^3 p$ in the p-version whereas C_1 is always constant.

An example suited to the above setting is the weakly singular integral equation.

$$V_k v(x) := \frac{i}{2} \int_\Gamma H_0^1(k|x - y|) v(y) ds_y = f(x), \quad x \in \Gamma, \tag{7.48}$$

from time harmonic acoustic scattering on a screen Γ where H_0^1 is the Hankel function of the first kind and order 0 with $k \in \mathbb{C}$. While (7.48) corresponds to a Dirichlet problem in $\mathbb{R}^2 \backslash \bar{\Gamma}$ the corresponding Neumann problem leads to

$$W_k v(x) := \frac{-i}{2} \frac{\partial}{\partial n_x} \int_\Gamma \frac{\partial}{\partial n_y} \left[H_0^1(k|x-y|) \right] v(y) \, ds_y = f(x), \quad x \in \Gamma. \tag{7.49}$$

7.4 ADDITIVE SCHWARZ METHODS FOR POSITIVE DEFINITE INTEGRAL OPERATORS IN THREE DIMENSIONS

In this section we briefly summarize results on the preconditioning of the hypersingular and weakly singular integral operators in $\mathbb{R}^3$ which are defined by

$$W u(x) := \frac{1}{4\pi} \frac{\partial}{\partial n_x} \int_\Gamma u(y) \frac{\partial}{\partial n_y} \frac{1}{|x-y|} \, dS_y, \quad x \in \Gamma$$

and

$$V u(x) := \frac{1}{4\pi} \int_\Gamma \frac{u(y)}{|x-y|} \, ds_y, \quad x \in \Gamma,$$

respectively. Here, for ease of presentation, $\Gamma \subset \mathbb{R}^3$ is a plane open surface and all the theoretical results are also valid for polyhedrons with rectangular sides. Again, W and V are positive definite operators of orders 1 and -1, respectively. As in the two–dimensional case we solve first kind integral equations involving W and V by the Galerkin method. For the hypersingular operator we have to use boundary element spaces $S^1(\Gamma) \subset \tilde{H}^{1/2}(\Gamma)$ consisting of continuous functions and for the weakly singular operator it suffices to take boundary element spaces $S^0(\Gamma) \subset \tilde{H}^{-1/2}(\Gamma)$ of discontinuous functions.

7.4.1 H VERSION FOR THE HYPERSINGULAR OPERATOR

We use quasi–uniform rectangular meshes of size h on Γ and take piecewise bilinear functions corresponding to these meshes to construct $S^1(\Gamma) = S_h^1(\Gamma)$. Just as described in §7.2.1 we perform 2–level and multilevel additive Schwarz methods in an analogous way, i.e. we decompose

$$S_h^1 = S_H^1 + S_{h,1}^1 + \cdots + S_{h,N_h-1}^1 \tag{7.50}$$

in the 2–level case and

$$S_h^1 = S_{h_1}^1 + \sum_{l=2}^L \left(S_{h_l,1}^1 + \cdots + S_{h_l,N_{h_l}-1}^1 \right) \tag{7.51}$$

in the multilevel case. We obtain a result similar to the two–dimensional case.

Theorem 7.5 [Heu97a] *The additive Schwarz operator corresponding to* (7.50) *or* (7.51) *has a condition number which is bounded by*

$$\kappa(P) \leq Ch^{-\epsilon}$$

with arbitrary $\epsilon > 0$. The constant C is independent of h, the mesh size of the finest level, and of the number of the levels L.

To get rid of the constraint $H = 2h$ one can also consider general 2–level methods where one has a coarse mesh Γ_H which is almost independent of the fine mesh Γ_h, the only restriction being the compatibility. The decomposition used is given by

$$S_h^1(\Gamma) = S_H^1(\Gamma) \cup S_{h,H}^1(\Gamma) \cup \cup_{j=1}^J S_h^1(\Gamma_j). \tag{7.52}$$

The space $S_H^1(\Gamma)$ consists of the usual continuous piecewise bilinear functions on the mesh Γ_H of size H. $S_{h,H}^1(\Gamma)$ is the so–called wire basket space which is spanned by the piecewise bilinear hat functions of $S_h^1(\Gamma)$ which are concentrated at the nodes lying on the element boundaries of the mesh of size H. The spaces $S_h^1(\Gamma_j)$ are spanned by the piecewise bilinear hat functions concentrated at the nodes interior to the restricted meshes $\Gamma_h|_{\Gamma_j} = \Gamma_{j,h}$, $j = 1, \ldots, J$. The result is the following.

Theorem 7.6 [HS97] *The condition number of the additive Schwarz operator P which is defined by the decomposition (7.52) is bounded by*

$$\kappa(P) \leq C(1 + \log \frac{H}{h})$$

where the constant $C > 0$ is independent of the coarse and fine mesh sizes H and h.

Thus, for this preconditioner, we have bounded condition numbers if the ratio H/h is fixed.

7.4.2   P VERSION FOR THE HYPERSINGULAR OPERATOR

We consider a fixed rectangular mesh Γ_h and take tensor products of piecewise linear functions and antiderivatives of Legendre polynomials as basis functions. For the following overlapping decomposition of the boundary element space $S^1(\Gamma) = S_p^1(\Gamma)$ we obtain bounded condition numbers of the corresponding additive Schwarz operator:

$$S_p^1 = S_1^1 \cup S_{p,1}^1 \cup \cdots \cup S_{p,N_h}^1. \tag{7.53}$$

The so–called coarse grid space is that of the h version, $S_1^1 = S_h^1(\Gamma)$, and the remaining spaces are subspaces localized at the neighborhood of each interior node. More precisely $S_{p,j}^1$ is the space of piecewise polynomials of degree p which are globally continuous and which have support contained in the elements adjacent to the node with number j. Therefore, subspaces for adjacent nodes have common functions.

Theorem 7.7 [Heu97a] *The condition number of the additive Schwarz operator P which is defined by the decomposition (7.53) has bounded condition number.*

Since the subspaces $S_{p,j}^1$ are rather large for large polynomial degree p one is interested in further splitting the decomposition. Due to the tensor product structure of the basis functions one has a natural decomposition into subspaces which consists of functions concentrated at the nodes, edges, and interior to the elements, separately. However, it is well known that one cannot take the usual nodal hat functions separately for such a splitting in higher dimensions. This would result in large condition numbers. Therefore, in order to use a non–overlapping decomposition, one has to consider well behaved basis functions, i.e. functions with small energy. In [Heu97c] the minimizing polynomial of degree p defined by

$$\|\varphi_0\|_{L^2(-1,1)} = \min_\varphi \|\varphi\|_{L^2(-1,1)}, \quad \varphi_0(1) = 1, \ \varphi_0(-1) = 0.$$

was taken to construct the nodal basis functions and as the component of the edge functions perpendicular to the edges. Further, basis functions for the edges and for the functions interior to the elements are defined as discrete tensor product solutions in the weak sense of the Laplacian. The decomposition is as follows.

$$S_p^1(\Gamma_h) = X_0 \oplus X_1 \oplus \cdots \oplus X_{J_h}. \tag{7.54}$$

Here $X_j = S_p^1(\Gamma_h) \cap \tilde{H}^{1/2}(\Gamma_j)$, $j = 1, \ldots, J_h$, where Γ_j is an element of the mesh. X_0 is the global space of the remaining functions which are concentrated at the nodes and the edges of the mesh. This space is called the wire basket space.

Theorem 7.8 [Heu97c] *The condition number of the additive Schwarz operator P defined by the decomposition (7.54) is bounded by*

$$\kappa(P) \le C(1 + \log p)^3.$$

The constant C is independent of the mesh size h and the polynomial degree p.

As shown in [Heu97c] a similar result holds even for the diagonal preconditioner if one takes the special discretely harmonic basis functions.

7.4.3   P VERSION FOR THE WEAKLY SINGULAR OPERATOR

We use quasi–uniform rectangular meshes of size h on Γ and take discontinuous piecewise polynomials of degree p for the boundary element space $S^0(\Gamma) = S_p^0(\Gamma)$. We decompose

$$S_p^0(\Gamma) = S_p^0(\Gamma_1) \cup \cdots \cup S_p^0(\Gamma_J). \tag{7.55}$$

The spaces $S_p^0(\Gamma_j)$ are the restriction of $S_p^0(\Gamma)$ onto a subdomain Γ_j where $\bar{\Gamma} = \cup_{j=1}^J \bar{\Gamma}_j$ is a, possibly overlapping, decomposition of Γ. We obtain

Theorem 7.9 [Heu97b] *For any $\epsilon > 0$ there exists a constant $C > 0$ such that the condition number of the additive Schwarz operator defined by the decomposition (7.55) is bounded with arbitrary $\epsilon > 0$ by*

$$\kappa(P) \le C p^\epsilon.$$

7.5   ADAPTIVE TWO–LEVEL BOUNDARY ELEMENT METHODS

In this section we show how to derive *a posteriori* error estimates from the properties of two-level additive Schwarz operators. If the corresponding subspace decomposition is chosen carefully we may obtain local error indicators which are easily computable and which can be used for adaptive refinement strategies. We derive the main result for general two-level decompositions and for brevity apply the theory to the hypersingular integral operator with a subspace decomposition which was analyzed in [Cao95] and [TSM97] in the hierarchical basis.

Let S_H be a finite dimensional space and let $u_H \in S_H$ be the Galerkin solution, i.e.

$$a(u_H, v) = \langle g, v \rangle \qquad \forall v \in S_H.$$

Let S_h be a refinement of S_H with discretization parameter $h = h(H)$ and let $u_h \in S_h$ be the Galerkin solution in S_h. We assume the following saturation condition:

There exists a constant $0 < \gamma < 1$ independent of H such that

$$\|u - u_h\| \leq \gamma \|u - u_H\| \tag{7.56}$$

where u is the exact solution of the problem under consideration and $\| \cdot \|^2 = a(\cdot, \cdot)$ is the energy norm.

By using the triangle inequality it can be shown from (7.56) that

$$(1 - \gamma) \|u - u_H\| \leq \|u_h - u_H\| \leq (1 + \gamma) \|u - u_H\|. \tag{7.57}$$

Let

$$S_h = S_H \oplus D_h, \qquad D_h = S_h^1 + S_h^2 + \ldots + S_h^N \tag{7.58}$$

be a subspace decomposition of the space S_h. If it can be shown that the correponding two-level additive Schwarz operator $P = P_{S_H} + \sum_{j=1}^{N} P_{S_h^j}$ has bounded condition number, i.e. if there are constants $\lambda_i > 0$ independent of H such that

$$\lambda_1 a(v, v) \leq a(Pv, v) \leq \lambda_2 a(v, v) \qquad \forall v \in S_h \tag{7.59}$$

then, by using

$$a(Pv, v) = a(P_{S_H} v, v) + \sum_{j=1}^{N} a(P_{S_h^j} v, v) = \|P_H v\|^2 + \sum_{j=1}^{N} \|P_{S_h^j} v\|^2$$

and the Galerkin property $\|P_{S_H}(u_h - u_H)\|^2 = 0$ we obtain the following result:

Theorem 7.10 *If (7.56) and (7.59) hold then*

$$\frac{1}{\lambda_2(1 + \gamma)} \sum_{j=1}^{N} \|P_{S_h^j}(u_h - u_H)\|^2 \leq \|u - u_H\|^2 \leq \frac{1}{\lambda_1(1 - \gamma)} \sum_{j=1}^{N} \|P_{S_h^j}(u_h - u_H)\|^2$$

In [TSM97], [Mun96] the h-version Galerkin method for the hypersingular integral operator W was analyzed. The spaces S_h and S_H were chosen as in Section 7.2.1 and

$$D_h = S_{h,1} \oplus S_{h,3} \oplus S_{h,5} \oplus \ldots \oplus S_{h,N_h-1}$$

with the same notation as in (7.14). It is shown in [Mun96] that (7.59) holds. Hence, from the above theorem, we obtain local error indicators

$$\eta_j = \|P_{S_{h,j}}(u_h - u_H)\| = \frac{|a(u_h - u_H, \phi_{h,j})|}{a(\phi_{h,j}, \phi_{h,j})^{1/2}} = \frac{|\langle W u_h - g, \phi_{h,j} \rangle|}{\langle W \phi_{h,j}, \phi_{h,j} \rangle^{1/2}}$$

with $j \in \{1, 3, \ldots, N_h - 1\}$ which can be used in an adaptive algorithm [Cao95]. Corresponding results for the weakly singular operator were obtained in [Fun96]. For an extension to surfaces in $\mathbb{R}^3$ see [MS96].

Table 1   Weakly singular integral equation (7.4) with $g_2(x) \equiv 1$: h-version, using CG and
2-level and multilevel additive Schwarz and BPX preconditioner

N	A_N	Condition number			Number of iterations			
		2-level	multilevel	BPX	CG	2-lev	m-lev	BPX
16	15.7545	2.1310	5.1005	4.0543	8	8	8	8
32	32.6847	2.6175	6.0171	4.8520	20	12	11	12
64	65.1292	2.9767	6.9188	8.8454	33	16	14	17
128	129.6566	3.1926	7.8182	10.4593	45	19	16	19
256	259.8891	3.3153	8.7199	12.1835	66	21	17	22
512	517.1460	3.3777	9.6253	14.0330	85	21	19	23
1024	1036.1733	3.4108	10.5344	16.0186	125	22	19	24
2048	2066.6248	3.4255	11.4468	18.1483	168	22	19	26
4096	4119.1740	3.4338	12.3619	20.4271	226	23	19	29
8192		3.4361	13.2791	22.8595		23	19	30
16384		3.4370	14.1980	25.4481		23	19	31
32768		3.4379	15.1182	28.1890		23	19	32
65536		3.4406	16.0395	31.1033		24	19	33

Table 2   Weakly singular integral equation (7.4) with $g_2(x) \equiv 1$: p-version with $N_0 = 2$,
using CG and PCG with 2-level and multilevel additive Schwarz methods.

p	A_N	Condition number		Number of iterations		
		2-level	multilevel	CG	2-level	multilevel
1	6.22	1.89	1.86	4	2	2
2	19.32	2.09	2.08	6	3	3
3	41.01	2.22	2.41	9	4	4
4	72.47	4.89	2.64	12	5	5
5	115.18	4.12	2.94	15	6	6
6	170.82	2.33	3.20	17	6	7
7	241.21	2.37	3.51	21	6	8
8	328.23	2.38	3.81	24	6	9
9	433.82	2.40	4.14	27	6	10
10	559.88	8.01	4.46	33	6	11

Table 3 Hypersingular integral equation (7.3) with $g_1(x) \equiv 1$: h-version, using CG and PCG with 2-level and multilevel additive Schwarz preconditioners

N	A_N	Condition number 2-level	Condition number multilevel	CG	Number of iterations 2-level	Number of iterations multilevel
15	7.7629	2.1475	3.0353	8	7	8
31	15.5445	2.2072	3.4613	11	11	11
63	31.1092	2.2162	3.7561	17	12	14
127	62.4163	2.2276	3.9714	26	13	16
255	125.0924	2.2299	4.1335	38	13	17
511	250.4733	2.2262	4.2578	55	12	17
1023	501.2394	2.2250	4.3545	78	12	17
2047	1002.7757	2.2236	4.4308	109	12	17
4095	2005.8634	2.2225	4.4917	154	12	18
8191		2.2160	4.5408		11	18
16383		2.2153	4.5808		11	18
32767		2.2150	4.6138		11	18
65535		2.2067	4.6413		10	18

Table 4 Hypersingular integral equation (7.3) with $g_1(x) \equiv 1$: p-version with $N_0 = 2$, using CG and PCG with 2-level and multilevel additive Schwarz methods

p	A_N	Condition number 2-level	Condition number multilevel	CG	Number of iterations 2-level	Number of iterations multilevel
2	4.44	2.31	2.31	2	2	2
3	12.75	3.55	3.55	3	3	3
4	26.04	4.16	4.85	4	4	4
5	45.88	4.84	6.14	5	5	5
6	73.32	5.31	7.41	7	6	6
7	109.94	5.84	8.69	8	7	7
8	156.98	6.21	9.95	9	7	8
9	215.97	6.61	11.24	12	7	9
10	288.23	6.95	12.50	13	7	10

7.6   IMPLEMENTATION AND NUMERICAL RESULTS.

For the implementation of the additive Schwarz method we have to distinguish two
basically different cases, namely when the subspaces form a direct sum or when they
are nested.

(a) When the subspace decomposition is given as a direct sum we proceed as follows.
Let
$$S = \mathrm{span}\,\{\phi_j : j = 1, \ldots, \dim S\},$$
and
$$S_i = \mathrm{span}\,\{\phi_j : j = N_{i-1} + 1, \ldots, N_i\}.$$

Let A_i denote the Galerkin matrix corresponding to S_i, $i = 0, \ldots, N$. Then we
have $P = \sum_i P_i = \sum_i C_i^T A_i^{-1} C_i A_S$, where the projection matrices $C_i = (c_{l,k}^{(i)})$
are diagonal matrices with $c_{l,k}^{(i)} = 1$ when $l = k = N_{i-1} + 1, \ldots, N_i$ and $c_{l,k}^{(i)} = 0$
otherwise. We also have $A_i = C_i^T A_S C_i$, where A_S is the Galerkin matrix
corresponding to S. The simple form of the projection matrices C_i is due to
the fact that the same basis functions are used in S and in the subspaces S_i.
Note this is exactly the situation of the p-version. If the matrices A_i are simply
the diagonal blocks of A_S, the preconditioner and P can be written as

$$P = \begin{pmatrix} A_1 & & & \\ & A_2 & & \\ & & \ddots & \\ & & & A_N \end{pmatrix}^{-1} A_S.$$

(b) When the subspaces form a nested sequence we proceed with the implementation
as follows. Again let A_i denote the Galerkin matrix belonging to S_i, $i = 0, \ldots, N$.
Note that now the subspaces S_i are spanned by different basis functions,
$S_i = \mathrm{span}\,\{\phi_j^i : j = 1, \ldots, \dim S_i\}$. Here the projection matrices C_i are defined
by $C_i = (c_{j,l}^i)$. Hence the projection to the lower level is given by

$$\phi_j^i = \sum_{l=1}^{\dim S_{i+1}} c_{j,l}^i \phi_l^{i+1},$$

and P can be written in matrix form

$$\left\{ (\mathrm{diag}\, A_N)^{-1} + \ldots + C_2^T ((\mathrm{diag}\, A_2)^{-1} \right.$$
$$\left. + C_1^T ((\mathrm{diag}\, A_1)^{-1} + C_0^T A_0^{-1} C_0) C_1) C_2 \cdots C_{N-1} \right\} A_S$$

where $A_i = C_i A_{i+1} C_i^T$ and $i = 0, \ldots, N-1$ and $A_N := A_S$.

Note that C_i are sparse matrices and therefore the action of C_i and C_i^T can be
implemented in a very efficient way; therefore their multiplication with a vector
costs only $\mathcal{O}(N)$ operations in contrary to $\mathcal{O}(N^2)$ in case of the full matrix.

The four tables give the results for the additive Schwarz methods described above. The numerical experiments were performed by Dr. M. Maischak on a SUN-Sparcstation 4/470 at the Institute for Applied Mathematics at University of Hannover.

Note that in the case of the hypersingular equation the multilevel additive Schwarz preconditioner differs only by a scaling factor from the BPX preconditioner (cf.(7.19)) whereas for the weakly singular operator these preconditioners behave completely differently, since in this case $a(\chi_{h_i}, \chi_{h_i}) := \langle V\chi_{h_i}, \chi_{h_i} \rangle$ depends on the step size h_i of the Haar basis function χ_{h_i} (cf. Table 1).

7.7 ACKNOWLEDGEMENTS

The author was partly supported by the German Research Foundation under Grant Nr. Ste 238/25-9.

References

[BCMP91] Babuška I., Craig A., Mandel J., and Pitkäranta J. (1991) Efficient preconditioning for the p-version finite element method in two dimensions. *SIAM J. Numer. Anal.* 28: 624–661.

[BLP94]   Bramble J. H., Leyk Z., and Pasciak J. E. (1994) The analysis of multigrid algorithms for pseudodifferential operators of order minus one. *Math. Comp.* 63: 461–478.

[BP93]   Bramble J. H. and Pasciak J. E. (1993) New estimates for multilevel algorithms including the V-cycle. *Math. Comp.* 60: 447–471.

[BPX90]   Bramble J. H., Pasciak J. E., and Xu J. (1990) Parallel multilevel preconditioners. *Math. Comp.* 55: 1–22.

[Cao95]   Cao T. (1995) *Adaptive and Additive Multilevel Methods for Boundary Integral Equations.* PhD thesis, School of Mathematics, University of New South Wales, Sydney, Australia.

[DW91]   Dryja M. and Widlund O. (1991) Multilevel additive methods for elliptic finite element problems. In Hackbusch W. (ed) *Parallel Algorithms for PDE's (Proc. of the 6. GAMM-Seminar, Kiel, Germany, January 19–21, 1990)*, pages 58–69. Vieweg, Braunschweig.

[EES83]   Eisenstat S. C., Elman H. C., and Schultz M. H. (1983) Variational iterative methods for non symmetric systems of linear equations. *SIAM J. Numer. Anal.* 20: 345–357.

[Fun96]   Funken S. A. (1996) *Schnelle Lösungsverfahren für FEM-BEM Kopplungsgleichungen.* PhD thesis, Institut für Angewandte Mathematik, Universität Hannover.

[Heu92]   Heuer N. (1992) *hp-Versionen der Randelementmethode.* PhD thesis, Hannover, Germany.

[Heu97a]   Heuer N. (1997) Additive Schwarz methods for hypersingular integral equations in $\mathbb{R}^3$. (to appear).

[Heu97b]   Heuer N. (1997) Additive Schwarz methods for weakly singular integral equations in $\mathbb{R}^3$ – the p-version. (to appear).

[Heu97c]   Heuer N. (1997) An iterative substructuring method for the p-version of the boundary element method for hypersingular integral equations in three dimensions. (to appear).

[HS96]   Hahne M. and Stephan E. P. (1996) Schwarz iterations for the efficient solution of screen problems with boundary elements. *Computing* 56: 61–85.

[HS97]   Heuer N. and Stephan E. P. (1997) Iterative substructuring for hypersingular integral equations in $\mathbb{R}^3$. (to appear).

[HTS97]   Heuer N., Tran T., and Stephan E. P. (1997) Multilevel additive schwarz method for the p and $h-p$ version boundary element method. (to appear).

[Lio88]   Lions P. L. (1988) On the Schwarz alternating method I. In Glowinski R., Golub G. H., Meurant G. A., and Périaux J. (eds) *Domain Decomposition Methods for Partial Differential Equations.* SIAM, Philadelphia.

[MS96]   Mund P. and Stephan E. P. (1996) Adaptive multilevel boundary element methods for the single layer potential in $\mathbb{R}^3$. Preprint, Institut für Angewandte Mathematik, Universität Hannover, Germany.

[Mun96]   Mund P. (1996) *Adaptive Zwei-Level-Verfahren für die Randelementmethode und für die Kopplung mit Finiten Elementen.* PhD thesis, Hannover, Germany.

[Pav94]   Pavarino L. (1994) Additive schwarz methods for the p-version finite element method. *Numer. Math.* 66: 493–515.

[ST95]   Stephan E. P. and Tran T. (1995) Additive Schwarz method for the p-version boundary element method. Applied math. report amr95/13, University of New South Wales, Sidney, Australia.

[ST97a]   Stephan E. P. and Tran T. (1997) Domain decomposition algorithms for indefinite hypersingular integral equations. the h and p-versions. SIAM J. Sci. Stat. COmput., to appear.

[ST97b]   Stephan E. P. and Tran T. (1997) Domain decomposition algorithms for indefinite weakly singular integral equations. the h and p-versions. (to appear).

[STZ97]   Stephan E. P., Tran T., and Zaprianov S. (1997) Wavelet-based preconditioners for first kind integral equations. (to appear).

[SW90]   Stephan E. P. and Wendland W. L. (1990) A hypersingular boundary integral method for two-dimensional screen and crack problems. *Arch. Rational Mech. Anal.* 112: 363–390.

[TS96]   Tran T. and Stephan E. P. (1996) Additive Schwarz method for the *h*-version boundary element method. *Applicable Analysis* 60: 63–84.

[TS97]   Tran T. and Stephan E. P. (1997) Preconditioners for the $h - p$ version of the galerkin boundary element method. (to appear).

[TSM97]   Tran T., Stephan E. P., and Mund P. (1997) Hierarchical basis preconditioners for first kind integral equations. (to appear).

[vP89]   von Petersdorff T. (1989) *Randwertprobleme der Elastizitätstheorie für Polyeder – Singularitäten und Approximation mit Randelementmethoden.* PhD thesis, Technische Hochschule Darmstadt, Darmstadt.

[vPS90]   von Petersdorff T. and Stephan E. P. (1990) On the convergence of the multigrid method for a hyper–singular integral equation of the first kind. *Numer. Math.* 57: 379–391.

[vPS92]   von Petersdorff T. and Stephan E. P. (1992) Multigrid solvers and preconditioners for first kind integral equations. *Numer. Meth.Part.Diff.Eqns.* 8: 443–450.

[Wid89]   Widlund O. (Jan 14–16 1989) Optimal iterative refinement methods. In Chan T., Glowinski R., Periaux J., and Widlund O. (eds) *Second Intern. Symp. on Domain Decomposition Methods*, pages 114–125. SIAM, LA California.

[Xu92]   Xu J. (1992) Iterative methods by space decomposition and subspace correction. *SIAM Review* 34: 581–613.

[Yse88]   Yserentant H. (1988) On the multi-level splitting of finite element spaces. *Numer. Math.* 54: 719–734.

[Zha92]   Zhang X. (1992) Multilevel schwarz methods. *Numer.Math.* 63: 521–539.

8

Non–overlapping Domain Decomposition with BEM and FEM

E. Schnack, Sz. Szikrai and K. Türke

Institute of Solid Mechanics,
Karlsruhe University,
D–76128 Karlsruhe, Germany

8.1 INTRODUCTION

This paper presents a coupling algorithm of the Finite Element Method and the Boundary Element Method within a domain decomposition procedure for solving mixed boundary value problems in 2D and 3D elastostatics. A mixed variational formulation on the coupling interfaces leads to the use of non–conforming grids for each substructure (cf. [Sch84], [Sch87], [Sch90]). The local BEM problems are solved iteratively by expanding the boundary integral equation in a Neumann series. As a consequence the properties of the coupling matrix concerning symmetry and definiteness can be controlled automatically within an adaptive algorithm. In particular the proposed FE/BE technique handles stress concentration problems very efficiently, providing locally high resolution of the desired stress fields. The BE macro–elements can be incorporated into standard FEM codes.

8.2 DOMAIN DECOMPOSITION WITH NON–OVERLAPPING SUBSTRUCTURES

We consider a bounded, isotropic interior domain $\bar{\Omega} = \Omega \cup \Gamma$ in $\mathbb{R}^m$, $(m = 2, 3)$, with a Lipschitz boundary $\Gamma = \partial\Omega$. The boundary value problem is defined by the Navier equation without body forces

$$u_{j,kk} + \frac{1}{1-2\nu} u_{k,kj} = 0 \qquad \text{or } \Delta^* u = 0 \qquad \text{for } u(\xi) \in \left(H^1(\Omega)\right)^m \qquad (8.1)$$

The Mathematics of Finite Elements and Applications
Edited by J. R. Whiteman © 1997 John Wiley & Sons Ltd.

with given displacements on the Dirichlet boundary Γ_u and given tractions on the Neumann boundary Γ_t:

$$u(\xi) \;\; = \;\; u^*(\xi) \qquad \text{for } \xi \in \Gamma_u = \Gamma \setminus \Gamma_t, \tag{8.2}$$

$$t(\xi) \;\; = \;\; \Pi\, u(\xi) = t^*(\xi) \qquad \text{for } \xi \in \Gamma_t. \tag{8.3}$$

Π denotes the Steklov–Poincaré traction operator. We first decompose $\bar{\Omega}$ into p non–overlapping subdomains Ω_i ($i = 1, \ldots, p$) according to

$$\bar{\Omega} = \bar{\Omega}_F \cup \bar{\Omega}_B, \quad \bar{\Omega}_F = \bigcup_{i=1}^{q} \bar{\Omega}_i, \quad \bar{\Omega}_B = \bigcup_{i=q+1}^{p} \bar{\Omega}_i, \quad \Omega_i \cap \Omega_j = \emptyset \;\; \forall\, i \neq j. \tag{8.4}$$

The global coupling boundary is defined by

$$\Gamma_C := \bigcup_{i=1}^{p} \Gamma_i \setminus \Gamma, \quad \Gamma_i := \partial\Omega_i, \quad i = 1, \ldots, p. \tag{8.5}$$

We also define the FE/BE coupling boundary Γ_0, the skeleton Γ_B of the BE substructures (also called macro–elements) and the coupling boundary Γ_F of the FE subdomains:

$$\Gamma_0 := \partial\Omega_F \bigcap \partial\Omega_B, \quad \Gamma_B := \bigcup_{i=q+1}^{p} \Gamma_i, \quad \Gamma_F := \Gamma_C \setminus \Gamma_B. \tag{8.6}$$

The indices F and B denote quantities corresponding to the FE substructures and the BE macro–elements, respectively. Later on, we suppose for simplicity that

$$\Gamma_B \bigcap \Gamma_t = 0 \tag{8.7}$$

for the simplicity. (For the general case, see [HSW95].) The energy test space corresponding to Ω_F will be denoted by

$$H^1_{\Gamma_u}(\Omega_F) = \{v \in H^1(\Omega_F) : v|_{\partial\Omega_F \bigcap \Gamma_u} = 0\}. \tag{8.8}$$

Let us define the product space

$$\mathcal{U}_F = \{(u_F, \tilde{u}) \in \left(H^1(\Omega_F)\right)^m \times \left(H^{1/2}(\Gamma_B)\right)^m, \; u_F|_{\Gamma_0} = \tilde{u}|_{\Gamma_0}\} \tag{8.9}$$

and the subspaces

$$\left(H_0^{-1/2}(\Gamma_i)\right)^m := \left\{\chi \in \left(H^{-1/2}(\Gamma_i)\right)^m : \int_{\Gamma_i} \chi \, d\Gamma = 0, \int_{\Gamma_i} r \times \chi \, d\Gamma = 0\right\}, \tag{8.10}$$

where r denotes the distance vector to the origin. The test function space associated with the skeleton is defined by

$$\mathcal{V}_F = \{(v_F, \tilde{v}) \in \mathcal{U}_F : v_F \in \left(H^1_{\Gamma_u}(\Omega_F)\right)^m, \; \tilde{v} = w|_{\Gamma_B} \text{with } w \in \left(H^1(\Omega)\right)^m, \; w|_{\Gamma_u} = 0\}. \tag{8.11}$$

The coupled domain and boundary integral variational formulation now reads:

Find

$$((u_F, \tilde{u}), u_{Bj}, t_{Bj}) \in \mathcal{U}_F \times \prod_{j=q+1}^{p} \left(H^{1/2}(\Gamma_j) \times H_0^{-1/2}(\Gamma_j) \right) \qquad (8.12)$$

with

$$u_F|_{\Gamma_u \cap \partial\Omega_F} = u^*|_{\Gamma_u \cap \partial\Omega_F} \quad and \quad \tilde{u}|_{\Gamma_u \cap \Gamma_B} = u^*|_{\Gamma_u \cap \Gamma_B} \qquad (8.13)$$

such that the global equations

$$a_F(u_F, v) + \sum_{j=q+1}^{p} < t_{Bj}, \tilde{v} >_{\Gamma_j} = f(v) \quad for\ all \quad (v_F, \tilde{v}) \in V_F, \qquad (8.14)$$

the weak coupling conditions

$$< \tilde{u} - u_{Bj}, \tau_{Bj} >_{\Gamma_j} = 0 \quad for\ all \quad \tau_{Bj} = \Pi\, v_{Bj} \in \left(H^{-1/2}(\Gamma_i) \right)^m, \qquad (8.15)$$

and the boundary integral equations

$$< (\tfrac{1}{2} I_j + K_j) u_{Bj}, \chi_j >_{\Gamma_j} = < V_j t_{Bj}, \chi_j >_{\Gamma_j} \quad for\ all \quad \chi_j \in \left(H_0^{-1/2}(\Gamma_j) \right)^m \qquad (8.16)$$

are satisfied. In (8.14) the following notations are used:

$$a_F(u_F, v) = \sum_{i=1}^{q} \int_{\Omega_i} \sigma_{kl}(u_F)\, e_{kl}(v)\, d\Omega, \qquad (8.17)$$

$$f(v) = \int_{\Gamma_t} t^* v\, d\Gamma. \qquad (8.18)$$

Equation (8.16) defines the local problem on the boundary Γ_j of a BEM substructure with the weak singular integral operator V_j and the singular integral operator K_j derived from Kelvin's fundamental solution, while I_j is the identity operator. The Lax–Milgram theorem implies the existence of a unique solution of the variational problem (8.12)–(8.16), see [HSW95].

8.3 THE FE/BE DISCRETIZATION

In order to solve the variational problem numerically, we discretize the subdomains and approximate the solution in finite dimensional subspaces. The finite element approximation is based on the triangulation of Ω_F, with the mesh–size H characterizing the largest distance between neighboring grid points, on which we choose a conforming finite element space $H_H^1(\Omega_F) \subset H^1(\Omega_F)$ having the approximation degree $d_F \geq 2$:

$$u_F^H = \Phi_F^H \underline{u}_F^H, \quad u_F^H \in \left(H_H^1(\Omega_F) \right)^m \subset \left(H^1(\Omega_F) \right)^m \qquad (8.19)$$

with

$$\Phi_F^H = [\Phi_I^H, \Phi_{C_F}^H, \Phi_{C_B}^H], \quad \underline{u}_F^H = \left[\underline{u}_I^{H^T}, \underline{u}_{C_F}^{H^T}, \underline{u}_{C_B}^{H^T}\right]^T. \tag{8.20}$$

Here and in the following the indices I, C_F and C_B denote quantities corresponding to the interior nodes of the FE–subdomains and to the coupling boundaries Γ_F, and Γ_B respectively. On the individual macro–element boundaries we define local quasi–regular boundary element grids with mesh parameter h. We approximate the traction field on the coarse grid of the macro–elements with finite element spaces of polynomial degree $d_T = d_F$ or $d_T = d_F - 1$, in such a way that the traction field $t_{B_j}^H$ fulfills the condition of equilibrium on Γ_j (see [Tür95]):

$$t_{B_j}^H = \Phi_{B_j}^H \underline{t}_{B_j}^H, \quad \text{span } \Phi_{B_j}^H \subset \left(H_0^{-1/2}(\Gamma_j)\right)^m, \quad j = q+1, ..., p. \tag{8.21}$$

Next, we map the traction ansatz (8.21) to the fine grid h:

$$t_{B_j}^h(x) = \Phi_{B_j}^h(x)\underline{t}_{B_j}^h = \Phi_{B_j}^h(x)P_{Hj}^h \underline{t}_{B_j}^H, \quad x \in \Gamma_j, \tag{8.22}$$

where $\Phi_{B_j}^h \in \left(S_{jh}^{d_B}(\Gamma_j)\right)^m$ consists of trial functions of polynomial degree $d_B \geq 1$ on the fine grid triangulation of the boundary Γ_j. The mapping

$$\underline{t}_{B_j}^h = P_{Hj}^h \underline{t}_{B_j}^H \tag{8.23}$$

is defined by the requirement:

$$t_{B_j}^h(x_i{}^j) = t_{B_j}^H(x_i{}^j), \quad i = 1, \ldots, n_{B_j}, \tag{8.24}$$

where $x_i{}^j$ denotes the BE nodes on the fine grid of Γ_j. The displacement field u_{B_j} on Γ_j is also approximated with piecewise polynomial trial functions of degree d_B on the fine grid:

$$u_{B_j}^h = \Phi_{B_j}^h \underline{u}_{B_j}^h, \quad \text{span } \Phi_{B_j}^h = \left(S_{jh}^{d_B}(\Gamma_j)\right)^m \subset \left(H^{1/2}(\Gamma_j)\right)^m. \tag{8.25}$$

In order to formulate the discretized version of the variational problem (8.12)–(8.16), we first consider the approximation of the Poincaré–Steklov mapping

$$u_{B_j} = \Pi_j^{-1} t_{B_j} = (\tfrac{1}{2}I_j + K_j)^{-1} V_j t_{B_j} \tag{8.26}$$

by an iterative algorithm.

8.4   A NEUMANN SERIES FOR THE LOCAL BOUNDARY INTEGRAL PROBLEM

We reformulate the Neumann problem (8.26) for a general interior domain Ω_+ with the Calderon projector

$$u(\xi) = (\tfrac{1}{2}I - K)u(\xi) + V t_+(\xi) \quad \text{for all} \quad \xi \in \Gamma = \partial\Omega_+ \tag{8.27}$$

with given tractions $t_+(\xi)$ and unknown displacements $u(\xi)$. If the solution $u(\xi)$ in Ω_+ is determined, it can be continuously extended to the exterior domain $\Omega_- = \mathbb{R}^m \backslash \overline{\Omega_+}$ with boundaries at infinity. In the construction of an algorithm to calculate the solution of (8.27) by iteration we start with some definitions:

The linear function space X is defined as the space of elastic states that includes all stress and deformation states in the interior domain Ω_+ and in the exterior domain Ω_- caused by a loading of Ω_+ on $\partial\Omega_+$ with an admissible traction field t_+. Admissible loadings are defined by the side condition in (8.10).

We call the static field $\sigma(\xi)$ and the kinematic field $u(\xi)$ with $\xi \in \overline{\Omega}_+ \cup \Omega_-$ the representatives of the elastic states: $g \in X$.

A linear manifold $[g]$ of these elements g is defined by adding an arbitrary constant c, in our problem given by rigid body motions: $[g] = \{g+c \ : \ c \in X'\}$. As a consequence the solution is defined in the factor space $\overset{\circ}{X} = X / X'$, if the space of elastic states is represented by the displacement field $u(\xi)$. Rieder [Rie62] has shown that the linear spaces X and $\overset{\circ}{X}$ are real Banach spaces. For the explicit algorithmic construction of the factor space $\overset{\circ}{X}$ we refer to [Tür95].

A general formulation of the Calderon projector (8.27) is given by

$$g - \lambda T g = V t_+ \tag{8.28}$$

with $\lambda = +1$ and $T = \left(\frac{1}{2} I - K\right)$. According to a fundamental theorem from functional analysis for the Banach space $\overset{\circ}{X}$ (cf. [Tay58]) we obtain convergence for the *Neumann* series

$$g = (1 - T)^{-1} g^* = \sum_{k=1}^{\infty} T^{k-1} g^* \quad , \tag{8.29}$$

if the spectral radius $r_\sigma(T)$ is smaller than "1" or if all eigenvalues λ_i of the homogeneous part of (8.28) are larger than "1". The properties of the eigenvalues of the integral operator T were investigated by Kupradze et.al. [KGBB79] and can be summarized in

Theorem 8.1 • *The eigenvalues λ_i are real.*
• *The eigenvalues λ_i are simple poles of the resolvent.*
• *The eigenvalues λ_i are not smaller than unity in absolute value:*

$$|\lambda_i| = \frac{a_+(u, u) + a_-(u, u)}{|a_-(u, u) - a_+(u, u)|} \geq 1 \quad . \tag{8.30}$$

According to the representation formula (8.30) the value $\lambda = +1$ can only be obtained, if the stress–strain energy a_+ or a_- in Ω_+ or Ω_- is zero, i.e. in the case of a rigid body motion. In the factor space $\overset{\circ}{X}$ the rigid body motion is defined as a zero element, so that $\lambda = +1$ is a regular value of our problem. Because the eigenvalues are isolated, the inequality

$$\lambda_i \geq \lambda_{i-1} \geq \ldots \geq \lambda_1 > 1 \tag{8.31}$$

is valid and the convergence for the Neumann series (8.29) is guaranteed.

As a consequence we can calculate the displacement vector u on $\partial\Omega_+$ as the limit of the approximations

$$u(\xi) = \lim_{m\to\infty} u^{(m)}(\xi) = \sum_{k=1}^{\infty} \left(\frac{1}{2}I - K\right)^{k-1} V\, t_+(\xi) \quad \forall \quad \xi \in \partial\Omega_+ \quad , \tag{8.32}$$

where an improved approximation $u^{(m)}(\xi)$ can be calculated by the recursive scheme

$$u^{(m)}(\xi) = \left(\frac{1}{2}I - K\right) u^{(m-1)}(\xi) + V\, t_+(\xi) \tag{8.33}$$

with the trivial start solution

$$u^{(0)} = 0 \quad . \tag{8.34}$$

8.5   THE NUMERICAL PROCEDURE IN MATRIX FORM

Once the nodal FE/BE basis

$$\Phi = \left[\Phi_F^H, \ \left(\Phi_{B_j}^H, \ \Phi_{B_j}^h\right)_{j=q+1}^p\right] \tag{8.35}$$

is chosen, we apply the standard conforming FEM and the BEM collocation with Neumann series as described in Section 8.4 to formulate the discretized version of the variational problem (8.12)–(8.16). This results in the block system

$$\begin{pmatrix} K_I & K_{IC_F} & K_{IC_B} & 0 \\ K_{C_FI} & K_{C_F} & K_{C_FC_B} & 0 \\ K_{C_BI} & K_{C_BC_F} & K_{C_B} & T^T \\ 0 & 0 & -T & \tilde{H} \end{pmatrix} \begin{pmatrix} u_I^H \\ u_{C_F}^H \\ u_{C_B}^H \\ t_B^H \end{pmatrix} = \begin{pmatrix} f_I \\ f_{C_F} \\ f_{C_B} \\ 0 \end{pmatrix} \tag{8.36}$$

of coupled linear FE/BE–equations with

$$\tilde{H} := \tilde{S}\tilde{L} = P_H^{h^T} SLP_H^h = P_H^{h^T} HP_H^h. \tag{8.37}$$

Besides the FE stiffness matrix K and nodal force vector f we have the block matrices S and T that result from the discretization of the weak FE/BE coupling conditions (8.15):

$$T = diag\,(T_i)_{i=\overline{q+1,p}}, \quad T_i = \int_{\Gamma_i} \Phi_{B_i}^{H^T} \Phi_{C_B}|_{\Gamma_i}\, d\Gamma, \tag{8.38}$$

and

$$S = diag\,(S_i)_{i=\overline{q+1,p}}, \quad S_i = \int_{\Gamma_i} \Phi_{B_i}^{h^T} \Phi_{B_i}^h\, d\Gamma, \tag{8.39}$$

and the matrix L that results from the approximation of the Poincaré–Steklov mapping:

$$L = diag\,(L_i)_{i=\overline{q+1,p}}, \quad u_{B_i}^h = L_i\, t_{B_i}^h. \tag{8.40}$$

One can show, that the coupling matrix $\tilde{H}$ is symmetric and positive definite for $h \longrightarrow 0$, i.e. if we solve the local Neumann problems exactly, see [Tür95], [Szi96].

8.6 STABILITY AND CONVERGENCE OF THE NUMERICAL SOLUTION

The unique solvability of the complete algebraic system (8.36) depends on how accurately we approximate the exact Poincaré–Steklov operator. Here we recall the main results of the error analysis of the method, for more details see [HSW95].

Theorem 8.2 *Under some assumptions of the approximation properties of a whole family of finite, skeleton mortar and boundary elements (see [HSW95]) there exist positive constants c_0 and c, such that $\forall\, h \leq c_0 H \leq c_0$ we have the uniform stability estimate*

$$\{\|u_F^H\|^2_{(H^1(\Omega_F))^m} + \|\tilde{u}^H\|^2_{(H^{1/2}(\Gamma_B))^m}\}^{\frac{1}{2}} \leq c\,(\|t^*\|_{(H^{-1/2}(\Gamma_t))^m} + \|u^*\|_{(H^{1/2}(\Gamma_u))^m})$$
$$\leq \|u\|_{(H^1(\Omega))^m}. \qquad (8.41)$$

Remark 8.1 *Theorem 8.2 guarantees the uniqueness and existence of the numerical solution $(u_F^H, \tilde{u}^H)$ together with $u_{B_j}^h$ and $t_{B_j}^h$.*

Lemma 8.1 *Under the choice of c_0 according to Theorem 8.2 the asymptotic estimate of optimal order for the global error holds:*

$$\{\|u_F - u_F^H\|^2_{(H^1(\Omega_F))^m} + \|\tilde{u} - \tilde{u}^H\|^2_{(H^{1/2}(\Gamma_B))^m}\}^{1/2} \leq c\,H^s\|u\|_{(H^{1+s}(\Omega))^m} \qquad (8.42)$$

provided $s \leq d_F - 1$.

Corollary 8.1 *Under the assumptions of Theorem 8.2 the following asymptotic error estimate for the macro–element displacement and traction field approximations holds:*

$$\sum_{j=q+1}^{p} \|u_{B_j}^h - \tilde{u}\|^2_{(H^{1/2}(\Gamma_j))^m} + \sum_{j=q+1}^{p} \|t_{B_j}^h - \Pi_j\tilde{u}\|^2_{(H^{-1/2}(\Gamma_j))^m} \leq$$
$$c_1 h^\delta \|u\|_{(H^{1+\delta}(\Omega))^m} + c_2 H^s\|u\|_{(H^{1+s}(\Omega))^m}, \qquad (8.43)$$

where $\delta \in (0, \frac{1}{2})$ is defined by the inverse assumption, see [HSW95].

The constant c_0 in the stability condition

$$h < c_0\,H \qquad (8.44)$$

is determined in the numerical implementation indirectly by defining error bounds ε_1 and ε_2 for the relative symmetry defect

$$MSD = \frac{\sum_{i=2}^{n}\sum_{j=1}^{i-1}|H_{ij} - H_{ji}|}{\sum_{i=2}^{n}\sum_{j=1}^{i-1}|H_{ij} + H_{ji}|}\,100\,(\%) \qquad (8.45)$$

and the smallest eigenvalue $\lambda_{\min}$ of the coupling matrix H. Here, n denotes the dimension of H. By an adaptive scheme using the expansion (8.33) (cf. [Tür95]) and refinement of the mesh h we can guarantee that the conditions

$$0 < MSD < \varepsilon_1 \quad \text{and} \quad \lambda_{\min} > \varepsilon_2 > 0 \tag{8.46}$$

for any given positive $\varepsilon_1, \varepsilon_2$ are fulfilled. Then we define the matrix

$$H_{i_{new}} := \frac{1}{2}\left(H_i + H_i^T\right), \tag{8.47}$$

and with this symmetric and positive definite matrix we compute the equivalent stiffness matrix $K_{M_i} = T_i^T H_i^{-1} T_i$ of the macro–element, which is positive semi–definite and symmetric, like the finite element stiffness matrices.

8.7   NUMERICAL EXPERIMENTS

We analyze a Neumann problem with the proposed FE/BE method under consideration of the Neumann series developed in Section 8.4. Figure 1 shows a plate with an elliptical cutout under uniform tension $\sigma_0 = 1$ MPa. In the same figure the decomposition of the domain is depicted with two macro–elements in the notch area and a very coarse FE mesh in the area with small stress gradients. We compute the

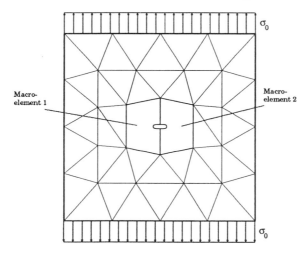

Figure 1   Neumann problem with FE/BE discretization (here: FE mesh H_2)

stress concentration factor $\alpha = \sigma_{\max}/\sigma_0$ and define the approximation error by

$$\varepsilon_\alpha = \frac{|\alpha_{analy} - \alpha_{num}|}{\alpha_{analy}} \cdot 100[\%]. \tag{8.48}$$

The exact value is given by $\alpha_{analy} = 4.7632$. In order to investigate the global convergence we define a coarse FE mesh H_1 with 14 finite elements and a refined mesh

H_2 with 46 finite elements. The fine mesh width h is kept constant with 135 boundary elements on both macro–elements. Table 1 shows the dependence of MSD, $\lambda_{\min}$ and ε_α from the number of Neumann iterations. One observes a decreasing approximation error ε_α for both discretizations. After 50 iterations there remains an approximation error of $\varepsilon_\alpha = 0.53\%$ for mesh H_1 and $\varepsilon_\alpha = 0.025\%$ for the refined mesh H_2. Finally,

Table 1   Numerical results of the Neumann problem

No. of iterations	1	10	20	30	40	50
MSD $[10^{-3}]$	1.903	2.144	2.269	2.317	2.336	2.344
$\lambda_{\min}$ $[10^{-6}]$	1.1049	1.4325	1.4316	1.4318	1.43195	1.4320
ε_α [%] (H_1)	51.7	10.6	3.80	1.66	0.86	0.53
ε_α [%] (H_2)	51.9	10.3	3.36	1.15	0.34	0.025

Figure 2 is a visualization of the stress distribution computed by the proposed FE/BE coupling technique. One can see the locally high resolution of the stress concentration.

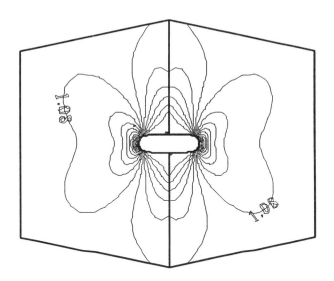

Figure 2   The distribution of the stress component σ_y

References

[HSW95]   Hsiao G., Schnack E., and Wendland W. (1995) A hybrid coupled finite-boundary element method. *Preprint 95–11 of the Mathematical Institute A, Stuttgart University* .

[KGBB79] Kupradze V., Gegelia T., Basheleishvili M., and Burchuladze T. (1979) Three dimensional problems of the mathematical theory of elasticity and thermoelasticity. In *North–Holland Series in Applied Mathematics and Mechanics*, volume 25. North–Holland Publishing Company, Amsterdam, New York, Oxford.

[Rie62]   Rieder G. (1962) Iterationsverfahren und Operatorgleichungen in der Elastizitätstheorie. In *Abhandlungen der Braunschweigischen Wissenschaftlichen Gesellschaft*, volume XIV. Fried. Vieweg & Sohn Verlag, Braunschweig.

[Sch84]   Schnack E. (1984) Stress analysis with a combination of HSM and BEM. In Whiteman J. R. (ed) *MAFELAP Conference on "The Mathematics of Finite Elements and Applications"*, pages 273–281. Academic Press 1985.

[Sch87]   Schnack E. (1987) A hybrid BEM–model. *Int. J. Num. Meth. Engng.* 24: 1015–1025.

[Sch90]   Schnack E. (1990) Macro-elements for 2D- and 3D-elasticity with BEM. In Tanaka M., Brebbia C. A., and Honma T. (eds) *Boundary Elements XII*, volume 2, pages 21–31. Springer-Verlag, Berlin, Heidelberg.

[Szi96]   Szikrai S. (1996) *Parallel coupling of FEM and BEM via domain decomposition for three dimensional linear elastostatic problems.* PhD thesis, University of Karlsruhe.

[Tay58]   Taylor A. E. (1958) *Introduction to functional analysis.* John Wiley & Sons, New York, London, Sydney.

[Tür95]   Türke K. (1995) *Eine Zweigitter–Methode zur Kopplung von FEM und BEM.* PhD thesis, University of Karlsruhe.

9

Fictitious Domain Methods for Incompressible Viscous Flow around Moving Rigid Bodies

R. Glowinski[1], T. W. Pan[1] and J. Periaux[2]

[1] Department of Mathematics, University of Houston, Houston, Texas 77204, U. S. A.

[2] Dassault Aviation, 92214 Saint-Cloud, France

9.1 INTRODUCTION

In this article we discuss the application of a Lagrange multiplier based fictitious domain method to the numerical simulation of incompressible viscous flow modelled by the Navier-Stokes equations around moving bodies. The solution method combines finite element approximations, time discretization by operator splitting and conjugate gradient algorithms for the solution of the linearly constrained quadratic minimization problems coming from the splitting method. The results of several numerical experiments for two-dimensional flow around a moving disk are presented.

9.2 INTRODUCTION: PRINCIPLE, HISTORICAL FACTS AND SYNOPSIS

Suppose that $\Omega \subset \mathbb{R}^d$ ($d=1, 2, 3$) is a connected open set (a domain) containing an inclusion ω, as shown in Figure 1; we denote by Γ and γ the boundaries of Ω and ω, respectively. We consider now the following *boundary value problem*

$$A(u) \ = \ f \ \text{in} \ \Omega \setminus \bar{\omega}, \tag{9.1}$$

$$B_0(u) \ = \ g_0 \ \text{on} \ \Gamma, \tag{9.2}$$

$$B_1(u) \ = \ g_1 \ \text{on} \ \gamma, \tag{9.3}$$

where, in (9.1)-(9.3), the functions f, g_0, g_1, and operators A, B_0, B_1, are given.

Assuming that the shape of Ω is *simple* (which is clearly the case for the example

The Mathematics of Finite Elements and Applications
Edited by J. R. Whiteman © 1997 John Wiley & Sons Ltd.

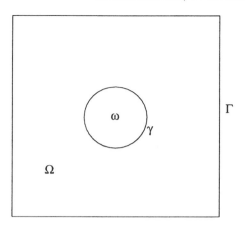

Figure 1   Geometry of Problem (9.1)-(9.3)

of Figure 1) it is reasonable to want to take advantage of that simplicity when solving problem (9.1)-(9.3) numerically; indeed, it may allow, among other things, the use of *regular finite difference or finite element meshes* and consequently of *fast solvers* for the finite dimensional systems approximating problem (9.1)-(9.3) on these grids. In order to address this goal a reasonable idea is to replace problem (9.1)-(9.3) by the following problem:

Find $\tilde{u}$ defined over Ω, S_γ a measure supported by γ, so that

$$\tilde{A}(\tilde{u}) \;=\; \tilde{f} + S_\gamma \quad \text{in} \quad \Omega, \tag{9.4}$$

$$\tilde{B}_0(\tilde{u}) \;=\; g_0 \quad \text{on} \quad \Gamma, \tag{9.5}$$

$$\tilde{B}_1(\tilde{u}|_{\Omega\setminus\bar{\omega}}) \;=\; g_1 \quad \text{on} \quad \gamma; \tag{9.6}$$

in (9.4)-(9.6), operator $\tilde{A}$ is an operator of the same type as A, which concides - in some sense - with A on $\Omega\setminus\bar{\omega}$, $\tilde{f}$ is some extension of f over Ω and $\tilde{B}_0$, $\tilde{B}_1$ are extensions of B_0, B_1.

If S_γ is *well-chosen*, so that the corresponding solution of the boundary value problem (9.4), (9.5) satisfies relation (9.6) we can expect to have $\tilde{u}|_{\Omega\setminus\bar{\omega}} = u$, where u is the solution of problem (9.1)-(9.3). At that stage, several comments are in order:

Remark 9.1 *There are other ways to "embed" domain $\Omega \setminus \bar{\omega}$, in the larger domain Ω. We can use a penalty, for example, as shown in [Glo90].*

Remark 9.2 *Fictitious domain methods can also be applied to time dependent problems as shown in this article (see also [Col97]- [Glo95a] [Col97, Glo94a, Glo94b, Glo95b, Glo95a]).*

Remark 9.3 *There is no particular difficulty in replacing ω by a finite number of "holes", $\omega_1, \omega_2, \ldots, \omega_q$, with $q \geq 2$.*

Remark 9.4 *Most references on fictitious domain methods are concerned with applications to linear problems. Actually, these methods are also well-suited to the solution of nonlinear problems as shown, for example, in [Glo94a, Glo94b, Glo95b, Glo95a](and in the present article).* □

To our knowledge fictitious domain techniques for the solution of partial differential equations were advocated for the first time, more than thirty years ago, by various investigators of the Marchuk-Yanenko school of Numerical Mathematics, at Novossibirsk. These methods belong, essentially, to the class of *boundary fitted* fictitious domain methods, since the discretization is taking place on a mesh which is regular with the exception of a neighborhood of γ where the mesh is locally distorted in order to fit accurately the boundary γ. This approach has motivated a *very* large number of publications; we shall limit our references to [Ast78, Mat93, Nep95] which are typical examples of the Novossibirsk fictitious domain methodology (see also the references therein). In the early seventies Golub and collaborators introduced a *domain embedding* technique for elliptic problems where, once again, the mesh has to follow the boundary γ (see ref. [Buz71] for details). The fictitious domain methods discussed in the present article are closer to the method advocated by Peskin in [Pes89]; indeed, in [Pes89] Peskin uses a fictitious domain method to simulate blood flow around heart valves (natural or artificial), the flow being modelled by the incompressible Navier-Stokes equations. An important analogy between the work in [Pes89] and the present article is that in both cases one uses a mesh which is nonfitted to γ, and which therefore *can stay fixed even if γ moves*. More recently, Leveque and collaborators have developed in [Lev94] a method closely related to that of Peskin.

In this article, motivated by the simulation of *Navier-Stokes flow around moving rigid bodies*, we shall follow the Peskin philosophy in the sense that we shall not use body fitted meshes; also - unlike Peskin - we shall make a systematic use of *variational principles* and of a *Lagrange multiplier* to enforce the boundary condition on γ. In fact the Lagrange multiplier will be the measure S_γ in equation (9.4). The content of this article is as follows:

In Section 9.2 we shall formulate a model flow problem governed by the incompressible Navier-Stokes equations. A Lagrange multiplier/fictitious domain based variational formulation of the above problem will be given in Section 9.3. In Section 9.4, we describe a finite element approximation of the above variational problem, while in Section 9.5 we discuss its time discretization by operator splitting methods à la Marchuk-Yanenko. In Section 9.6 we discuss the solution of the various sub-problems associated to the splitting method, and finally, in Section 9.7, we present the results of numerical experiments.

Remark 9.5 *The present article is not a close repetition of the Navier-Stokes/fictitious domain methods related parts of [Glo94b, Glo95b, Glo95a].Indeed in the above articles the incompressibility condition was forced via a Stokes solver à la Cahouet-Chabard (see [Cah88, Bri87, Glo91, Glo89, Glo92a, Glo92b]), while in this chapter we shall use a L^2-projection method, closely related to the one used in, e.g., [Tur96] (see also the references therein).*

9.3   FORMULATION OF A MODEL PROBLEM

With the geometrical situation as in Figure 1 (with $d = 2, 3$) with $\omega = \omega(t)$ a *moving rigid body* we consider for $t \geq 0$ the solution of the following system of Navier-Stokes equations

$$\frac{\partial \mathbf{u}}{\partial t} - \nu \Delta \mathbf{u} + (\mathbf{u} \cdot \nabla) \mathbf{u} + \nabla p \; = \; \mathbf{f} \quad \text{in} \quad \Omega \setminus \overline{\omega(t)}, \tag{9.7}$$

$$\nabla \cdot \mathbf{u} \; = \; 0 \quad \text{in} \quad \Omega \setminus \overline{\omega(t)}, \tag{9.8}$$

$$\mathbf{u}(\mathbf{x}, 0) \; = \; \mathbf{u}_0(\mathbf{x}), \quad \mathbf{x} \in \Omega \setminus \overline{\omega(0)}, \; (\text{with } \nabla \cdot \mathbf{u}_0 = 0), \tag{9.9}$$

$$\mathbf{u} \; = \; \mathbf{g}_0 \quad \text{on} \quad \Gamma, \tag{9.10}$$

$$\mathbf{u} \; = \; \mathbf{g}_1 \quad \text{on} \quad \gamma(t). \tag{9.11}$$

In (9.7)-(9.11), $\mathbf{u}$ and p denote as usual *velocity* and *pressure*, respectively; $\nu (> 0)$ is a *viscosity* coefficient, $\mathbf{f}$ a density of *external forces*, $\mathbf{x}$ the *generic point* of $\mathbb{R}^d$ ($\mathbf{x} = \{x_i\}_{i=1}^d$), $\gamma(t) = \partial \omega(t)$ and

$$(\mathbf{u} \cdot \nabla) \mathbf{u} = \left\{ \sum_{j=1}^{j=d} u_j \frac{\partial u_i}{\partial x_j} \right\}_{i=1}^{i=d}$$

We suppose that

$$\int_\Gamma \mathbf{g}_0 \cdot \mathbf{n} \, d\Gamma = 0, \quad \int_{\gamma(t)} \mathbf{g}_1 \cdot \mathbf{n} \, d\gamma = 0, \tag{9.12}$$

where, in (9.12), $\mathbf{n}$ is the outer normal unit vector at $\partial(\Omega \setminus \overline{\omega(t)})$; if $\mathbf{g}_1$ is the velocity associated to a rigid body motion the second condition in (12) is automatically satisfied. In the following, we shall use, if necessary, the notation $\phi(t)$ for the function

$$\mathbf{x} \rightarrow \phi(\mathbf{x}, t).$$

The existence of solution for problem (9.7)-(9.11) is a nontrivial mathematical issue *when ω is moving*; we shall not address it in this article (see however [Ami93] and the references therein).

9.4   A LAGRANGE MULTIPLIER/FICTITIOUS DOMAIN VARIATIONAL FORMULATION OF PROBLEM (9.7)-(9.11)

We introduce first the following functional spaces

$$\mathbf{V}_{\mathbf{g}_0(t)} \; = \; \{\mathbf{v} | \; \mathbf{v} \in (H^1(\Omega))^d, \mathbf{v} = \mathbf{g}_0(t) \quad \text{on} \quad \Gamma\}, \tag{9.13}$$

$$\mathbf{V}_0 \; = \; (H_0^1(\Omega))^d, \tag{9.14}$$

$$L_0^2(\Omega) = \left\{ q \mid q \in L^2(\Omega); \int_\Omega q \, dx = 0 \right\}, \tag{9.15}$$

$$\Lambda(t) = (\mathbf{H}^{-1/2}(\gamma(t)))^d. \tag{9.16}$$

We consider next $\mathbf{U}_0$ (resp., $\tilde{\mathbf{f}}$) such that

$$\nabla \cdot \mathbf{U}_0 = 0, \quad \mathbf{U}_0|_{\Omega \backslash \overline{\omega(0)}} = \mathbf{u}_0, \tag{9.17}$$

(resp., $\tilde{\mathbf{f}}|_{\Omega \backslash \overline{\omega}} = \mathbf{f}$).

It can be shown - at least formally - that problem (9.7)-(9.11) is *equivalent* to:

For $t \geq 0$, find $\{\mathbf{U}(t), P(t), \lambda(t)\} \in \mathbf{V}_{\mathbf{g}_0(t)} \times L_0^2(\Omega) \times \Lambda(t)$ such that

$$\int_\Omega \frac{\partial \mathbf{U}}{\partial t} \cdot \mathbf{v} \, dx + \nu \int_\Omega \nabla \mathbf{U} \cdot \nabla \mathbf{v} \, dx + \int_\Omega (\mathbf{U} \cdot \nabla) \mathbf{U} \cdot \mathbf{v} \, dx - \int_\Omega P \nabla \cdot \mathbf{v} \, dx$$

$$= \int_\Omega \tilde{\mathbf{f}} \cdot \mathbf{v} \, dx + \int_{\gamma(t)} \lambda \cdot \mathbf{v} \, d\gamma, \quad \forall \mathbf{v} \in \mathbf{V}_0, \tag{9.18}$$

$$\nabla \cdot \mathbf{U}(t) = 0, \tag{9.19}$$

$$\mathbf{U}(0) = \mathbf{U}_0, \tag{9.20}$$

$$\mathbf{U}(t) = \mathbf{g}_1(t) \quad \text{on} \quad \gamma(t), \tag{9.21}$$

in the sense that

$$\mathbf{U}(t)|_{\Omega \backslash \overline{\omega}} = \mathbf{u}, \quad P|_{\Omega \backslash \overline{\omega}} = p; \tag{9.22}$$

it is very easily shown that

$$\lambda = \left[\nu \frac{\partial \mathbf{U}}{\partial \mathbf{n}} - \mathbf{n} P \right]_\gamma. \tag{9.23}$$

where $[\quad]_\gamma$ denotes the jump at γ.

Remark 9.6 *For $\tilde{\mathbf{f}}$ we can take an L^2-extension of $\mathbf{f}$ (by 0 for example).*

Remark 9.7 *We observe that the actual geometry, i.e. $\omega(t)$ and $\gamma(t)$, occurs "only" through the $\gamma(t)$-integral in (9.18) and in (9.21).*

9.5 FINITE ELEMENT APPROXIMATION OF PROBLEM (9.18)-(9.21)

We suppose that $\Omega \subset \mathbb{R}^2$ ($d = 2$). With h a *space discretization step* we introduce a finite element triangulation $\mathcal{T}_h$ of $\overline{\Omega}$ and then $\mathcal{T}_{2h}$ a triangulation twice coarser (in practice we should construct $\mathcal{T}_{2h}$ first and then $\mathcal{T}_h$ by joining the midpoints of the edges of $\mathcal{T}_{2h}$, dividing thus each triangle of $\mathcal{T}_{2h}$ into 4 similar subtriangles). We define then the following finite dimensional spaces which approximate $\mathbf{V}_{\mathbf{g}_0(t)}$, $\mathbf{V}_0$, $L^2(\Omega)$, $L_0^2(\Omega)$, respectively:

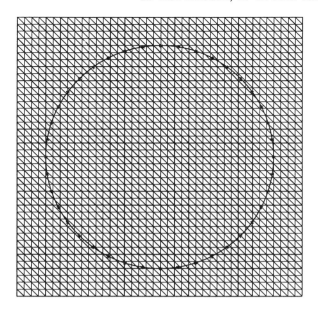

Figure 2 Part of the triangulation of Ω with mesh points
indicated by "$*$" on the disk boundary

$$\mathbf{V}_{g_{0h}} = \{\mathbf{v}_h| \quad \mathbf{v}_h \in C^0(\bar{\Omega})^2, \quad \mathbf{v}_h|_T \in P_1 \times P_1, \quad \forall T \in \mathcal{T}_h, \quad \mathbf{v}|_\Gamma = g_{0h}\}, \quad (9.24)$$
$$\mathbf{V}_{0h} = \{\mathbf{v}_h| \quad \mathbf{v}_h \in C^0(\bar{\Omega})^2, \quad \mathbf{v}_h|_T \in P_1 \times P_1, \quad \forall T \in \mathcal{T}_h, \quad \mathbf{v}|_\Gamma = 0\}, \quad (9.25)$$
$$L_h^2 = \{q_h| \quad q_h \in C^0(\bar{\Omega}), \quad q_h|_T \in P_1, \forall T \in \mathcal{T}_{2h}\}, \quad (9.26)$$
$$L_{0h}^2 = \left\{q_h| \quad q_h \in L_h^2, \quad \int_\Omega q_h\, d\mathbf{x} = 0\right\}; \quad (9.27)$$

in (9.24), g_{0h} is an approximation of g_0 satisfying $\int_\Gamma g_{0h} \cdot \mathbf{n}\, d\Gamma = 0$.
Concerning the space $\Lambda_h(t)$ approximating $\Lambda(t)$, we define it by

$$\Lambda_h(t) = \{\mu_h| \quad \mu_h \in (L^\infty(\gamma(t)))^2, \mu_h \quad \text{is constant on the arc joining}$$
$$\text{2 consecutive mesh points on } \gamma(t) \}. \quad (9.28)$$

A particular choice for the mesh points on γ is visualized on Figure 2 where ω is a disk.

Let us mention that the mesh points on γ *do not have to be at the intersection of* γ with the triangle edges; as shown in [Gir95] one still has convergence properties (for elliptic problems at least) if $h_\gamma \geq Ch_\Omega$ with C of the order of 3 (numerical experiments suggest that $C = 2\sqrt{2}$). This kind of decoupling between the Ω and γ meshes makes the fictitious domain approach very attractive for problems with moving boundaries, like those addressed in this article.

With the above spaces it is quite natural to approximate problem (9.18)-(9.21) (with obvious notation) by:

$$\int_\Omega \frac{\partial \mathbf{U}_h}{\partial t} \cdot \mathbf{v}\, dx + \nu \int_\Omega \nabla \mathbf{U}_h \cdot \nabla \mathbf{v}\, dx + \int_\Omega (\mathbf{U}_h \cdot \nabla)\mathbf{U}_h \cdot \mathbf{v}\, dx - \int_\Omega P_h \nabla \cdot \mathbf{v}\, dx$$

$$= \int_\Omega \tilde{\mathbf{f}}_h \cdot \mathbf{v}\, dx + \int_{\gamma(t)} \lambda_h \cdot \mathbf{v}\, d\gamma, \quad \forall \mathbf{v} \in \mathbf{V}_{0h}, (9.29)$$

$$\int_\Omega q \nabla \cdot \mathbf{U}_h(t)\, dx = 0, \quad \forall q \in L_h^2, \tag{9.30}$$

$$\mathbf{U}_h(0) = \mathbf{U}_{0h}, \tag{9.31}$$

$$\int_{\gamma(t)} (\mathbf{U}_h(t) - \mathbf{g}_1(t)) \cdot \mu\, d\gamma = 0, \quad \forall \mu \in \Lambda_h(t), \tag{9.32}$$

$$\{\mathbf{U}_h(t), P_h(t), \lambda_h(t)\} \in \mathbf{V}_{\mathbf{g}_0(t)h} \times L_{0h}^2 \times \Lambda_h(t); \tag{9.33}$$

in (9.31), $\mathbf{U}_{0h}$ is an approximation of $\mathbf{U}_0$ so that

$$\int_\Omega q \nabla \cdot \mathbf{U}_{0h}\, dx = 0, \quad \forall q \in L_h^2.$$

The time discretization of problem (9.29)-(9.33) by operator splitting methods will be discussed in Section 9.5.

9.6   TIME DISCRETIZATION OF PROBLEM (9.29)-(9.33) BY OPERATOR SPLITTING METHODS

Most "modern" Navier-Stokes solvers are based on operator splitting algorithms in order to force the incompressibility condition via a Stokes solver or an L^2- projection method (see [Glo92b], [Tur96] for details). This approach still applies to the initial value problem (9.29)-(9.33). Indeed this problem contains three numerical difficulties with each of which can be associated a specific operator, namely

1. The incompressibility condition and the related unknown pressure.
2. An advection-diffusion term.
3. The boundary condition on $\gamma(t)$ and the related multiplier $\lambda(t)$.

The operators in (9.1) and (9.3) are essentially *projection operators*.

   From an abstract point of view problem (9.29)-(9.33) is a particular case of the following class of initial value problems

$$\begin{cases} \dfrac{d\phi}{dt} + A_1(\phi) + A_2(\phi) + A_3(\phi) = f, \\[2mm] \phi(0) = \phi_0, \end{cases} \tag{9.34}$$

where operators A_i can be *multivalued*. Among the many operator splitting schemes which can be employed to solve (9.34) we shall advocate the very simple one below (analyzed in, e.g., [Mar90]); it is only first order accurate, but its low order accuracy is compensated by good stability and robustness properties.

A fractional step scheme à la Marchuk-Yanenko:

In the following, Δt is a time discretization step.

$$\phi^0 = \phi_0, \tag{9.35}$$

for $n \geq 0$, we obtain ϕ^{n+1} from ϕ^n via the solution of

$$\frac{\phi^{n+1/3} - \phi^n}{\Delta t} + A_1(\phi^{n+1/3}) = f_1^{n+1}, \tag{9.36}$$

$$\frac{\phi^{n+2/3} - \phi^{n+1/3}}{\Delta t} + A_2(\phi^{n+2/3}) = f_2^{n+1}, \tag{9.37}$$

$$\frac{\phi^{n+1} - \phi^{n+2/3}}{\Delta t} + A_3(\phi^{n+1}) = f_3^{n+1}, \tag{9.38}$$

with $\sum_{i=1}^{3} f_i^{n+1} = f^{n+1}$.

Applying the above scheme to problem (9.29)-(9.33) we obtain (with $0 \leq \alpha, \beta \leq 1$, $\alpha + \beta = 1$ and after dropping some of the subscripts h):

$$\mathbf{U}^0 = \mathbf{U}_{0h}; \tag{9.39}$$

for $n \geq 0$, we compute $\{\mathbf{U}^{n+1/3}, P^{n+1/3}\}$, $\mathbf{U}^{n+2/3}$, $\{\mathbf{U}^{n+1}, \lambda^{n+1}\}$ via the solution of

$$\begin{cases} \int_\Omega \frac{\mathbf{U}^{n+1/3} - \mathbf{U}^n}{\Delta t} \cdot \mathbf{v}\, d\mathbf{x} - \int_\Omega P^{n+1/3} \nabla \cdot \mathbf{v}\, d\mathbf{x} = 0, \quad \forall \mathbf{v} \in \mathbf{V}_{0h}, \\[4mm] \int_\Omega q\nabla \cdot \mathbf{U}^{n+1/3}\, d\mathbf{x} = 0, \quad \forall q \in L_h^2; \quad \mathbf{U}^{n+1/3} \in \mathbf{V}_{g0h}^{n+1}, \; P^{n+1/3} \in L_{0h}^2, \end{cases} \tag{9.40}$$

$$\begin{cases} \int_\Omega \frac{\mathbf{U}^{n+2/3} - \mathbf{U}^{n+1/3}}{\Delta t} \cdot \mathbf{v}\, d\mathbf{x} + \alpha\nu \int_\Omega \nabla \mathbf{U}^{n+2/3} \cdot \nabla \mathbf{v}\, d\mathbf{x} \\[4mm] + \int_\Omega (\mathbf{U}^{n+1/3} \cdot \nabla)\mathbf{U}^{n+2/3} \cdot \mathbf{v}\, d\mathbf{x} = \alpha \int_\Omega \tilde{\mathbf{f}}^{n+1} \cdot \mathbf{v}\, d\mathbf{x}, \quad \forall \mathbf{v} \in \mathbf{V}_{0h}; \\[4mm] \mathbf{U}^{n+2/3} \in \mathbf{V}_{g0h}^{n+1}, \end{cases} \tag{9.41}$$

$$
\begin{cases}
\displaystyle\int_{\Omega} \frac{\mathbf{U}^{n+1} - \mathbf{U}^{n+2/3}}{\Delta t} \cdot \mathbf{v}\, dx + \beta\nu \int_{\Omega} \nabla \mathbf{U}^{n+1} \cdot \nabla \mathbf{v}\, dx \\[2mm]
\displaystyle = \beta \int_{\Omega} \tilde{\mathbf{f}}^{n+1} \cdot \mathbf{v}\, dx + \int_{\gamma^{n+1}} \lambda^{n+1} \cdot \mathbf{v}\, d\gamma, \quad \forall \mathbf{v} \in \mathbf{V}_{0h}, \\[2mm]
\displaystyle \int_{\gamma^{n+1}} (\mathbf{U}^{n+1} - \mathbf{g}_{1h}^{n+1}) \cdot \mu\, d\gamma = 0, \quad \forall \mu \in \Lambda_h^{n+1}; \\[2mm]
\mathbf{U}^{n+1} \in \mathbf{V}_{g0h}^{n+1}, \lambda^{n+1} \in \Lambda_h^{n+1}.
\end{cases}
\tag{9.42}
$$

In (9.40)-(9.42) we have used the following notation

$$
\mathbf{V}_{g0h}^{n+1} = \mathbf{V}_{g0((n+1)\Delta t)h}, \quad \Lambda_h^{n+1} = \Lambda_h((n+1)\Delta t).
$$

Remark 9.8 *Many other splitting schemes are possible, some more complicated (and accurate) than the one above; on the other hand, scheme (9.35)-(9.38) is the simplest splitting scheme for those situations involving three operators and the results obtained with it compare favorablely with those obtained by more sophisticated schemes (for the particular problem considered here, at least).*

9.7 SOLUTION OF THE SUBPROBLEMS (9.40), (9.41) AND (9.42)

9.7.1 *SOLUTION OF THE SUBPROBLEM (9.40): L^2-PROJECTION ON $\mathbf{V}_{g0h}$*

The subproblems (9.40) can be viewed as *degenerated (zero viscosity) quasi-Stokes problems* of the following form (some h and n have been dropped):

$$
\alpha \int_{\Omega} \mathbf{U} \cdot \mathbf{v}\, dx - \int_{\Omega} P\nabla \cdot \mathbf{v}\, dx = \int_{\Omega} \mathbf{f} \cdot \mathbf{v}\, dx, \quad \forall \mathbf{v} \in \mathbf{V}_{0h}, \tag{9.43}
$$

$$
\int_{\Omega} q\nabla \cdot \mathbf{U}\, dx = 0, \quad \forall q \in L_h^2, \tag{9.44}
$$

with $\{\mathbf{U}, P\} \in \mathbf{V}_{g0h} \times L_{0h}^2$ (and $\alpha = 1/\Delta t$).

It is very easy to interpret $\mathbf{U}$ in (9.43), (9.44); it is the L^2-projection of $\mathbf{f}/\alpha$ on the subspace of $\mathbf{V}_{g0h}$ consisting of those functions satisfying

$$
\int_{\Omega} q\nabla \cdot \mathbf{v}\, dx = 0, \quad \forall q \in L_h^2. \tag{9.45}
$$

The pressure P is the Lagrange multiplier associated to the linear constrains (9.45); P is *nonunique* unless we specify an additional relation, like - for example - $\int_{\Omega} P\, dx = 0$,

i.e. $P \in L_{0h}^2$.

The *saddle-point problem* (9.43), (9.44) can be solved by an *Uzawa/ Preconditioned Conjugate Gradient algorithm* operating in the space L^2_{0h}. This algorithm is as follows:

Step 0: Initialization

$$P^0 \in L^2_{0h} \quad \text{is given;} \tag{9.46}$$

solve the projection problem:

$$\begin{cases} \alpha \int_\Omega \mathbf{U}^0 \cdot \mathbf{v} \, d\mathbf{x} = \int_\Omega \mathbf{f} \cdot \mathbf{v} \, d\mathbf{x} + \int_\Omega P^0 \nabla \cdot \mathbf{v} \, d\mathbf{x} \quad \forall \mathbf{v} \in \mathbf{V}_{0h}; \\ \mathbf{U}^0 \in \mathbf{V}_{g_{0h}}, \end{cases} \tag{9.47}$$

then

$$\begin{cases} \int_\Omega r^0 q \, d\mathbf{x} = \int_\Omega q \nabla \cdot \mathbf{U}^0 \, d\mathbf{x}, \quad \forall q \in L^2_{0h}, \\ r^0 \in L^2_{0h}, \end{cases} \tag{9.48}$$

and finally

$$\begin{cases} \int_\Omega \nabla g^0 \cdot \nabla q \, d\mathbf{x} = \int_\Omega r^0 q \, d\mathbf{x}, \quad \forall q \in L^2_h, \\ g^0 \in L^2_{0h}. \end{cases} \tag{9.49}$$

Take

$$w^0 = g^0. \tag{9.50}$$

Then for $k \geq 0$, assuming that P^k, r^k, g^k, w^k are known, compute P^{k+1}, r^{k+1}, g^{k+1}, w^{k+1} as follows:

Step 1: Descent

Solve:

$$\begin{cases} \alpha \int_\Omega \overline{\mathbf{U}}^k \cdot \mathbf{v} \, d\mathbf{x} = \int_\Omega w^k \nabla \cdot \mathbf{v} \, d\mathbf{x}, \quad \forall \mathbf{v} \in \mathbf{V}_{0h}; \\ \overline{\mathbf{U}}^k \in \mathbf{V}_{0h}, \end{cases} \tag{9.51}$$

then

$$\begin{cases} \int_\Omega \bar{r}^k q \, d\mathbf{x} = \int_\Omega q \nabla \cdot \overline{\mathbf{U}}^k \, d\mathbf{x}, \quad \forall q \in L^2_{0h}, \\ \bar{r}^0 \in L^2_{0h}, \end{cases} \tag{9.52}$$

and finally

$$\begin{cases} \int_\Omega \nabla \bar{g}^k \cdot \nabla q \, d\mathbf{x} = \int_\Omega \bar{r}^k q \, d\mathbf{x}, \quad \forall q \in L_h^2, \\ \bar{g}^k \in L_{0h}^2. \end{cases} \tag{9.53}$$

Compute

$$\rho_k = \int_\Omega r^k g^k \, d\mathbf{x} \Big/ \int_\Omega \bar{r}^k w^k \, d\mathbf{x}, \tag{9.54}$$

and then

$$\begin{aligned} P^{k+1} &= P^k - \rho_k w^k, & (9.55) \\ g^{k+1} &= g^k - \rho_k \bar{g}^k, & (9.56) \\ r^{k+1} &= r^k - \rho_k \bar{r}^k. & (9.57) \end{aligned}$$

Step 2: Convergence Testing and Construction of the New Decent Direction

    If

$$\int_\Omega r^{k+1} g^{k+1} \, d\mathbf{x} \Big/ \int_\Omega r^0 g^0 \, d\mathbf{x} \le \epsilon, \tag{9.58}$$

take $P = P^{k+1}$ and compute $\mathbf{U}$ from relation (9.43). Else, compute

$$\gamma_k = \int_\Omega r^{k+1} g^{k+1} \, d\mathbf{x} \Big/ \int_\Omega r^k g^k \, d\mathbf{x} \tag{9.59}$$

and set

$$w^{k+1} = g^k + \gamma_k w^k. \tag{9.60}$$

Do $k = k + 1$ and go back to (9.51).
The choice of ϵ in the stopping test (9.58) will be discussed in Section 9.7.

Remark 9.9 *Use of the trapezoidal rule to evaluate the various $L^2(\Omega)$-integrals in (9.40)-(9.42) and in the algorithm (9.46)-(9.60) makes the implementation of the above algorithm very easy and economical.*

9.7.2 *SOLUTION OF THE SUBPROBLEMS (9.41)*

The solution by least squares/preconditioned conjugate gradient methods of linear or nonlinear advection-diffusion problems such as (9.41) has been discussed at length in

[Bri87], [Glo91], [Glo92b] (see also [Glo84]). Due to page limitation we shall skip it in the present article. Let us mention that iterative methods like GMRES can also be used to solve problems such as (9.40).

9.7.3 SOLUTION OF THE SUBPROBLEMS (9.42): FORCING THE DIRICHLET CONDITIONS ON γ

If $\beta > 0$, problem (9.42) is indeed a *saddle-point problem* whose solution has been discussed in [Glo94a], [Glo94b]. It can be solved by an Uzawa/conjugate gradient algorithm operating in the space Λ_h^{n+1}. For *two-dimensional* problems an efficient preconditioning operator is provided by a discrete version of the boundary operator $(\frac{\Delta t}{\beta\nu}I - \frac{\partial^2}{\partial s^2})^{-1/2}$, where s is the arc-length along γ (see [Glo94a] for details and computational experiments).

In the particular case where $\beta = 0$, problem (9.42) reduces to an $L^2(\Omega)$-projection over the subspace of $\mathbf{V}_{g_{0h}}^{n+1}$ of the functions satisfying the condition

$$\int_{\gamma^{n+1}} (\mathbf{v} - \mathbf{g}_{1h}^{n+1}) \cdot \mu \, d\gamma = 0, \quad \forall\mu \in \Lambda_h^{n+1}.$$

It follows from the above observation that if $\beta = 0$, problem (9.42) can be solved by an Uzawa/conjugate gradient algorithm operating in the space Λ_h^{n+1}, which has many similarities with algorithm (9.46)-(9.60). If one uses the trapezoidal rule to compute the various $L^2(\Omega)$-integrals in (9.42), taking $\beta = 0$ brings further simplification since in that particular case $\mathbf{U}^{n+1}$ will coincide with $\mathbf{U}^{n+2/3}$ at those vertices of $\mathcal{T}_h$ such that the support of the related shape function does not intersect γ^{n+1}; from the above observation it follows that to obtain $\mathbf{U}^{n+1}$ and λ^{n+1} we have to solve a linear system of the following form

$$\begin{cases} A\mathbf{x} + B^t\mathbf{y} & = \mathbf{b}, \\ B\mathbf{x} & = \mathbf{c}, \end{cases} \tag{9.61}$$

where A is an $N \times N$ matrix, symmetric and positive definite and where B is an $N \times M$ matrix; we have M and N both of the order of $1/h$. The saddle-point problem (9.61) can be solved also by an Uzawa/conjugate gradient algorithm operating in $\mathbb{R}^M$ (other methods are possible).

9.8 NUMERICAL EXPERIMENTS

For the test problem that we consider, we shall simulate a two-dimensional flow with $\Omega = (-0.35, 0.9) \times (-0.5, 0.5)$ (see Figure 3) and ω a moving disk of radius 0.125. The center of the disk is moving between $(0,0)$ and $(0.5,0)$ along a prescribed trajectory $(x(t), y(t))$ (see Figure 3) given as follows

$$x(t) = 0.25(1 - \cos(\frac{\pi t}{2})), \quad y(t) = -0.1\sin(\pi(1 - \cos(\frac{\pi t}{2})));$$

we have thus a *periodic* motion of period 4. Several different positions of the disk have

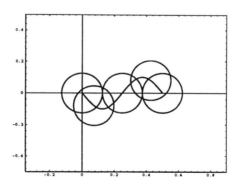

Figure 3

been shown on Figure 3. The boundary conditions are $\mathbf{u} = 0$ on Γ and $\mathbf{u}$ on $\partial\omega(t)$ coinciding with the disk velocity.

We suppose that the disk rotates counterclockwise at angular velocity π. Since we are taking $\nu = 10^{-2}$ the maximum Reynolds number based on the disk diameter as characteristic length is 25.584. On Ω we have used a regular triangulation $\mathcal{T}_h$ to approximate the velocity, like the one in Figure 2, the pressure grid $\mathcal{T}_{2h}$ being twice coarser. Concerning $\Lambda_h(t)$, $\gamma(t)$ has been divided into M subarcs of equal length.

We have run two series of tests: For the first series we have taken $h = 1/128$, $\Delta t = 0.0025$ and $M = 80$. For the second we have taken $h = 1/256$, $\Delta t = 0.00125$ and $M = 160$. With stopping criteria of the order of 10^{-12} we need around 10 iterations at most to have convergence of the conjugate gradient algorithms used to solve the problems at each step of scheme (9.39)-(9.42).

In Figures 4 and 5 (resp., 6 and 7) we show the isobar contours, the vorticity density and the streamlines obtained at $t = 4.5, 5, 5.5, 6, 6.5, 7, 7.5, 8$ for $h = 1/128$, $\Delta t = 0.0025$ and $M = 80$ (resp., $h = 1/256$, $\Delta t = 0.00125$ and $M = 160$). There is a good agreement between those results, concerning particularly streamlines and vorticity density. In order to improve pressure convergence we intend to consider more sophisticated methods with a stronger coupling between the steps of scheme (9.39)-(9.42). The corresponding results will be reported in a forthcoming publication.

9.9   ACKNOWLEDGEMENTS

We would like to acknowledge the helpful comments and suggestions of E. J. Dean, J.W. He, D.D. Joseph, Y. Kuznetsov, B. Maury, A.H. Sameh and F.J. Sanchez.

The support of the following institutions is acknowledged: University of Houston and Department of Computer Science, University of Minnesota. We also benefited from the support of NSF (Grants DMS 8822522, DMS 9112847, DMS 9217374), DRET (Grant 89424), DARPA (Contracts AFOSR F49620-89-C-0125 and AFOSR-90-0334), the Texas Board of Higher Education (Grants 003652156ARP and 003652146ATP), University of Houston (PEER grant 1-27682) and again NSF under the HPCC Grand Challenge Grant ECS-952.

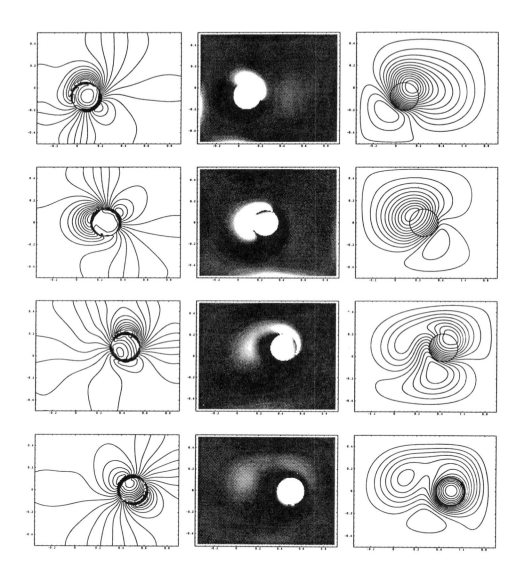

Figure 4 Isobar contours (at left), vorticity density (at middle) and streamlines (at right) at time $t = 4.5, 5, 5.5, 6$ during first half of one period of the disk motion. The disk is moving from the left to the right. The mesh size for velocity is $h = 1/128$ and the mesh size for pressure is $h = 1/64$.

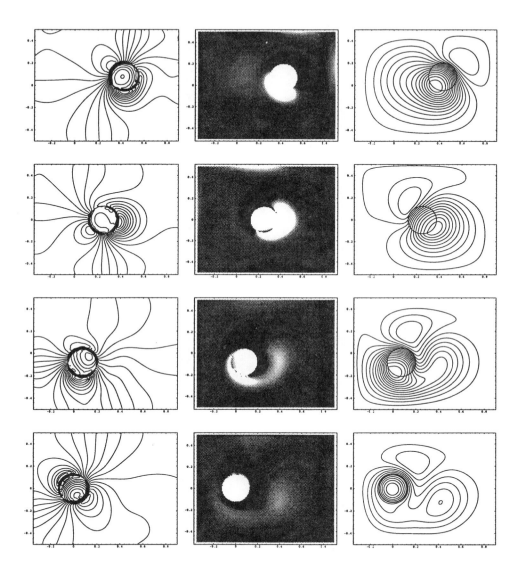

Figure 5   Isobar contours (at left), vorticity density (at middle) and streamlines (at right) at time $t = 6.5, 7, 7.5, 8$ during second half of one period of the disk motion. The disk is moving from the right to the left. The mesh size for velocity is $h = 1/128$ and the mesh size for pressure is $h = 1/64$.

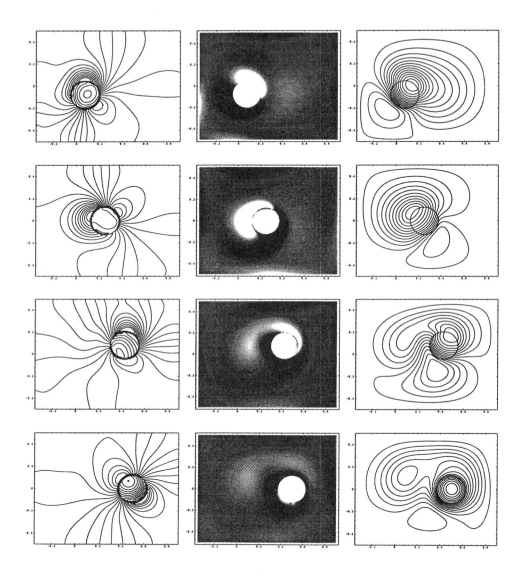

Figure 6 Isobar contours (at left), vorticity density (at middle) and streamlines (at right) at time $t = 4.5, 5, 5.5, 6$ during first half of one period of the disk motion. The disk is moving from the left to the right. The mesh size for velocity is $h = 1/256$ and the mesh size for pressure is $h = 1/128$.

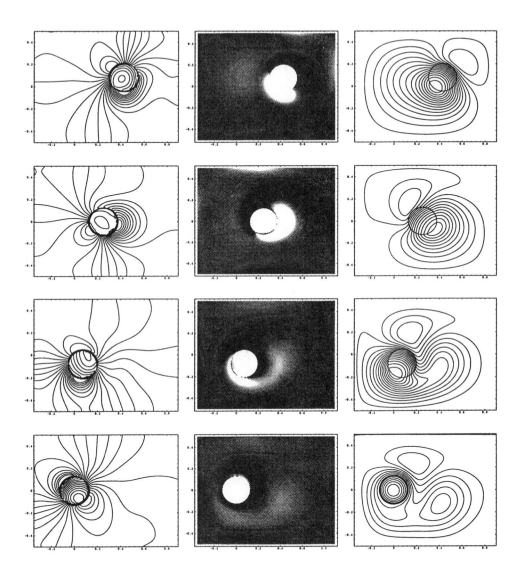

Figure 7 Isobar contours (at left), vorticity density (at middle) and streamlines (at right) at time $t = 6.5, 7, 7.5, 8$ during second half of one period of the disk motion. The disk is moving from the right to the left. The mesh size for velocity is $h = 1/256$ and the mesh size for pressure is $h = 1/128$.

References

[Ami93] Amiez G. and Gremaud P. A. (1993) On a penalty method for the Navier-Stokes problem in regions with moving boundaries. *Comp. Appl. Math.* 12: 113–122.

[Ast78] Astrakhantsev G. P. (1978) Fictitious domain methods for second order elliptic equations with natural boundary conditions. *Zh. Vychisl. Mat. Mat. Fiz.* 18: 118–125.

[Bri87] Bristeau M. O. and Glowinski R. and and Periaux J. (1987) Numerical methods for the Navier-Stokes equations. Applications to the simulation of compressible and incompressible viscous flow. *Comp. Phys. Rep.* 6: 73–187.

[Buz71] Buzbee B. L. and Dorr F. W. and George J. A. and Golub G. H. (1971) The direct solution of the discrete Poisson equation on irregular regions. *SIAM J. Num. Anal.* 8: 722–736.

[Cah88] Cahouet J. and Chabard J. P. (1988) Some fast 3-D finite element solvers for the generalized Stokes problem. *Int. J. Num. Methods in Fluids* 8: 869–895.

[Col97] Collino F. and Joly P. and and Millot F. (1996/1997) Fictitious domain method for unsteady problems: application to Electro-Magnetic scattering. to appear.

[Gir95] Girault V. and Glowinski R. (1995) Error analysis of a fictitious domain method applied to a Dirichlet problem. *Japan J. Ind. Appl. Math.* 12: 487–514.

[Glo84] Glowinski R. (1984) *Numerical methods for nonlinear variational problems.* Springer-Verlag, New York.

[Glo89] Glowinski R. and Le Tallec P. (1989) *Augmented Lagrangian and operator splitting methods in nonlinear mechanics.* SIAM, Philadelphia.

[Glo90] Glowinski R. and Periaux J. and Ravachol M. and Pan T. W. and Wells R. O. and Zhou X. (1990) Wavelet methods in computational fluid dynamics. In Y. H. M. (ed) *Algorithmic Trends in Computational Fluid Dynamics*, pages 259–276. Springer-Verlag, N. Y.

[Glo91] Glowinski R. (1991) Finite element methods for the numerical simulation of incompressible viscous flow. Introduction to the control of the Navier-Stokes equations. In *et. al.* A. C. R. (ed) *Vortex Dynamics and Vortex Methods*, volume 28 of *Lectures in Applied Mathematics*, pages 219–301. AMS, Providence, R.I.

[Glo92a] Glowinski R. (1992) Ensuring well-posedness by analogy; Stokes problem and boundary control for the wave equation. *J. Comp. Phys.* 103: 189–221.

[Glo92b] Glowinski R. and Pironneau O. (1992) Finite element methods for Navier-Stokes equations. *Annu. Rev. Fluid Mech.* 24: 167–204.

[Glo94a] Glowinski R. and Pan T. W. and and Periaux J. (1994) A fictitious domain method for Dirichlet problems and applications. *Comp. Meth. Appl. Mech. Eng.* 111: 283–303.

[Glo94b] Glowinski R. and Pan T. W. and and Periaux J. (1994) A fictitious domain method for external incompressible viscous flow modeled by Navier–Stokes equations. *Comp. Meth. Appl. Mech. Eng.* 112: 133–148.

[Glo95a] Glowinski R. and Kearsley A. J. and Pan T. W. and Periaux J. (1995) Numerical simulation and optimal shape for viscous flow by fictitious domain method. *Int. J. Num. Methods in Fluids* 20: 695–711.

[Glo95b] Glowinski R. and Pan T. W. and and Periaux J. (1995) A Lagrange multiplier/fictitious domain method for the Dirichlet problem. Generalization to some flow problems. *Japan J. Ind. Appl. Math.* 12: 87–108.

[Lev94] Leveque R. and Li Z. (1994) The immersed interface method for elliptic equations with discontinuous coefficients and singular sources. *SIAM J. Num. Anal.* 31: 1019–1044.

[Mar90] Marchuk G. I. (1990) Splitting and alternate direction methods. In *et al.* C. P. G. (ed) *Handbook of Numerical Analysis, Vol. I*, pages 197–462. North-Holland, Amsterdam.

[Mat93] Matsokin A. M. and Nepomnyaschikh S. V. (1993) The fictitious domain method and explicit continuation operators. *Zh. Vychisl. Mat. Mat. Fiz.* 33: 45–59.

[Nep95] Nepomnyaschikh S. V. (1995) Fictitious space method on unstructured meshes. *East-West J. Num. Math.* 3: 71–79.

[Pes89] Peskin C. Y. and McQueen D. M. (1989) A three–dimensional computational method for blood flow in the heart : I. Immersed elastic fibers in a viscous incompressible fluid. *J. Comp. Phys.* 81: 372–405.

[Tur96] Turek S. (1996) A comparative study of time-stepping techniques for the incompressible Navier-Stokes equations: from fully implicit non-linear schemes to semi-implicit projection methods. *Int. J. Num. Math. in Fluids* 22: 987–1011.

10

Theory and Applications of Numerical Schemes for Nonlinear Convection–Diffusion Problems and Compressible Navier–Stokes Equations

Miloslav Feistauer and Jiří Felcman

Institute of Numerical Mathematics, Faculty of Mathematics and Physics, Charles University Prague, Malostranské nám. 25, 118 00 Praha 1, Czech Republic

10.1   INTRODUCTION

Numerical simulation of viscous compressible high-speed flow is one of the most difficult areas of Computational Fluid Dynamics because of the facts that convection effects dominate those of diffusion, that shock waves and boundary layers are present and that there is interaction between shocks and boundary layers. In addition there is a significant lack of theory for problems of this type. Our goal is to develop a robust theoretically based numerical method for the solution of viscous compressible flow applied on unstructured meshes.

Since the viscosity and heat conduction coefficients of gases are small, the viscous dissipative terms are often considered as perturbations of the inviscid Euler system. This leads us to the conclusion that any efficient method for the solution of viscous gas flow should be based on a numerical method which is effective for inviscid flow. The most popular technique for the solution of inviscid compressible flow at present is the finite volume (FV) method (cf., e. g., [Fei93, Par. 7.3]).

Some authors ([FHKV95], [HHK92], [VH92]) carry out the FV discretization for the complete viscous system. However, the character of second order viscous terms also offers the possibility of discretization by the finite element (FE) method. In this way we arrive at a combined finite volume – finite element method.

The Mathematics of Finite Elements and Applications
Edited by J. R. Whiteman © 1997 John Wiley & Sons Ltd.

We have developed several combined FV – FE procedures for the numerical solution of nonlinear convection – diffusion problems and compressible viscous flow based on a general class of cell centred flux vector splitting FV schemes for the discretization of inviscid terms together with the discretization of viscous terms by the FE method over a triangular grid. Substantial attention has been paid to the theoretical analysis of the resulting combined FV – FE schemes applied to a simplified scalar nonlinear convection – diffusion conservation law equation. In the present paper we discuss the convergence and error estimates of the method and present some applications to the solution of viscous compressible transonic flow. The comparison of numerical results with experimental data shows the applicability and robustness of the method.

10.2   CONTINUOUS MODEL PROBLEM

We will denote by $\mathbb{R}^n$ the n-dimensional Euclidean space equipped with the norm $|\cdot|$. By x_1, x_2 and t we will denote the Cartesian coordinates of points $x \in \mathbb{R}^2$ and time, respectively. Let $\Omega \subset \mathbb{R}^2$ be a bounded polygonal domain. In the space-time cylinder $Q_T = \Omega \times (0, T)$ $(0 < T < \infty)$ we consider the following *initial-boundary value problem*: Find $u : \overline{Q_T} \to \mathbb{R}$, $u = u(x, t)$, $x \in \overline{\Omega}$, $t \in [0, T]$, such that

$$\frac{\partial u}{\partial t} + \sum_{s=1}^{2} \frac{\partial f_s(u)}{\partial x_s} - \nu \Delta u = g \quad \text{in } Q_T, \tag{10.1}$$

$$u|_{\partial\Omega \times (0,T)} = 0, \tag{10.2}$$

$$u(x, 0) = u^0(x), \quad x \in \Omega, \tag{10.3}$$

where $\nu > 0$ is a given constant and $f_s : \mathbb{R} \to \mathbb{R}$, $s = 1, 2$, $g : Q_T \to \mathbb{R}$, $u^0 : \Omega \to \mathbb{R}$ are given functions.

We assume that the following *assumptions on data* are satisfied: $f_s \in C^1(\mathbb{R})$, $g \in C([0, T]; W^{1,q}(\Omega))$ for some $q > 2$ and $u^0 \in W_0^{1,p}(\Omega)$ for some $p > 2$. (For the definition of these function spaces, see, e.g., [KJF77].)

Let us set $V = H_0^1(\Omega)$,

$$(u, v) \equiv \int_\Omega uv \, dx, \quad u, v \in L^2(\Omega), \tag{10.4}$$

$$a(u, v) \equiv \nu \int_\Omega \nabla u \cdot \nabla v \, dx, \quad u, v \in H^1(\Omega), \tag{10.5}$$

$$b(\varphi, v) \equiv -\int_\Omega \sum_{s=1}^{2} f_s(\varphi) \frac{\partial v}{\partial x_s} \, dx \quad \text{for } \varphi \in L^\infty(\Omega), \quad v \in V. \tag{10.6}$$

We say that a function u is a *weak solution* of problem (10.1)–(10.3), if it satisfies the following conditions

a) $u \in L^2(0, T; V) \cap L^\infty(Q_T)$,

b) $\dfrac{d}{dt}(u(t), v) + b(u(t), v) + a(u(t), v) = (g(t), v) \quad \forall v \in V,$ \hfill (10.7)

in the sense of distributions on $(0, T)$,

c) $u(0) = u^0$.

This problem has a unique solution.

Equation (10.1) is a strongly simplified prototype of equations describing viscous gas flow. It also appears in such other areas as hydrology and environmental protection. Convection-diffusion problems have been extensively investigated, mainly in a linear case. Let us mention the papers [Joh81], [AFRS89], [AR93], [Zho95], [ZR96], monographs [Ike83], [Mor96], [RST96] and the references therein.

In what follows we will be concerned with numerical methods for the solution of problem (10.1) – (10.3) based on the combination of the finite volume method with the finite element method. The numerical schemes so developed will be applied to solve problems of viscous gas flow.

10.3   DISCRETE PROBLEM

The discretization of problem (10.7) is carried out with the aid of a generally unstructured triangular mesh. By T_h we denote a triangulation of Ω with standard properties ([Cia79]). The triangulation T_h is called a *basic mesh*. Let $\mathcal{P}_h = \{P_i; i \in J\}$ be the set of all vertices of all $T \in T_h$. (J is a suitable index set.)

Now, over the basic mesh T_h we construct the *dual mesh* $\mathcal{D}_h = \{D_i; i \in J\}$ covering $\overline{\Omega}$. The *dual finite volume* D_i associated with a vertex $P_i \in \mathcal{P}_h$ is a closed polygon obtained in the following way (see Figure 1): We join the centre of gravity of every triangle $T \in T_h$ that contains the vertex P_i with the centre of every side of T containing P_i. If $P_i \in \mathcal{P}_h \cap \partial\Omega$, then we complete the contour so obtained with the additional straight segments joining P_i to the centres of boundary sides that contain P_i. In this way we get the boundary ∂D_i of the finite volume D_i.

Two different finite volumes D_i and D_j are called *neighbours*, if their boundaries contain a common straight segment. In this case we write

$$\Gamma_{ij} = \bigcup_{\alpha=1}^{\beta_{ij}} \Gamma_{ij}^\alpha = \partial D_i \cap \partial D_j = \Gamma_{ji}, \tag{10.8}$$

where $\Gamma_{ij}^\alpha = \Gamma_{ji}^\alpha$ are straight segments. We have $\beta_{ij} = 2$ for D_i or $D_j \subset \Omega$. If both D_i and D_j are adjacent to $\partial\Omega$, then $\beta_{ij} = 1$.

For $i \in J$, let $s(i) = \{j \in J; D_j$ is a neighbour of $D_i\}$. If $P_i \in \mathcal{P}_h \cap \partial\Omega$, then we denote by $\Gamma_{i,-1}^\alpha$, $\alpha = 1, 2 =: \beta_{i,-1}$, the segments that form $\partial D_i \cap \partial\Omega$. In this case we set $S(i) = s(i) \cup \{-1\}$, otherwise (for $P_i \in \mathcal{P}_h \cap \Omega$) we put $S(i) = s(i)$. Obviously, for every $D_i \in \mathcal{D}_h$ we have

$$\partial D_i = \bigcup_{j \in S(i)} \Gamma_{ij} = \bigcup_{j \in S(i)} \bigcup_{\alpha=1}^{\beta_{ij}} \Gamma_{ij}^\alpha. \tag{10.9}$$

(Dual finite volumes described above were applied to the numerical solution of inviscid gas flow in [Vij86], [ADF+89], [SS94], [FLWW95], [FFL95].)

Further, we introduce the following *notation*: $|D_i|$ = area of $D_i \in \mathcal{D}_h$, $|T|$ = area of $T \in T_h$, $\mathbf{n}_{ij}^\alpha = (n_{1ij}^\alpha, n_{2ij}^\alpha)$ = unit outward normal to ∂D_i on Γ_{ij}^α, ℓ_{ij}^α = length of Γ_{ij}^α, $|\partial D_i|$ = the length of ∂D_i. Moreover, we consider a partition $0 = t_0 < t_1 < \cdots$ of the time interval $[0, T]$ and set $\tau_k = t_{k+1} - t_k$ for $k = 0, 1, \ldots$.

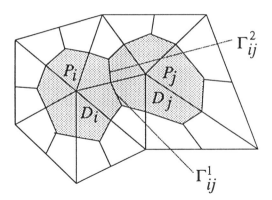

Figure 1   Dual finite volumes over a triangular mesh

Let us define the following spaces:

$$X_h \;=\; \{v_h \in C(\overline{\Omega}); v_h|_T \text{ is linear for each } T \in \mathcal{T}_h\} \subset H^1(\Omega) \qquad (10.10)$$

(the space of linear conforming finite elements),

$$V_h \;=\; \{v_h \in X_h; v_h = 0 \text{ on } \partial\Omega\}.$$

By r_h we denote the operator of the Lagrange X_h- interpolation. Further, by L_h we denote the so-called *lumping operator*:

$$v \in C(\overline{\Omega}) \to L_h v : L_h v|_{D_i} = v(P_i), \; i \in J. \qquad (10.11)$$

In order to derive the discrete form of problem (10.7), we put

$$(u, v)_h \;=\; \int_\Omega r_h(uv) \, dx, \qquad u, v \in C(\overline{\Omega}), \qquad (10.12)$$

and for $u, v \in V_h$ we write

$$b(u, v) \;=\; \int_\Omega \sum_{s=1}^{2} \frac{\partial f_s(u)}{\partial x_s} v \, dx \approx \int_\Omega \sum_{s=1}^{2} \frac{\partial f_s(u)}{\partial x_s} L_h v \, dx = \qquad (10.13)$$

$$= \sum_{i \in J} v(P_i) \int_{D_i} \sum_{s=1}^{2} \frac{\partial f_s(u)}{\partial x_s} \, dx =$$

$$= \sum_{i \in J} v(P_i) \int_{\partial D_i} \sum_{s=1}^{2} f_s(u) \, n_s \, dS \; =$$

$$= \sum_{i \in J} v(P_i) \sum_{j \in s(i)} \sum_{\alpha=1}^{\beta_{ij}} \int_{\Gamma_{ij}^\alpha} \sum_{s=1}^{2} f_s(u) \, n_s \, dS \; \approx$$

$$\approx \; b_h(u, v) = \sum_{i \in J} v(P_i) \sum_{j \in s(i)} \sum_{\alpha=1}^{\beta_{ij}} H\big(u(P_i), u(P_j), \mathbf{n}_{ij}^\alpha\big) \, \ell_{ij}^\alpha.$$

Here $\mathbf{n} = (n_1, n_2)$ denotes the unit outward normal to ∂D_i. The function H defined on $\mathbb{R}^2 \times \mathcal{S}$, where $\mathcal{S} = \{\mathbf{n} \in \mathbb{R}^2; |\mathbf{n}| = 1\}$, is called a *numerical flux*, and has the following properties:

a) $H = H(y, z, \mathbf{n})$ is *Lipschitz-continuous* with respect to y, $z \in [-M, M]$ with a constant $c(M)$.

b) H is *consistent*: $H(u, u, \mathbf{n}) = \sum_{s=1}^{2} f_s(u) n_s \quad \forall u \in \mathbb{R}, \forall \mathbf{n} = (n_1, n_2), |\mathbf{n}| = 1$.

c) H is *conservative*: $H(y, z, \mathbf{n}) = -H(z, y, -\mathbf{n}) \quad \forall y, z \in \mathbb{R}, \forall \mathbf{n}$ with $|\mathbf{n}| = 1$.

d) H is *monotone*, i.e., for a given fixed number $M > 0$ the function $H(y, z, \mathbf{n})$ is nonincreasing with respect to the second variable z on the set $\{(y, z, \mathbf{n}); y, z \in [-M, M], |\mathbf{n}| = 1\}$.

We consider the following *numerical schemes*: We set $u_h^0 = r_h u^0$, $g^k = g(\cdot, t_k)$. Provided u^k has beeen computed, we determine u^{k+1} such that either

$$\frac{1}{\tau_k}(u_h^{k+1} - u_h^k, v_h)_h + b_h(u_h^k, v_h) + a(u_h^{k+1}, v_h) = (g^{k+1}, v_h)_h \tag{10.14}$$

$$\forall v_h \in V_h, \ t_k \in [0, T),$$

(semiimplicit scheme) or

$$\frac{1}{\tau_k}(u_h^{k+1} - u_h^k, v_h)_h + b_h(u_h^k, v_h) + a(u_h^k, v_h) = (g^k, v_h)_h \forall v_h \in V_h, \ t_k \in [0, T) \tag{10.15}$$

(purely explicit scheme). The function u_h^k is the approximate solution at time t_k.

In view of [Hac89], the above schemes can be completely derived via the FV approach. We speak however about a combined FV–FE method, because the convergence analysis and the discretization of viscous flow are based on the FE methodology.

10.4   CONVERGENCE

Let us consider a system $\{\mathcal{T}_h\}_{h\in(0,h_0)}$ ($h_0 > 0$) of triangulations of the domain Ω, set $\tau = T/r$ for an integer $r > 1$ and define the partition of the interval $[0, T]$ formed by $t_k = k\tau$, $k = 0, \ldots, r$. Hence, $\tau_k = \tau$ for $k = 0, \ldots, r - 1$.

We denote by h_T and σ_T the length of the longest side of the triangle $T \in \mathcal{T}_h$ and the radius of the largest circle inscribed in T, respectively, and set $h = \max_{T \in \mathcal{T}_h} h_T$. We introduce the functions $u_{h\tau}$, $w_{h\tau} : (-\infty, +\infty) \to V_h$ associated with an approximate solution $\{u_h^k\}_{k=0}^r$:

$$u_{h\tau}(t) = u_h^0, \quad t \leq 0, \quad u_{h\tau}(t) = u_h^k, \quad t \in (t_{k-1}, t_k], \quad k = 1, \ldots, r, \tag{10.16}$$

$$u_{h\tau}(t) = u_h^r, \quad t \geq T;$$

$w_{h\tau}$ is a continuous, piecewise linear mapping of $[0, T]$ into V_h, $\quad$ (10.17)

$$w_{h\tau}(t_k) = u_h^k, \ k = 0, \ldots, r, \quad w_{h\tau}(t) = 0 \quad \text{for} \quad t < 0 \text{ or } t > T.$$

Our goal is to prove the convergence (in suitable spaces) of the mappings $u_{h\tau}$ and $w_{h\tau}$ (considered on $[0, T]$) to the exact solution u, if $h, \tau \to 0$ in an appropriate way. Here we summarize results from [FFL96] and [FS96].

By $c, c_1, \ldots, \hat{c}_1, \hat{c}_2, \ldots, c^*$ we denote constants independent of h, τ, ν, whilst $C, C_1, \ldots$ will denote constants that are independent of h, τ, but depend on ν.

Assumptions (A1).   a) The system $\{T_h\}_{h\in(0,h_0)}$ is *regular*, i.e., all angles of all $T \in T_h$ are bounded from below by a positive constant independent of T and h.
b) The triangulations T_h are of *weakly acute type*. This means that the magnitude of all angles of all $T \in T_h$, $h \in (0, h_0)$, is less than or equal to $\pi/2$.

In the convergence analysis we proceed in several steps.

I. L^∞-stability.   With the aid of the above assumptions and discrete maximum principle ([Ike83]) we find that the inequalities

$$\|u^0\|_{L^\infty(\Omega)} \leq \tilde{M}, \quad \|g\|_{L^\infty(Q_T)} \leq \tilde{K}. \tag{10.18}$$

imply that

$$\|u_{h\tau}\|_{L^\infty(Q_T)}, \quad \|w_{h\tau}\|_{L^\infty(Q_T)} \leq M = \tilde{M} + T\tilde{K}, \tag{10.19}$$

provided the following *stability conditions* are satisfied:

$$\tau c(M)|\partial D_i| \leq |D_i|, \; i \in J, \tag{10.20}$$

in the case of the semiimplicit scheme (hence, $\tau = O(h)$);

$$\frac{1}{3}|T| - \tau c(M)|\partial D_i \cap T| - \frac{1}{4}\tau\nu h_T/\sigma_T \geq 0, \tag{10.21}$$
$$T \in T_h, P_i \in \mathcal{P}_h \cap T,$$

in the case of the explicit scheme (hence, $\tau = O(h^2/(h + \nu))$).

II. Consistency   is represented by the estimates

a) $\hat{c}_1\|v\|_{L^2(\Omega)} \leq \|L_h v\|_{L^2(\Omega)} \leq \hat{c}_2\|v\|_{L^2(\Omega)}, \quad v \in X_h,$ \hfill (10.22)

b) $\|v - L_h v\|_{L^2(\Omega)} \leq c\,h\|v\|_{H^1(\Omega)}, \quad v \in X_h,$

c) $|(u, v) - (u, v)_h| \leq c\,h^2\|u\|_{H^1(\Omega)}\|v\|_{H^1(\Omega)}, \quad u, v \in X_h,$

d) $|(g^k, v) - (g^k, v)_h| \leq c\,h\|g^k\|_{W^{1,q}(\Omega)}\|v\|_{H^1(\Omega)}, \quad v \in V_h.$

e)   If $M > 0$, then there exist constants $\tilde{c} = \tilde{c}(M), \hat{c} = \hat{c}(M)$ such that
   $|b(u, v) - b_h(u, v)| \leq \tilde{c}\,h\|u\|_{H^1(\Omega)}\|v\|_{H^1(\Omega)},$

f) $|b_h(u, v)| \leq \hat{c}M|v|_{H^1(\Omega)},$
   $u \in V_h \cap L^\infty(\Omega), \|u\|_{L^\infty(\Omega)} \leq M, \; v \in V_h, \; h \in (0, h_0).$

III. A priori estimates   :

$$\|u_{h\tau}\|_{L^2(-1,T;V)}, \quad \|w_{h\tau}\|_{L^2(0,T;V)} \quad \leq C, \quad \|u_{h\tau} - w_{h\tau}\|_{L^2(Q_T)} \leq C\tau, \tag{10.23}$$

provided condition (10.20) is satisfied in the case of scheme (10.14), whereas in the case of scheme (10.15) in addition to condition (10.21), the *inverse assumption*

$$h \leq c_1\,h_T, \quad T \in T_h, \; h \in (0, h_0), \tag{10.24}$$

is satisfied and the *second stability condition* holds:

$$0 < \tau \leq \frac{c_2 h^2}{\nu + h^2 + h}. \tag{10.25}$$

IV. Limit process. The above results, the application of the Fourier transform with respect to time combined with suitable compactness arguments and uniqueness of the exact solution imply the following result:

Theorem 10.1 *If the assumptions introduced above are satisfied, then*

$$u_{h\tau}, w_{h\tau} \rightarrow u \quad \text{weakly in} \quad L^2(0,T;V),$$
$$u_{h\tau}, w_{h\tau} \rightarrow u \quad \text{weak-* in} \quad L^\infty(Q_T),$$
$$u_{h\tau}, w_{h\tau} \rightarrow u \quad \text{strongly in} \quad L^2(Q_T),$$

as $h, \tau \rightarrow 0$ so that the stability conditions (10.20) or (10.21) together with (10.24) and (10.25) are satisfied in the case of schemes (10.14) or (10.15), respectively, where u is the unique weak solution of problem (10.1)–(10.3) (i.e., u satisfies (10.7)).

10.5 ERROR ESTIMATES

Let us assume that the exact solution of the continuous problem (10.7) is regular in the following sense:

$$u \in L^\infty(0, T; H^{1+\bar\varepsilon}(\Omega)),$$
$$u' = \frac{\partial u}{\partial t} \in L^\infty(0, T; L^2(\Omega)), \tag{10.26}$$
$$u'' = \frac{\partial^2 u}{\partial t^2} \in L^\infty(0, T; V^*)$$

for some $\bar\varepsilon \in (0,1]$. By V^* we denote the dual of V. For $\bar\varepsilon \in (0,1)$, $H^{1+\bar\varepsilon}(\Omega)$ is the Sobolev–Slobodetskii space of functions with "noninteger derivatives of order $1 + \bar\varepsilon$" (see [KJF77]). For piecewise polynomial approximations of such functions, see, e.g. [Fei89]. Assumptions (10.26) imply that

$$u \in C([0,T]; L^2(\Omega)), \quad u' \in C([0,T]; V^*). \tag{10.27}$$

In a similar manner to that of the previous Section we consider a system $\{T_h\}_{h\in(0,h_0)}$ of triangulations of the domain Ω satisfying Assumptions (A1,a–b). Moreover, we use the inverse assumption (10.24).

If $\varphi \in V$, then by $P_h \varphi$ we denote the *Ritz projection* of φ:

$$\text{a)} \quad P_h \varphi \in V_h, \qquad \text{b)} \quad a(P_h \varphi - \varphi, v_h) = 0 \quad \forall v_h \in V_h. \tag{10.28}$$

Since Ω is a polygonal domain, the regularity results and duality technique imply the existence of such an $\varepsilon^* \in (0,1]$ that for $\varepsilon \in (0, \varepsilon^*]$ and $\varphi \in H^{1+\varepsilon}(\Omega) \cap V$ we have

$$|P_h \varphi|_{H^1(\Omega)} \leq c|\varphi|_{H^1(\Omega)},$$
$$\|\varphi - P_h \varphi\|_{L^2(\Omega)} \leq c h^{1+\varepsilon}\|\varphi\|_{H^{1+\varepsilon}(\Omega)}, \tag{10.29}$$
$$\|\varphi - P_h \varphi\|_{H^1(\Omega)} \leq c h^\varepsilon\|\varphi\|_{H^{1+\varepsilon}(\Omega)}, \qquad h \in (0, h_0).$$

The error estimates for the combined finite volume – finite element method have been derived in several steps. We demonstrate the process for the semiimplicit scheme (10.14). The detailed analysis is carried out in [FFLW96].

I. Additional a priori estimate: There exists a constant $C > 0$ $(C = O(\nu^{-1}))$ such that

$$\|u_h^k\|_{H^1(\Omega)} \leq C, \quad t_k \in [0, T], \tag{10.30}$$

$h \in (0, h_0), \tau > 0$ satisfy the stability condition (10.20).

II. Truncation error. Let us set $u^k = u(\cdot, t_k)$, $\hat{M} = \max\{\|u\|_{L^\infty(Q_T)}, M\}$. Assumptions (10.26) imply the following estimates:

a) $\left|(u^{k+1} - u^k, v) - \tau(u'(t_{k+1}), v)\right| \leq \tau^2 \|v\|_{L^2(\Omega)} \|u''\|_{L^\infty(0,T;V\bullet)}, \quad v \in V, \tag{10.31}$

b) $\left|b(u^{k+1}, v) - b(u^k, v)\right| \leq 2 \max_{\substack{\xi \in [-\hat{M}, \hat{M}] \\ i=1,2}} |f_i'(\xi)| \|u^{k+1} - u^k\|_{L^2(\Omega)} |v|_{H^1(\Omega)}, \quad v \in V,$

c) $\|u^{k+1} - u^k\|_{L^2(\Omega)} \leq \tau \|u'\|_{L^\infty(0,T;L^2(\Omega))},$

d) $\left|b(u^{k+1}, v) - b(u^k, v)\right| \leq c\tau |v|_{H^1(\Omega)}, \quad v \in V.$

Using estimates (10.22) and (10.30), we derive the following relation for the error of the method defined as $e^k = u_h^k - u^k$ for $t_k \in [0, T]$:

$$(e^{k+1}, v_h) - (e^k, v_h) + \tau a(e^{k+1}, v_h)) = \tag{10.32}$$
$$= -\tau \left[b(u_h^k, v_h) - b(u^k, v_h)\right] + \varepsilon_1(\tau; u, v_h) + \varepsilon_2(\tau, h; u_h^k, u_h^{k+1}, v_h),$$
$$v_h \in V_h, \quad t_k \in [0, T),$$

where

$$|\varepsilon_1(\tau; u, v_h)| \leq c\tau^2 \|v_h\|_{H^1(\Omega)} \tag{10.33}$$
$$|\varepsilon_2(\tau, h; u_h^k, u_h^{k+1}, v_h)| \leq c\|v_h\|_{H^1(\Omega)} \left[h^2(\|u_h^k\|_{H^1(\Omega)} + \|u_h^{k+1}\|_{H^1(\Omega)}) + \tau h\right].$$

Moreover, let us note that

$$|b(\varphi_1, v) - b(\varphi_2, s)| \leq \tilde{c}(\hat{M}) \|\varphi_1 - \varphi_2\|_{L^2(\Omega)} |v|_{H^1(\Omega)} \tag{10.34}$$
$$\text{for } \varphi_i, v \in V, \|\varphi_i\|_{L^\infty(\Omega)} \leq \hat{M}, i = 1, 2.$$

III. Error estimates. Substituting $v_h := P_h e^{k+1}$ in (10.32), assuming that $h = 0(\tau)$ and using (10.33)–(10.34) and Young's inequality, we find that

$$\|e^{k+1}\|_{L^2(\Omega)}^2 + \tau\nu|e^{k+1}|_{H^1(\Omega)}^2 \leq \tag{10.35}$$
$$\leq \left(1 + \frac{c\tau}{\nu}\right) \|e^k\|_{L^2(\Omega)}^2 + \tau(\hat{C}_1 h^2 + \hat{C}_2 h^{2\varepsilon} + \hat{C}_3 h^{1+2\varepsilon}),$$
$$t_k \in [0, T], \ h \in (0, h_0), \ \tau > 0 \text{ satisfy } (10.20),$$
$$\hat{C}_1 = O(\nu^{-3}), \hat{C}_2 = O(\nu), \hat{C}_3 = O(1).$$

From this we obtain:

Theorem 10.2 *If $h \in (0, h_0)$ and $\tau > 0$ satisfy the conditions (10.20) and $h = 0(\tau)$ and we set $\varepsilon = \min(\tilde{\varepsilon}, \varepsilon^*)$, then*

$$\max_{t_k \in [0,T]} \|e^k\|_{L^2(\Omega)} \leq$$

$$\leq \left[C_1 h^{1+\varepsilon} + C_2 h + C_3 h^\varepsilon + C_4 h^{1/2+\varepsilon} \right] exp\left(\frac{cT}{\nu}\right), \qquad (10.36)$$

$$\left\{ \nu\tau \sum_{t_k \in [0,T]} |e^k|^2_{H^1(\Omega)} \right\}^{1/2} \leq$$

$$\leq \left[C_1 h^{1+\varepsilon} + C_2 h + C_3 h^\varepsilon + C_4 h^{1/2+\varepsilon} \right] exp\left(\frac{cT}{\nu}\right) \nu^{-1/2},$$

$$C_1 = O(1),\ C_2 = O(\nu^{-1}),\ C_3 = O(\nu),\ C_4 = O(\nu^{1/2}).$$

Remark 10.1 *a) The expressions $\max_{t_k \in [0,T]} \|e^k\|_{L^2(\Omega)}$ and $\left\{ \nu\tau \sum_{t_k \in [0,T]} |e^k|^2_{H^1(\Omega)} \right\}^{1/2}$ are discrete analogues of the $L^\infty(0,T; L^2(\Omega))$–norm and of the energy norm, respectively, of the error. The above asymptotic error estimates are rather pessimistic for a very small diffusion parameter ν.*
b) If the polygonal domain Ω is convex, then it is possible to set $\varepsilon^* = 1$. Assuming also $\tilde{\varepsilon} = 1$, we have $\varepsilon = 1$ in (10.36).
c) The above convergence results and error estimates can be extended to the case when $\Omega \subset \mathbb{R}^3$ is a bounded polyhedral domain and $p, q > 3$.
d) With the aid of the theory of finite element variational crimes developed by Feistauer and Ženíšek ([FŽ87], [FŽ88]), the theoretical analysis presented here can be generalized to the case of nonhomogeneous mixed Dirichlet–Neumann boundary conditions on a piecewise–smooth boundary $\partial\Omega$. Moreover, it is possible to consider a nonlinear diffusion term on the right hand side of equation (10.1) provided some assumptions on monotonicity or pseudomonotonicity are satisfied.

10.6 SCHEMES OVER OTHER MESHES

The combined FV – FE schemes for the discretization of the convection – diffusion problem (10.1)–(10.3) can also be applied on other types of meshes. Let us mention two possibilities.

I) *Triangular finite volumes – triangular finite elements.* The convection terms represented by the form b are discretized by the FV method on a mesh $\mathcal{D}_h = \{D_i\}_{i \in J}$, where D_i are triangles and $\mathcal{D}_h$ is a triangulation of the domain Ω with standard properties from the FE method. The same notation as in Section 10.3, where $\beta_{ij} := 1$, can be used. We consider $\mathcal{D}_h$ as a *primary mesh*. (For the application of triangular finite volumes to the solution of inviscid transonic flow, see, e.g., [Fel92], [FDF94], [FFL95], [FLWW95], [FD96].)

The diffusion term is discretized using conforming piecewise linear finite elements on a triangulation $\mathcal{T}_h$ of Ω, compatible with the primary grid $\mathcal{D}_h$ in the sense that the set of all vertices of the triangles $T \in \mathcal{T}_h$ consists of the barycentres P_i, $i \in J$, of all

$D_i \in \mathcal{D}_h$ and the vertices of $D_i \in \mathcal{D}_h$ lying on $\partial\Omega$. We call the mesh $\mathcal{T}_h$ adjoint to the grid $\mathcal{D}_h$. See Figure 2.

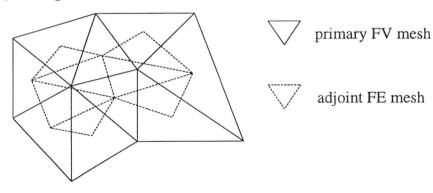

Figure 2   Adjoint finite elements $T \in \mathcal{T}_h$ to the primary finite volume mesh $\mathcal{D}_h$

In this case the spaces X_h, V_h and the lumping operator L_h are defined ·in the same way as in Section 10.3.

The convergence of this scheme remains open.

II) *Barycentric finite volumes – nonconforming finite elements.* Let $\mathcal{T}_h$ be a triangulation of the domain Ω. By $\{S_i; i \in J\}$ we denote the set of all sides of all $T \in \mathcal{T}_h$. (J is a suitable index set.) We construct the *barycentric finite volume mesh* $\mathcal{D}_h = \{D_i; i \in J\}$ over the primary mesh $\mathcal{T}_h$: Each side $S_i \not\subset \partial\Omega$ is associated with the quadrilateral D_i for which the boundary consists of segments connecting the endpoints of S_i with the barycentres of the triangles adjacent to the side S_i. For $S_i \subset \partial\Omega$, D_i is defined as the triangle with sides formed by the above segments and S_i. See Figure 3.

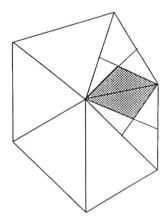

Figure 3   Barycentric finite volume over a triangular mesh

The finite volume approximation of the exact solution u at time $t = t_k$ is sought as a piecewise constant function in Ω with constant values on D_i, $i \in J$, whereas the finite element approximation uses the Crouzeix–Raviart nonconforming finite elements (cf. [Fei93, Par. 8.9]), linear on each $T \in \mathcal{T}_h$ and continuous at midpoints of S_i, $i \in J$.

10.7 NUMERICAL SOLUTION OF COMPRESSIBLE TRANSONIC VISCOUS FLOW

The complete system consisting of the continuity equation, the Navier-Stokes equations, the energy equation and the state equation, describing viscous compressible flow in a plane bounded domain Ω and time interval $(0, T)$ is written in the form

$$\frac{\partial w}{\partial t} + \sum_{s=1}^{2} \frac{\partial f_s(w)}{\partial x_s} = \sum_{s=1}^{2} \frac{\partial R_s(w, \nabla w)}{\partial x_s} \quad \text{in } Q_T = \Omega \times (0, T), \tag{10.37}$$

where

$$
\begin{aligned}
w &= (\rho, \rho v_1, \rho v_2, e)^{\mathrm{T}}, \quad w = w(x, t), \quad x \in \Omega, \ t \in (0, T), \\
f_s(w) &= (\rho v_s, \rho v_s v_1 + \delta_{s1} p, \rho v_s v_2 + \delta_{s2} p, (e + p) v_s)^{\mathrm{T}}, \\
R_s(w) &= (0, \tau_{s1}, \tau_{s2}, \tau_{s1} v_1 + \tau_{s2} v_2 + k \partial\theta/\partial x_s)^{\mathrm{T}}, \\
\tau_{ij} &= \lambda \operatorname{div} v \, \delta_{ij} + \mu(\partial v_i/\partial x_j + \partial v_j/\partial x_i), \\
p &= (\gamma - 1)(e - \rho|v|^2/2), \quad e = \rho(c_v \theta + |v|^2/2).
\end{aligned}
\tag{10.38}
$$

We use the following notation: t–time, x_s ($s = 1, 2$)–Cartesian coordinates, ρ–density, p–pressure, θ–absolute temperature, e–total energy, $v = (v_1, v_2)$–velocity, δ_{ij}–Kronecker delta, $\gamma > 1$, c_v, k, $\mu > 0$–given constants, $\lambda = -2\mu/3$.

The above system is equipped with *initial conditions* prescribing the state w on Ω at time $t = 0$ and *boundary conditions*: At inlet we prescribe ρ, v_1, v_2 and set $\partial\theta/\partial n = 0$, on fixed walls we assume that $v_1 = v_2 = 0$, $\partial\theta/\partial n = 0$ and at outlet we set $-pn_j + \sum_{i=1}^{2} \tau_{ij} n_i = 0$, $j = 1, 2$, and $\partial\theta/\partial n = 0$. (Here $\partial/\partial n$ is the derivative in the direction of the unit outer normal $n = (n_1, n_2)$ to the boundary $\partial\Omega$.) In the solution of *flow through a cascade of profiles*, the domain Ω represents one period of the cascade. Then $\partial\Omega$ consists of the inlet, outlet, impermeable profile and artificial boundary formed by two arcs Γ^+ and Γ^- on which periodicity conditions are considered. (Cf., e. g., [FDF94], [FFL95].)

The above problem is discretized with the aid of the numerical schemes described in Sections 10.3 and 10.6. These schemes can be written in a similar form to (10.14) (semiimplicit scheme) or (10.15) (purely explicit scheme), where the form a is replaced by its approximation a_h obtained with the aid of numerical integration (cf. (10.42)). The form b_h (see (10.13)) is now defined with the aid of the numerical flux H of an *upwind flux vector splitting scheme of Godunov type* ([Fei93, Chap. 7]).

Another method, which has been applied with success, is the *inviscid–viscous operator splitting* ([FFD96]). In this case we split (10.37) into the inviscid system

$$\frac{\partial w}{\partial t} + \sum_{s=1}^{2} \frac{\partial f_s(w)}{\partial x_s} = 0, \tag{10.39}$$

and the purely viscous system

$$\frac{\partial w}{\partial t} = \sum_{s=1}^{2} \frac{\partial R_s(w, \nabla w)}{\partial x_s}, \tag{10.40}$$

and discretize these separately.

The inviscid system (10.39) is discretized by the FV method on a mesh $\mathcal{D}_h = \{D_i\}_{i \in J}$, whereas the purely viscous system (10.40) is discretized using conforming piecewise linear finite elements on a triangulation $\mathcal{T}_h$ of Ω, as described above. One time step $t_k \to t_{k+1}$ is divided into two *fractional steps*:

Step I (*inviscid FV step on the mesh* $\mathcal{D}_h$): Assume that the values w_i^k, $i \in J$, approximating the solution on the finite volumes D_i at time t_k are known. Compute the values $w_i^{k+1/2}$, $i \in J$, from the FV formula

$$w_i^{k+1/2} = w_i^k - \frac{\tau_k}{|D_i|} \sum_{j \in S(i)} \sum_{\alpha=1}^{\beta_{ij}} H(w_i^k, w_j^k, n_{ij}^\alpha) \ell_{ij}^\alpha \tag{10.41}$$

equipped with inviscid boundary conditions. On $\Gamma^\pm$ periodicity conditions are considered. (Cf. [Fei93, Par. 7.3], [FFL95], [FDF94].)

Step II (*viscous FE step on the mesh* $\mathcal{T}_h$): Define the FE (i.e., continuous piecewise linear) approximation $w_h^{k+1/2}$ with values $w_h^{k+1/2}(P_i) = w_i^{k+1/2}$ at the vertices P_i, $i \in J$ of $\mathcal{T}_h$. On $\partial\Omega$, the viscous boundary conditions are used. Compute the FE approximation w_h^{k+1} as the solution of the following problem:

(i) w_h^{k+1} satisfies the viscous Dirichlet boundary conditions, $\qquad$ (10.42)

(ii) $(w_h^{k+1}, \varphi_h)_h = (w_h^{k+1/2}, \varphi_h)_h - \tau_k \, a_h(w_h^{k+1/2}, \varphi_h)$

for all test functions $\varphi_h = (\phi_1, \ldots, \phi_4)$ such that ϕ_j ($j = 1, \ldots, 4$) is continuous in Ω, linear on each $T \in \mathcal{T}_h$ and vanishes on the part of $\partial\Omega$ where the j-th component w_j of the state vector w satisfies the Dirichlet boundary condition.
Now set $w_i^{k+1} := w_h^{k+1}(P_i)$ for $i \in J$, $k := k+1$ and go to Step I.

In (10.42), $(w, \varphi)_h$ and $a_h(w, \varphi)$ denote the approximation of $\int_\Omega w \varphi \, dx$ and $\int_\Omega \sum_{s=1}^{2} R_s(w, \nabla w) \, \partial\varphi/\partial x_s \, dx$, respectively.

The computation of viscous flow can be realized with the aid of a purely explicit scheme similar to (10.15) and the operator splitting scheme (10.41)–(10.42), because their algorithmization is simple. The following *stability conditions* are used:

$$\frac{\tau_k}{|D_i|} |\partial D_i| \max_{j \in S(i), \alpha=1,\ldots,\beta_{ij}} \rho(P(w_i^k, n_{ij}^\alpha)) \leq \text{CFL} \approx 0.85, \quad i \in J, \tag{10.43}$$

$$\frac{3}{4} \frac{h_T}{\sigma_T} \frac{\tau_k}{|T|} \max(\mu, k) \leq \text{CFL}, \quad T \in \mathcal{T}_h, \tag{10.44}$$

where $P(w, n) = \sum_{s=1}^{2} (D f_s(w)/Dw) n_s$, $\rho(P) = $ spectral radius of the matrix P. Conditions (10.43) and (10.44) are obtained on the basis of linearization and in analogy with a scalar problem.

In order to increase the accuracy of the method, the term $H(w_i^k, w_j^k, n_{ij}^\alpha)$ is replaced by $H(w_{ij}^{k,\alpha}, w_{ji}^{k,\alpha}, n_{ij}^\alpha)$ where $w_{ij}^{k,\alpha}$ and $w_{ji}^{k,\alpha}$ are the values of the *"second order recovery"* of the piecewise constant FV solution combined with the use of a suitable flux limiter ([FFL95]).

The accuracy of the solution of transonic flow is also increased with the aid of *automatic adaptive mesh refinement* in the vicinity of shock waves, based on a shock indicator using divided differences of the density and taking into account the direction of the flow ([FDF94], [FFL95], [FD96]).

Computational results. The combined FV – FE method was applied to several test problems. The goal was to obtain steady viscous transonic flow via time stabilization for $t_k \to \infty$. The numerical fluxes by Steger – Warming ([SW81]), Van Leer ([Van75]), Vijayasundaram ([Vij86]) and Osher – Solomon ([OS82], [Spe88]) were used in our schemes applied on the types of meshes described above. The best results were obtained with aid of the method using the *Osher – Solomon numerical flux* (in the form described in [FFL95]) *applied on a primary triangular finite volume mesh and combined with the FE method on an adjoint triangulation*. Here we present two examples of the application of this method which were obtained in cooperation with V. Dolejší.

I) *Flow of air ($\gamma = 1.4$) through the GAMM channel* (10 % circular bump) with inlet Mach number $= 0.67$ and Reynolds number $\approx 10^5$. In Figure 4 and Figure 5, the primary FV and adjoint FE meshes are plotted. Figure 6 shows the convergence history to the steady state measured in the L_1-norm and the density distribution on the walls. In Figure 7, Mach number isolines are plotted. We obtained identical results with the aid of the purely explicit scheme and the inviscid-viscous operator splitting method.

II) *Flow past a cascade of profiles* was solved by the inviscid – viscous operator splitting method using the Osher – Solomon numerical flux, applied on a primary triangular FV mesh and an adjoint triangular FE mesh. The computational results are compared with a wind tunnel experiment (by courtesy of the Institute of Thermomechanics of the Czech Academy of Sciences in Prague, see [ŠŠ90]). The experiment and the computations were performed for the following data: angle of attack $= 19° 18'$, inlet Mach number $= 0.32$, outlet Mach number $= 1.18$, $\gamma = 1.4$, Reynolds number $\approx 10^5$. (Inviscid computations were presented in [FD96], [FDF94], [FFL95].) Figure 8 shows the primary FV mesh $\mathcal{D}_h$ and the adjoint FE mesh $\mathcal{T}_h$, refined a priori towards the profile and adaptively in the vicinity of the shock waves according to the algorithm described in [FDF94], [FFL95], [FD96]. In Figure 9, the convergence history to the steady state is shown. Figure 10, where density isolines are plotted and compared with an interferogram, indicates a very good agreement of computational and experimental results.

Acknowledgements. The present research has been supported under Grant No. 201/96/0313 of the Czech Grant Agency. The authors gratefully acknowledge this support.

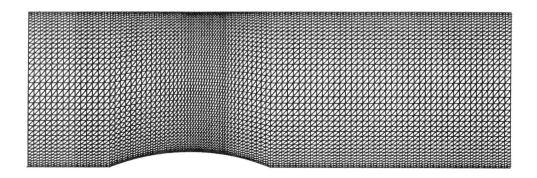

Figure 4 Primary FV mesh

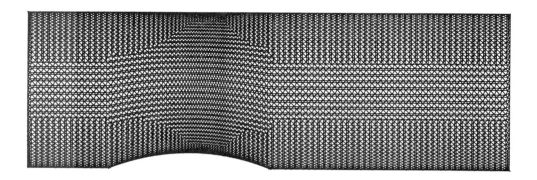

Figure 5 Adjoint FE mesh

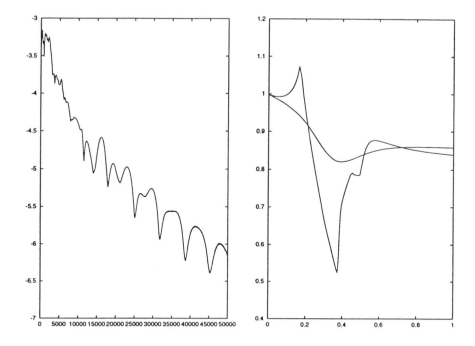

Figure 6   Convergence history $(\log\|(\rho^{k+1} - \rho^k)/\rho^k\|_{L^1(\Omega)}$ against the number of time steps); the density distribution on the walls against the relative arclength

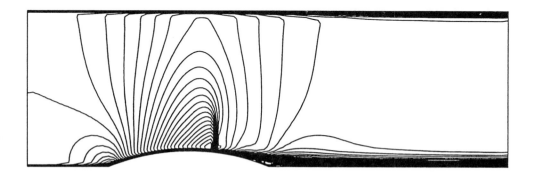

Figure 7   Mach number isolines

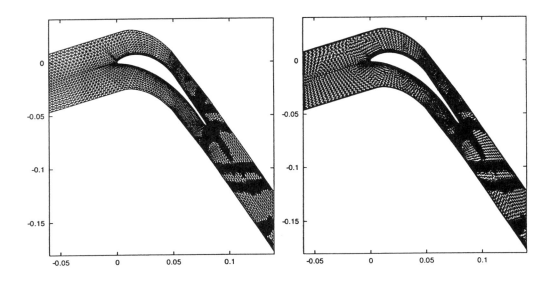

Figure 8   Basic and adjoint mesh

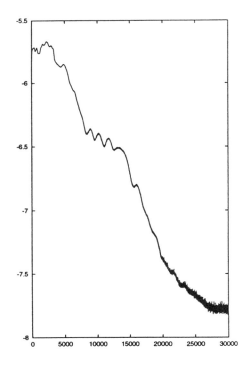

Figure 9   Convergence history

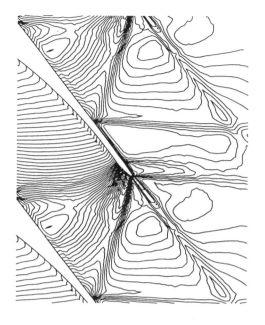

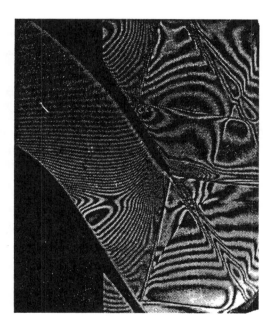

Figure 10   Density isolines compared with interferogram

References

[ADF+89] Arminjon P., Dervieux A., Fezoui L., Steve H., and Stoufflet B. (1989) Non-oscillatory schemes for multidimensional Euler calculations with unstructured grids. In Ballmann J. and Jeltsch R. (eds) *Notes on Numerical Fluid Mechanics Volume 24 (Nonlinear Hyperbolic Equations – Theory, Computation Methods, and Applications)*. Vieweg, Braunschweig – Wiesbaden.

[AFRS89] Adam D., Felgenhauer A., Roos H.-G., and Stynes M. (1989) A nonconforming finite element method for a singularly perturbed boundary value problem. Preprint 1992–10, Department of Mathematics, University College, Cork, Ireland.

[AR93] Adam D. and Roos H.-G. (1993) A nonconforming exponentially fitted finite element method I: The interpolation error. Preprint MATH-NM-06-1993, Technische Universität Dresden.

[Cia79] Ciarlet P. G. (1979) *The Finite Element Method for Elliptic Problems*. North – Holland, Amsterdam.

[FD96] Felcman J. and Dolejší V. (1996) Adaptive methods for the solution of the Euler equations in elements of blade machines. *ZAMM (to appear)* .

[FDF94] Felcman J., Dolejší V., and Feistauer M. (1994) Adaptive finite volume method for the numerical solution of the compressible Euler equations. In Wagner S., Hirschel E. H., Périaux J., and Piva R. (eds) *Computational Fluid Dynamics 94*, pages 894 – 901. Wiley, New York.

[Fei89] Feistauer M. (1989) On the finite element approximation of functions with noninteger derivatives. *Numer. Funct. Anal. and Optimiz.* 10: 91 – 110.

[Fei93] Feistauer M. (1993) *Mathematical Methods in Fluid Dynamics*. Pitman Monographs and Surveys in Pure and Applied Mathematics 67, Longman Scientific & Technical, Harlow.

[Fel92] Felcman J. (1992) Finite volume solution of the inviscid compressible fluid flow. *ZAMM* 72: 513 – 516.

[FFD96] Feistauer M., Felcman J., and Dolejší V. (1996) Numerical simulation of compressible viscous flow through cascades of profiles. *ZAMM (to appear)*

[FFL95] Feistauer M., Felcman J., and Lukáčová – Medviďová M. (1995) Combined finite element – finite volume solution of compressible flow. *Journal of Comput. and Appl. Math.* 63: 179 – 199.

[FFL96] Feistauer M., Felcman J., and Lukáčová – Medviďová M. (1996) On the convergence of a combined finite volume – finite element method for nonlinear convection – diffusion problems. *Numer. Methods for PDE's (to appear)* .

[FFLW96] Feistauer M., Felcman J., Lukáčová M., and Warnecke G. (1996) Error estimates of a combined finite volume–finite element method for nonlinear convection–diffusion problems. Preprint, Faculty of Mathematics and Physics, Charles Univ., Prague.

[FHKV95] Fořt J., Huněk M., Kozel K., and Vavřincová M. (1995) Numerical simulation of steady and unsteady flows through plane cascades. In Feistauer M., Rannacher R., and Kozel K. (eds) *Numerical Modelling in Continuum Mechanics II*, pages 95–102. Faculty of Mathematics and Physics, Charles Univ., Prague.

[FLWW95] Felcman J., Lukáčová–Medviďová M., Warnecke G., and Wendland W. L. (1995) Adaptive mesh refinement for Euler equations. Preprint 95-15, Universität Stuttgart.

[FS96] Feistauer M. and Stupka P. (1996) Convergence of the combined finite volume–finite element method for nonlinear convection–diffusion problems. Explicit schemes. Preprint, Faculty of Mathematics and Physics, Charles Univ., Prague.

[FŽ87] Feistauer M. and Ženíšek A. (1987) Finite element solution of nonlinear elliptic problems. *Numer. Math.* 50: 451–475.

[FŽ88] Feistauer M. and Ženíšek A. (1988) Compactness method in the finite element theory of nonlinear elliptic problems. *Numer. Math.* 52: 147–163.

[Hac89] Hackbusch W. (1989) On first and second order box schemes. *Computing* 41: 277–296.

[HHK92] Hůlek T., Huněk M., and Kozel K. (1992) Numerical solution of Euler and Navier–Stokes equations for 2D transonic flow. In Wagner S., Hirsch C., Périaux J., and Kordulla W. (eds) *Computational Fluid Dynamics–92, Vol. I*. Elsevier Science Publishers, Amsterdam.

[Ike83] Ikeda T. (1983) *Maximum Principle in Finite Element Models for Convection–Diffusion Phenomena*. Lecture Notes in Numerical and Applied Analysis Vol. 4, North–Holland, Amsterdam–New York–Oxford.

[Joh81] Johnson C. (1981) Finite element methods for convection–diffusion problems. In Glowinski R. and Lions J. L. (eds) *Computing Methods in Engineering and Applied Sciences V*. North-Holland, Amsterdam.

[KJF77] Kufner A., John O., and Fučík S. (1977) *Function Spaces*. Academia, Prague.

[Mor96] Morton K. W. (1996) *Numerical Solution of Convection-Diffusion Problems*. Chapman & Hall, London.

[OS82] Osher S. and Solomon F. (1982) Upwind difference schemes for hyperbolic systems of conservation laws. *Math. Comp.* 38: 339–374.

[RST96]   Roos H.-G., Stynes M., and Tobiska L. (1996) *Numerical Methods for Singularly Perturbed Differential Equations.* Springer Series in Computational Mathematics 24, Springer-Verlag, Berlin.

[Spe88]   Spekreijse S. P. (1988) *Multigrid Solution of Steady Euler Equations.* Centre for Mathematics and Computer Science, Amsterdam.

[ŠŠ90]   Šťastný M. and Šafařík P. (1990) Experimental analysis data on the transonic flow past a turbine cascade. ASME Publ. No. 90−GT−313, New York.

[SS94]   Sonar T. and Süli E. (1994) A dual graph norm refinement indicator for the DLR-τ -Code. Forschungsbericht 94-24, Institut für Strömungsmechanik, Göttingen.

[SW81]   Steger J. L. and Warming R. F. (1981) Flux vector splitting of the inviscid gas dynamics equations with applications to finite difference methods. *J. Comput. Phys.* 40: 263−293.

[Van75]   Van Leer B. (1975) Towards the ultimate conservative difference scheme III. upstream centered finite-difference schemes for ideal compressible flow. *J. Comp. Phys.* 23: 263−275.

[VH92]   Vilsmeier R. and Hänel D. (1992) Adaptive solutions of the conservation equations on unstructured grids. In Vos J. B. and A. Rizzi I. L. R. (eds) *Notes on Numerical Fluid Mechanics Vol. 35 (Proceedings of the Ninth GAMM-Conference on Numerical Methods in Fluid Mechanics)*, pages 321−330. Vieweg, Braunschweig−Wiesbaden.

[Vij86]   Vijayasundaram G. (1986) Transonic flow simulation using an upstream centered scheme of Godunov in finite elements. *J. Comput. Phys.* 63: 416−433.

[Zho95]   Zhou G. (1995) A local L^2-error analysis of the streamline diffusion method for nonstationary convection-diffusion systems. M^2AN 29: 577−603.

[ZR96]   Zhou G. and Rannacher R. (1996) Pointwise superconvergence of streamline diffusion finite-element method. *Numer. Methods for PDE's* 12: 123−145.

11

Towards a Finite Element Capability for the Modelling of Viscous Compressible Flows

K. Morgan, O. Hassan, M. T. Manzari, N. Verhoeven and N. P. Weatherill

Department of Civil Engineering,
University of Wales,
Swansea SA2 8PP,
United Kingdom.

11.1   INTRODUCTION

Aerospace companies have a considerable interest in developing tools which will enable the routine simulation of steady transonic viscous flows over general configurations [KGB96]. To meet these requirements, the algorithm developer must not only provide a capability for handling complex geometries, but also a method for modelling the turbulent flow regimes which are typically encountered.

Computational domains of complex geometrical shape are readily discretised by an unstructured mesh approach. In addition, if consideration is restricted to the use of cells which are tetrahedral in shape, fully automatic techniques are now available which are designed to generate reasonably isotropic meshes. These techniques are based upon either Delaunay [Wea92] or advancing front [PMP90] approaches, with the generated meshes being post processed to ensure that the mesh cells are all of good quality [PMH96]. A recent development in this area has seen the coupling of a mesh generator to a solid modeller [PM96] and the result is a robust, rapid meshing procedure. The current status of the unstructured mesh approach is that it has been fully validated for the simulation of a range of inviscid aerodynamic flows over general three dimensional configurations [MPPH91, PPM+93b].

It is natural, therefore, to consider the means by which this approach can be extended to provide a capability for the simulation of problems involving transonic viscous flows. At present, any successful implementation will require the adoption of an appropriate turbulence model. In addition, a method will have to be devised for generating the highly stretched anisotropic meshes which are typically employed

The Mathematics of Finite Elements and Applications
Edited by J. R. Whiteman © 1997 John Wiley & Sons Ltd.

for the efficient representation of viscous features in high speed compressible flow modelling. It can be expected that the computational demands, in terms of both cpu time and memory requirements, of an approach of this type will be large. However, these demands become more reasonable if the problem can be divided and the sub–problems solved on a distributed system of computers. For this reason, it will ultimately be essential to achieve a successful parallel implementation of both the mesh generator and the solution algorithm.

In this paper, we will describe the steps we are taking to provide an unstructured mesh capability of this type. It will be seen that the process is not yet complete, but that much of the required technology is now in place.

11.2 TURBULENCE MODELLING

For the simulation of three dimensional compressible turbulent flow, we work with a time averaged form of the governing equations. The averaging is made over a time scale τ, which is large in comparison to the time scale of the turbulence, but much less than the time scale of the mean flow. Following this procedure, each variable is decomposed into a time varying mean value and a fluctuation about the mean. The conventional time average

$$f = \bar{f} + f' \qquad \bar{f}(t) = \frac{1}{\tau} \int_{t-\tau/2}^{t+\tau/2} f(s)ds \qquad (11.1)$$

is employed for the density, pressure and temperature. The Favre (mass) average

$$f = \tilde{f} + f'' \qquad \tilde{f} = \frac{1}{\bar{\rho}\tau} \int_{t-\tau/2}^{t+\tau/2} \rho(s)f(s)ds \qquad (11.2)$$

is used for the velocity components and the enthalpy. To model the turbulent stress components in the resulting equations, we seek an approach which (a) results in the appearance of simple source terms and (b) requires the implementation of straightforward boundary conditions. These requirements will be met by accomplishing this closure by the use of the two equation k–ω model [Wil93], where k denotes the turbulent kinetic energy and

$$\omega = \frac{\epsilon}{\beta^* k} \qquad (11.3)$$

Here ϵ is the turbulence dissipation and β^* is a constant. In this approach, the basic set of conservation equations is supplemented by the addition of transport equations for k and ω.

11.3 GOVERNING EQUATIONS

The time averaged equations for the k–ω model are considered in the non–dimensional form [Man96]

$$\frac{\partial \mathbf{U}}{\partial t} + \frac{\partial \mathbf{F}^j}{\partial x_j} - \frac{\partial \mathbf{G}^j}{\partial x_j} = \mathbf{S} \qquad j = 1, 2, 3 \qquad (11.4)$$

where summation is implied over the repeated index j,

$$\mathbf{U} = \begin{bmatrix} \rho \\ \rho u_1 \\ \rho u_2 \\ \rho u_3 \\ \rho E \\ \rho k \\ \rho \omega \end{bmatrix} \qquad \mathbf{F}^j = \begin{bmatrix} \rho u_j \\ \rho u_1 u_j + p\delta_{1j} \\ \rho u_2 u_j + p\delta_{2j} \\ \rho u_3 u_j + p\delta_{3j} \\ (\rho E + p)u_j \\ \rho k u_j \\ \rho \omega u_j \end{bmatrix} \qquad (11.5)$$

and

$$\mathbf{G}^j = \begin{bmatrix} 0 \\ \sigma_{1j} \\ \sigma_{2j} \\ \sigma_{3j} \\ u_i \sigma_{ij} - q_j + D_{kj} \\ D_{kj} \\ D_{\omega j} \end{bmatrix} \qquad \mathbf{S} = \begin{bmatrix} 0 \\ 0 \\ 0 \\ 0 \\ 0 \\ P_k - D_k \\ P_\omega - D_\omega \end{bmatrix} \qquad (11.6)$$

It should be noted that the overbars have now been omitted, on the understanding that the equations are expressed in terms of the appropriate time averaged quantities. The equations have been written with respect to a cartesian coordinate system $Ox_1x_2x_3$ and u_j denotes the fluid velocity in direction x_j. The density, pressure and specific total energy of the fluid are denoted by ρ, p and E respectively, where

$$E = \frac{p}{(\gamma - 1)\rho} + \frac{u_i u_i}{2} + k \qquad (11.7)$$

and γ is the ratio of the fluid specific heats. The components of the stress tensor σ_{ij} are defined according to

$$\sigma_{ij} = \frac{(\mu + \mu_t)}{Re} \left\{ \left(\frac{\partial u_i}{\partial x_j} + \frac{\partial u_j}{\partial x_i} \right) - \frac{2}{3} \frac{\partial u_k}{\partial x_k} \delta_{ij} \right\} - \frac{2}{3} \rho k \delta_{ij} \qquad (11.8)$$

with the turbulent viscosity in the k–ω model being computed as

$$\mu_t = \alpha^* \frac{\rho k}{\omega} Re \qquad (11.9)$$

Here α^* is a constant and Re is the flow Reynolds number. The production and dissipation of k and ω are represented by the terms

$$P_k = \left[\frac{\mu_t}{Re} \left\{ \left(\frac{\partial u_i}{\partial x_j} + \frac{\partial u_j}{\partial x_i} \right) - \frac{2}{3} \frac{\partial u_k}{\partial x_k} \delta_{ij} \right\} - \frac{2}{3} \rho k \delta_{ij} \right] \frac{\partial u_i}{\partial x_j} \qquad D_k = \beta^* \rho \omega k \quad (11.10)$$

and

$$P_\omega = \frac{\alpha\omega}{k} P_k \qquad\qquad D_\omega = \beta\rho\omega^2 \qquad\qquad (11.11)$$

where α and β are constants. The quantities D_{kj} and $D_{\omega j}$ represent the components of the diffusive fluxes of k and ω respectively in the direction x_j and are taken in the form

$$D_{kj} = \frac{1}{Re}\left(\mu + \frac{\mu_t}{\sigma_k}\right)\frac{\partial k}{\partial x_j} \qquad\qquad D_{\omega j} = \frac{1}{Re}\left(\mu + \frac{\mu_t}{\sigma_\omega}\right)\frac{\partial\omega}{\partial x_j} \qquad (11.12)$$

where σ_k and σ_ω are constants. The component of the diffusive heat flux in the direction x_j is defined to be

$$q_j = -\frac{1}{Re}\left(\frac{\mu}{Pr} + \frac{\mu_t}{Pr_t}\right)\frac{\partial T}{\partial x_j} \qquad\qquad (11.13)$$

where T is the temperature in the fluid and Pr and Pr_t denote Prandtl numbers. Standard values of

$$\beta = \frac{3}{40} \qquad\qquad \beta^* = 0.09 \qquad\qquad \sigma_\omega = 2.0 \qquad\qquad (11.14)$$

$$\sigma_k = 2.0 \qquad\qquad \alpha = \frac{5}{9} \qquad\qquad \alpha^* = 1 \qquad\qquad (11.15)$$

are employed for the closure coefficients.

The equation set is completed by the addition of the perfect gas equation of state. The viscosity μ is assumed to vary with temperature according to Sutherland's relation [Sch84]. At a solid wall, the fluid velocity will be zero, the temperature or the heat flux will be specified and k will be equal to zero. The value of ω will tend to infinity as the wall is approached. It will be assumed that the distribution of all the variables is known at some initial time $t = 0$.

11.4   WEAK VARIATIONAL FORM

The starting point for the development of a numerical solution approach will be the replacement of the classical formulation of the problem by an equivalent weak variational form [Joh87]. In a region Ω, bounded by a closed surface Γ, the variational formulation takes the form: find $\mathbf{U}(\mathbf{x}, t)$ such that

$$\int_\Omega \frac{\partial\mathbf{U}}{\partial t}W\,d\Omega = \int_\Omega \mathbf{F}^j\frac{\partial W}{\partial x_j}\,d\Omega - \int_\Gamma \bar{\mathbf{F}}^j n_j W\,d\Gamma$$
$$-\int_\Omega \mathbf{G}^j\frac{\partial W}{\partial x_j}\,d\Omega + \int_\Gamma \bar{\mathbf{G}}^j n_j W\,d\Gamma + \int_\Omega \mathbf{S}W\,d\Omega \qquad (11.16)$$

for every suitable weighting function $W(\mathbf{x})$ and for every $t > 0$, subject to the initial condition $\mathbf{U}(\mathbf{x}, 0) = \mathbf{U}_0(\mathbf{x})$, for all $\mathbf{x}$ in Ω, where $\mathbf{U}_0$ is a prescribed function. In equation (11.16), $\mathbf{n} = (n_1, n_2, n_3)$ is the unit outward normal to Γ and the overbar is now to be interpreted as denoting a prescribed normal boundary flux.

11.5 APPROXIMATE SOLUTION

To obtain an approximate solution to this variational formulation, the problem domain Ω is discretised using triangular or tetrahedral elements. Nodes are located at the element vertices and are numbered from 1 to p. An approximate piecewise linear finite element solution is then sought in the form

$$\mathbf{U} \approx \mathbf{U}^{(p)} = \mathbf{U}_J(t) N_J(\mathbf{x}) \qquad\qquad J = 1, 2, \cdots, p \qquad\qquad (11.17)$$

where there is an implied summation over the repeated index J. Here N_J denotes the standard linear shape function associated with node J of the mesh and $\mathbf{U}_J(t)$ denotes the value of the approximation at node J. A Galerkin approximate solution [ZM83] follows from using the above weak variational statement in the form: find $\mathbf{U}^{(p)}$ such that

$$\sum_{E \in I} \int_{\Omega_E} \frac{\partial \mathbf{U}^{(p)}}{\partial t} N_I d\Omega = \sum_{E \in I} \int_{\Omega_E} \mathbf{F}^{j(p)} \frac{\partial N_I}{\partial x_j} d\Omega - \sum_{B \in I} \int_{\Gamma_B} \bar{\mathbf{F}}^{j(p)} n_j N_I d\Gamma$$

$$- \sum_{E \in I} \int_{\Omega_E} \mathbf{G}^{j(p)} \frac{\partial N_I}{\partial x_j} d\Omega + \sum_{B \in I} \int_{\Gamma_B} \bar{\mathbf{G}}^{j(p)} n_j N_I d\Gamma + \sum_{E \in I} \int_{\Omega_E} \mathbf{S}^{(p)} N_I d\Omega \qquad (11.18)$$

for each $I = 1, 2, \cdots, p$. In this equation, the summations extend over all elements E, and all boundary faces B, which contain node I. It can be observed that equation (11.18) has been expressed in a form which indicates how the discrete equations are normally constructed in the finite element approach, by looping over the elements. In practice, we choose to evaluate the discrete equation in an alternative form which, in the three dimensional case, can be achieved by writing equation (11.18) as

$$\left[\mathbf{M}_L \frac{d\mathbf{U}}{dt} \right]_I = - \sum_{s=1}^{m_I} \frac{C^j_{II_s}}{2} \{ (\mathbf{F}^j_I + \mathbf{F}^j_{I_s}) - (\mathbf{G}^j_I + \mathbf{G}^j_{I_s}) \}$$

$$+ \langle \sum_{f=1}^{\ell_I} D_f \{ (6\bar{\mathbf{F}}^n_I + 2\bar{\mathbf{F}}^n_{If_1} + \bar{\mathbf{F}}^n_{If_2}) - (6\bar{\mathbf{G}}^n_I + 2\bar{\mathbf{G}}^n_{If_1} + \bar{\mathbf{G}}^n_{If_2}) \} \rangle_I + [\mathbf{M}_L]_I \, \mathbf{S}_I \qquad (11.19)$$

where the consistent mass matrices have, for computational convenience, been replaced by the standard lumped mass matrix, $\mathbf{M}_L$, and the summations now extend over the m_I edges, and the ℓ_I boundary faces, in the mesh which are connected to node I. The term enclosed by the angle brackets will only appear when node I is located on the boundary. This method of constructing the discrete equations, involving looping over the edges in the mesh, is preferred because it offers improvements in computational efficiency over the standard approach [MPPH94]. In these expressions,

$$C^j_{II_s} = - \sum_{E \in II_s} \frac{\Omega_E}{2} \left[\frac{\partial N_I}{\partial x_j} \right]_E + \langle \frac{\Gamma_f}{12} n^j \rangle_{II_c} \qquad\qquad D_f = - \frac{\Gamma_f}{24} \qquad (11.20)$$

The nodal values of the viscous fluxes can be determined when the nodal values of both the solution and the solution gradients are known. Assuming the solution takes the form of equation (11.17), the solution gradients in the coordinate directions can be computed by Galerkin projection [PPM93a], leading to the expression

$$
\left[\mathbf{M}_L \frac{\partial \mathbf{U}}{\partial x_j} \right]_I = -\sum_{s=1}^{m_I} \frac{C_{II_s}^j}{2} (\mathbf{U}_I + \mathbf{U}_{I_s}) + \langle \sum_{f=1}^{\ell_I} D_f n^j (6\mathbf{U}_I + 2\mathbf{U}_{If_1}^n + \mathbf{U}_{If_2}) \rangle_I \quad (11.21)
$$

The resulting scheme is central difference in character and cannot be used successfully for the simulation of practical compressible flows. The standard remedy is to replace the inviscid physical flux function by a consistent numerical flux function to produce a modified scheme which has the desired properties [Hir90].

11.5.1 THE NUMERICAL FLUX FUNCTION

Ignoring boundary face contributions, the inviscid flux contribution on the right hand side of equation (11.19) for I can be re–written in the form

$$
\sum_{s=1}^{m_I} \frac{C_{II_s}^j}{2} (\mathbf{F}_I^j + \mathbf{F}_{I_s}^j) = \sum_{s=1}^{m_I} \frac{\mathcal{L}_{II_s}}{2} \left[(\mathbf{F}_I^j + \mathbf{F}_{I_s}^j) \mathcal{S}_{II_s}^j \right] = \frac{1}{2} \sum_{s=1}^{m_I} \mathcal{L}_{II_s} \mathbf{F}_{II_s} \quad (11.22)
$$

where

$$
\mathbf{F}_{II_s} = \mathbf{F}_I^j \mathcal{S}_{II_s}^j + \mathbf{F}_{I_s}^j \mathcal{S}_{II_s}^j \quad (11.23)
$$

and

$$
\mathbf{C}_{II_s} = (C_{II_s}^1, C_{II_s}^2, C_{II_s}^3) \qquad \mathcal{L}_{II_s} = |\mathbf{C}_{II_s}| \qquad \mathcal{S}_{II_s}^j = \frac{C_{II_s}^j}{|\mathbf{C}_{II_s}|} \quad (11.24)
$$

Here, it has been assumed that the typical edge s, of the set of m_I edges connected to node I, connects nodes I and I_s. To produce a practical numerical scheme, we replace the actual physical flux function $\mathbf{F}_{II_s}$ by a consistent numerical flux function $\mathcal{F}_{II_s}$, where

$$
\mathcal{F}_{II_s} = \mathbf{F}_{II_s} + \mathcal{D}_{II_s} \quad (11.25)
$$

and $\mathcal{D}_{II_s}$ represents an added dissipation

A scheme which is first order upwind in character will result from the use of the flux function of equations (11.23) and (11.25) with

$$
\mathcal{D}_{II_s} = -|\mathbf{A}_{II_s}(\mathbf{U}_I, \mathbf{U}_{I_s})|(\mathbf{U}_{I_s} - \mathbf{U}_I) \quad (11.26)
$$

The matrix $\mathbf{A}_{II_s}$ is the jacobian matrix of the edge flux function and is computed as

$$
\mathbf{A}_{II_s} = \frac{d\mathbf{F}^j}{d\mathbf{U}} \mathcal{S}_{II_s}^j \quad (11.27)
$$

and, in equation (11.26), this matrix is evaluated at the Roe interface state [Hir90] between $\mathbf{U}_I$ and $\mathbf{U}_{I_s}$. A high resolution extension is readily constructed, using MUSCL

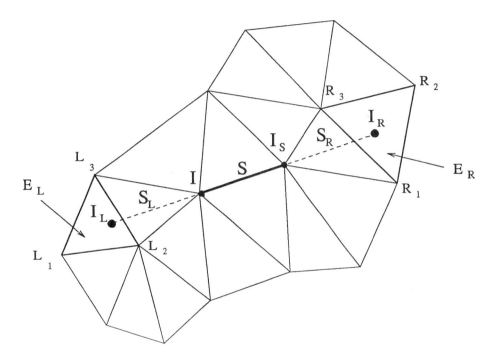

Figure 1   The four point stencil for the edge s

concepts [vL79, Man96]. On an unstructured mesh, a linear four point stencil is created for each edge, as illustrated in Figure 1. Interface states, $\mathbf{U}_l$ and $\mathbf{U}_r$, are determined at the midpoint of the edge s in the form

$$\mathbf{U}_l = \mathcal{I}(\mathbf{U}_{I_L}, \mathbf{U}_I, \mathbf{U}_{I_s}) \qquad\qquad \mathbf{U}_r = \mathcal{I}(\mathbf{U}_I, \mathbf{U}_{I_s}, \mathbf{U}_{I_R}) \qquad (11.28)$$

where $\mathcal{I}$ denotes a suitable limited interpolating function. The function proposed by Thomas and Venkatakrishnan [Tho91, Ven91] is employed in this work. The revised scheme then follows by replacing the numerical flux function of equations (11.23), (11.25) and (11.26) by

$$\mathcal{F}_{II_s} = \mathbf{F}^j(\mathbf{U}_l)\mathcal{S}^j_{II_s} + \mathbf{F}^j(\mathbf{U}_r)\mathcal{S}^j_{II_s} + \mathcal{D}_{(lr)II_s} \qquad (11.29)$$

where

$$\mathcal{D}_{(lr)II_s} = -|\mathbf{A}(\mathbf{U}_l, \mathbf{U}_r)|(\mathbf{U}_r - \mathbf{U}_l) \qquad (11.30)$$

This method of constructing the higher order flux function is not unique and other possibilities have also been investigated [PMVP94, PPM93a].

11.5.2 IMPLEMENTATION DETAILS

In this initial version of the compressible turbulent code, the solution is advanced by explicit Euler timestepping. This means that, with the discrete equation system expressed in the form

$$\left[\mathbf{M}_L \frac{d\mathbf{U}}{dt}\right]_I = \mathbf{R}_I \qquad (11.31)$$

the solution is advanced, over a timestep Δt, from time t_n to time t_{n+1}, according to

$$\mathbf{U}_I^{n+1} = \mathbf{U}_I^n + \Delta t[\mathbf{M}_L]_I^{-1}\mathbf{R}_I^n \qquad (11.32)$$

where a superscript n denotes an evaluation at time $t = t_n$. The time advancement process begins with the density and fluid velocity set equal to their free stream values, while ω is initialised within the range $1 \le \omega \le 10$ and k is assigned the value 0.001.

At a solid wall boundary, the value of ω is normally set according to the expression

$$\omega_w = \frac{60\mu_w}{\beta\rho_w\delta^2 Re} \qquad (11.33)$$

where the subscript w denotes wall values. In this expression, δ denotes the height of the first layer of computational cells adjacent to the wall.

An alternative treatment of a solid wall boundary follows from the use of wall function boundary conditions. In this approach, we avoid the requirement for accurately representing the viscous sub–layer by imposing on the computed solution the velocity distribution prescribed by the law of the wall [Sch84]. This approach, when applicable, can result in a considerable reduction in the number of grid points necessary to produce an accurate solution. In this case, see Figure 2, with the definitions

$$Y^+ = \frac{\rho u_\tau y}{\mu_w} \qquad U^+ = \frac{u}{u_\tau} \qquad u_\tau = \sqrt{\frac{\tau_w}{\rho}} \qquad (11.34)$$

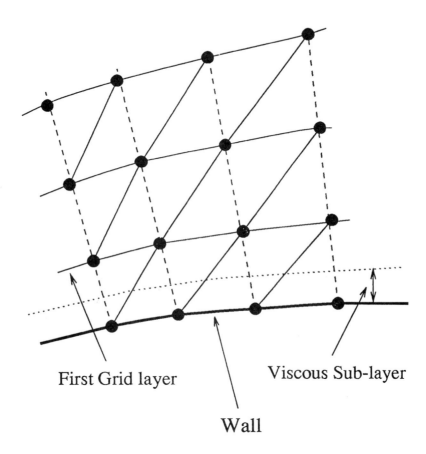

Figure 2 The form of the mesh for the application of the law of the wall

the first layer of nodal points away from the wall is constructed in such a manner that $10 < Y^+ < 100$. The value of u_τ, at the end of a time step, is obtained from the law of the wall as

$$U^+ = \frac{1}{0.41} \log(Y^+) + 5.5 \qquad (11.35)$$

The values of k and ω at the first layer of points adjacent to the wall are then corrected according to

$$k = \frac{u_\tau^2}{\sqrt{\beta^*}} \qquad \omega = \frac{u_\tau}{0.41\sqrt{\beta^*}y} \qquad (11.36)$$

before proceeding to the next advancement of the solution.

11.5.3 EXAMPLE

The numerical performance of the algorithm is illustrated by considering its application to the simulation of turbulent supersonic flow over a flat plate. The free stream Mach number is 2, the plate length is 1 metre and the Reynolds number is 10^7 per metre. The plate is taken to be isothermal, with $T_w = 222^0K$, while the Prandtl numbers are constant and set according to $Pr_t = 0.9$ and $Pr = 0.72$. The mesh employed is a triangulation of a structured 101×101 grid, with the first layer of nodes located at a distance of 3×10^{-6} metres above the wall. The aspect ratio of the elements adjacent to the wall varies from 1,600 at the leading edge to 5,200 at the trailing edge. Figure 3 shows the computed variation of the velocity component u_1, with distance x_2 away from the wall, at the section $x_1 = 0.93$ metres. The result of Prandtl's power law [Whi91] is also shown on this figure, for comparison. More detail of this comparison in the near wall region is apparent in the representation of the profiles in the plot of U^+ against Y^+, which is given in Figure 4. A comparison between the computed skin friction distribution and the closed form solution of White–Cristoph [Whi91] is shown in Figure 5. These comparisons form the initial validation of the proposed solution procedure.

11.6 MESH GENERATION

The simulation of the flow over a flat plate has demonstrated that the efficient computation of turbulent flows normally requires the use of highly stretched meshes. For configurations of general shape, anisotropic meshes of this form cannot be constructed by standard unstructured mesh generators, which have generally been designed with the generation of isotropic meshes in mind. For this reason, initial attempts at employing unstructured mesh algorithms for the simulation of viscous flows tended to adopt a structured mesh in the immediate vicinity of solid surfaces and an unstructured mesh elsewhere. This approach is viable for most two–dimensional simulations [Nak86, HMP89, HC89] and also for three–dimensional simulations where the geometry is not too complex [HMP91a]. However, this method tends to negate the geometrical advantages of the fully unstructured mesh approach. Another possibility is to begin with an isotropic unstructured mesh over the domain of interest and to

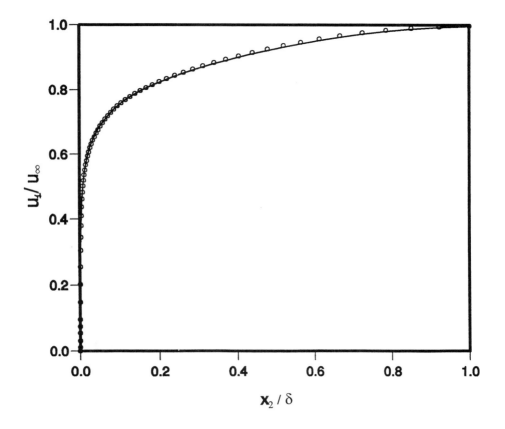

Figure 3 Comparison between the computed velocity profile (circles) and Prandtl's power law (solid line) for the case of supersonic turbulent flow over a flat plate

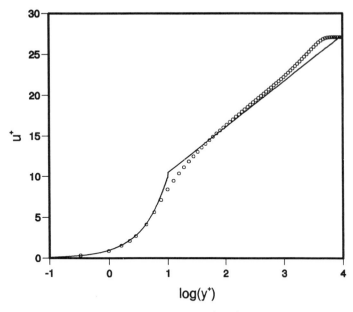

Figure 4 Comparison between the computed (U^+, Y^+) profile (circles) and the law of the wall (solid line) for the case of supersonic turbulent flow over a flat plate

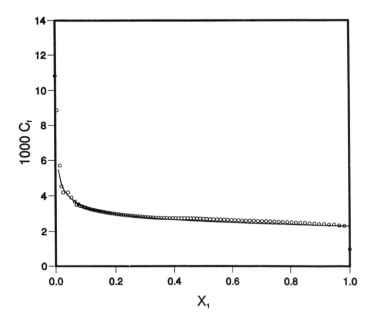

Figure 5 Comparison between the computed skin friction distribution along the wall (circles) with the closed form solution of White–Cristoph (solid line)

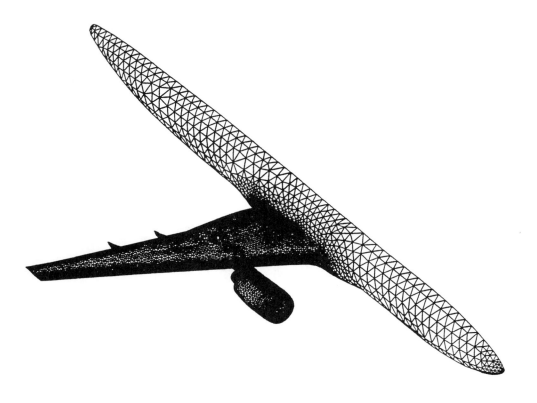

Figure 6 Surface discretisation for the wing/body/pylon/nacelle configuration

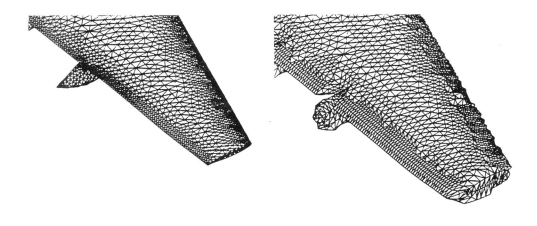

Figure 7 Comparison of details of the initial surface discretisation with the corresponding mesh on the upper surface of the generated layers for the wing configuration

adapt this in some manner to produce an anisotropic mesh suitable for the viscous flow analysis. Following this method, a degree of success has been achieved in both two–dimensional [HMP$^+$91b, HPMP94, Wea94] and, more recently, three dimensional simulations [PM96]. In this paper, we have followed an alternative approach in which we attempt to generate a suitable mesh directly [Pir96, HMPP96].

The problem of directly generating meshes suitable for viscous flow analysis is divided into three separate phases. Firstly, the boundary surface components are triangulated using the advancing front method. The element size distribution is controlled by a user–specified mesh control function. Secondly, unstructured layers of highly stretched tetrahedral elements are generated adjacent to those boundary surface components which represent solid walls. The height and number of these layers is specified by the user, who will normally attempt to ensure that the distribution of points in the final mesh will be such that the boundary layer in the simulated flow can be adequately represented. The layers are constructed by generating points along normals emanating from surface nodes and connecting the points to form tetrahedral elements. When normals begin to intersect, the process is locally terminated. Thirdly, the remainder of the domain is discretised using a Delaunay procedure [WH92].

This approach has been successfully employed to produce meshes on a number of different aerospace configurations. For example, starting from the surface discretisation shown in Figure 6, a mesh consisting of a maximum of eight layers of stretched elements was generated. The initial surface discretisation consisted of 27,924 triangles and the height of the first layer was set at 0.0003 of the root chord. Details of the initial surface mesh and the corresponding details of the mesh on the upper surface of the generated layers are shown in Figures 7 and 8. There are 621,511 elements and 118,947 points

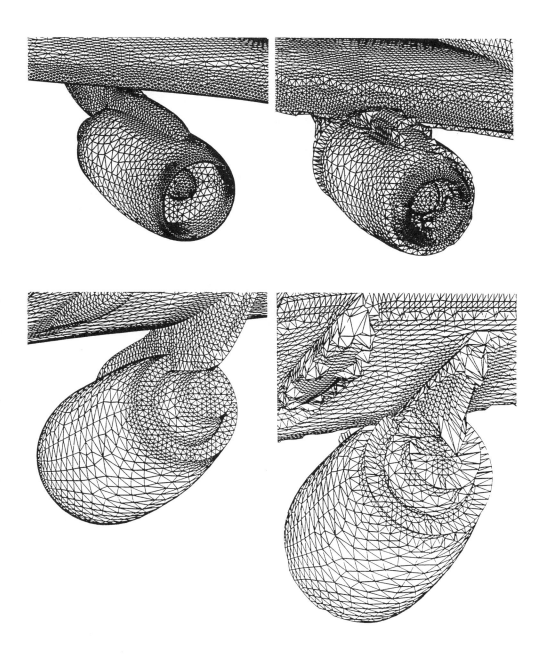

Figure 8 Comparison of details of the initial surface discretisation with the corresponding mesh on the upper surface of the generated layers for the body/pylon/nacelle configuration

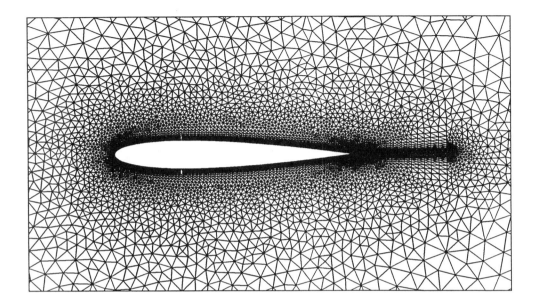

Figure 9   Detail of the mesh employed for the simulation of subsonic flow over a
NACA0012 aerofoil

in the generated layers, while the complete mesh consists of 793,805 elements and
138,071 points. When it is realised that this mesh is far too coarse to be employed
for the simulation of a realistic turbulent flow over a configuration of this type, it is
apparent that the problems of interest are going to be characterised by the requirement
for using very large meshes, which will include high aspect ratio elements.

11.6.1   EXAMPLE

An example involving the two dimensional simulation of subsonic turbulent flow, at a
free stream Mach number of 0.502, over a NACA0012 aerofoil is chosen to illustrate
the coupled use of the mesh generator and the flow solver. The Reynolds number,
based upon the aerofoil chord, is $Re = 2.91 \times 10^6$ and the angle of attack is 1.77^o.
The Prandtl numbers are $Pr = 0.72$ and $Pr_t = 0.9$ and the free–stream temperature
is $116.43^o R$.

   For this problem, computations were performed using both the standard and the
wall function boundary conditions. For the standard computation, a grid containing
$14,097$ nodes and $27,939$ triangular elements, with $\delta = 0.00001$ of the chord length,
was generated. A detail of this grid is given in Figure 9. The computation employing
the wall function boundary condition was performed on a grid containing $3,831$ nodes

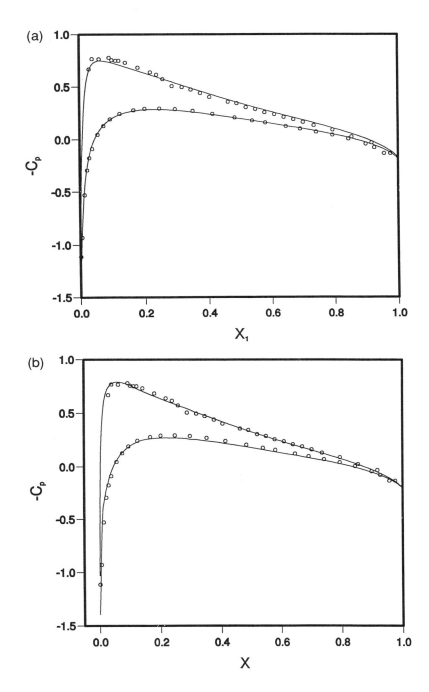

Figure 10 Subsonic flow over a NACA0012 aerofoil. Comparison between the computed (line) and the experimental (circles) distribution of the wall pressure (a) when the standard wall boundary condition is employed (b) when the wall function boundary condition is employed

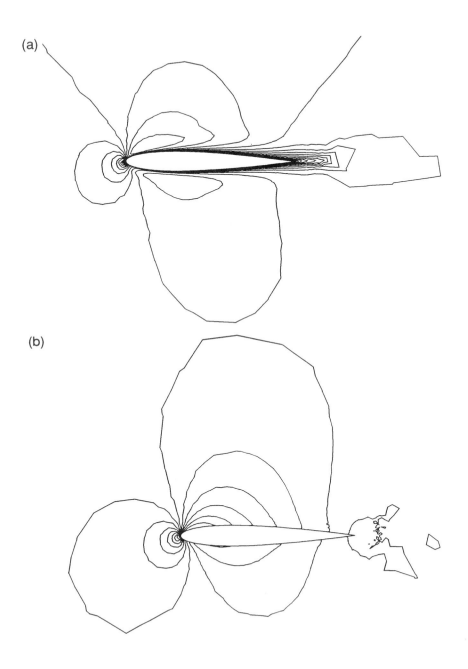

Figure 11   Subsonic flow over a NACA0012 aerofoil. Detail of the computed contours, showing (a) Mach number (b) pressure coefficient

and $7,426$ triangular elements, with $\delta = 0.001$ of the chord length. The computed wall pressure distributions are compared with experimental measurements in Figure 10. It is apparent that both procedures have produced results which are in close agreement with experimental data. Details from the distribution of computed Mach number and pressure contours, when the wall function boundary conditions were applied, are shown in Figure 11. The corresponding plots for the other simulation were found to be very similar.

11.7   PARALLEL FLOW SOLVER

The topics discussed in this paper lead to the inevitable conclusion that the simulation of three dimensional turbulent flows over realistic aerospace configurations can be expected to require considerable computational resources. However, it is apparent that existing memory and cpu limitations may be overcome if the problem can be subdivided and solved on a system of distributed processors [Cab95, MM94, GWZ94]. In this section, we briefly describe the results of some initial experiments in implementing the flow algorithm on a computer system of this type. At this stage, the experiments have been confined to the simulation of inviscid flows only. Portability of the resulting code has been accorded prime importance and no attention has been paid to attempting optimisation for particular computer platforms. For this reason, only the standard PVM message passing libraries [GBD+94] have been used.

The approach which has been followed begins by decomposing the mesh for the original computational domain into a number of smaller meshes of sub–domains. Here, this is achieved by employing the recursive spectral bisection software of Simon [Sim91]. Following this decomposition, the nodal points and the mesh edges are locally renumbered within each sub–domain. The governing equations are then solved on the individual sub–domains using the standard algorithm, with the sub–domain solutions being patched together to produce the solution of the original problem.

To achieve this, we recall that, for the sequential version of the edge based flow solver described above, the main data structure takes the form of a pointer from edges to points. In the main computation, consisting of loops over edges in the mesh, information from points is gathered to edges and edge information is scattered and added to points.

In the parallel implementation, edges in the mesh are owned by only one domain and are not duplicated. For a typical sub–domain $\mathcal{I}$, interior edges and neighbour edges can then be defined. An interior edge will be an edge which belongs to sub–domain $\mathcal{I}$ and which is such that both nodes connected by the edge also belong to $\mathcal{I}$. In this case, data locality is achieved during the gather and scatter processes, from the nodes to the edge and vice versa, and there is no need for inter–domain communication. A neighbour edge will be an edge for which one node belongs to sub–domain $\mathcal{I}$ while the other belongs to a different sub–domain $\mathcal{J}$. The convention employed will assign this edge to sub–domain $\mathcal{I}$ only if $\mathcal{J} > \mathcal{I}$. This edge will obviously contribute to the accumulation of contributions to a node in sub–domain $\mathcal{J}$ and, in this case, inter–domain communication is required.

Nodal points are regarded as being owned by one domain, but are duplicated at sub–domain boundaries, creating a halo of dummy points, to enforce data locality.

Table 1 Computational performance on a CRAY T3D for a mesh of approximately two million elements

Processors	Edges(min/max)	Cut Edges	T3D(m)	OSU	ASU
4	469771/484465	31122	650	1	1
8	230459/244827	54824	345	2	1.88
16	114139/125040	84347	159	4	4.09
32	55144/64926	118506	92	8	7.07
64	26760/33475	168030	44	16	14.77
128	12826/17323	220279	27	32	24.07
256	6224/8613	292210	14	64	46.42

For a typical sub–domain $\mathcal{I}$, interior points and neighbour points are defined. A point will be an interior point for sub–domain $\mathcal{I}$ provided that all the edges surrounding the point belong to sub–domain $\mathcal{I}$. For an interior point, data locality is achieved during the scatter process from edges to points and there is no need to communicate. A point which does not belong to sub–domain $\mathcal{I}$ will be regarded as a neighbour point if at least one surrounding edge belongs to sub–domain $\mathcal{I}$. In this case, inter–domain communication is required.

The parallel version of the code has been used in the simulation of the inviscid flow over the complex aircraft configuration shown in Figure 6. The volume mesh employed consisted of approximately two million tetrahedral elements. The simulation was undertaken on a CRAY T3D computer and 1000 time steps were computed, using initially 4 processors. The simulation was then repeated several times, each time with an increased number of processors. The performance of the code can be judged from Table 1, which displays for each simulation the optimum speed up attainable (OSU) and the actual speed up attained (ASU). It can be observed that excellent efficiency is obtained as the number of processors in increased to 64. For a mesh of this size, the performance shows evidence of degradation when 128 and 256 processors are employed, as communication is then beginning to dominate over computation.

These results are very encouraging and show that the chosen algorithm can be efficiently and effectively parallelised. Work is currently underway on the parallelisation of the turbulent viscous flow solver and there is no reason to suggest that the performance of this code will be markedly different.

11.8 PARALLEL MESH GENERATION

Unstructured mesh generators can be expensive in terms of computer memory requirements and the normal user may experience problems in generating the large meshes which are required for the accurate solution of complex turbulent flows. We would like to investigate if these memory requirements can be overcome by generating the meshes in parallel, using distributed processors. As a by product, a parallel implementation would also lead to a significant reduction in the cpu time needed for the mesh generation.

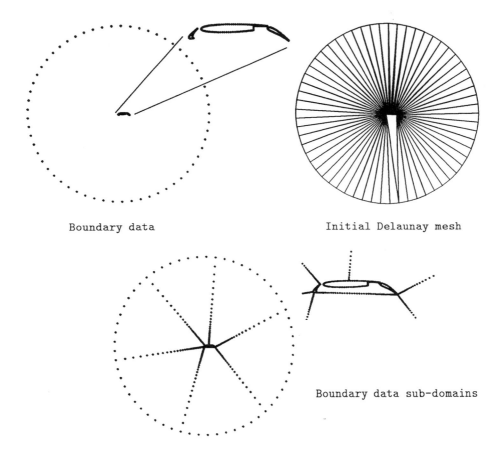

Boundary data   Initial Delaunay mesh

Boundary data sub-domains

Figure 12   Decomposition of a domain into a number of sub–domains

(a) (b)

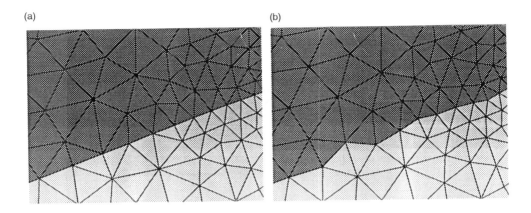

Figure 13 Mesh at inter-domain boundary (a) without smoothing (b) with smoothing

Parallelisation of the Delaunay mesh generation procedure in two dimensions has already been investigated [VWM95]. The approach which is followed employs a manager/worker structure, in which the workers execute the mesh generation routines. Following a Delaunay triangulation of the boundary points, a greedy–type algorithm is employed to decompose the computational domain into a number of sub–domains, as shown in Figure 12. This stage is accomplished by the manager and the discretisation of the sub–domains is accomplished by the workers. With standard Delaunay mesh generation, a post–processing step is normally applied in which the generated mesh is smoothed. The smoothing has the effect of moving points locally in a way that the quality of the mesh is improved. When the mesh is generated in parallel, the inter-domain boundaries generated by the initial decomposition are straight lines and these remain in the final mesh. In this case, post–processing is also possible at the inter–domain boundaries following the generation of the complete mesh. To achieve this, the workers communicate the neighbour points to the manager and the manager executes the inter–domain smoothing as a post–processing step. Figure 13 illustrates the form of the mesh at an inter-domain boundary with and without smoothing.

In this type of implementation, there is no communication between the individual workers. With a dynamic loading implementation, when a worker has completed its current meshing task, it is allocated a new sub–domain. The size of the mesh which can be generated on any sub–domain is limited only by the memory available on the

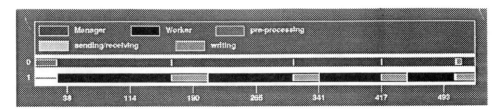

Figure 14   MPI/Upshot screen timing snapshot with a single worker

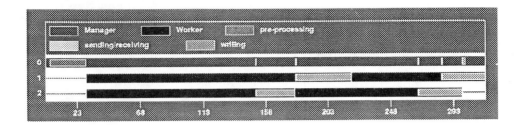

Figure 15   MPI/Upshot screen timing snapshot with two workers

worker. An important practical result of this procedure is that large global meshes can be obtained, without paging, by combining several small sub–domain meshes.

Figures 14 and 15 represent MPI/Upshot snapshots of an example of a mesh generated with four sub–domains. The time bar 0 represents the manager, while the other time bars represent the workers.

The meshes were generated on a network of SGI Indy workstations, each with an R4600 processor, a data cache size of 16 kbytes and 32 Mbytes of main memory. Apart from the pre–processing and file writing, the manager is only active when a worker asks for a new task (sending/receiving). The worker memory was set for meshes of up to 30,000 points. In Figure 14, one worker discretises all four sub–domains, producing a mesh with a total number of 82,547 points, 247099 edges, 164552 elements and 542 boundary points. Table 2 represents the number of points, edges, and elements per sub-domain generated. It also shows, for the case in which two workers are employed (see Figure 15), which sub–domain was meshed by which worker.

Table 2   Sub–domain mesh sizes generated by workers 1 and 2

worker	domain	points	edges	elements	boundary
1	1	25551	75654	50104	996
2	2	21744	64345	42602	884
2	4	21083	62382	41300	864
1	3	15601	46146	30546	654

11.9 CONCLUSIONS

The results presented have demonstrated the validation of an explicit sequential $k - \omega$ turbulence model implementation on unstructured triangular and tetrahedral meshes. A method for generating anisotropic unstructured tetrahedral meshes suitable for the simulation of turbulent compressible flows over complex three dimensional aerospace configurations has also been discussed. The computer memory and cpu requirements of the approach are considerable and would currently prohibit its routine use. For this reason, the possibility of employing distributed memory computers has been investigated and initial experiences with both parallel mesh generation and parallel flow solver have been shown to be encouraging. To bring this to fruition, further work is now needed to enable the rapid generation of grids of the size necessary to perform accurate turbulent flow simulations over realistic geometries. Although parallelisation of the flow solver does not appear to be a major difficulty, the analyst would certainly benefit from the additional incorporation of some form of convergence acceleration. In this context, multigrid or implicit methods are worthy of further investigation.

ACKNOWLEDGEMENTS

N. Verhoeven thanks the UK Engineering and Physical Sciences Research Council for the support provided under Research Grant GR/J91234. K. Morgan and N.P. Weatherill thank the UK Engineering and Physical Sciences Research Council for the support provided under Research Grant GR/J12321. The authors thank the UK Engineering and Physical Sciences Research Council for providing access to the CRAY T3D at Edinburgh under Research Grant GR/K42264.

References

[Cab95]   Cabello J. (1995) Parallel explicit unstructured grid solvers on distributed memory computers. Technical report, Department of Civil Engineering, University of Wales, Swansea.

[GBD$^+$94] Geist A., Beguelin A., Dongara J., Jiang W., Manchek R., and Sunderam V. (1994) *PVM3 User's Guide and Reference Manual.* Oak Ridge National Laboratory, Tennessee.

[GWZ94]   Grant P. W., Webster M. F., and Zhang X. (1994) Distributed parallel processing of a CFD code on workstation clusters. Technical report, Department of Computer Science, University of Wales, Swansea.

[HC89]   Holmes D. G. and Connell S. D. (1989) Solution of the 2–D Navier–Stokes equations on unstructured adaptive grids. *AIAA Paper 89–1932–CP* .

[Hir90]   Hirsch C. (1990) *Numerical Computation of Internal and External Flows. Volume 2.* John Wiley, Chichester.

[HMP89]   Hassan O., Morgan K., and Peraire J. (1989) An implicit/explicit scheme
             for compressible viscous high speed flow. *Computer Methods in Applied
             Mechanics and Engineering* 76: 245–258.

[HMP91a]   Hassan O., Morgan K., and Peraire J. (1991) An implicit finite element
             method for high speed flows. *International Journal for Numerical
             Methods in Engineering* 32: 183–205.

[HMP+91b]   Hassan O., Morgan K., Peraire J., Probert E. J., and Thareja R. R. (1991)
             Adaptive unstructured mesh methods for steady viscous flow. *AIAA
             Paper 91–1538* .

[HMPP96]   Hassan O., Morgan K., Probert E. J., and Peraire J. (1996) Unstructured
             tetrahedral mesh generation for three dimensional viscous flows.
             International Journal for Numerical Methods in Engineering 39: 549–
             567.

[HPMP94]   Hassan O., Probert E. J., Morgan K., and Peraire J. (1994) Unstructured
             mesh generation for viscous high speed flows. In Weatherill N. P.,
             Eiseman P. R., Häuser J., and Thompson J. F. (eds) *Proceedings of the
             4th International Conference on Numerical Grid Generation in CFD and
             Related Fields*, pages 779–793. Pineridge Press, Swansea.

[Joh87]   Johnson C. (1987) *Numerical Solution of Partial Differential Equations
             by the Finite Element Method*. Cambridge University Press.

[KGB96]   Kalitzin G., Gould A. R. B., and Benton J. J. (1996) Application of
             two–equation turbulence models in aircraft design. *AIAA Paper 96–0327*

[Man96]   Manzari M. T. (1996) *An unstructured grid finite element algorithm
             for compressible turbulent flow computations*. PhD thesis, University of
             Wales Swansea, C/Ph/191/96.

[MM94]   Morano E. and Mavriplis D. (1994) Implementation of a parallel
             unstructured Euler solver on the CM–5. *AIAA Paper* .

[MPPH91]   Morgan K., Peraire J., Peiró J., and Hassan O. (1991) The computation
             of three dimensional flows using unstructured grids. *Computer Methods
             in Applied Mechanics and Engineering* 87: 335–352.

[MPPH94]   Morgan K., Peraire J., Peiró J., and Hassan O. (1994) Unstructured Grid
             Methods for High Speed Compressible Flows. In Whiteman J. R. (ed)
             The Mathematics of Finite Elements and Applications: Highlights 1993,
             pages 215–241. John Wiley, Chichester.

[Nak86]   Nakahashi K. (1986) FDM–FEM zonal approach for compuations of
             compressible viscous flows. In *Lecture Notes in Physics 264*, pages 494–
             498. Springer Verlag, Berlin.

[Pir96]   Pirzadeh S. (1996) Progress towards a user–orientated unstructured
             viscous grid generator. *AIAA Paper 96–0031* .

[PM96]   Peraire J. and Morgan K. (1996) Unstructured mesh generation including directional refinement for aerodynamic flow simulation. *Finite Element Methods for Engineering Analysis and Design* In press.

[PMH96]   Peraire J., Morgan K., and Hassan O. (1996) Edge based unstructured mesh procedure for 3D time domain electromagnetic scattering. *AIAA Paper 96–0830* .

[PMP90]   Peraire J., Morgan K., and Peiró J. (1990) Unstructured finite element mesh generation and adaptive procedures for CFD. In *Conference Proceedings No: 464—Applications of Mesh Generation to Complex 3D Configurations*, pages 18.1–18.12. AGARD, Paris.

[PMVP94]   Peraire J., Morgan K., Vahdati M., and Peiró J. (1994) The construction and behaviour of some unstructured grid algorithms for compressible flows. In Morton K. W. and Baines M. J. (eds) *Numerical Methods for Fluid Dynamics IV*, pages 221–229. Oxford Science Publications.

[PPM93a]   Peraire J., Peiró J., and Morgan K. (1993) Finite element multigrid solution of Euler flows past installed aero–engines. *Computational Mechanics* 11: 433–451.

[PPM$^+$93b] Peraire J., Peiró J., Morgan K., Vahdati M., and Molina R. C. (1993) Hypersonic flow computations around re–entry vehicles. In *Conference Proceedings No: 514—70th FDP Meeting on Theoretical and Experimental Methods in Hypersonic Flows*, pages 39.1–39.13. AGARD, Paris.

[Sch84]   Schetz J. A. (1984) *Foundations of Boundary Layer Theory for Momentum, Heat and Mass Transfer*. Prentice–Hall, Englewood Cliffs, N.J.

[Sim91]   Simon H. D. (1991) Partitioning of unstructured problems for parallel processing. *Computing Systems in Engineering* 2: 135–148.

[Tho91]   Thomas J. L. (1991) An implicit multigrid scheme for hypersonic strong–interaction flowfields. In *Proceedings of the Fifth Copper Mountain Conference on Multigrid Methods*.

[Ven91]   Venkatakrishnan V. (1991) Preconditioned conjugate gradient methods for the compressible Navier–Stokes equations. *AIAA Journal* 29: 1092–1100.

[vL79]   van Leer B. (1979) Towards the ultimate conservative difference scheme. V: A second order sequel to Godunov's method. *Journal of Computational Physics* 32: 101–136.

[VWM95]   Verhoeven N., Weatherill N. P., and Morgan K. (1995) Dynamic load balancing in a 2D parallel Delaunay mesh generator. In Ecer A., Periaux J., Satofuka N., and Taylor S. (eds) *Parallel Computational Fluid Dynamics: Implementations and Results Using Parallel Computers*, pages 641–648. Elsevier Science, Amsterdam.

[Wea92]   Weatherill N. P. (1992) Delaunay triangulation in computational fluid
           dynamics. *Computer Mathematics and Applications* 24: 129–150.

[Wea94]   Weatherill N. P. (1994) Grid generation by the Delaunay triangulation.
           In *Lecture Series 1994–02.* von Karman Institute for Fluid Mechanics,
           Brussels.

[WH92]   Weatherill N. P. and Hassan O. (1992) Efficient three dimensional grid
           generation using Delaunay triangulation. In Hirsch C. (ed) *Proceedings
           of the 1st European Computational Fluid Dynamics Conference.* Elsevier,
           Amsterdam.

[Whi91]   White F. M. (1991) *Viscous Fluid Flow.* McGraw Hill, New York, second
           edition.

[Wil93]   Wilcox D. C. (1993) *Turbulence Modelling for CFD.* DCW Industries
           Inc, La Cañada, Calfornia.

[ZM83]   Zienkiewicz O. C. and Morgan K. (1983) *Finite Elements and
           Approximation.* John Wiley, New York.

12

Multicomponent, Multiphase Flow and Transport in Porous Media

Mary Wheeler, Clint Dawson and Carlos Celentano

Texas Institute for Computational and Applied Mathematics
The University of Texas at Austin
Taylor Hall 2.400
Austin, Texas, 78712, U.S.A.

12.1   INTRODUCTION

The contamination of groundwater is one of the most serious environmental problems. The characterization and remediation of contaminated sites is difficult and expensive and only now is technology emerging to cope with this severe and widespread problem. Modeling of multiphase flow in permeable media plays a central role in these technologies and is essential for risk assessment, cost reduction and the rational and effective use of resources. The research needed in this area applies to both subsurface environmental remediation and to problems associated with the environmentally prudent production of hydrocarbon energy from existing oil and gas fields.

In this paper we will describe the general structure of the mathematical and physical models required for the simulation of subsurface flow and transport in heterogeneous porous media and indicate areas of present and future research. In Section 12.2 we present a model formulation for multicomponent, multiphase flow that includes mass transfer, geochemical and biochemical effects. In Section 12.3 we briefly discuss some of the computational issues arising in the numerical solution of subsurface flow equations. Section 12.4 is concerned with a black-oil model formulation, and is intended to illustrate in a simple way the coupling of fluid phase behavior with flow and transport. In Section 12.5 we show how geostatistics can be used to help in the estimation of uncertainty in the performance of remediation strategies and oil production. Finally, in Section 12.6, we point out some of the current problems that demand further research.

The Mathematics of Finite Elements and Applications
Edited by J. R. Whiteman © 1997 John Wiley & Sons Ltd.

12.2   MODEL FORMULATION

Below we outline a model for multicomponent, multispecies, multiphase flow. The model includes equilibrium and nonequilibrium mass transfer between phases, geochemistry, biochemistry and radionuclide decay.

In the discussion below, we use the terms *components* and *species*. A component refers to a basic chemical entity such that every species can be uniquely represented as a combination of components. A species is a product of a chemical reaction involving components as reactants.

Let n_c denote the number of components in the system. Let n_p denote the total number of flowing and stationary phases in the system, and n_s the maximum number of species in the system.

The flow of species k within phase j can be written as [LPCS84, Lak89]

$$\frac{\partial N_{kj}}{\partial t} + \nabla \cdot \mathbf{F}_{kj} = R_{kj},\tag{12.1}$$

where N_{kj} is the molar density (moles/unit volume of phase j), $\mathbf{F}_{kj}$ is the species flux and R_{kj} describes kinetic source and sink terms.

The species flux can be written as

$$\mathbf{F}_{kj} = \xi_j x_{kj} \mathbf{u}_j - \phi \xi_j S_j \mathbf{K}_{kj} \nabla x_{kj},\tag{12.2}$$

where ξ_j is the molar density of phase j, x_{kj} is the mole fraction of species k in phase j, $\mathbf{u}_j$ is the Darcy velocity of phase j, ϕ is the porosity of the medium, S_j is the saturation of phase j and $\mathbf{K}_{kj}$ is the diffusion/dispersion tensor for the species. We note that $\xi_j x_{kj} = N_{kj}$, $\sum_{k=1}^{n_s} x_{kj} = 1$, and $\sum_j S_j = 1$. We also note that (12.1) does not preclude changes in the porosity of the media due to, for example, biomass growth.

The flowing phase velocity is given by Darcy's Law:

$$\mathbf{u}_j = -\frac{K k_{rj}}{\mu_j} \left(\nabla P_j - \rho_j g \nabla D \right),\tag{12.3}$$

where k_{rj} is the relative permeability, μ_j is the viscosity, P_j is the pressure and ρ_j is the mass density for phase j, and $g \nabla D$ is a gravitational force vector. Here

$$\frac{K k_{rj}}{\mu_j} \equiv \lambda_j = \lambda_j(S, x, \mathbf{x}, T),\tag{12.4}$$

where S is a vector of phase saturations, x is a vector of species mole fractions, $\mathbf{x}$ is spatial location, and T is temperature . The capillary pressure relation gives

$$P_j - P_n = P_{cjn}(S, x, \mathbf{x}, T), \quad j \neq n.\tag{12.5}$$

This relation may also be hysteric. Molar and mass densities, viscosity and porosity are assumed to be functions of composition, pressure and temperature, with porosity also varying spatially:

$$\xi_j = \xi_j(x, P_j, T), \quad \rho_j = \rho_j(x, P_j, T),\tag{12.6}$$

$$\mu_j = \mu_j(x, P_j, T), \quad \phi = \phi(\mathbf{x}, x, P_j, T).\tag{12.7}$$

For nonmobile species (such as microorganisms attached to the pore surface), the flux term $\mathbf{F}_{kj} = 0$.

The reaction term R_{kj} can be quite complicated, involving numerous types of reactions. In particular,

$$
\begin{aligned}
R_{kj} \quad = \quad & \text{mass transfer of species } k \text{ into and out of phase } j \\
& + \text{ geochemical reactions } + \text{ biochemical reactions} \\
& + \text{ radionuclide decay } + \text{ external sources and sinks .} \quad (12.8)
\end{aligned}
$$

The geochemical reactions include reactions within the phase (complexation and redox), sorption (adsorption and ion exchange) and precipitation/dissolution.

In many cases of interest, the reactions occur at such fast rates relative to transport that they are assumed to be in local equilibrium. In the case of equilibrium mass transfer, the species balance equations are generally summed over phases (noting that the mass transfer terms sum to zero), giving an equation for the total species molar density:

$$
\frac{\partial N_k}{\partial t} + \sum_{j=1}^{n_p} \nabla \cdot F_{kj} = \sum_{j=1}^{n_p} R_{kj}. \quad (12.9)
$$

In practice, one solves for the total species by some means and performs a "flash calculation" to determine the mole fraction of the species within each phase.

When modeling geochemistry, similar manipulations can be performed to form equations for total component densities. Multiplying (12.1) by stoichiometric coefficients a_{ik}, determined by

$$
\sum_{k=1}^{n_s} a_{ik} N_{kj} = \tilde{N}_{ij}, \quad (12.10)
$$

where $\tilde{N}_{ij}$ is the molar density of component i in phase j, summing on species and phases, we obtain an equation for the total molar density of the component:

$$
\frac{\partial \tilde{N}_i}{\partial t} + \sum_{k=1}^{n_s} a_{ik} \sum_{j=1}^{n_p} \nabla \cdot F_{kj} = \sum_{k=1}^{n_s} a_{ik} \sum_{j=1}^{n_p} R_{kj}. \quad (12.11)
$$

Since the total reaction rate must sum to zero, the right hand side in (12.11) reduces to a sum of external sources and sinks and radionuclide decay terms only.

For both equilibrium mass transfer and geochemistry, determining mole fractions of each species in each phase involves minimizing the Gibbs free energy, subject to constraints. Equivalent methods are available for specialized forms of phase equilibrium such as needed for microemulsions [DPS]. For mass transfer, simplified methods such as Henry's Law are often used for dilute solutions or trace components

Nonequilibrium reactions require returning to (12.1) to determine the mole fraction of a species within a phase. A functional form for the reaction term must be specified. For nonequilibrium mass transfer, for example, the reaction term which would appear on the right side of (12.1) is generally of the form

$$
\kappa_k (x_{kj}^{eq} - x_{kj}), \quad (12.12)
$$

where κ_k is an effective mass transfer coefficient and x_{kj}^{eq} is obtained through an equilibrium calculation   [PLAW91, MPMM90, PAW94]. In chemical reactions, a similar though more complicated expression is commonly used. Thus nonequilibrium reactions may also require solving a constrained minimization problem.

In order to complete the model, we need an energy balance. The molar energy balance for the control volume using internal energy rather than temperature as a primary variable, assuming local equilibrium heat transfer between phases, can be expressed as [BPS91]

$$\frac{\partial u}{\partial t} + \nabla \cdot \sum_{j=1}^{n_p} \xi_j h_j \mathbf{u}_j - \nabla \cdot (\lambda_T \nabla T) - q_H + q_L = 0. \tag{12.13}$$

Here $u = u_r + u_f$ is the sum of the internal energies of the minerals (rock or soil) and the fluids per unit bulk volume, h_j denotes the molar enthalpy of phase j, λ_T is the effective thermal conductivity, q_H is the enthalpy injection rate per unit bulk volume and q_L is heat loss to the over- and under-burdens per unit bulk volume.

In summary, equations for total species molar density or total component molar density can be formed using (12.9) or (12.11). Within these equations are mole fractions of species within a phase. These mole fractions are determined by an equilibrium relationship or the nonequilibrium equation (12.1). These equations, together with Darcy's Law, the constitutive relationships (12.4)-(12.7), volume balance, energy balance and appropriate boundary and initial conditions comprise the mathematical model.

12.3   BLACK OIL MODEL

As a special case of the general transport model described above, we present here in some detail the basic equations corresponding to the black oil model, which has been widely used in the petroleum industry.

In the black oil model, it is assumed that there are only three components and three phases, both denoted by oil, gas and water. All fluids are at constant temperature and in thermodynamic equilibrium throughout the reservoir. Furthermore, mass transfer between phases is allowed in the following way: the gas component can exist in all three phases, the oil component can be present in the oil and gas phases and the water component only exists in the water phase.

For each component, the mass conservation equation can be written as

$$\frac{\partial}{\partial t} [\phi N_I] + \nabla \cdot \left[\sum_{J=1}^{N_P} C_{IJ} \rho_J \mathbf{u}_J \right] + q_I = 0 \tag{12.14}$$

where ϕ is porosity, N_I is the total mass density of component I, with $I = O, G, W$ for oil, gas and water, respectively. The mass fraction coefficients C_{IJ} represent the amount of component I in phase J, and will have to be determined by phase equilibrium as described below. They are related to the component densities by

$$N_I = \sum_{J=1}^{N_P} C_{IJ} S_J \rho_J$$

where S_J are phase saturations and ρ_J phase densities. Phase velocities $\mathbf{u}_J$ are given by Darcy's law (12.3). Finally, q_I represents a source term for component I.

In addition to the three mass conservation equations for each component, we have a volume constraint so that the fluids fill the porous space:

$$S_o + S_w + S_g = 1 \tag{12.15}$$

(Following standard notation, we use capital letter subscripts for component variables $(I = O, G, W)$ and lower-case letter subscripts for phase variables $(J = o, g, w)$.) Moreover, capillary pressures are assumed in the form

$$p_{Cow} = p_o - p_w = f(S_o, S_g, S_w) \tag{12.16}$$

$$p_{Cog} = p_g - p_o = f(S_o, S_g, S_w) \tag{12.17}$$

Additional relations for relative permeabilities $k_{rJ} = k_{rJ}(S_o, S_g, S_w)$ and phase viscosities $\mu_J = \mu_J(p_J, C_{IJ})$ complete the system of equations.

Thus, the flow and transport problem consists of six equations (three equations as given by (12.14), plus (12.15),(12.16) and (12.17)) with six unknowns: three mass components N_O, N_G, N_W and three phase pressures p_o, p_g, p_w. Using (12.16)-(12.17), the system can be reduced to 4×4 with primary variables given by, for instance, N_O, N_G, N_W and p_w.

Coupled with the flow description, there is the phase equilibrium problem that governs the thermodynamic conditions under which the components combine to form phases. By introducing certain functions of pressure and temperature obtained through experimental measurements, it is possible to obtain mass fractions C_{IJ}, phase densities ρ_J and saturations S_J given a state with known primary variables N_O, N_G, N_W and p_w. For example, when all three phases are present, the component densities and saturations are related through the following expressions [Pea77]:

$$\begin{bmatrix} N_O/N_{OS} \\ N_G/N_{GS} \\ N_W/N_{WS} \end{bmatrix} = \begin{bmatrix} B_o^{-1} & R_v B_g^{-1} & 0 \\ R_{so} B_o^{-1} & B_g^{-1} & R_{sw} B_w^{-1} \\ 0 & 0 & B_w^{-1} \end{bmatrix} \begin{bmatrix} S_o \\ S_g \\ S_w \end{bmatrix} \tag{12.18}$$

where the subscript S in the component densities denotes standard conditions. The functions B_o, B_g and B_w are the formation volume factors for each phase. R_{so} defines the solubility of gas in oil, R_{sw} is the solubility of gas in water and R_v gives the volatility of oil in gas. All these quantities are known functions of pressure and temperature. Similar equations can be derived for the mass fractions C_{IJ} and phase densities ρ_J.

It should be noted that a thermodynamic test is required to define the number of phases present at a given state. In correspondence with our mass transfer assumptions, it is possible to have either the gas phase missing or the oil phase missing. In the first case, all gas is dissolved in oil and/or water (undersaturated oil), while in the second case, all oil is dissolved into gas (undersaturated gas). For these cases, slightly different equations have to be employed in the phase equilibrium problem [TB89]. Notice, however, that the flow equations (1.14) are still valid when one phase is missing. This is why we use the component densities N_I as primary variables instead of saturations S_J, which would appear to be a better choice in view of the volume constraint (1.15), but would require to modify the flow equations as one phase disappears.

12.4   MIXED FINITE ELEMENT APPROXIMATION

We now describe a time-implicit mixed finite element approximation for the black oil system described above. We consider rectangular elements and the lowest-order Raviart-Thomas approximating spaces. Thus the mixed method reduces, through the use of numerical quadrature, to a cell-centered finite difference scheme.

Let Ω denote the computational domain, and for simplicity assume no-flow boundary conditions on $\partial\Omega$; i.e.,

$$\mathbf{u}_J \cdot \eta = 0, \quad J = O, W, G, \qquad (12.19)$$

where η is the outward normal. Let $(.,.)$ denote the $L^2(\Omega)$ inner product, scalar and vector.

We will approximate the L^2 inner product with various quadrature rules, denoting these approximations by $(.,.)_R$, where R = M, T and TM are application of the midpoint, trapezoidal and trapezoidal by midpoint rules, respectively.

Let $0 = t^0 < t^1 < \cdots < t^N = T$ be a given sequence of time steps, $\Delta t^n = t^n - t^{n-1}$, $\Delta t = \max_n \Delta t^n$, and for $\phi = \phi(t, .)$, let $\phi^n = \phi(t^n, .)$ with

$$d_t \phi^n = \frac{\phi^n - \phi^{n-1}}{\Delta t^n}.$$

Let $V = H(\Omega, div) = \{\mathbf{v} \in (L^2(\Omega))^d : \nabla \cdot \mathbf{v} \in L^2(\Omega)\}$ and $W = L^2(\Omega)$.

We will consider a quasi-uniform triangulation of Ω with mesh size h denoted by $\mathcal{T}$ and consisting of rectangles in two dimensions or parallelepipeds in three dimensions. We consider the lowest order Raviart-Thomas-Nedelec space on bricks ([RT77], [Ned80]). Thus, on an element $E \in \mathcal{T}$, we have

$$
\begin{aligned}
V_h(E) &= \{(\alpha_1 x_1 + \beta_1, \alpha_2 x_2 + \beta_2, \alpha_3 x_3 + \beta_3)^T : \alpha_i, \beta_i \in \mathbb{R}\}, \\
W_h(E) &= \{\alpha : \alpha \in \mathbb{R}\}.
\end{aligned}
$$

For an element on the boundary, $\partial E \subset \partial \Omega$, we have the edge space,

$$\Lambda_h(\partial E) = \{\alpha : \alpha \in \mathbb{R}\}.$$

We use the standard nodal basis, where for V_h the nodes are at the midpoints of edges or faces of the elements, and for W_h the nodes are at the centers of the elements. The nodes for Λ_h are at midpoints of edges.

For a given phase, the expanded mixed finite element method approximates all scalar quantities, N_I, ρ_J, C_{IJ}, S_J, p_J, etc. at each time level in the space W_h. In order to more easily handle full tensors we introduce the vector quantity $\tilde{\mathbf{u}}_J = -\nabla p_J + \rho_J g \nabla D$ and define $F_I = \sum_{J=1}^{N_p} C_{IJ}\rho_J \mathbf{u}_J$. We approximate all vector quantities in the space V_h. Denoting approximations with a subscript h, the method with numerical quadrature is as follows:

$$(d_t(\phi N_{I,h}^n), w) = -(\nabla \cdot \sum_{J=1}^{N_p} F_{I,h}^n, w) - (q_I^n, w)_M, \ \forall w \in W_h, \tag{12.20}$$

$$(\tilde{\mathbf{u}}_{J,h}^n, \mathbf{v})_{TM} = (p_{J,h}^n, \nabla \cdot \mathbf{v}) - (\alpha_{J,h}^n, \mathbf{v} \cdot \eta)_\Gamma + (\rho_{J,h}^n g \nabla D, \mathbf{v})_{TM}, \ \forall \mathbf{v} \in V_h, \tag{12.21}$$

$$(F_{I,h}^n, \mathbf{v})_{TM} = \sum_{J=1}^{N_p} (C_{IJ,h}^n \rho_{J,h}^n \lambda_{J,h}^n \tilde{\mathbf{u}}_{J,h}^n, \mathbf{v})_T, \ \forall \mathbf{v} \in V_h, \tag{12.22}$$

$$(F_{I,h}^n \cdot \eta, \mu)_\Gamma = 0, \ \forall \mu \in \Lambda_h, \tag{12.23}$$

The system (12.20)-(12.23) coupled with the algebraic equations (12.15-12.18) reduces to a finite difference scheme for one pressure and the three component densities.

12.5 GEOSTATISTICS AND UPSCALING

It is known that some of the parameters entering into the flow and transport equations can have large variations throughout the reservoir. Due to the range of spatial lengths present in actual fields, these variations cannot be measured properly, making unrealistic the assumption that precise values can be assigned to those parameters. As a means to quantify this uncertainty, some key parameters, like porosity and absolute permeability, are viewed instead as random variables with known statistics and spatial correlation. The flow simulation will now be understood as one realization of a random system whose probabilistic behavior we want to estimate. To this end, many flow simulations must be performed on different realizations of the geological data.

In order to capture the influence of fine scale heterogeneities in the flow field, it has become common practice to generate random permeability fields comprising millions of grid blocks. As flow simulators cannot perform practically on that scale, techniques are needed to upscale those parameters so that they can be used in computationally feasible grids.

We describe here a technique that is primarily concerned with the upscaling problem in oil and gas reservoirs. Within that context, the following assumptions are made: 1) the governing flow equations on the coarse scale are identical to those at the fine scale level; and 2) the absolute permeability tensor $k(\mathbf{x})$ is interpreted as a random variable with known second order statistics and spatial correlation. The goal is then to determine effective permeabilities such that major features of the flow field on the fine scale are replicated on an arbitrary chosen coarse scale.

The present upscaling technique has three main components. First, a layered model for the permeability field is constructed on each large cell using tools from geostatistics. Second, we solve the single phase pressure equation on this layered model in order to calculate the velocity field on the cell boundaries. Finally, the effective permeability tensor K_{eff} is obtained by minimizing the difference between the flow through the layered and uniform cells. We comment briefly on each part of the method.

In general, there is no unique way to define an effective permeability tensor K_{eff} for a uniform cell that will produce the same average flow as that of a fine scale $k(\mathbf{x})$ cell when subject to arbitrary boundary conditions. The value of K_{eff} calculated to match average fluxes on a large cell will depend on the applied boundary conditions and on the position at which we do the matching. By using an appropriate layered model k_L instead of the original fine scale geological model $k(\mathbf{x})$, we can define K_{eff} in a *unique* way, at least with respect to the class of uniform pressure boundary conditions on the cell surface. The layered permeability model is obtained through kriging [JH78], thus incorporating explicitly the spatial structure of $k(\mathbf{x})$.

The single phase pressure equation containing the layered model k_L is solved subject to Dirichlet boundary conditions (i.e., constant pressure on one face, zero pressure on all the others). Hence, by avoiding no-flow or periodic boundary conditions, we obtain a cell model closer to the real one when embedded in the reservoir.

To illustrate, suppose we want to calculate $K_{eff_{xx}}$ for a large cell of dimensions (a, b, c), containing many small blocks with $k(x, y, z)$ piece-wise constant. We have to solve the following pressure equation:

$$\nabla \cdot (k_L \nabla p) \;=\; 0$$

with $p(a, y, z) = 1$ and $p = 0$ on all other faces. Then, $K_{eff_{xx}}$ will be chosen so as to minimize the residual

$$R_x \;=\; \left\{ \int_0^b \int_0^c \left[k_L \frac{\partial p}{\partial x} - K_{eff_{xx}} \frac{\partial \phi}{\partial x} \right]^2 dy \, dz \right\}_{x=0}^{x=a}$$

where $\{f(x)\}_{x=0}^{x=a} = f(a) - f(0)$. $\phi(x, y, z)$ represents the pressure field on the uniform cell and therefore is the solution of Laplace's equation subject to the same boundary conditions. It can be easily shown that, for a *consistent* $K_{eff_{xx}}$, that is, for a $K_{eff_{xx}}$ independent of any particular uniform boundary value, the layered model has to be of the form $k_L(z)$ or $k_L(y)$. From these two possibilities, we will pick up the layered field that produces the minimum residual R_x. A similar procedure is repeated for $K_{eff_{yy}}$ and $K_{eff_{zz}}$. The method can be extended to handle effective permeabilities in full tensor form, which in general will be non-symmetric. For convenience, in this report we will only work with effective diagonal tensors. Notice that even for isotropic fine scale $k(\mathbf{x})$, the method will give anisotropic diagonal K_{eff}.

There are three sources of error in the above procedure. The first one has to do with the error in constructing a layered model k_L. This error can be estimated through the kriging variance σ_L^2. In general, σ_L^2 can be reduced by enlarging the point support employed in estimating k_L.

The second type of error arises when there exists a large contrast in the values k_L among different layers within a cell. This error can be monitored through the value of the residual R. Large values of R may indicate that the coarse grid has to be modified to better replicate the effect of fine scale heterogeneities.

Finally, we also have the error associated with the fact that, in general, the real boundary values affecting the large cell when embedded in the reservoir are not uniform. We can control this error, at least in a qualitative way, by a reasonable definition of the coarse scale grid.

We applied the proposed upscaling method to several water flooding cases having different fine scale permeability fields $k(\mathbf{x})$. In all cases, the geological reservoir contained 32^3 blocks and four wells, one injector and three producers, located on the corners. This uncommon flooding pattern represents a more stringent test for the uspcaling technique than the usual quarter 5-spot. The coarse scale model comprises 8^3 blocks. To avoid discretization errors, we ran both fine and coarse scale models on the 32^3 fine grid. The permeabilities are log-normally distributed, with mean 5.0 and standard deviation 1.0, so that 95% of the values lie in the interval 20 - 1000 mD. For brevity, only one case with correlation lengths $16 \times 16 \times 8$ fine grid blocks is presented here.

For this and other test cases, good agreement is observed between fine and coarse scale models in terms of fractional flow of oil at producers and total oil recovery (see Figure 1). In general, accurate breakthrough predictions are obtained for all wells, even though their fine scale values differ considerably between producers. Thus, the geostatistically-based upscaling method seems to capture well the average displacement patterns of fine scale flows whose permeability fields have a wide range of correlation lengths. More details can be found in [Cel96].

12.6 FURTHER RESEARCH

From this overview, it is clear that the simulation of complex subsurface phenomena on multiple realizations requires a flexible yet extremely powerful computational approach. To cope with this challenge, over the last years two main modeling approaches have emerged, which in the near future will be tied into a single problem solving environment.

The first modeling approach will emphasize computational speed at the expense of model complexity. This set of simulators will be based on a sequential decoupling of the flow and reactive transport processes. Problems involving two and three flowing phases with weakly compositional phase behavior will be solved within this framework (i.e., dilute mixtures and low pressure applications).

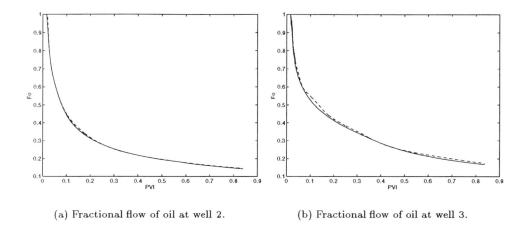

(a) Fractional flow of oil at well 2.   (b) Fractional flow of oil at well 3.

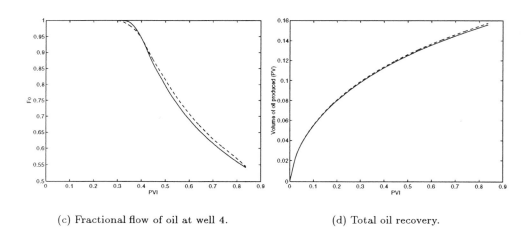

(c) Fractional flow of oil at well 4.   (d) Total oil recovery.

Figure 1   Water flooding test: fractional flow of oil at producer wells and total oil recovery as functions of pore volumes injected (PVI), for fine scale (continuous line) and coarse scale (dashed line) models, comprising 32^3 and 8^3 grid blocks, respectively.

The second modeling approach will be appropriate for problems with a tight coupling between flow and reactive transport. It will require a fully compositional formulation for phase behavior, with great demands in computational resources. Problems with complex phase behavior and high pressure fields fall into this group, such as petroleum and groundwater spill and DNAPL applications.

For this unified problem solving environment to be successful, more research is needed in the following areas:

• numerical algorithms, in particular for tightly coupled multicomponent, multiphase flow models;
• parallel linear and nonlinear solvers, which are crucial for the overall efficiency and robustness of the simulators;
• efficient techniques for solving the phase behavior problem, which in its most general form involves the minimization of Gibbs free energy;
• geostatistical modeling and scale up techniques for uncertainty estimation and risk assessment;
• incorporation of equilibrium and nonequilibrium geochemical reactions and biodegradation modules into the flow model.

References

[BPS91]   Brantferger K., Pope G. A., and Sepehrnoori K. (1991) Development of a thermodynamically consistent, fully implicit, compositional, equation-of-state steamflood simulator. In *paper SPE 21253 presented at the 11th SPE Symposium on Reservoir Simulation*. Anaheim, CA.

[Cel96]   Celentano C. (1996) A geostatistically-based upscaling technique for multiphase flow in heterogeneous reservoirs. Advanced Computational Technology Initiative. Annual Report, The University of Texas at Austin, Austin, TX.

[DPS]   Delshad M., Pope G. A., and Sepehrnoori K.A compositional simulator for modeling surfactant enhanced aquifer remediation 1: Formulation. *J. Contaminant Hydrology, to appear* .

[JH78]   Journel A. G. and Huijbregts C. J. (1978) *Mining Geostatistics*. Academic Press, London.

[Lak89]   Lake L. W. (1989) *Enhanced Oil Recovery*. Prentice Hall, Englewood Cliffs, NJ.

[LPCS84] Lake L. W., Pope G. A., Carey G. F., and Sepehrnoori K. (1984) Isothermal, multiphase, multicomponent fluid flow in permeable media. *In Situ* 8(1): 1–40.

[MPMM90] Miller C. T., Poirier-McNeill M. M., and Mayer A. S. (1990) Dissolution of trapped nonaqueous phase liquids: Mass transfer characteristics. *Water Resources Research* 26: 2783–2796.

[Ned80]   Nedelec J. C. (1980) Mixed finite elements in $\mathbb{R}^3$. *Numer. Math.* 35: 315–341.

[PAW94]   Powers S. E., Abriola L. M., and Weber, W. J. Jr. (1994) An experimental investigation of nonaqueous phase liquid dissolution in saturated subsurface systems: Transient mass transfer rates.   *Water Resources Research* 30: 321–332.

[Pea77]   Peaceman D. W. (1977) *Fundamentals of Numerical Reservoir Simulation*. Elsevier, Amsterdam.

[PLAW91]   Powers S. E., Louriero C. O., Abriola L. M., and Weber, W. J.   Jr. (1991) Theoretical study of the significance of nonequilibrium dissolution of nonaqueous phase liquids in subsurface systems.   *Water Resources Research* 27: 463–477.

[RT77]   Raviart P. A. and Thomas J. M. (1977) A mixed finite element method for second order elliptic problems.   In Galligani I. and Magenes E. (eds) *Mathematical Aspects of Finite Element Methods: Lecture Notes in Mathematics 606*, pages 292–315. Springer-Verlag, Berlin.

[TB89]   Trangestein J. A. and Bell J. B. (1989) Mathematical structure of the black-oil model for petroleum reservoir simulation. *SIAM J. Appl. Math.* 49: 749–783.

13

Method of characteristics implemented in relaxation schemes for convection diffusion problems with degenerate diffusion

J. Kačur

Department of Numerical Analysis, Faculty of Mathematics and Physics,
Comenius University, Mlynska dolina, 842 15 Bratislava, Slovakia

13.1 INTRODUCTION

In [AL83], [JK95], [KHK93], [Kač94], approximation schemes have recently been developed to solve double nonlinear degenerate parabolic problems of the form

$$\partial_t b(x, u) - \nabla a(t, x, u, \nabla \beta(x, u)) = f(t, x, u)$$

$$\text{in} \quad I \times \Omega \equiv Q_T, \quad (\Omega \quad \text{bounded domain in} \quad R^N), \quad I \equiv (0, T), \quad T < \infty \quad (13.1)$$

with the mixed boundary conditions

$$\beta(x, u) = 0 \quad \text{on} \quad I \times \Gamma_1,$$
$$a(t, x, u, \nabla \beta(x, u)).\nu = g(t, x, \beta(x, u)) \quad \text{on} \quad I \times \Gamma_2 \quad (13.2)$$

and initial conditions

$$b(x, u(x, 0)) = b(x, u_0) \quad \text{on} \quad \Omega \quad (13.3)$$

where $\partial \Omega = \Gamma_1 \cup \Gamma_2 \cup \Lambda$ (Γ_1, Γ_2 are open in $\partial \Omega, \Gamma_1 \cap \Gamma_2 = \emptyset$ and $mes_{N-1} \Lambda = 0$). Here $u = (u_1, \ldots, u_m)$ is a vector function and $b(x, u) \equiv (b^1(x, u^1), \ldots, b^m(x, u^m)), \beta(x, u) \equiv (\beta^1(x, u^1), \ldots, \beta^m(x, u^m))$. We assume that $b(x, s), \beta(x, s)$ are strictly increasing in s, and that $a(t, x, \eta, \xi)$ is monotone, coercive in ξ, e.g., $a(t, x, \eta, \xi) = \xi |\xi|^{p-2}, p > 1$.
In [Kač94], nondecreasing $b(x, s), \beta(x, s)$ have also been considered under the restrictions:

The Mathematics of Finite Elements and Applications
Edited by J. R. Whiteman © 1997 John Wiley & Sons Ltd.

(A) If b is nondecreasing in s then $\beta(x,s) \equiv s$. In this case $a \equiv a(t,x,b(x,u),\nabla u)$,
$f \equiv f(t,x,b(x,u)), q \equiv g(t,x)$;

(B) If $\beta(x,s)$ is nondecreasing in s then $b(x,s) \equiv q(x)s, q(x) \geq q_0 > 0$. In this case the terms a, f, g are considered in the form $a \equiv a(t,x,\beta(x,u),\nabla\beta(x,u))$, $f \equiv f(t,x,\beta(x,u)), g \equiv g(t,x,\beta(x,u))$.

We shall here apply the method of characteristics to relaxation schemes for the more special system

$$\partial_t u - A(x)\nabla u - \nabla a(t,x,u,\nabla\beta(x,u)) = f(t,x,u) \tag{13.4}$$

where A is an $m \times N$ matrix and $\beta(x,s)$ is nondecreasing in s. For simplicity we shall consider scalar equations in (13.1) and (13.4).

Relaxation schemes for (13.1) are based on a nonstandard time discretization and can be introduced in the following way. Let $\tau = \frac{T}{n}$ be a time step $(n \in N)$, $u_i \approx u(t_i, x)$ and $\Theta_i \approx \beta(x, u_i)$. At the time level $t_i = i\tau$ we determine Θ_i from the regular elliptic problem

$$\lambda_i(\Theta_i - \beta(x, u_{i-1})) - \tau\nabla a(t_i, x, u_{i-1}, \nabla\Theta_i) = \tau f(t_i, x, u_{i-1}) +$$
$$\tau\tilde{f}_i \equiv \tau f_i + \tau\tilde{f}_i \tag{13.5}$$
$$\Theta_i = 0 \quad \text{on} \quad \Gamma_1, \quad a(t_i, x, u_{i-1}, \nabla\Theta_i).\nu = g(t_i, x, \Theta_{i-1}) \equiv g_i \quad \text{on} \quad \Gamma_2$$

where $\lambda_i \in L_\infty(\Omega)$ has to satisfy the "convergence condition"

$$\frac{1}{2}\tau^d \leq \lambda_i \leq \min\left\{\tau^{-d}, \frac{b(x, u_{i-1} + \mu_i(\Theta_i - \beta_n(x, u_{i-1}))) - b(x, u_{i-1})}{\Theta_i - \beta_n(x, u_{i-1})}\right\} + \tau^d \tag{13.6}$$

with a relaxation function $\mu_i \in L_\infty(\Omega)$ satisfying

$$\frac{1}{2}\tau^d \leq \mu_i \leq min\left\{\frac{\beta_n^{-1}(x, \beta_n(x, u_{i-1}) + \alpha(\Theta_i - \beta_n(x, u_{i-1}))) - u_{i-1}}{\Theta_i - \beta_n(x, u_{i-1})}\right\} \tag{13.7}$$

where $d \in (0,1), 0 < \alpha < 1(\alpha \approx 1)$ are parameters of the method. We determine u_i from

$$b_n(x, u_i) := b_n(x, u_{i-1}) + \lambda_i(\Theta_i - \beta(x, u_{i-1})) \tag{13.8}$$

where b_n, β_n are regularizations of b, β such that respectively

$$b_n(x, s) := b(x, s) + \tau^d s \tag{13.9}$$

$$\beta_n(x, s) \quad \text{is absolutely continuous in} \quad s \quad \forall x \in \Omega$$

satisfying:

i) $\beta_n(x,s) \rightrightarrows \beta(x,s)$ locally uniformly in s
ii) $\tau^d \leq \partial_s\beta_n(x,s) \leq \tau^{-d}$
iii) $\min\{\partial_s\beta(x,s), 1\} \leq C\partial_s\beta_n(x,s)$

If we put $\mu_i = \lambda_i = \frac{1}{2}\tau^d$ then (13.6) and (13.7) are satisfied. However, a good approximation requires that the terms on the R.H.S in the inequalities (13.6), (13.7) are attained and this is comfirmed in the numerical experiments, see [JK91], [KHK93], [Kač94].

By means of the relaxation functions λ_i, μ_i we control the degeneracy of $\partial_s b(x, s) = 0, +\infty$ and $\partial_s \beta(x, s) = 0, +\infty$. When $b(x, s) \equiv s$, we can take $\lambda_i \equiv \mu_i$, and when $\beta(x, s) \equiv s$ we can take $\mu_i \equiv \alpha$.

If $\beta(x, s)$ is Lipschitz continuous in s then it is sufficient to use the approximation

$$\beta_n(x, s) := \beta(x, s) + \tau^d s$$

The term $\tilde{f}_i$ in (13.5) represents a small term (error) satisfying

$$|\tilde{f}_i|_2 = o(1), \quad |\tilde{f}_i|_{q'} \le c, \quad \forall n, i = 1, \ldots, n,$$
$$q' > q \quad (p^{-1} + q^{-1} = 1) \tag{13.10}$$

where $o(1)$ is the Landau symbol denoting any term converging to 0 for $\tau \to 0$ $(n \to \infty)$.

In the paper [JK95] a more simple approximation of (13.1) is used where the relaxation function λ_i has to satisfy

$$\left| \lambda_i - \frac{b_n(x, u_{i-1} + \mu_i(\Theta_i - \beta_n(x, u_{i-1}))) - b_n(x, u_{i-1})}{\Theta_i - \beta_n(x, u_{i-1})} \right| < \tau \tag{13.11}$$

and μ_i satisfies (13.7). Then we define

$$u_i := u_{i-1} + \mu_i(\Theta_i - \beta_n(x, u_{i-1})), \quad i = 1, \ldots, n. \tag{13.12}$$

In this case the restriction on λ_i is stronger than in the case (13.5)-(13.8). Moreover the growth of a in ξ is superlinear, i.e. $p \ge 2$. On the other hand we obtain u_i in (13.11) without inverting b.

To guarantee that the convergence conditions (13.6), (13.7), resp. (13.11),(13.7), are satisfied we use the iteration process as follows

$$(\lambda_{i,k-1}(\Theta_{i,k} - \beta(u_{i-1})), v) + \tau(a(t_i, u_{i-1}, \nabla\Theta_{i,k}), \nabla v) +$$
$$\tau(g_i, v)_{\Gamma_2} = \tau(f_i, v) \quad \forall v \in V \tag{13.13}$$

where

$$\bar{\lambda}_{i,k} := \min \left\{ \frac{b(u_{i-1} + \mu_{i,k}(\Theta_{i,k} - \beta_n(u_{i-1}))) - b(u_{i-1})}{\Theta_{i,k} - \beta_n(u_{i-1})}, \tau^{-d} \right\} + \frac{2}{3}\tau^d$$
$$\lambda_{i,k} := \min\{\bar{\lambda}_{i,k}, \lambda_{i,k-1}\} \quad \text{for} \quad k = L, L+1, \ldots \quad \text{and} \tag{13.14}$$
$$\lambda_{i,k} := \bar{\lambda}_{i,k} \quad \text{for} \quad k = 1, \ldots, L$$

with

$$\bar{\mu}_{i,k} := \frac{\beta_n^{-1}(\beta_n(u_{i-1}) + \alpha(\Theta_{i,k} - \beta_n(u_{i-1}))) - u_{i-1}}{\Theta_{i,k} - \beta_n(u_{i-1})}$$
$$\mu_{i,k} := \min\{\bar{\mu}_{i,k}, \mu_{i,k-1}\} \quad \text{for} \quad k = L, L+1, \ldots \tag{13.15}$$
$$\mu_{i,k} := \bar{\mu}_{i,k} \quad \text{for} \quad k = 1, \ldots, L$$

and the starting values $\mu_{i,0} := \min \frac{\alpha}{\partial_s \beta_n(u_{i-1})}, \lambda_{i,0} := \min\{\{\tau^{-d}, \partial_s b_n(u_{i-1})\}\mu_{i,0}, \}$.

13.2   ASSUMPTIONS AND THE CONVERGENCE OF ITERATIONS (13.12)-(13.14)

Let

$$B(x, s) := b(x, \beta^{-1}(x, s))s - \int_0^s b(x, \beta^{-1}(x, z))dz \quad \text{for} \quad s \in \{y \in R : y = \beta(x, z)\}$$

where by $\beta^{-1}(x, s)$ we understand the inverse of $\beta(x, s)$ (x is fixed) with respect to s. If $\beta(x, R) = (r, s)$ and r or s are finite then we put $\beta^{-1}(x, y) = +\infty(-\infty)$ for $y \geq s$ ($y \leq r$). We then make the following assumptions:

H_1) b is increasing, absolutely continuous in s ($b(x, 0) = 0$) and $|\partial_s b(x, s) - \partial_s b(y, s)| \leq \omega(|x - y|)\partial_s b(x, s)$ where $\omega : R_+ \to R_+$ is continuous, $\omega(0) = 0$ and

$$|b_n(x, s)|^2 \leq c_1 B_n(x, \beta_n(x, s)) + c_2 \qquad \forall s \in R; n$$

H_2) $\beta(x, s)$ is absolutely continuous in s, $\beta(x, 0) = 0$ for $x \in \Omega$ and $|\partial_s \beta(x, s) - \partial_s \beta(y, s)| \leq \omega(|x - y|)\partial_s \beta(x, s)$

H_3) $a(t, x, \eta, \xi) : I \times \Omega \times R \times R^N \to R^N$ satisfies:

$$(a(t, x, \eta, \xi_1) - a(., \xi_1)) - a(., \xi_2)).(\xi_1 - \xi_2) \geq 0;$$
$$|a(t, x, \eta, \xi)| \leq C(1 + (B(x, \beta(x, s)))^{1/\gamma} + |\xi|^{p-1});$$
$$a(t, x, \eta, \xi).\xi \geq C_1|\xi|^\gamma - C_2$$
$$\text{where} \quad p > 1 \quad \text{and} \quad 0 < \gamma < \frac{p-1}{p};$$

H_4) $|f(t, x, \eta)| \leq C(1 + B(x, \beta(x, \eta))^\gamma)$;

H_5) $|g(t, x, s)| \leq C(1 + |s|^\gamma)$;

H_6) $u_0 \in L_2(\Omega)$;

We shall use the standard function spaces $L_\infty \equiv L_\infty(\Omega)$, $L_p = L_p(\Omega), W_p^1(\Omega)$ (Sobolev space), $V = \{u \in W_p^1; v/\Gamma_1 = 0\}$ $L_p(I, V)$. We denote $x.y = \sum_{i=1}^N x_i y_i$, $(u, v) = \int_\Omega uv$ $(u, v)_{\Gamma_2} = \int_{\Gamma_2} uv$ and let V^* be the dual space of V with the duality $< u, v >$ for $u \in V^*, v \in V$.

Let $|.|_\infty, |.|_p, \|.\|, \|.\|_*, |.|_{\Gamma,p}$ denote the norms in $L_\infty, L_p, V, V^*, L_p(\Gamma)$. By C we denote a generic positive constant.

Definition 13.1 *A measurable function $u : Q_T \to R$ is a variational solution of (13.1)-(13.3) if*

i) $b(u) \in L_2(Q_T), \partial_t b(u) \in L_q(I, V^*)$ $(p^{-1} + q^{-1} = 1)$ $\beta(u) \in L_p(I, V)$;

ii) $\int_I < \partial_t b(u), v >= - \int_{Q_T}(b(u) - b(u_0)).\partial_t v$ $\forall v \in V \cap L_\infty(Q_T)$ with $\partial_t v \in L_\infty(Q_T), v(x, T) = 0$;

iii) $\int_I < \partial_t b(u), v > + \int_I (a(t, u, \nabla\beta(u)), \nabla v) + \int_I (g(t, \beta(u)), v)_{\Gamma_2} = \int_I (f(t, u), v)$, $\forall v \in L_p(I, V)$.

Similarly in (13.5) Θ_i has the sense of a variational solution. To guarantee that the convergence conditions are satisfied we have the following result:

Theorem 13.1 *Let the assumptions H_1-H_6, $u_0 \in L_\infty(\Omega)$ and $p > \frac{2N}{N+2}$ be satisfied. Let $\{\lambda_{i,k}\}, \{\mu_{i,k}\}, \{\Theta_{i,k}\}$ arise from (13.12)-(13.14). Then $\lambda_{i,k} \to \lambda_i, \mu_{i,k} \to \mu_i$ in $L_r(\Omega)$, $\forall r > 1$ and $\Theta_{i,k} \to \Theta_i$ in $L_2(\Omega)$ for $k \to \infty$ where λ_i, μ_i and Θ_i satisfy (13.1), (13.5), (13.6) and (13.7) with $\tilde{f}_i \equiv 0$. If $|\Theta_{i,k}|_{q'} \le c_i$ with $q' > \max\{2, q\}$ then there exists $\tilde{f}_i, k_0 = k_0(i, \tau) < \infty$ such that $\lambda_i := \lambda_{i,k_0}, \Theta_i := \Theta_{i,k_0}$ satisfy (13.5)-(13.7) where $\tilde{f}_i$ satisfies (13.10).*

By means of Θ_i, μ_i we construct Rothe's functions
$$\Theta^n(t) = \Theta_{i-1} + (t - t_{i-1})\tau^{-1} (\Theta_i - \Theta_{i-1}) \text{ for } t \in (t_{i-1}, t_i), i = 1, \ldots, n$$
and the corresponding step function $\bar{\Theta}^n(t) = \Theta_i$ for $t \in (t_{i-1}, t_i), i = 1, \ldots, n$,
$\bar{\Theta}^n(0) = \Theta_0 = B_n(x, u_0)$.
   Similarly we define $u^n, \bar{u}^n$. Denote by $\{\bar{n}\}$ a subsequence of $\{n\}$.
   Our main result is

Theorem 13.2 *Let the assumptions H_1-H_6 and (13.9),(13.10) be fulfilled. Then there exists a variational solution u of (13.1)-(13.3). Moreover $b_{\bar{n}}(x, \bar{u}^{\bar{n}}) \to b(x, u)$ in $L_s(Q_T)$ $s < 2$ and $\bar{\Theta}^{\bar{n}} \to \beta(x, u)$ in $L_{p_0}(Q_T), p_0 < \min\{p, 2\}$. If the variational solution u is unique, then the original sequences $\{b_n(x, \bar{u}^n)\}, \{\Theta^n\}$ are convergent.*

The proof is based on energy type *a priori* estimates

$$\max_{1 \le i \le n} \int_\Omega B_n(x, \beta_n(x, u_i)) \le C, \quad \sum_{i=1}^n \|\nabla \Theta_i\|_\tau^2 \le C.$$

$$\sum_{i=1}^n \int_\Omega \frac{1}{\lambda_i}(b_n(x, u_i) - b_n(x, u_{i-1}))^2 \le C, \quad \sum_{i=1}^n |u_i - u_{i-1}|^2 \le C\tau^{-2d}$$

$$\sum_{i=1}^n |\Theta_i - \beta(x, u_{i-1})|^2 \le C\tau^{-d}, \quad \sum_{i=1}^n |u_i - u_{i-1}|_2^2 \le C\tau^{-d}$$

To verify these *a priori* estimates we replace the first term in (13.5) using (13.8) and then we test it with $v = (\Theta_i - u_{i-1}) + u_i - (u_i - u_{i-1})$, and sum for $i = 1, \ldots, j$. The parabolic term gives us

$$\sum_{i=1}^j (b_n(x, u_i) - b_n(x, u_{i-1}), u_i) \ge \int_\Omega B_n(x, u_j) - \int_\Omega B_n(x, u_0),$$

$$J_1 \equiv \sum_{i=1}^j (b_n(x, u_i) - b_n(x, u_{i-1}), \Theta_i - u_{i-1}) \ge \sum_{i=1}^j \int_\Omega \frac{1}{\lambda_i}(b_n(x, u_i) - b_n(x, u_{i-1}))^2$$

and

$$J_3 = \left| \sum_{i=1}^j (b_n(x, u_i) - b_n(x, u_{i-1}), u_i - u_{i-1}) \right| \le \alpha J_1$$

where (13.6),(13.7) have been used substantially. The elliptic term, R.H.S and the boundary terms are estimated in a standard way. Then Gronwall's argument implies that the required *a priori* estimates hold.

To guarantee compactness we need some more information (with respect to the t variable). For this purpose we sum (13.5) for $i = j + 1, \ldots, j + k$ and then we test the result with $v = (\Theta_{j+k} - \Theta_j)\tau$. Summing this for $j = 1, \ldots, n - k$ and using the *a priori* estimates already obtained we conclude that

$$\sum_{j=1}^{n-k-1} (b_n(x, u_{j+k}) - b_n(x, u_j), \Theta_{j+k+1} - \Theta_{j+1})\tau \leq Ck\tau.$$

$$\int_0^{T-z-\tau} (b_n(x, \bar{u}^n(t + z)) - b_n(x, \bar{u}^n(t)), \beta_n(x, \bar{u}^n(t + z)) - \beta_n(x, \bar{u}^n(t))) \leq$$

$$C(z + \tau^{(1-d)/2}), \quad \text{for} \quad k\tau < z \leq (k + 1)\tau$$

Then, a strictly increasing and Lipschitz continuous function (in s) $W(x, s)$ can be constructed so that

$$\int_0^{T-z-\tau} \int_\Omega |W(x, \bar{u}^n(x, t + z) - W(x, \bar{u}^n(x, t)|^2 \leq C(z + \tau^{(1-d)/2}).$$

From this and the estimates obtained

$$\int_I |\bar{\Theta}^n - \bar{\beta}(x, \bar{u}^n(t - \tau)|_2^2 \leq \frac{c}{n^{1-d}}, \quad \int_I |\nabla \bar{\Theta}^n|_2^2 \leq C$$

we conclude the compactness of $\{W(x, \bar{u}^n)\}$ in $L_1(Q_T)$ and hence that $\bar{u}^{\bar{n}} \to u$ a.e. in Q_T.

By the duality argument in

$$\left(\frac{b(\bar{u}^n(t)) - b(\bar{u}^n(t - \tau))}{\tau}, v \right) + (a_n(t, \bar{u}^n(t - \tau), \nabla \bar{\Theta}^n), v) +$$

$$+ (g_n(t, \bar{\Theta}^n(t - \tau)), v) = (f_n(t, \bar{u}^n(t - \tau)), v) \qquad \forall v \in V$$

(here $a_n(t, x, \eta, \xi) = a(t_i, x, \eta, \xi)$ for $t \in (t_{i-1}, t_i >, \quad i = 1, \ldots, n)$ we find that

$$\left\{ \frac{b(\bar{u}^n(t)) - b(\bar{u}^n(t - \tau))}{\tau} \right\} \quad \text{is bounded in} \quad L_p(I, V^*)$$

and hence

$$\frac{b(\bar{u}^n(t)) - b(\bar{u}^n(t - \tau))}{\tau} \rightharpoonup \partial_t b(u) \quad \text{in} \quad L_q(I, V^*).$$

To prove that u is a variational solution of (13.1)-(13.3) we use the "integration by parts formula"; see [AL83]

$$\int_0^t \int_\Omega < \partial_t b(x, \bar{u}), \beta(x, u) >= \int_\Omega B(x, \beta(x, \bar{u})) - \int_\Omega B(x, \beta(x, x_0))$$

for a.e. $t \in I$ and follow the argument of Minty-Browder. For details see [Kač94] where (13.1) has been considered with Lipschitz continuous b, β in s.

Assuming the strong ellipticity we obtain

Theorem 13.3 *Let the assumptions of Theorem 13.2 be satisfied. If*

$$(a(t, x, \eta, \xi_1) - a(t, x, \eta, \xi_2)).(\xi_1 - \xi_2) \geq C|\xi_1 - \xi_2|^p \quad p \geq 2.$$

(13.16)

is fulfilled then $\bar{\Theta}^{\bar{n}} \to u$ in $L_p(I, V)$.

The analogous convergence results can be obtained for the cases (A) and (B). In case (B) we only obtain $\bar{u}^n \rightharpoonup u$ in $L_2(Q_T)$.

Remark 13.1 *The uniqueness of the variational solution of (13.1)-(13.3) has been studied in [Ott] under some additional structural restrictions.*

13.3 IMPLEMENTATION OF THE METHOD OF CHARACTERISTICS

Implementation of the method of characteristics in convection-diffusion problems has been discussed intensively in the last decade, [Pir82], [DR82], [WEC95], [DRW89] and [Ber]., and the effectiveness of this method has been demonstrated for convection dominated problems. Stability and error estimates have been analysed in [Ber] and [MPS88]. Here we consider the scalar convection-diffusion equation of the form (13.4) where $A(x)$ is a vector function $A(x) \equiv (a_1(x), \ldots, a_N(x))$ and $\beta(x, s)$ is Lipschitz continuous, strictly increasing in s. We assume that $A \in C^1(\bar{\Omega})$. The implementation of the method of characteristics in the corresponding relaxation schemes (13.5)-(13.8) reads as follows.

At the time level $t_i = i\tau$ we solve the elliptic problem (variable x is omitted)

$$\mu_i(\Theta_i - \beta(u_{i-1}^*)) - \tau\nabla a(t_i, u_{i-1}, \nabla\Theta_i) = \tau f_i$$
$$\Theta_i = 0 \quad \text{on} \quad \Gamma_1, \quad a(t_i, u_{i-1}, \nabla\Theta_i).\nu = g_i \quad \text{on} \quad \Gamma_2 \quad (13.17)$$

where the relaxation function has to satisfy the "convergence condition"

$$\frac{\varepsilon}{\beta_n'(u_{i-1}^*)} \leq \mu_i \leq \min\left\{\frac{\beta_n^{-1}(\beta_n(u_{i-1}^*) + \alpha(\Theta_i - \beta_n(u_{i-1}^*))) - u_{i-1}^*}{\Theta_i - \beta_n(u_{i-1}^*)}, \frac{1}{\beta_n'(u_{i-1}^*)}\right\}$$

(13.18)

with $0 < \varepsilon < 1$ and

$$|\nabla\mu_i|_\infty \leq c\tau^{-d/2}$$

(13.19)

where

$$u_{i-1}^*(x) := \tilde{u}_{i-1}(x - \tau A(x)) \quad (\tilde{u} \quad \text{is an extension of} \quad u).$$

(13.20)

Then we define

$$u_i := u_{i-1}^* + \mu_i(\Theta_i - \beta_n(u_{i-1}^*)).$$

(13.21)

successively for $i = 1, \ldots, n$. Here $\tilde{u}_{i-1} \in W_p^1(\Omega^*)$ is an extension of $u_{i-1} \in W_p^1(\Omega)$ so that $\bar{\Omega} \subset \Omega^*$ and $x - \tau A(x) \in \Omega^*$, $\forall x \in \Omega$ and $\tau \leq \tau_0$. If $\partial\Omega$ is Lipschitz continuous, then

$$\|\tilde{u}_{i-1}\|_{W_p^1(\Omega^*)} \leq C\|u\|_{W_p^1(\Omega)} \qquad \forall u \in W_p^1(\Omega).$$

The function β_n is a regularization of β as in Section 13.1. The convergence conditions in (13.18) are more restrictive than in the relaxation schemes for (13.5)-(13.8) where

$$b(x, s) = s \quad \text{and} \quad \lambda_i \equiv \mu_i.$$

Remark 13.2 *If $\tau \leq \tau_0$ is sufficiently small then one can expect that $\Theta_i - \beta_n(u_{i-1}^*)$ is small (in $L_\infty(\Omega)$). Then the difference quotient on the R.H.S in (13.18) is close to $\frac{\alpha}{\beta_n'(u_{i-1}^*)}$. Thus, when α is close to 1 and ε is close to 0 then the interval for μ_i in (13.18) is not empty. These facts has been proved for linear elliptic operators and under some smoothness assumptions on the data $g, f, \partial\Omega$, see [Kač96].*

To determine Θ_i, μ_i in (13.17)-(13.21) we propose the following iterations

$$\mu_{i,k-1}(\Theta_{i,k} - \beta(u_{i-1}^*)) - \tau\nabla a(t_i, u_{i-1}, \nabla\Theta_{i,k}) = \tau f_i$$
$$\Theta_{i,k} = 0 \quad \text{on} \quad \Gamma_1, \quad a(t_i, u_{i-1}, \nabla\Theta_{i,k})\nu = g_i \quad \text{on} \quad \Gamma_2 \qquad (13.22)$$

and

$$\mu_{i,k} \in \left(\frac{\varepsilon}{\beta_n'(u_{i-1}^*)}, \min\left\{\frac{\beta_n^{-1}(\beta_n(u_{i-1}^*) + \alpha(\Theta_{i,k} - \beta_n(u_{i-1}^*))) - u_{i-1}^*}{\Theta_{i,k} - \beta_n(u_{i-1}^*)}, \frac{1}{\beta_n'(u_{i-1}^*)}\right\}\right) \qquad (13.23)$$

so that

$$|\nabla\mu_{i,k}|_\infty \leq c\tau^{-d/2} \qquad (13.24)$$

Remark 13.3 *If $|\Theta_i - \beta_n(u_{i-1}^*)|_\infty$ is sufficiently small (uniformly for $i = 1, \ldots, n$) for τ small (see Remark 13.2) then the length of the interval in (13.23) is bounded from below uniformly for $x \in \Omega$ by a positive constant. Then $\mu_{i,k}$ can be chosen to satisfy (13.24) because of the Lipschitz continuity of β.*

The convergence of (13.22),(13.23), cannot generally be guaranteed. Some practical modifications can be chosen to force the convergence of $\{\mu_{i,k}\}, \{\Theta_{i,k}\}$ in (13.22),(13.23) similarly to those in Section 13.1, e.g., we can put for $k \leq L > 1$

$$\bar{\mu}_{i,k} = \min\left\{\frac{\beta_n^{-1}(\beta_n(u_{i-1}^*) + \alpha(\Theta_{i,k} - \beta_n(u_{i-1}^*))) - u_{i-1}^*}{\Theta_{i,k} - \beta_n(u_{i-1}^*)}, \frac{1}{\beta_n'(u_{i-1}^*)}\right\}$$

and then

$$\mu_{i,k} = \min\{\mu_{i,k-1}, \bar{\mu}_{i,k}\} \qquad \text{where} \qquad \alpha + \beta = 1, \quad \alpha, \beta > 0.$$

The convergence of the method proposed in (13.17)-(13.21) can be formulated in the following way.

Theorem 13.4 *Let the assumptions of Theorem 13.3 be satisfied. Let $\beta(x, s)$ be Lipschitz continuous in s and $\beta(x, u_0) \in W_p^1(\Omega)$. Then $\bar{\Theta}^{\bar{n}} \rightarrow \beta(x, u)$ in $L_p(I, W_p^1(\Omega))$ and $\bar{u}^{\bar{n}} \rightharpoonup u$ in $L_2(Q_T)$. If β is strictly increasing in s, then $\bar{u}^{\bar{n}} \rightarrow u$ in $L_2(Q_T)$.*

For the proof we proceed as in Section 13.2. The stronger restriction (13.18) here guarantees that $u^n \in L_2(I, W_2^1(\Omega))$ and $\{\bar{u}^n\}_{n=1}^{\infty}$ is uniformly bounded in $L_2(I, W_2^1(\Omega))$. This enables us to prove the convergence of (13.17) for $n \rightarrow \infty$ to (13.4) in the setting of the variational formulation.

References

[AL83]   Alt H. W. and Luckhaus S. (1983) Quasilinear elliptic-parabolic differential equations. *Math. Z* 183: 311–341.

[Ber]   Bermejo R.A Galerkin-characteristics algorithm for transport-diffusion equations. *SIAM J. Numer. Anal* 32(2): 426–454.

[DR82]   Douglas J. and Russel T. F. (1982) Numerical methods for convection dominated diffusion problems based on combining the method of the characteristics with finite elements or finite differences. *SIAM J. Numer. Annal* 19: 871–885.

[DRW89]   Dawson C. N., Russel T. F., and Wheeler M. F. (1989) Some improved error estimates for the modified method of characteristics. *SIAM J. Numer. Anal* 26: 1487–1512.

[JK91]   Jäger W. and Kačur J. (1991) Solution of porous medium systems by linear approximation scheme. *Num.Math* 60: 407–427.

[JK95]   Jäger W. and Kačur J. (1995) Solution of doubly nonlinear and degenerate parabolic problems by relaxation schemes. *M^2AN Mathematical modelling and numerical analysis* 29(5): 605–627.

[Kač94]   Kačur J. (1994) Solution to strongly nonlinear parabolic problems by a linear approximation scheme. Technical Report Mathematics Preprint No. IV-M1-94, Comenius University Faculty of Mathematics and Physics.

[Kač96]   Kačur J. (1996) Solution to strongly nonlinear parabolic problems by a linear approximation scheme. Technical Report Mathematics Preprint No. IV-M1-96, Comenius University Faculty of Mathematics and Physics.

[KHK93]   Kačur J., Handlovičova A., and Kačurova M. (1993) Solution of nonlinear diffusion problems by linear approximation schemes. *SIAM J. Numer. Anal* 30: 1703–1722.

[MPS88]   Morton K. W., Priestly A., and Suli E. (1988) Stability of the Lagrangian-Galerkin method with nonexact integration. *RAIRO, M^2AN* 22(123).

[Ott]   Otto F.L^1-contraction principle and uniqueness for quasilinear elliptic-parabolic equations. to appear.

[Pir82]   Pironneau O. (1982) On the transport-diffusion algorithm and its applications to the Navier-Stokes equations. *Numer. Math* 38: 309–332.

[WEC95] Wang H., Ewing R. E., and Celia M. A. (1995) Eulerian-Lagrangian localized adjoint method for reactive transport with biodegradation. *Numerical Methods for PDE* 11: 229–254.

14

On Nodal Transport Methods

J. P. Hennart[a1] and E. del Valle[b]

[a] INRIA - Domaine de Voluceau - B.P. 105,
78153 LE CHESNAY Cedex (FRANCE),
On sabbatical leave from IIMAS - UNAM,
Apdo Postal 20.726, 01000 México D.F.,
MEXICO

[b] Instituto Politécnico Nacional
Escuela Superior de Física y Matemáticas
Unidad Profesional "Adolfo López Mateos"
07738 México D.F.,
MEXICO

14.1   INTRODUCTION

One of the most important partial differential equations in the nuclear engineering
field is the neutron transport equation in its discrete-ordinates approximation S_N
form which in $x - y$ geometry reads:

$$L\psi_k \equiv \mu_k \frac{\partial \psi_k}{\partial x} + \nu_k \frac{\partial \psi_k}{\partial y} + \sigma_t \psi_k = \sigma_s \sum_{l=1}^{M} \omega_l \psi_l + S_k \equiv Q_k, k = 1, \ldots, M, \qquad (14.1)$$

where the unknown is ψ_k, the angular neutron flux corresponding to the k-th ray of
the S_N approximation, M being the total number of rays considered; in this case this
is given by $N(N + 2)/2$. The domain to be considered is a union of rectangles on the
boundary of which boundary conditions must also be imposed.

Classically, with nodal methods, the domain of interest is decomposed into relatively
large homogeneous regions or "nodes", over which each angular flux ψ is approximated
by a generalized interpolant, with interpolation parameters which are cell and/or edge
Legendre moments.

This unique interpolant is piecewise continuous, and uses polynomial or exponential
shape functions in the case of the so-called *analytical* nodal methods which depend on

---

[1] Consultant to the "Comisión Nacional de Seguridad Nuclear y Salvaguardias" - MEXICO

The Mathematics of Finite Elements and Applications
Edited by J. R. Whiteman © 1997 John Wiley & Sons Ltd.

transverse integration procedures; see e.g. [Bad90]. For a ray in the first quadrant, the possible left and bottom edge parameters are known from the boundary conditions or from the neighboring left and bottom cells. The unknowns are thus the right and top edge parameters as well as the cell parameters.

In this paper, we present two "new" (in a sense to be defined later) classes of polynomial nodal methods. In essence, both classes of methods lead to discontinuous approximations as they at most conserve some edge moments between adjacent nodes, as in the case of the first class of methods which we call *weakly discontinuous*. In this case one or more moments of the angular flux are conserved between a given cell and its upstream neighbors. The second class of methods, called *strongly discontinuous*, is fully discontinuous and has only outgoing (at top and right) edge moments as parameters, in addition to possible cell moments.

Before dealing with these methods in detail, in the next section we present some notation and the basic formalism. The two classes of methods are then developed in the following two sections. Some numerical results are then given after which some conclusions are presented in the last section.

14.2   NOTATION AND BASIC FORMALISM

14.2.1   NOTATION

Assuming that the domain Ω consisting of the union of rectangles has been discretized into N_e nodes or rather *cells* or *elements*, i.e. $\Omega \equiv \Omega_h = \cup_{e=1}^{N_e} \Omega_e$, each cell Ω_e is mapped onto a reference cell $\widehat{\Omega} \equiv [-1, +1] \times [-1, +1]$, as is traditional with finite element methods. A particular finite element is then defined by a set of degrees of freedom D and a space of functions S with $card(D) = dim(S)$. With degrees of freedom which are cell and/or edge moments as in this paper, we shall speak of *nodal finite elements*. For practical purposes, these moments will be taken as Legendre moments.

Some notation will be helpful for describing D and S in a compact way in the *nodal* case.

Let P_i be the normalized Legendre polynomial of degree i over $[-1, +1]$ which satisfies

$$P_i(+1) = 1, \quad P_i(-1) = (-1)^i, \quad \text{and} \quad \int_{-1}^{+1} P_i(x)P_j(x)dx = N_i\delta_{ij}. \qquad (14.2)$$

with $N_i = 2/(2i + 1)$. Define moreover $P_{ij}(x, y)$ as $P_i(x)P_j(y)$. Assuming that $L\psi = Q$ is the given equation, ψ is approximated by ψ_h and, over $\widehat{\Omega}$, cell moments of $\psi_h(x, y) \in S$ are defined as follows

$$\psi_C^{ij} = \int_{-1}^{+1} \int_{-1}^{+1} P_{ij}(x, y)\psi_h(x, y)\, dxdy/N_i \cdot N_j. \qquad (14.3)$$

Edge moments are moreover given by

$$\psi_E^i = \int_{-1}^{+1} P_i(s_E)\psi_h(x_E, y_E)ds_E/N_i, \qquad (14.4)$$

where E is a generic symbol corresponding to L, R, B, and T for the left, right, bottom, and top edges respectively, x_E or y_E is ± 1 depending on the particular edge considered, the other coordinate being s_E, the coordinate along that edge.

S is a space of functions, which in this paper are taken to be *polynomials*. To describe these in a systematic way, let us introduce the spaces of polynomials of degree i in x and j in y, $\mathcal{Q}_{ij}(x, y) \equiv \{x^a y^b \mid 0 \leq a \leq i, 0 \leq b \leq j\}$, with in particular $\mathcal{Q}_i \equiv \mathcal{Q}_{ii}(x, y)$ and also the spaces of polynomials of degree i in x and y, $\mathcal{P}_i(x, y) \equiv \{x^a y^b \mid 0 \leq a + b \leq i\}$. For each nodal finite element, we shall call $N_p = dim(D)$ the total number of parameters and N_u the number of unknowns which is less than N_p in the weakly discontinuous case, where the interpolation parameters on the left and bottom edges are taken from the neighboring cells or given by the boundary conditions. In the strongly discontinuous case, there are no left and bottom parameters and we have $N_p = N_u$. In the following, each particular method will be assigned a symbol consisting of two capital letters, WD in the weakly discontinuous case and SD in the strongly discontinuous one; indexed by the two numbers N_p and N_u in the first case, and by N_p or N_u indifferently in the second case. In the WD case, $(N_p - N_u)/2$ is the number of edge moments conserved between adjacent cells. In most practical cases, this number is one or two. In both cases, we have programmed all the methods from two to eight unknowns per cell and applied them to multiplicative and nonmultiplicative problems of the nuclear literature. In the weakly discontinuous case, we have given a constructive algorithm to deduce S if D is known [HJR88]. In that paper, we always assumed that we had the same number of edge moments on each pair of opposite edges. This is clearly not true in the strongly discontinuous case and we had to adapt the earlier algorithm to that situation for applications in neutron transport problems. For second order elliptic equations (the diffusion case), extensions of the basic procedure were also studied to provide transition elements of the p type with progressively more edge moments. These two classes of extensions will be presented elsewhere in a paper in preparation. In two dimensional applications as in this paper, the spaces S can be conveniently described with a Pascal triangle in which the different entries ab correspond to the polynomials $P_{ab}(x, y)$ instead of the more usual monomials $x^a y^b$. The definition of the basic polynomial spaces introduced earlier will be modified accordingly, for instance $\mathcal{P}_i(x, y) \equiv \{span(P_{ab}) \mid 0 \leq a + b \leq i\}$. With these conventions, it is extremely easy to check the unisolvency with respect to the set D of the space S provided by the algorithm. Due to the normalizations adopted, the basis functions can practically be determined by hand and they are given in a very compact way in terms of the P_{ab}'s. For the details, at least for the WD case, the readers are referred to [HJR88].

We shall give a concrete example, namely the element WD_{53}. In this case, $D \equiv \{\psi_L^0, \psi_R^0, \psi_B^0, \psi_T^0, \psi_C^{00}\}$ and correspondingly $S \equiv \{P_{00}, P_{01}, P_{10}, P_{02}, P_{20}\}$. On each cell, ψ is approximated by

$$\psi_h = \sum_E \psi_E^0 u_E^0 + \psi_C^{00} u_C^{00} \tag{14.5}$$

where the basis functions have very compact expressions in terms of the P_{ab}'s. For instance, $u_L^0 = -\frac{1}{2}(P_{10} - P_{20})$ and $u_C^{00} = P_{00} - P_{20} - P_{02}$, the other edge basis functions being obtained by changing the sign of x and/or y or by permuting them.

14.2.2 BASIC FORMALISM

Replacing ψ by ψ_h in each cell, a local residual $L\psi_h - Q_h$ arises where Q_h is evaluated from a previous iterate in the standard source iteration procedure which can be accelerated or not. If the ray (μ_k, ν_k) is in the first quadrant, one proceeds cell by cell by a standard *diagonal sweeping*, beginning with the first cell in Ω_h seen by the particular ray considered. Consequently, we know from the boundary conditions at the left or bottom of the domain, or from the left or bottom neighbors which have been processed previously, the edge moments on the left and bottom edges. If as it is the case with the weakly discontinuous methods, some of these moments are conserved between neighbors, they are directly known in either cell. The moment equations we shall now mention produce edge moments which are not interpolation parameters and it is indispensable to know precisely if we must evaluate these in the current cell or in the previous cells. Locally, Legendre moments of the residual $L\psi_h - Q_h$ are taken to obtain as many equations as unknowns. Since L is a first order partial differential operator, its application to a *discontinuous* approximation ψ_h generates delta distributions and the correct way to take these into account is to derive the moment equations over a cell Ω_e which is shifted upstream by ϵ, in the limit of a vanishing ϵ. For a ray in the first quadrant this means that the cell considered is moved slightly downward and to the left. As $\epsilon \to 0$, boundary terms arise at the left and the bottom of the cell, connecting it to its upstream neighbors or boundaries.

Returning to the WD_{53} example, there are actually two posibilities: in these two cases the zeroth order moment of the residual should be taken , which ensures balance; i.e. *particle conservation*. Let p_{ij} be P_{ij} scaled to the particular cell considered $[x_L, x_R] \times [y_B, y_T]$. The balance equation corresponds to taking the moment of the residual with respect to p_{00}. The missing equations correspond to taking that moment with respect to p_{10} and p_{01} by symmetry or with respect to p_{20} and p_{02}. Both methods work but the first is more satisfactory, and in fact better, as it fills in the Pascal triangle from the top by taking moments with respect to $\mathcal{P}_1$.

After taking the limit for ϵ, the general final equations read

$$\int_{y_B}^{y_T} \{\mu p_{ij}(x_L, y)[\psi_h(x_L + 0, y) - \psi_h(x_L - 0, y)]\}\, dy$$

$$+ \quad \int_{x_L}^{x_R} \{\nu p_{ij}(x, y_B)[\psi_h(x, y_B + 0) - \psi_h(x, y_B - 0)]\}\, dx$$

$$+ \quad \int_{\Omega_e} p_{ij}(x, y)(L\psi_h - Q_h)\, dx dy \quad = \quad 0. \tag{14.6}$$

The first two integrals in (14.6) correspond to the boundary terms at the left and bottom of the cell. For the first of these, since $p_{ij}(x_L, y) = (-1)^i p_j(y)$, we get after some algebra

$$\mu(2i + 1)(-1)^i[\psi_{L+}^j - \psi_{L-}^j]/\Delta x, \tag{14.7}$$

where $\Delta x \equiv x_R - x_L$ while $\psi_{L\pm}^j$ is the jth left edge moment of ψ_h evaluated at $x_L \pm 0$, that is in the cell considered or in the adjacent one on the left. The corresponding expression at the bottom is easily obtained by permuting i and j, x and y, and L and B, to give

$$\nu(2j+1)(-1)^j[\psi^i_{B+} - \psi^i_{B-}]/\Delta y, \tag{14.8}$$

The third line in (14.6) corresponds to cell moments of the residual. For $ij = 00$ for instance, we get the cell balance equation

$$\frac{\mu}{\Delta x}[\psi^0_{R-} - \psi^0_{L+}] + \frac{\nu}{\Delta y}[\psi^0_{T-} - \psi^0_{B+}] + \sigma_t\psi^{00}_C = Q^{00}_C. \tag{14.9}$$

The corresponding $ij = 10$ and $ij = 01$ equations are respectively given by

$$\frac{3\mu}{\Delta x}[\psi^0_{R-} + \psi^0_{L+} - 2\psi^{00}_C] + \frac{\nu}{\Delta y}[\psi^1_{T-} - \psi^1_{B+}] + \sigma_t\psi^{10}_C = Q^{10}_C \tag{14.10}$$

and

$$\frac{\mu}{\Delta x}[\psi^1_{R-} - \psi^1_{L+}] + \frac{3\nu}{\Delta y}[\psi^0_{T-} + \psi^0_{B+} - 2\psi^{00}_C] + \sigma_t\psi^{01}_C = Q^{01}_C. \tag{14.11}$$

These last two equations have been obtained using the expression for ψ_h given by (14.5) with the explicit forms of u^0_E and u^{00}_C, as well as the fact that $P'_3 = 3P_1$. Notice that as soon as i or j are equal to 1, some edge and cell moments of ψ_h appear, which are not among the original interpolation parameters.

With classical nodal methods (see e.g. [Bad90] and the references it contains) polynomial (or exponential) shapes are assumed for the angular flux at the surface and in the interior of each cell *separately*, leading to schemes typically described (in part at least) by two capital letters, like for instance C-Q for Constant-surface-Quadratic-interior, L-L for Linear-surface-Linear-interior, etc.

Here we present "new" nodal methods: with respect to the classical nodal schemes, their main distinctive feature is that all the information contained at the surface *and* in the interior of the cells is integrated into a unique interpolant which is piecewise continuous. One of the advantages of this approach is that *all* the possible edge and cell moments of the resulting representation are *unambiguously* defined in the cell considered (and its upward neighbors if necessary). If we combine (14.7) and (14.8) with (14.9), (14.10), or (14.11) following (14.6), we get a final set of equations which are identical to the former ones except that everywhere ψ^i_{L+} and ψ^i_{B+} with $i = 0$ or 1 are replaced by ψ^i_{L-} and ψ^i_{B-} respectively. It then remains to evaluate in the cell considered or in its upward neighbors the cell and edge moments which are not interpolation parameters, in terms of the interpolation parameters of the common cell. This is a straightforward operation and we have for instance $\psi^{10}_C = (\psi^0_{R-} - \psi^0_{L+})/2$ and $\psi^1_{T-} = (\psi^0_{R-} - \psi^0_{L+})/2$.

Before leaving this section, we want to add a few words about the selection of the N_u moment equations needed. The possible moments are with respect to any P_{ij} included in S. In the strongly discontinuous case, since we have $N_u = N_p = dim(S)$, there is only one possibility but in the weakly discontinuous case several choices of equations are possible as mentioned above for the WD_{53} example. The criteria used are the following ones:

1. The moment (00) must always be included to ensure particle balance.
2. If the moment (ij) is taken, then by symmetry the moment (ji) must also be considered.

3. For some of the spaces S considered where it may happen that the P_{ab}'s do not all appear independently (see [HJR88]), some equations are built by taking the difference between moments (ab) and (ba) of the neutron transport equation (14.1).
4. The algebraic system obtained for a given set of moments must not be singular in the case of an infinite medium without scattering, i.e. for the equation $\sigma_t \psi_k = Q_k$.
5. The moments of the neutron transport equation considered should ideally fill the Pascal triangle from the top.

If these features are taken into account, the final system generally provides good numerical results as we shall see.

14.3   WEAKLY DISCONTINUOUS METHODS

In the case of the weakly discontinuous family of schemes, we have programmed all the methods from two to eight unknowns per cell and applied them to different benchmark problems of the nuclear literature. In the following we shall describe them briefly by giving in each case the set D of interpolation parameters and the corresponding space S. In all the following examples, E stands for L, R, B, and T and we have:

$$
\begin{aligned}
WD_{42} &= [D \equiv \{\psi_E^0\}, S \equiv \{\mathcal{P}_1 \oplus (P_{20} - P_{02})\}], \\
WD_{53} &= [D \equiv \{\psi_E^0, \psi_C^{00}\}, S \equiv \{\mathcal{P}_1 \oplus (P_{20}, P_{02})\}], \\
WD_{64} &= [D \equiv \{\psi_E^0, \psi_C^{00}, \psi_C^{11}\}, S \equiv \{\mathcal{P}_2\}], \\
WD_{95} &= [D \equiv \{\psi_E^0, \psi_E^1, \psi_C^{00}\}, S \equiv \{\mathcal{P}_2 \oplus (P_{21}, P_{12}, P_{31} - P_{13})\}], \\
WD_{86} &= [D \equiv \{\psi_E^0, (\psi_C^{ij}, i, j = 0, 1)\}, S \equiv \{\mathcal{P}_2 \oplus (P_{30}, P_{03})\}], \\
WD_{11,7} &= [D \equiv \{\psi_E^0, \psi_E^1, (\psi_C^{ij}, i + j = 0, 1)\}, S \equiv \{\mathcal{P}_3 \oplus (P_{31} - P_{13})\}], \\
WD_{12,8} &= [D \equiv \{\psi_E^0, \psi_E^1, (\psi_C^{ij}, i, j = 0, 1)\}, S \equiv \{\mathcal{P}_3 \oplus (P_{31}, P_{13})\}].
\end{aligned}
$$

With respect to the approximation properties of the schemes described above, all depends on the greatest k such that $\mathcal{P}_k \in S$. With the notation used to describe the different spaces S, it is clear that WD_{42} and WD_{53} contain $\mathcal{P}_1$ but not $\mathcal{P}_2$. Similarly WD_{64}, WD_{95}, and WD_{86} contain $\mathcal{P}_2$ while the last two spaces contain $\mathcal{P}_3$.

A particular scheme is completely defined if we specify the moments of the residual which are taken in (14.6) to yield a system of linear algebraic equations of order N_u. As mentioned above, these choices are not always unique. In the following, we give the ones chosen to produce the numerical results exhibited later which essentially satisfy the criteria formulated at the end of the previous section. In each case, before the symbol of the method, we give the space M spanned by the Legendre polynomials used to weight the residual in (14.6), following the conventions defined above. The result is:

$$
\begin{aligned}
M &\equiv \{\mathcal{P}_0, \oplus(P_{10} - P_{01})\} \quad \text{for} \quad WD_{42}, \\
M &\equiv \{\mathcal{P}_1\} \quad \text{for} \quad WD_{53}, \\
M &\equiv \{\mathcal{Q}_1\} \quad \text{for} \quad WD_{64},
\end{aligned}
$$

$$M \equiv \{\mathcal{Q}_1 \oplus (P_{31} - P_{13})\} \quad \text{for} \quad WD_{95},$$
$$M \equiv \{\mathcal{Q}_1 \oplus (P_{20}, P_{02})\} \quad \text{for} \quad WD_{86},$$
$$M \equiv \{\mathcal{P}_2 \oplus (P_{21} - P_{12})\} \quad \text{for} \quad WD_{11,7},$$
$$M \equiv \{\mathcal{P}_2 \oplus (P_{21}, P_{12}\} \quad \text{for} \quad WD_{12,8}.$$

14.4 STRONGLY DISCONTINUOUS METHODS

In the case of the strongly discontinuous methods, we also programmed all the methods with 2 to 8 (actually 10) unknowns per cell. With the notations of the previous section these methods are very easy to describe. To each weakly discontinuous method $WD_{N_p N_u}$, corresponds a strongly discontinuous method SD_{N_u} with the same number of unknowns. Here $N_u = N_p$ as there are no edge moments on the ingoing edges, that is the left and bottom edges for a ray in the first quadrant. The set D of interpolation parameters for each SD_{N_u} is the same as for $WD_{N_p N_u}$, provided E stands for R and T only, corresponding to the outgoing edges. This set will be called $D^*_{N_p N_u}$. For each case, we also need to specify the space S which turns out to be identical to the space M of the weighting Legendre polynomials and we have:

$$
\begin{aligned}
SD_2 &= [D^*_{42}, S \equiv \{\mathcal{P}_0 \oplus (P_{10} - P_{01})\}], \\
SD_3 &= [D^*_{53}, S \equiv \{\mathcal{P}_1\}], \\
SD_4 &= [D^*_{64}, S \equiv \{\mathcal{Q}_1\}], \\
SD_5 &= [D^*_{95}, S \equiv \{\mathcal{Q}_1 \oplus (P_{21} - P_{12})\}], \\
SD_6 &= [D^*_{86}, S \equiv \{\mathcal{P}_2\}], \\
SD_7 &= [D^*_{11,7}, S \equiv \{\mathcal{P}_2 \oplus (P_{21} - P_{12})\}], \\
SD_8 &= [D^*_{12,8}, S \equiv \{\mathcal{P}_2 \oplus (P_{21}, P_{12})\}].
\end{aligned}
$$

With the notation used, it is clear that SD_2 only contains $\mathcal{P}_0$ and is therefore not very interesting. SD_3, SD_4, and SD_5 all contain $\mathcal{P}_1$ as do the last three spaces with $\mathcal{P}_2$.

14.5 NUMERICAL RESULTS

All these methods have been tested on a series of multiplicative and nonmultiplicative benchmark problems of the nuclear literature. A first report on the weakly discontinuous methods appeared in the masters thesis of Filio [Fil93]. The strongly discontinuous methods were developed in the masters thesis of Delfin [Del96]. All these results in their final correct form will be a part of the doctoral thesis of one of us (E. del Valle) which is in preparation.

In this paper, some results will be given for a test problem proposed by Azmy [Azm88], in fact problem 2 of his paper. As described by Azmy, it is a modification of Khalil's steel and water problem [Kha85] with a smaller average scattering ratio to avoid excessively slow convergence of the standard source iteration. The geometric configuration can be described by introducing four square domains $\Omega_i \equiv [0, L_i] \times [0, L_i], i = 0, \ldots, 4$ with $L_1 = 40, L_2 = 50, L_3 = 70$, and $L_4 = 100$. Then four regions I to IV can be defined as follows: $I \equiv \Omega_1, II \equiv \Omega_2 \backslash \Omega_1, III \equiv \Omega_3 \backslash \Omega_2$, and $IV \equiv \Omega_4 \backslash \Omega_3$. Vacuum boundary conditions are specified at the top and right edges, while reflection is imposed at the left and bottom edges. The nuclear data σ_t, σ_s, and the neutron source S are given in Table 1.

Table 1   Nuclear data for Azmy's test problem 2

Region	σ_t	σ_s	S
I	1.0	0.50	1.0
II	0.1	0.01	0.0
III	0.3	0.10	0.0
IV	0.1	0.01	0.0

This problem was solved by Azmy on a sequence of uniform meshes using an S_4 EQN-type angular quadrature, and a pointwise, relative convergence criterion of 10^{-4} on each one of the calculated nodal flux moments. The converged solution was used to calculate the quadrant-averaged scalar fluxes over the four regions. Different methods were used, in particular the Linear Nodal one (LN), which has 7 unknowns ($N_u = 7$) and 11 parameters ($N_p = 11$) per angular direction and node. The method of solution proposed relies on *transverse integration* where zeroth and first order moments of the neutron transport equation (14.1) are taken on a set of horizontal and vertical slices. The final result is a set of one-dimensional coupled equations, two per horizontal and per vertical slice. These equations are solved *analytically* and exponentials appear in the final expressions which are considerably more complicated than in the purely polynomial case we have considered. Finally no attempt is made to combine the linear edge and cell behaviors, which in the case of the $WD_{11,7}$ method for which results are given below lead to a global interpolant containing P_3 and for which all the extra moments not contained in D are perfectly defined in every cell. The numerical results obtained for different meshes are given in Table 2. The results for the 80×80 mesh can be taken as the reference result.

Table 2   Numerical results obtained for Azmy's test problem 2 using the LN method

Mesh	I	II	III	IV
10×10	1.953	3.650E-1	1.626E-2	2.222E-5
20×20	1.955	3.504E-1	1.556E-2	2.727E-5
40×40	1.957	3 388E-1	1.516E-2	2.696E-5
80×80	1.957	3.339E-1	1.504E-2	2.682E-5

Using different weakly and strongly continuous methods presented in this paper, the Azmy sample problem 2 was run with the same angular quadrature and stopping criterion. The corresponding numerical results are shown in Table 3. The motivation behind our choice of the results presented is that WD_{64} and SD_4 have the same number of unknowns. On the other hand WD_{64} and SD_6 have the same number of parameters. SD_4 includes Q_1 and thus P_1 but no more. WD_{64}, SD_6, and SD_8

include $\mathcal{P}_2$. $WD_{11,7}$ finally has the same number of parameters and unknowns as LN and includes $\mathcal{P}_3$.

Table 3   Numerical results obtained for Azmy's test problem 2 using the $WD_{64}, WD_{11,7}, SD_4, SD_6$, and SD_8 methods

Mesh	Region	WD_{64}	$WD_{11,7}$	SD_4	SD_6	SD_8
10×10	I	1.957	1.957	1.953	1.955	1.955
	II	3.272E-1	3.329E-1	3.776E-1	3.501E-1	3.493E-1
	III	1.557E-2	1.500E-2	1.463E-2	1.544E-2	1.567E-2
	IV	1.399E-5	2.569E-5	******	3.636E-5	3.699E-5
20×20	I	1.957	1.957	1.955	1.957	1.956
	II	3.330E-1	3.328E-1	3.546E-1	3.390E-1	3.390E-1
	III	1.508E-2	1.502E-2	1.552E-2	1.511E-2	1.518E-2
	IV	2.873E-5	2.677E-5	2.062E-5	2.722E-5	2.724E-5
40×40	I	1.957	1.957	1.956	1.957	1.957
	II	3.331E-1	3.328E-1	3.418E-1	3.340E-1	3.341E-1
	III	1.504E-2	1.502E-2	1.524E-2	1.503E-2	1.505E-2
	IV	2.695E-5	2.685E-5	2.614E-5	2.684E-5	2.685E-5

14.6   CONCLUSIONS

From Table 3, it is clear that all the methods proposed converge to the reference solution when the mesh is refined. The difference between them is more pronounced in a low flux region like IV when coarse meshes are used. In the case of SD_4, which is clearly the least accurate method, we even get negative (non physical) results in that region. If we compare the results given by WD_{64} and SD_6, with the same number of parameters, 4 unknowns for the first one and 6 for the second one, it turns out that SD_6 is clearly better.

$WD_{11,7}$ on a 40×40 mesh yields much better results than LN on the same mesh, quite comparable with the reference solution on a 80×80 mesh. These results may even be better as LN, at least in regions II and III, does not seem to have converged.

On the same 40×40 mesh, SD_6 and SD_8 which have slightly less or more unknowns are remarkably good, with less parameters and only $\mathcal{P}_2$ included in S: they are actually slightly better than $WD_{11,7}$ with an S containing $\mathcal{P}_3$.

In conclusion, most of the schemes developed are quite valuable. They provide a global representation of the angular fluxes, taking into account all the edge and cell information available. Being polynomial, this representation is clearly simpler than the ones based on the exact anlytical solution of the transverse integrated equations. Finally, if moments are needed which do not belong to D, they can always be retrieved in a completely consistent way in terms of the parameters belonging to D.

References

[Azm88]   Azmy Y. Y. (1988) Comparison of three approximations to the linear-linear nodal transport method in weighted diamond-difference form. *Nuclear Science and Engineering* 100: 190–200.

[Bad90]   Badruzzaman A. (1990) Nodal methods in transport theory. *Advances in Nuclear Science and Technology* 21: 1–119.

[Del96]   Delfin A. (1996) *Solución numérica de la ecuación de transporte de neutrones usando métodos nodales discontinuos en geometría X-Y.* Tesis de Maestría, ESFM-IPN, México.

[Fil93]   Filio C. (1993) *Solución numérica de la ecuación de transporte de neutrones en geometría X-Y.* Tesis de Maestría, ESFM-IPN, México.

[HJR88]   Hennart J.-P., Jaffré J., and Roberts J. E. (1988) A constructive method for deriving finite elements of nodal type. *Numerische Mathematik* 53: 701–738.

[Kha85]   Khalil H. (1985) A nodal diffusion technique for synthetic acceleration of nodal S_n calculations. *Nuclear Science and Engineering* 90: 263–280.

15

Some Recent Advances in Computational Strategies for Metal Forming Processes

D. R. J. Owen, D. Perić, Y. T. Feng and M. Dutko

Department of Civil Engineering,
University of Wales Swansea,
Swansea,
SA2 8PP, U. K.

15.1   INTRODUCTION

Over the last two decades, the numerical analysis of forming processes has been based almost exclusively on the finite element method and a strong impact on the computational community is evident from numerous conferences (see Shen and Dawson [SD95] for the most recent meeting).

The demands of industry have also increased: Apart of the technical requirements, including developments in material, structural and contact models, the finite element based simulation must also be appropriately integrated within a CAD/CAM environment both to reduce design to product time and minimise costs. In such a *decision support system* a crucial link between numerical analysis and other phases of the design process is provided by *automated, adaptive finite element strategies.*

A general feature of finite element simulation of forming operations is that the distribution of the optimal mesh refinement changes continually throughout the forming process. Considerable benefits may accrue from adaptive analysis in terms of robustness and efficiency. At the same time, error estimation procedures have a crucial role in quality assurance of the decision making process by providing a reliable finite element solution.

A present barrier to the simulation of industrial scale forming problems by implicit methods is the inability to efficiently solve the resulting large systems of simultaneous equations. The use of direct equation solution techniques for even moderately sized 3-

The Mathematics of Finite Elements and Applications
Edited by J. R. Whiteman © 1997 John Wiley & Sons Ltd.

D problems is precluded by computer memory requirements and consequently there is current intense interest in the development of appropriate iterative methods in which the storage needs are substantially reduced.

The paper examines the above two issues in detail and presents computational strategies suitable for the modelling of industrial scale forming problems. The algorithmic and computational links between the adaptive remeshing approach and the non-nested multigrid method are demonstrated. A set of numerical examples is provided that illustrate the effectiveness and robustness of the techniques developed.

15.2   ADAPTIVE STRATEGIES FOR INDUSTRIAL FORMING OPERATIONS

Important issues related to *finite strain plasticity* implementation are described in Section 15.2.1.

Due to the nature of deformation in forming operations, which are very large and mostly plastic, the problem of the proper treatment of *incompressibility* of plastic flow becomes important. This applies critically for bulk forming operations, but also becomes important in thin sheet forming due to the significance of the need for correct modelling of the double sided contact occurring under the blank holder area. Aspects of technology of elements capable of the treatment of incompressibility are discussed in Section 15.2.2.

The history dependent nature of the process necessitates transfer of all relevant problem variables from the old mesh to the new one, as successive remeshing is applied during the process simulation. As the mesh is adapted, with respect to an appropriate error estimator, the solution procedure, in general, cannot be re-computed from the initial state, but has to be continued from the previously computed state. Hence, some suitable means for transferring the state variables between meshes, or *transfer operators*, needs to be defined. A class of transfer operators for large strain elasto-plastic problems occurring in forming processes is defined in Section 15.2.3.

15.2.1   FINITE STRAIN PLASTICITY

Multiplicative decomposition

The main hypothesis underlying the approach employed for finite strain elasto-plasticity is the multiplicative split of the deformation gradient into elastic and plastic parts, i.e.,

$$\boldsymbol{F} := \boldsymbol{F}^e \, \boldsymbol{F}^p \tag{15.1}$$

This assumption , firstly introduced by Lee   [Lee69], admits the existence of a local unstressed *intermediate configuration*. Due to its suitability for the computational treatment of finite strain elasto-plasticity, the hypothesis of multiplicative decomposition is currently widely employed in the computational mechanics literature [EB90, Sim92, POH92].

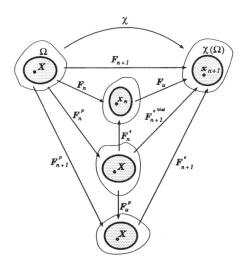

Figure 1 Multiplicative decomposition of deformation gradient.

Stress update for finite strains

In the context of finite element analysis of path dependent problems, the load path is followed incrementally and a numerical approximation to the material constitutive law is needed to update stresses as well as the internal variables of the problem within each increment. Then, given the values of the variables $\{\sigma_n, q_n\}$ at the beginning of a generic increment $[t_n, t_{n+1}]$ an algorithm for integration of the evolution equations is required to obtain the updated values $\{\sigma_{n+1}, q_{n+1}\}$ at the end of the increment.

In the present work, the *backward Euler scheme* is employed for time integration and the *Newton-Raphson algorithm* is used in the solution of the resulting set of nonlinear equations. For convenience, the operations, on the kinematic level, of the algorithm for integration of the constitutive equations at finite strains are summarized in Box 1 while Box 2 describes stress update for small strains.

15.2.2 FINITE ELEMENT TECHNOLOGY

It is a well known fact that the performance of low order kinematically based finite elements is extremely poor near the incompressible limit. Problems of elasto-plastic forming simulations under plastic dominant deformations and the assumption of isochoric plastic flow are included in this class of analysis. In such situations, spurious *locking* frequently occurs as a consequence of the inability of low order interpolation polynomials to adequately represent general volume preserving displacement fields. However, due to their simplicity, low order elements are often preferred in large scale computations and several formulations have been proposed to allow their use near the incompressible limit. Within the context of geometrically linear theory, the class of assumed enhanced strain methods described by Simo and Rifai [SR90], which incorporates popular procedures such as the classical incompatible modes formulation

Box 1. *Algorithm for integration of constitutive equations*

(i) For given displacement $\boldsymbol{u}$ and increment of displacement $\boldsymbol{\Delta u}$, evaluate total and incremental deformation gradient

$$\boldsymbol{F}_{n+1} := \mathbf{1} + \text{Grad}_{\varphi_{n+1}}[\boldsymbol{u}_{n+1}], \qquad \boldsymbol{F_u} := \mathbf{1} + \text{Grad}_{\varphi_{n+1}}[\boldsymbol{\Delta u}_{n+1}]$$

(ii) Evaluate elastic trial deformation gradient and elastic trial Fingers tensor

$$\boldsymbol{b}_{n+1}^{e \text{ trial}} := \boldsymbol{F_u}(\boldsymbol{b}_n^e)(\boldsymbol{F_u})^T$$

(iii) Compute eigenvalues (principal stretches $\boldsymbol{\lambda}^{e \text{ trial}}$) and eigenvectors (rotation tensor $\boldsymbol{R}_{n+1}^{e \text{ trial}}$) of elastic trial Fingers tensor $\boldsymbol{b}_{n+1}^{e \text{ trial}}$

(iv) Evaluate elastic trial left strain tensor and its logarithmic strain measure

$$\boldsymbol{V}_{n+1}^{e \text{ trial}} = (\boldsymbol{R}_{n+1}^{e \text{ trial}})^T (\boldsymbol{\lambda}^{e \text{ trial}}) \boldsymbol{R}_{n+1}^{e \text{ trial}}$$

$$\boldsymbol{e}_{n+1}^{e \text{ trial}} = (\boldsymbol{R}_{n+1}^{e \text{ trial}})^T \ln[\boldsymbol{\lambda}^{e \text{ trial}}] \boldsymbol{R}_{n+1}^{e \text{ trial}}$$

(v) Perform stress updating procedure for small strain (see **Box 2**)

(vi) Update Cauchy stress and internal variable

$$\boldsymbol{\sigma}_{n+1} := \det[\boldsymbol{F}_{n+1}]^{-1}\boldsymbol{\tau}_{n+1}$$

$$\boldsymbol{V}_{n+1}^e := \exp[\boldsymbol{e}_{n+1}^e], \qquad \boldsymbol{b}_{n+1}^e := (\boldsymbol{V}_{n+1}^e)^2$$

Box 2. *Stress updating procedure – Small Strains*

(i) Elastic predictor

- Evaluate trial elastic stress
$$\boldsymbol{\tau}_{n+1}^{\text{trial}} := \mathbf{h} : \boldsymbol{e}_{n+1}^{e \text{ trial}}$$

- Check plastic consistency condition

IF $\Phi^{\text{trial}} := J_2(\boldsymbol{\tau}_{n+1}^{\text{trial}}) - K(R_n) - \tau_y \leq 0$ THEN

Set $(\cdot)_{n+1} = (\cdot)_{n+1}^{\text{trial}}$ and RETURN

ELSE go to (ii)

(ii) Plastic corrector (solve the system for $\boldsymbol{\tau}_{n+1}$, and $\Delta\gamma$)

$$\left\{ \begin{array}{l} J_2(\boldsymbol{\tau}_{n+1}) - K(R_n + \Delta\gamma) - \tau_y \\ \boldsymbol{\tau}_{n+1} - \mathbf{h} : (\boldsymbol{e}_{n+1}^{e \text{ trial}} - \Delta\gamma\, \partial_\tau\phi_{n+1}) \end{array} \right\} = \left\{ \begin{array}{c} 0 \\ 0 \end{array} \right\}$$

with

$$\partial_\tau\phi_{n+1} = \frac{3}{2}\frac{\text{dev}\boldsymbol{\tau}_{n+1}}{J_2(\boldsymbol{\tau}_{n+1})}$$

(iii) Update elastic part of the strain $\boldsymbol{e}^e$ and plastic consistency parameter R

$$\boldsymbol{e}_{n+1}^e := \mathbf{h}^{-1} : \boldsymbol{\tau}_{n+1}, \qquad R_{n+1} := R_n + \Delta\gamma$$

(iv) RETURN

[TBW76] and *B-bar* methods [Hug80], is well established and is employed with success in a number of existing commercial finite element codes. In the geometrically non-linear regime, however, the enforcement of incompressibility is substantially more demanding and the development of robust and efficient low order finite elements is by no means trivial. Different approaches have been introduced in the computational literature; the class of mixed variational methods developed by Simo *et al.* [STP85], the mixed u/p formulation proposed by Sussman and Bathe [SB87], the non-linear B-bar methodology adopted by Moran *et al.* [MOS90] and the family of enhanced elements of Simo and Armero [SA92] are particularly important. However, due to the occurrence of pathological hourglassing patterns, a serious limitation on the applicability of enhanced elements has been identified by de Souza Neto *et al.* [dPHO95] for the elasto-plastic finite strain case. An element which does not suffer from such drawbacks is the F-bar element presented in [dPDO96] based on the concept of multiplicative deviatoric/volumetric split in conjuction with the replacement of the compatible deformation gradient with an assumed modified counterpart.

15.2.3   ADAPTIVE SOLUTION UPDATE

Error indicators

The extension of the error estimation based on the *plastic dissipation functional* and the rate of *plastic work* described by Perić *et al.* [PYO94] for large strain elasto-plasticity has been found to be the most appropriate solution to the adaptive solution of metal forming processes.

Mesh regeneration

An unstructured meshing approach is used for the mesh generation and subsequent mesh adaptation. The algorithm employed is based on the *Delaunay triangulation* technique which is particularly suited to local mesh regeneration. An extension of the Delaunay scheme to quadrilateral elements is also available and creates the possibility of employment of the low order elements described in Section 15.2.2.

Transfer operations for evolving meshes

After creating a new mesh, the transfer of displacement and history–dependent variables from the old mesh to a new one is required. Several important aspects of the transfer operation have to be addressed ([PHDO96], [LB94]):

(i) consistency with the constitutive equations,

(ii) requirement of equilibrium (which is fundamental for implicit FE simulation),

(iii) compatibility of the history–dependent internal variables transfer with the displacement field on the new mesh,

(iv) compatibility with evolving boundary conditions,

(v) minimisation of the numerical diffusion of transferred state fields.

To describe the transfer operation, let us define a state array ${}^h\Lambda_n = ({}^h u_n, {}^h e_n, {}^h \sigma_n, {}^h q_n)$ where ${}^h u_n, {}^h e_n, {}^h \sigma_n, {}^h q_n$ denote values of the displacement, strain tensor, stress tensor and a vector of internal variables at time t_n for the mesh h. Assume, furthermore, that the estimated error of the solution $^h\Lambda_n$ respects the prescribed criteria, while these are violated by the solution $^h\Lambda_{n+1}$. In this case a new mesh $h + 1$ is generated and a new solution $^{h+1}\Lambda_{n+1}$ needs to be computed. As the backward Euler scheme is adopted the internal variables $^{h+1}q_n$ for a new mesh $h + 1$ at time t_n need to be evaluated. In this way the state $^{h+1}\tilde{\Lambda}_n = (\bullet, \bullet, \bullet, {}^{h+1}q_n)$ is constructed, where a symbol $\sim$ is used to denote a reduced state array. It should be noted that this state characterises the history of the material and, in the case of a fully implicit scheme, provides sufficient information for computation of a new solution $^{h+1}\Lambda_{n+1}$.

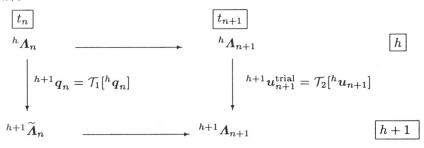

Figure 2   Transfer operator diagram

Conceptually, Figure 2 summarises a typical transfer operation that includes both, the mapping of the internal variables and mapping of the displacement field. The implementation of the given general transfer operation is performed for the case of evolving finite element meshes composed of constant strain triangles in the following fashion:

• Mapping of the internal variables - Transfer operator $\mathcal{T}_1$

Algorithm of the transfer operator $\mathcal{T}_1$ comprises the following steps:

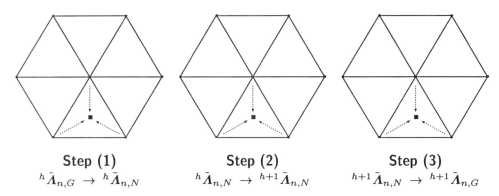

Figure 3   A procedure illustrating the implementation of the transfer operator $^{h+1}q_{n,G} = \mathcal{T}_1[^hq_{n,G}]$ for finite element meshes composed of three noded triangles.

(1) The Gauss point components of the old mesh $^h q_{n,G}$ are projected to nodes $^h q_{n,N}$ using the finite element shape functions. The nodal point averages are then performed resulting in $^h q_{n,N}^*$.

(2) In the second step the nodal components of the state variables $^h q_{n,N}^*$ will be transferred from the old mesh h to a new mesh $h+1$ resulting in $^{h+1} q_{n,N}^*$. This step of the transfer operation is the most complex one and can be subdivided as follows:

- *Construction of the background triangular mesh*
  For each node A of the new mesh $h+1$ with known coordinates $^{h+1} x_{n,A}$ the so-called background element is found in the old mesh h, i.e. element $^h \Omega^{(e)}$ for which $^{h+1} x_{n,A} \in {}^h \Omega^{(e)}$.

- *Evaluation of the local coordinates*
  Knowing the coordinates of all nodal points for both old and new meshes, the local coordinates $(^h r_A, {}^h s_A)$ within the background element corresponding to the position of nodal point A in the new mesh can be found by solving the following equation

$$^{h+1} x_{n,A} = \sum_{b=1}^{3} {}^h N_b({}^h r_A, {}^h s_A) {}^h x_{n,b} \tag{15.2}$$

  where $^h N_b$ represent shape functions of the element $^h \Omega^{(e)}$. Since three noded elements are used for the background mesh, local coordinates for each node of the new mesh $h+1$ can be obtained by resolving the linear system (15.2).

- *Transfer of the nodal values*
  By using the shape functions $^h N_b({}^h r_A, {}^h s_A)$ the state variables $^h \tilde{A}_{n,B} = {}^h q_{n,B}^*$ are mapped from the nodes B of the old mesh h to the nodes A of the new mesh $h+1$. This mapping may be expressed as

$$^{h+1} \tilde{A}_{n,A} = \sum_{b=1}^{3} {}^h N_b({}^h r_A, {}^h s_A) {}^h \tilde{A}_{n,b} \tag{15.3}$$

(3) The state variables at the Gauss points of the new mesh $^{h+1} q_{n,G}^*$ can be easily obtained by employing the shape functions of the element $^{h+1} \Omega^{(e)}$, i.e.

$$^{h+1} \tilde{A}_{n,A} = \sum_{a=1}^{3} {}^{h+1} N_a(r_G, s_G) {}^{h+1} \tilde{A}_{n,a} \tag{15.4}$$

where (r_G, s_G) are the Gauss point coordinates.

- **Mapping of the displacements - Transfer operator $\mathcal{T}_2$**
Since the displacement field over new mesh $h+1$ is fully prescribed by the nodal values $^{h+1} u_{n,a}$ and element shape functions of the new mesh $^{h+1} N_a(^{h+1} r_a, {}^{h+1} s_a)$ defined for each element $^{h+1} \Omega^{(e)}$ of the new mesh, the task of transferring displacements (i.e. transfer operator $\mathcal{T}_2$) is performed by repeating the step *Mapping of the nodal values*, in the procedure describing the transfer operator $\mathcal{T}_1$.

15.3 NUMERICAL VERIFICATION

To illustrate the adaptive strategy its application to examples of industrial metal forming operations are presented in this section.

EXAMPLE 15.3.1   Axisymmetric piercing. The finite element simulation of the axisymmetric piercing of a cylindrical workpiece is presented. The geometry of the problem is shown in Figure 4 and the initial mesh in Figure 5. The workpiece is assumed to be made of an elastic-plastic material with Young's modulus $E = 210\text{GPa}$, Poisson ratio $\nu = 0.3$, yield stress $\sigma = 100\text{MPa}$ and linear hardening with hardening modulus $H = 900\text{MPa}$, while the punch is assumed to be rigid. Frictional contact between workpiece and tool is defined by a Coulomb law with coefficient of friction $\mu = 0.1$.

In analysis an error indicator based on the rate of plastic work is used. The initial mesh consists of 101 quadrilateral elements and the final mesh contains 426 elements. Convergence of the finite element solution is established on the basis of the standard Euclidean norm of the out-of-balance forces with a tolerance of 10^{-3}. No difficulties related to the convergence have been observed during the simulation despite frequent remeshings.

Distribution of effective plastic strain on deformed meshes at various stages of the process is shown in Figure 6. The deformed meshes show no hourglassing patterns, which is in agreement with analyses of a similar class of problems carried out in [dPDO96].

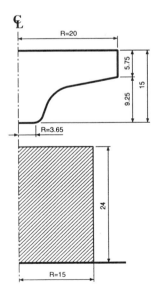

Figure 4   Axisymmetric
piercing. Geometry.

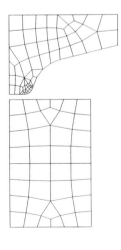

Figure 5   Axisymmetric
piercing. Initial mesh.

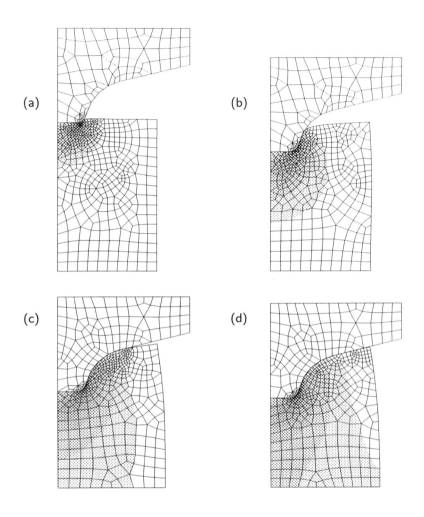

Figure 6   Axisymmetric piercing. Evolution of effective plastic strain.
(a) U = 0.62, (b) U = 5.66, (c) U = 9.83, (d) U = 11.00.

EXAMPLE 15.3.2   Plane strain spike forming. The finite element simulation of plane strain spike forming is presented. The geometry of the problem is shown in Figure 7 and the initial mesh in Figure 8. The workpiece is assumed to be made of an elastic-plastic material with Young's modulus $E = 125$GPa, Poisson ratio $\nu = 0.3$ and yield stress $\sigma = 40$MPa with an exponential hardening law. In the present FE analysis the punch is assumed to be rigid. A Coulomb law with a coefficient of friction $\mu = 0.2$ defines the frictional contact between workpiece and tool. A convergence tolerance of 10^{-3} of the finite element solution is established on the basis of the standard Euclidean norm of the out-of-balance forces.

   In total 12 mesh adaptions were performed, which involve complete new mesh

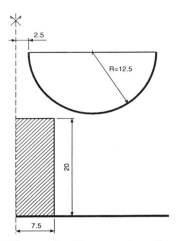
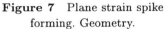

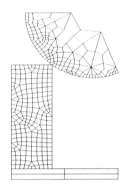

| **Figure 7** Plane strain spike | **Figure 8** Plane strain spike |
| forming. Geometry. | forming. Initial mesh. |

definitions related to the error indicator based on the rate of plastic work. The number of elements varies in subsequent meshes from initially 302 quadrilateral elements to finally 353 elements. Deformed meshes obtained after adaptive remeshing are shown in Figure 9. Similarly to the previous analysis, meshes at various stages show no hourglassing patterns despite very large deformation.

15.4   ITERATIVE EQUATION SOLUTION TECHNIQUES

The FE simulation of 3-D sheet and bulk forming operations within an implicit scheme gives rise to large algebraic equation systems

$$Ax = b \tag{15.5}$$

where A is a $n \times n$ nonsingular matrix, b is the known vector, and x is the solution to be found.

In solution of the algebraic system (15.5), iterative methods offer compelling promise over direct methods with regard to the following aspects:

- Much easier to exploit system sparsity and thus the computer memory required may be substantially reduced, especially for large problems;
- Relatively simple in implementation;
- Accuracy may be more controllable and thus computing time may be saved in cases where only a lower level accuracy is required;
- They may lead to a more synergistic incorporation in the solution of evolving nonlinear problems;
- These methods may prove more conducive to effective implementation in emerging computer systems with vector and parallel processing facilities.

Symmetric positive definite problems could arise if forming operations without

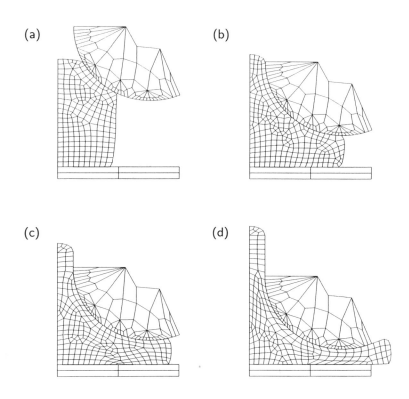

Figure 9 Plane strain spike forming. Evolution of deformed mesh.
(a) U = 6.3, (b) U = 12.1, (c) U = 13.5, (d) U = 15.0.

friction are simulated, or only the thermal aspects of the problem are considered. Otherwise, a set of nonsymmetric algebraic equations can be expected if a true forming process is analysed where frictional contact is essential. In the former case, it seems quite clear that the *conjugate gradient method (CG)* is the most efficient iterative solver to date. In the past decade, a number of *Krylov-type methods* have been proposed that are applicable to nonsymmetric cases. It is not apparent, however, which Krylov algorithm is to be preferred for solving nonsymmetric equations.

Along with the development of Krylov-type methods, another important progress is the development of *multi-grid (MG) methods*, which have been established as an efficient iterative approach for obtaining solutions of a wide variety of practical problems.

In this paper, we will focus on Krylov-type iterative solvers and the multi-grid method for solving 3-D metal forming operations which are characterised by contact-friction, material and geometrical nonlinearities.

15.5 PRECONDITIONING TECHNIQUES

Although a considerable number of successful applications of iterative methods in many areas have been reported, iterative methods suffer from the following shortcomings:

- Convergence rates may be very low; or more significantly, convergence cannot always be guaranteed;
- Performance of many iterative algorithms is problem-dependent.

These difficulties may be partially solved by employing preconditioning techniques. A suitable preconditioner is crucial to obtaining a rapid convergence of iterative methods and choosing good preconditioners for general matrices is an important research issue. An efficient preconditioner should be a close approximation of matrix A, easily computed and inverted (explicitly or implicitly) as well as being reasonably sparse. Moreover, all relevant computation processes should be concurrent if the iterative methods need to be highly parallelised or vectorized.

In sequential computers, a commonly used preconditioning technique that possesses these properties (except concurrency) is the *incomplete Cholesky decomposition (IC) with no fill-in scheme*, termed IC(0).

In order to improve the quality of the scheme, many strategies for altering the pattern of the nonzeros in the incomplete factorisation, normally classified as *high-level fill-in schemes*, have been proposed. Unfortunately, these schemes have disadvantages: (i) they are not suitable for a "black-box" implementation; (ii) it is difficult to choose *a priori* the effective parameters and to estimate the required storage; (iii) the numerical operations involved in the incomplete decomposition phase and in the following preconditioning steps are usually much larger than IC(0).

On the contrary, the IC(0) technique may be enhanced in most cases by a very simple strategy, namely bandwidth optimisation. In the example reported in this paper, Lewis' implementation version of the Gibbs-King algorithm [Lew82], which has proved to be relatively effective in terms of performance, is actually employed.

The performance of IC(0) may be degraded if a negative diagonal term is encountered during the incomplete decomposition procedure. Although several approaches have been proposed, our experience suggests the following scheme to be the most successful one: if a negative diagonal term occurs, modify A to $\bar{A} = A + \lambda D$, where $D = \mathrm{diag}(A)$ and λ is a small positive parameter. The value of λ is selected as small as possible but should make the diagonal of the incomplete decomposition of $\bar{A}$ positive. This strategy is applied to both symmetric and nonsymmetric cases and appears to be essential to obtain reasonable convergence of the corresponding preconditioned iterative algorithms for the nonsymmetric case.

In a parallel environment, however, IC(0) is hardly employed due to its intrinsic sequential nature. In this case, diagonal scaling (DS) is the most popular used preconditioning scheme. Its high concurrency may compensate for its slow convergence and thus it may outperform IC(0) in many cases. Note, however, that for very ill-conditioned problems, DS may not perform well and the corresponding preconditioned iterative algorithm may even fail to converge.

An important issue which has not, at present, received adequate attention in the literature concerns the effects of the preconditioning patterns on the convergence of the iterative algorithms. In applying IC(0) preconditioning to linear systems of equations, there are three slightly different versions from which to choose, which are referred to as left, right and left-right preconditioning respectively [FHOP95]. Different forms of preconditioning patterns lead to different condition numbers and different distributions of eigenpairs of the preconditioned matrix, and consequently have different effects on the convergence of iterative methods. No theoretical results, however, are at present available to compare the efficiencies of these three preconditioning schemes. Our previous experience [FHOP95] indicates that the left preconditioning pattern exhibits slightly better behaviour, but the right preconditioning pattern is actually used in the current work.

15.6 ITERATIVE METHODS FOR NONSYMMETRIC PROBLEMS

The commonly used Krylov-type iterative approaches include:

- CGS, *Conjugate Gradient Squared method* [Son89]
- BiCGStab, *Bi-Conjugate Gradient Stabilised method* [vdV92]
- GMRES, *Generalised Minimum Residual method* [SS86]

These methods require only three operations in their implementation: saxpy operations, inner products, and matrix-vector multiplications. As a result they can be efficiently implemented on scalar, vector and parallel computers. In this work some aspects of CGS and BiCGStab iterative approaches to the solution of nonsymmetric problems are presented which are relevant for metal forming simulations.

The CGS approach is an accelerated version of the *biconjugate gradient method*, which is an extension of CG to nonsymmetric cases. CGS sometimes suffers from severe numerical instability, even though it often exhibits good convergence. BiCGStab is the stabilised version of CGS. Numerical experiments indicate that BiCGStab appears to be a far more stable method. For this reason, only the BiCGStab algorithm is considered below. For more details concerning the algorithm see [vdV92].

Algorithm 1: BiCGStab with right preconditioning

---

1. Initial guess x_0, and $r_0 = b - Ax_0$; Choose $\bar{r}_0$ such that $r_0^T \bar{r}_0 \neq 0$, e.g. $\bar{r}_0 = r_0$;
   Set $\rho_0 = \alpha_0 = \omega_0 = 1$ and $p_0 = v_0 = 0$
2. For $i = 1, 2, \cdots$ Do

   - $\rho_i = \bar{r}_0^T r_{i-1}$ $\beta_i = (\rho_i/\rho_{i-1})(\alpha_{i-1}/\alpha_i)$
   - $p_i = r_{i-1} + \beta_i(p_{i-1} - \omega_{i-1} v_{i-1})$
   - Solve: $My = p_i$
   - $v_i = Ay$; $\alpha_i = \rho_i/(\bar{r}_0^T v_i)$
   - $s = r_{i-1} - \alpha_i v_i$
   - Solve: $Mz = s$
   - $t = Az$; $\omega_i = (t^T s)/(t^T t)$;
   - $x_i = x_{i-1} + \alpha_i y + \omega_i z$
   - $r_i = s - \omega_i t$
   - Convergence Check: if $\|r_i\|/\|r_0\| \leq \epsilon$ (given tolerance) then stop

   End Do

In the above algorithm, M denotes the preconditioning matrix. The BiCGStab algorithm involves two matrix-vector multiplications and two preconditioning operations at each iteration, and requires seven working vectors.

15.7   GALERKIN MULTI-GRID METHOD

The essential multi-grid principle is based on the observation that the smooth (or long-wavelength) part of the error, which may not be efficiently swept out by iterative methods such as CG and BiCGStab, can be substantially reduced by a coarse mesh correction. The success of MG strategies lies primarily in (i) their excellent convergence characteristics, which theoretically should not depend on the size of the finite element mesh; (ii) their high efficiency whereby solutions of problems with n_{eq} unknown are obtained with $O(n_{eq})$ in terms of work and storage for large classes of problems. Several different schemes of multi-grid techniques have been put forward in the last decade [Bra77, RS87, Hac85, Wes92]. In this paper we focus on one particular scheme termed the Galerkin Multi-Grid (GMG) method proposed in [FOP96].

To illustrate the basic idea of the Galerkin multi-grid scheme we consider its two grid form. Suppose that G_c and G are, respectively, coarse and fine grids which discretise the same geometrical domain Ω, and that the fine grid is supposed to represent the current problem considered. We use subscript c to distinguish the quantities of the coarse grid from those of the fine grid. Let A_c be the coarse grid matrix, and P and Q be, respectively, the matrix representations of the interpolation and projection operators. In the GMG method, the coarse grid matrix A_c is constructed by direct projection of the fine mesh matrix as

$$A_c = QAP \tag{15.6}$$

Here projection operator Q is taken as $Q = P^T$. Therefore

$$A_c = P^T A P \tag{15.7}$$

Efficient computation of A_c is crucial to achieve an overall high performance of the complete GMG method. Such an implementation can be found in [FPO96]. Let $S(x, \mu)$ denote the smoother with x as the initial guess and μ the maximum number of iterations, and μ_1 and μ_2 be the maximum iterations of the pre- and post- smoothing procedures performed respectively before and after the coarse grid correction which is accomplished by a profile solver.

The standard Galerkin multi-grid algorithm for solving (15.5) can then be outlined as follows

Algorithm 2: Standard Two-Level Galerkin Multi-Grid Algorithm

---

1. Set up: Define interpolation operator P, and construct $A_c = P^T A P$; Choose initial guess x^0, smoothing algorithm S and parameters μ_1, μ_2, and compute $r^0 = b - A x^0$
2. For $l = 1, ..., m_l$ Do

   - One cycle MG iteration:
     - — Smoothing on fine mesh:
       $\Delta x^l \leftarrow S(0, \mu_1)$
     - — Compute the new residual:
       $\Delta r^l = r_{l-1} - A \Delta x^l$
     - — Coarse mesh correction:
       $A_c \Delta x_c^l = P^T \Delta r^l$
     - — Update: $\Delta x^l \leftarrow \Delta x^l + P \Delta x_c^l$
     - — Smoothing on fine mesh:
       $\Delta x^l \leftarrow S(\Delta x^l, \mu_2)$
   - Solution update : $x^l = x^{l-1} + \Delta x^l$
   - Residual update : $r^l = r^{l-1} - A \Delta x^l$
   - Convergence check: if $\|r^l\| / \|r^0\| \leq \epsilon$ stop.

   End Do

Obviously, the efficiency of GMG is dependent on the quality of the coarse grid and the appropriate selection of interpolation and projection operators. Once A_c, P and Q are determined, the performance of GMG will entirely depend on the smoother S and the numbers of iterations μ_1, μ_2. The practical selection of smoothers can range from very simple Jacobi (or DS), Gauss-Seidel, SOR, SSOR to incomplete decomposition, and even to any iterative algorithm. In our case, preconditioned BiCGStab is chosen as the smoother.

In order to enhance the performance of GMG, the outer loop of the MG iterations is further accelerated by GMRES and it is equivalent to nonlinear GMRES with MG as its preconditioning scheme. For more details regarding this enhancement, we refer to [FPO96]

As the Galerkin strategy has been fully adopted for the generation of coarse mesh equations and no material and loading information for coarse meshes are utilised, the

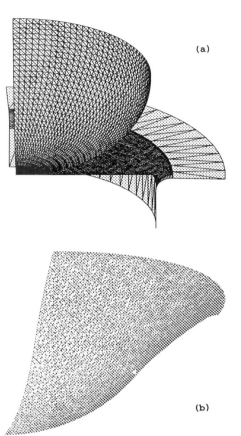

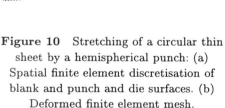

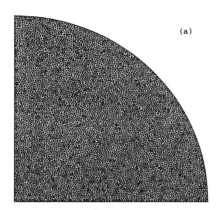

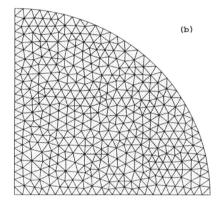

Figure 10 Stretching of a circular thin sheet by a hemispherical punch: (a) Spatial finite element discretisation of blank and punch and die surfaces. (b) Deformed finite element mesh.

Figure 11 Finite element discretisation of the blank: (a) Fine mesh: 9098 nodes, 17666 elements, 26298 active d.o.f. (b) Coarse mesh: 464 nodes, 841 elements, 1230 active d.o.f.

GMG approach is relatively easy to incorporate into the existing solution procedures, and is particularly suitable for implementation in material nonlinear cases, including frictional contact. For geometrical nonlinear cases, the approach uses the constant transfer operator throughout the whole solution processing without significantly influencing the convergence property. Another important feature of the GMG method is that coarse and fine meshes can be non-nested and unstructured which not only allows for easy treatment of complex geometry problems, but also provides a possibility to easily combine with adaptive mesh refinement techniques such as those described in Section 15.2.

15.8 NUMERICAL EXAMPLE

Stretching of a circular thin sheet by a hemispherical punch. The geometry for this example is as follows: blank thickness is $t = 1$ mm, blank radius is $R = 59.18$ mm, while the punch and die radii are $R_p = 50.8$ mm and $R_d = 6.83$ mm respectively. It is assumed that the blank material is described by a large strain elasto-plastic constitutive model with Young's modulus given by $E = 2.1 \times 10^5$ N/mm^2, Poisson's ratio $\nu = 0.3$ and isotropic hardening, which relates the equivalent stress $\bar{\sigma}$ and the equivalent plastic logarithmic strain $\bar{\varepsilon}^p$ as $\bar{\sigma} = 589 \cdot (10^{-4} + \bar{\varepsilon}^p)^{0.216}$ N/mm^2. The analysis is performed employing a membrane formulation for a quarter of the problem with appropriate boundary conditions. From a numerical point of view this problem is considered as a full three-dimensional analysis with appropriate algorithmic treatment of the frictional contact problem. Results are obtained for a coefficient of friction between tools and blank of $\nu_F = 0.30$.

To solve this problem the blank is discretised with 17666 constant strain triangular finite elements resulting in 26298 active d.o.f. The surfaces of the punch and die are respectively represented by 2145 and 612 triangular flat elements. Spatial discretisation of the problem is depicted in Figure 10a. Figure 10b shows the deformed mesh at punch displacement of $D_p = 30$ mm.

Within the multi-grid strategy the above mesh represents the fine mesh, while the coarse mesh contains 841 constant strain triangular finite elements with 1230 active d.o.f. Both meshes are depicted in Figure 11. It should be emphasised that the fine and coarse meshes are fully unstructured meshes and have been obtained independently from each other.

A full Newton-Raphson method, with an unsymmetric tangent stiffness arising from the non-associated frictional contact law, is employed in all computations. Convergence of the finite element solution is established on the basis of the standard Euclidian norm of the out-of-balance forces.

Three solution algorithms are employed in the solution of the algebraic system of equations: profile (direct) solver, Bi-CGStab iterative solution and multigrid algorithm. A comparison of performances of these algorithms, in terms of CPU time (seconds) and memory requirements (Mbytes), is presented in Figure 12. It is clear that both BiCGStab and multigrid solution provide a significant improvement over the standard direct solver. Memory requirements and CPU time have both been reduced by approximately 7-8 times in comparison to the direct solution. It should be emphasised that, in comparison to the BiCGStab solution, the multigrid strategy requires some small additional storage, but shows almost two-fold reduction in the CPU time. In addition, the multigrid solution is expected to have a more stable behaviour in the solution of complex 3-D problems that may arise in the simulation of forming processes.

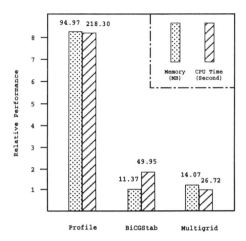

Figure 12 A comparison of performances, in terms of CPU time (seconds) and memory requirements (Mbytes), for three equation solution algorithms: profile (direct) solver, Bi-CGStab iterative solution and multi-grid algorithm.

15.9   CONCLUSIONS

A reliable and efficient automated, adaptive finite element strategy based on effective error estimates and automatic remeshing techniques has been developed for the FE analysis of sheet and bulk metal forming operations. A set of numerical examples is provided that illustrate the effectiveness and robustness of the developed approach.

By simulation of a sheet forming example, it has been shown that iterative methods combined with multigrid solutions offer a significant improvement in performance in comparison to direct solutions. It has been proved that the appropriate choice and implementation of the iterative solutions within an implicit solution strategy, may lead to successful simulation of metal forming processes, governed by complex large strain elasto-plastic material models and frictional contact interface laws.

It is expected that for general 3-D problems that may arise in simulation of forming processes, iterative methods combined with multigrid solutions can potentially offer substantial savings in terms of CPU time and memory requirements, in comparison to direct solution methods, thus offering a viable alternative to explicit solution strategies for large classes of problems.

ACKNOWLEDGEMENT

This work is funded by the EPSRC of UK under grant No. GR/J47008. This support is gratefully acknowledged.

References

[Bra77]   Brandt A. (1977) Multi-level adaptive solution to boundary-value problems. *Math. Comput.* 31: 333–390.

[dPDO96]   de Souza Neto E. A., Perić D., Dutko M., and Owen D. R. J. (1996) Design of simple low order finite elements for large strain analysis of nearly incompressible solids. *Int. J. Solids Struct.* 33: 3277–3296.

[dPHO95]   de Souza Neto E. A., Perić D., Huang G. C., and Owen D. R. J. (1995) Remarks on the stability of enhanced strain elements in finite elasticity and elastoplasticity. *Comm. Num. Meth. Engng.* 11: 951–961.

[EB90]   Eterovic A. and Bathe K.-J. (1990) A hyperelastic based large strain elasto-plastic constitutive formulation with combined isotropic-kinematic hardening using the logarithmic stress and strain measures. *Int. J. Num. Meth. Engng.* 30: 1099–1114.

[FHOP95]   Feng Y. T., Huang G. J., Owen D. R. J., and Perić D. (1995) An evaluation of iterative methods in the solution of a convection-diffusion problem. *Int. J. Num. Meth. Heat Fluid Flow* 5: 213–223.

[FOP96]   Feng Y. T., Owen D. R. J., and Perić D. (1996) A Multi-grid enhanced GMRES algorithm for elasto-plastic problems. *Technical Report CR/921/96, Dept. Civil Eng., University of Wales Swansea* .

[FPO96]   Feng Y. T., Perić D., and Owen D. R. J. (1996) A non-nested Galerkin Multi-Grid method for solving linear and nonlinear solid mechanics problems. *Comp. Meth. Appl. Mech. Engng.* (in press).

[Hac85]   Hackbuch W. (1985) *Multi-grid Methods And Applications.* Springer, Berlin, Germany.

[Hug80]   Hughes T. J. R. (1980) Generalization of selective integration procedures to anisotropic and nonlinear media. *Int. J. Num. Meth. Engng.* 15: 1413–1418.

[LB94]   Lee N.-S. and Bathe K.-J. (1994) Error indicators and adaptive remeshing in large deformation finite element analysis. *Fin. Elem. Anal. Design* 16: 99–139.

[Lee69]   Lee E. H. (1969) Elastic-plastic deformation at finite strains. *J. Appl. Mech.* 36: 1–6.

[Lew82]   Lewis J. G. (1982) Implementation of the Gibbs-Poole-Stockmeyer and Gibbs-King algorithm. *ACM Trans. Math. Softw.* 8: 180–189.

[MOS90]   Moran B., Ortiz M., and Shih F. (1990) Formulation of implicit finite element methods for multiplicative finite deformation plasticity. *Int. J. Num. Meth. Engng.* 29: 483–514.

[PHDO96]   Perić D., Hochard C., Dutko M., and Owen D. R. J. (1996) Transfer operators for evolving meshes in small strain elasto–plasticity. *Comp. Meth. Appl. Mech. Engng.* (in press).

[POH92]   Perić D., Owen D. R. J., and Honnor M. (1992) A model for finite strain elasto-plasticity based on logarithmic strains: Computational issues. *Comp. Meth. Appl. Mech Engng.* 94: 35–61.

[PYO94]   Perić D., Yu J., and Owen D. R. J. (1994) On error estimates and adaptivity in elastoplastic solids: Application to the numerical simulation of strain localization in classical and cosserat continua. *Int. J. Num. Meth. Engng.* 37: 1351–1379.

[RS87]   Ruge J. W. and Stuben K. (1987) Chapter 4: Algebraic multigrid. In McCormick S. F. (ed) *Multigrid Methods*, pages 73–130. SIAM, Philadelphia, Pennsylvania, USA.

[SA92]   Simo J. C. and Armero F. (1992) Geometrically non-linear enhanced strain mixed methods and the method of incompatible modes. *Int. J. Num. Meth. Engng.* 33: 1413–1449.

[SB87]   Sussman T. and Bathe K.-J. (1987) A finite element formulation for nonlinear incompressible elastic and inelastic analysis. *Comp. Struct.* 26: 357–409.

[SD95]   Shen S.-F. and Dawson P. R. (eds) (1995) *Proceedings of the Fifth International Conference on Numerical Methods in Industrial Forming processes*, Rotterdam. A.A.Balkema.

[Sim92]   Simo J. C. (1992) Algorithms for static and dynamic multiplicative plasticity that preserve the classical return mapping schemes of the infinitesimal theory. *Comp. Meth. Appl. Mech. Engng.* 99: 61–112.

[Son89]   Sonneveld P. (1989) CGS: a fast Lanczos-type solver for nonsymmetric linear systems. *SIAM J. Sci. Stat. Comp.* 10: 36–52.

[SR90]   Simo J. C. and Rifai S. (1990) A class of mixed assumed strain methods and the method of incompatible modes. *Int. J. Num. Meth. Engng.* 29: 1595–1638.

[SS86]   Saad Y. and Schultz M. (1986) GMRES: A generalized minimal residual algorithm for solving nonsymmetric linear systems. *SIAM J. Sci. Stat. Comp.* 7: 856–869.

[STP85]   Simo J. C., Taylor R. L., and Pister K. S. (1985) Variational and projection methods for the volume constraint in finite deformation elasto-plasticity. *Comp. Meth. Appl. Mech. Engng.* 51: 177–208.

[TBW76]   Taylor R. L., Beresford P. J., and Wilson E. L. (1976) A non-conforming element for stress analysis. *Int. J. Num. Meth. Engng.* 10: 1211–1219.

[vdV92]   van der Vorst H. (1992) Bi-CGSTAB: a fast and smoothly converging variant of Bi-CG for the solution of nonsymmetric linear equations. *SIAM J. Sci. Stat. Comp.* 13: 631–644.

[Wes92]   Wesseling P. (1992) *An Introduction to Multigrid Methods*. John Wiley & Sons, Chichester, England.

16

An Integrated Approach for the Geometric Design and Finite Element Method Formability Analysis of Sheet Metal Parts

J.C. Cavendish and S.P. Marin

Manufacturing and Design Systems Department
General Motors Research and Development Center,
Warren, Michigan 48090-9055, USA

16.1   INTRODUCTION

At MAFELAP 1990 we introduced a new feature-based computer-aided design method for designing functional parts, such as automobile sheet metal inner panels. With this method, a designer can directly and smoothly assemble and represent a complex, multi-featured surface using simpler component surfaces and information about feature shape (see   [Cav91b] and   [Cav92]). At MAFELAP 1990 we also described a companion feature-based finite element mesh generation algorithm for triangulating multi-featured surface designs (see also [Cav91a]).

   In this paper we will focus on manufacturing issues related to downstream processing of inner panel surfaces designed by our feature-based approach. In Section 16.2 we briefly review the basic elements of our approach to feature-based surface design. Issues related to finite element formability analysis are taken up in Section 16.3. We conclude with summary remarks in Section 16.4.

16.2   A REVIEW OF FEATURE-BASED SURFACE DESIGN

We have developed a feature-based CAD methodology for use in the design of complex, multi-featured surfaces. A single feature is defined as a pocket by first inputting plan view feature boundaries along with definitions of primary and secondary component surfaces. For example, the simple pocket shown in Figure 1 was designed as a

The Mathematics of Finite Elements and Applications
Edited by J. R. Whiteman © 1997 John Wiley & Sons Ltd.

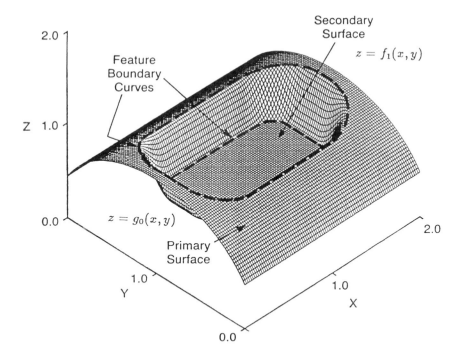

Figure 1   A simple pocket feature

modification to the primary surface $z = g_0(x, y)$ by inputting the definition of the pocket bottom, the secondary surface $z = f_1(x, y)$, a plan view inner feature boundary curve, C_1, and a plan view outer feature boundary curve C_0, and then computing a new surface $z = g_1(x, y)$ according to the blending formula

$$g_1(x, y) = (1 - \phi(x, y))g_0(x, y) + \phi(x, y)f_1(x, y) \qquad (16.1)$$

In (16.1), $\phi(x, y)$ is what we term a *transition function*. It may be any continuously differentiable function $\phi : R^2 \to [0, 1]$ which is identically 0 for points (x, y) that are outside the outer curve C_0 and identically 1 for points inside the inner curve C_1.

Although there are a variety of ways to define such transition functions between the inner and outer curves for arbitrary plan view features boundaries, we have found one specialization to be of significant practical value. This case arises when the outer curve is a constant normal offset of the inner boundary. Here, $\phi(x, y)$ is defined in the *transition region* (the region between C_0 and C_1) as follows. Let (x, y) be a point in the transition region and define $\rho(x, y)$ to be the distance from (x, y) to the inner curve C_1 and let $\eta(x, y) = \frac{(D - \rho(x, y))}{D}$, where D is the constant offset distance. Then $\phi(x, y)$ is given by

$$\phi(x, y) = \eta^2(x, y)(3 - 2\eta(x, y)). \qquad (16.2)$$

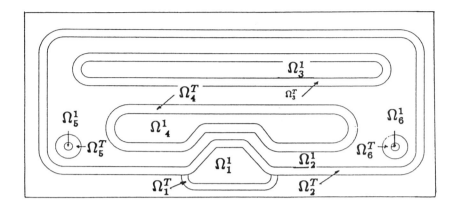

Figure 2   Feature-boundaries for a six-feature design

Additional features are added by treating the surface $z = g_1(x, y)$ as a new primary surface and constructing another pocket as described above. The result is a sequence of blending steps, each of which adds one pocket-like feature to the surface. If the base surface (that is, the initial primary surface) is denoted by $z = g_0(x, y)$ and we wish to design a surface with N features, we begin by identifying the secondary surfaces $z = f_1(x, y), z = f_2(x, y), ..., z = f_N(x, y)$ involved in the construction of each feature. With these secondary surfaces specified we then input feature boundary curves $C_{0,i}, C_{1,i}, i = 1, 2, ...N$ at each step. Our feature-based methodology automatically constructs the transition functions $\phi_i(x, y)$ for the ith feature. As each blending step is completed, a continuously differentiable surface

$$z = g_i(x, y) = (1 - \phi_i)g_{i-1} + \phi_i f_i \tag{16.3}$$

is defined which contains the first i features of the planned design. The last step yields the desired surface $z = g_N(x, y)$ containing all N features.

We conclude this section with a simple 6-feature surface designed using the procedural algorithm defined by (16.3). We begin with Figure 2 which defines plan view outlines of all feature boundaries. Also shown in Figure 2 are feature transition regions, Ω_i^T and feature interior regions, Ω_i^1. Figure 3 represents a shaded image of the 6-feature surface $z = g_6(x, y)$ determined by a six-fold execution of (16.3). For this exercise, the primary and all secondary surfaces are sections of parallel planes.

16.3   FINITE ELEMENT FORMABILITY ANALYSIS

Part formability evaluation is an essential step in making the transition from design concept to part construction in the creation of sheet metal parts. Formability

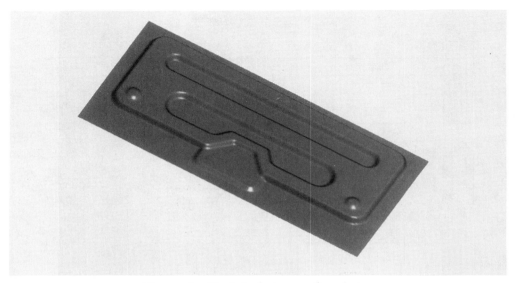

Figure 3   Final six-feature surface design

evaluations have traditionally been performed by first constructing an evaluation or *proof die*, usually from an easily modified material like zinc alloy or plastic. The proof die is then tried, modified and tried again until a good part is stamped. This die tryout process is typically time consuming and costly. Recently, analytic techniques have become available which provide the capability to conduct finite element computer simulations of the sheet metal forming process and make formability evaluations at panel design time. This approach allows changes in geometry and stamping conditions to be quickly evaluated and incorporated in the math model, and eliminates much of the costly iteration with a proof die.

A particularly efficient formability analysis technique arises from the finite element method applied to a membrane theory of shells for one-dimensional plane strain or axisymmetric deformations. The basic geometry for a typical punch stretching problem is shown in Figure 4. The sheet is assumed initially flat and is stretched over a die by a punch with curved edges. For a plane strain problem, no quantity varies in the direction perpendicular to the paper, therefore, the strain component in this direction is zero. Points on the sheet material are identified by their initial distances, ξ from the z-axis and the sheet initially lies in the interval $\xi_L < \xi < \xi_R$. The current horizontal and vertical displacements are u_1 and u_2 respectively and are functions of ξ and time t (see Figure 4).

The principal logarithmic strain components for the plane strain problem are

$$\epsilon_1 = ln((1 + \frac{\partial u_1}{\partial \xi})^2 + (\frac{\partial u_2}{\partial \xi})^2)^{\frac{1}{2}} \tag{16.4}$$

$$\epsilon_2 = 0 \tag{16.5}$$

where subscript 1 is tangential to the sheet (i.e., in the ξ direction) and 2 is normal to the paper. For axisymmetric problems, ϵ_1 is as in (16.4), but the circumferential

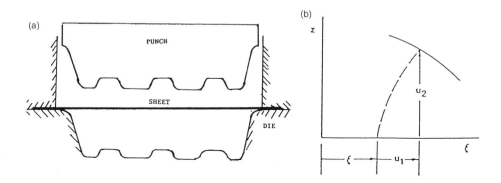

Figure 4   4(a) punch stretching, (b) notation and geometry

component of strain is defined by

$$\epsilon_2 = ln(1 + \frac{u_1}{\xi})$$ (16.6)

During punch stretching, the sheet may be divided into a punch contact region, a die contact region and a non-contact region. The shape $S_P(\xi, t)$ of the moving punch is a specified function of time determined by first cutting a user-specified cross section in a multi-featured geometry defined by (16.3) and then translating that cross section at a constant rate in time in a direction parallel to the z-axis (the punch progression direction). In a similar fashion, the shape $S_D(\xi)$ of the fixed die is specified. The interface conditions between the sheet and the punch or die are modeled by Coulomb friction(see [Fre87]) for details).

The basic mathematical equations used to model the sheet metal deformation process are the *equilibrium equations* of shell membrane theory (partial differential equations expressed in rate form which represent the Euler equations of the principle of virtual work) and the *constitutive equations* (rate-sensitive, elastic-plastic equations relating temporal derivatives of stress and temporal derivatives of strains for the particular sheet material being formed). Combining the rate forms of the equilibrium and constitutive equations leads to a variational form whose Euler equations are differential equations involving the displacements, u_1 and u_2, and their spatial and time derivatives. Galerkin's method is applied (the sheet domain is decomposed into an assembly of piecewise linear membrane finite elements in which nodal variables are horizontal and vertical displacements) to produce a finite element discretization in the spatial dimension, leaving the time variable continuous. Finally, the resulting system of nonlinear ordinary differential equations is numerically integrated using a low order explicit modified Euler method (see [Fre87] for complete details).

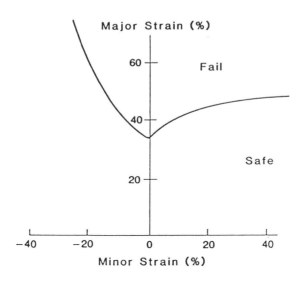

Figure 5   Forming limit diagram

An important concept in sheet metal forming analysis is the notion of a *forming limit diagram*. This diagram partitions the $\epsilon_1 \epsilon_2$ - strain plane into safe and failure regions, see Figure 5. Whenever the principal logarithmic strains enter the failure region, it is presumed that fracture occurs in the sheet and the part fails to form. The forming limit diagram is empirically derived and varies in shape with the type of sheet steel being formed.

We illustrate this one-dimensional formability analysis capability first for the surface designed in Figure 3. The location and shape of the chosen cross section is shown in Figure 6.

It was judged that sheet metal deformation near this cross section was close enough to a one-dimensional problem that a plane strain analysis would be approximately valid. The plane strain forming limit for this material is about 34%. After each time step in the finite element formability analysis, the strains, ϵ_1, in each element are computed and compared with the forming limit value. If at any time during the forming process the strain exceeds 34%, it is presumed that fracture occurs, the location of the failure and punch depth at that time are recorded, and the simulation is terminated. Figure 7 shows plots of part shape and strain distribution in the cross section displayed in Figure 6 at a punch depth of 11 mm. The strain spike that appears in this figure occurs in the transition region of the second feature in Figure 2 (that is, in Ω_2^T). This spike exceeds 34%, hence the computations suggest that the part would fail to be made successfully. The top graph in Figure 7 displays the sheet metal shape (thick, bold line) at the time of failure. Here, the sheet is pinned between the punch and the die surfaces (as they are outlined by the fine lines in Figure 7). The cause of the strain excess is the steep transition surface defined on Ω_2^T. This in turn is the result of a small offset distance ($D = 10$ mm) used in equation (16.2) to define the transition function of the second feature. If D is increased to 13 mm, the plane strain finite

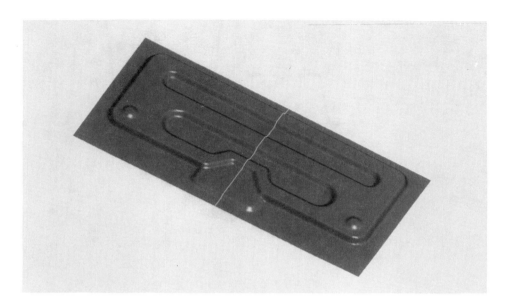

Figure 6 Cross section defining punch geometry

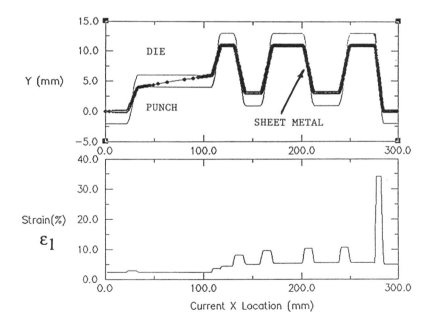

Figure 7 Plot of part shape and strain distribution, forming limit exceeded

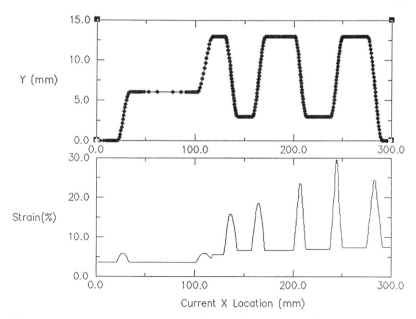

Figure 8   Part shape and strain distribution, forming limit not exceeded

element formability analysis produces the results of Figure 8. For this case the punch successfully bottoms out with peak strain less than 30%

It is well worth noting that the time scales associated with the design/analysis/modify cycle described above are quite different from those that attend current computer-aided technology for part design and analysis. Most commercial CAD systems rely on patch-based parametric surface mathematics for representing part geometry. It would require about 60 separate patches and at least 4 hours to create a parametric equivalent of the part depicted in Figure 3. Once this were done, tests would be carried out to confirm that adjacent patches join with acceptable continuity and smoothness. Additional tests would be used to verify that there are no holes or missing surface patches in the design. By contrast, the feature-based method described in Section 16.2 yields a single, continuously differentiable surface defining the entire part: it is not necessary to carry out the surface verification steps refered to above. Use of the feature-based approach required approximately 15 minutes to define the input parameters needed to generate the surface geometry displayed in Figure 3. One complete plane strain or axisymmetric formability analysis is carried out in less than a minute (on a HP 715 workstation). The modification needed to produce a makeable part (namely, the change of a single parameter, D, from 10 mm to 13 mm) is accomplished in seconds. The time scale associated with feature-based design/modification with one dimensional formability analysis is measured in minutes, while conventional approaches require hours or days. It is this real time design/analysis capability which offers high potential for sheet metal product development processes.

Figure 9 illustrates an axisymmetric part with attendant formability analysis section.

The results of an axisymmetric finite element formability analysis are also shown in Figure 9. These figures display the sheet shape at a punch depth of 40 mm (full

(a)

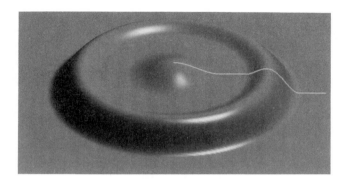

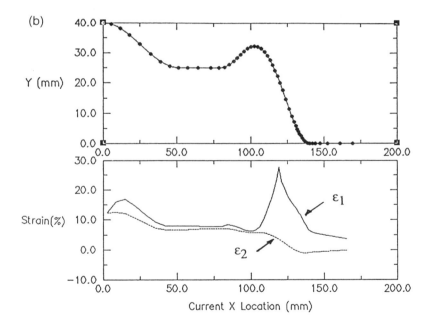

Figure 9   (a) axisymmetric part, (b) part shape and strain distribution.

penetration), and the distribution of radial (ϵ_1) and circumferential (ϵ_2) strains. These strains all reside in the safe region of the forming limit diagram, which suggests that the part would form.

16.4 SUMMARY

We have illustrated the capability of a feature-based design methodology for defining smooth surface representations of automobile inner panels. We have also illustrated the application of finite element analysis and processing tools for formability analysis of inner panels designed by these methods. Although the formability formulation is strictly valid only for plane strain or axisymmetric problems, experience has shown that this restriction can be relaxed in practice. One of the benefits of a rapid geometry creation capability coupled to an economical formability analysis capability is that varying combinations of material, processs parameters and geometric configurations can be studied at the time of part design. Such studies are necessary in order that the final manufacturing process be insensitive to day-to-day variations in press settings, material properties and die/punch lubricants.

References

[Cav91a] Cavendish, J.C. and Frey, W.H., and Marin, S.P. (1991) Feature-Based Design and Finite Element Mesh Generation for Functional Surfaces. *Engineering Software* 13,(5/6): 226–237.

[Cav91b] Cavendish, J.C. and Marin, S.P. (1991) Feature-Based Design and Finite Element Analysis of Functional Surfaces. In Whiteman J. (ed) *The Mathematics of Finite Elements and Applications (MAFELAP VII)*, pages 129–139. Academic Press, London.

[Cav92] Cavendish, J.C. and Marin, S.P. (1992) A Procedural Feature-Based Approach for Designing Functional Surfaces. In Hagen H. (ed) *Topics in Surface Modeling*, pages 145–168. Society for Industrial and Applied Mathematics, Philadelphia.

[Fre87] Frey, W.H. and Wenner, M.L. (1987) Development and Application of a One-Dimensional Finite Element Code for Sheet Metal Forming Analysis. Interdisciplinary Issues in Materials Processing and Manufacturing - Volume One, pages 307–319. The American Society of Mechanical Engineers.

17

Multiphysics Modelling of Materials Processing Using Finite Volume Unstructured Mesh Procedures

C. Bailey, M.Cross, P. Chow and K. Pericleous

Centre for Numerical Modelling and Process Analysis,
University of Greenwich,
Woolwich, London
SE18 6PF, U. K.

17.1   INTRODUCTION

This paper will outline techniques used for modelling the phenomena governing the processing of materials into products. Such phenomena can be termed multiphysics, where the formation of material products in manufacturing involves controlling a whole range of physics. A classical example of this is the casting process, where a metal is melted and poured into a mould cavity. The molten metal then solidifies and behaves as a solid in interacting with the mould. Modelling the effects of such a process involves coupling the equations of fluid flow, solidification and solid mechanics in order that accurate modelling of the full process can be accomplished. Other manufacturing processes such as plastic extrusions, welding and soldering also need to be viewed in a multiphysics framework by modellers.

Over the last twenty years numerous software products have evolved that solve problems of fluid flow, electromagnetics and solid mechanics. Until recently such codes only addressed specific phenomena, for example fluid flow (PHOENICS, CFX, FLOW3D etc.) or solid mechanics (ANSYS, ABACUS, LUSAS etc.). This has been the general trend in computational mechanics where models are developed which are specifically for fluid flow or solid mechanics, etc. . Obviously the majority of manufacturing processes, particularly in materials processing, are not governed by a distinct set of physical laws, but will involve numerous physical relationships as the material in question undergoes a phase change.

The Mathematics of Finite Elements and Applications
Edited by J. R. Whiteman © 1997 John Wiley & Sons Ltd.

In the past the technique used to predict multiphysics processes has been to couple different codes together (i.e PHOENICS - ABACUS). This has proved to be very time consuming and prone to extensive errors, especially when each code is using different types of meshes. Currently there is a great need for multiphysics software that can predict the behaviour of materials in both the solid and fluid state, plus the effects of forces such as electromagnetic forces on the material. It would be beneficial for such software to allow the inclusions of new physics easily and that the coupling between such physics should be undertaken on the same mesh within the same computational environment. To achieve this aim the group at Greenwich have produced the PHYSICA multiphysics software environment which it is hoped will provide modellers with a computational toolkit that will allow easy inclusion of relevant physics.

Some interesting questions, which will be discussed in this paper, for multiphysics modellers are :

- What discretisation methods should be used for Multiphysics modelling?
- Can parallel computing help speed up Multiphysics simulations.?
- How can we link up the length scales (Nano-Continuum Level)?

Answers to the above questions are needed to help modellers produce techniques and appropriate software which will aid industry in designing higher quality products with much shorter lead times. At present computational models, especially in the materials processing arena, can be very slow in producing solutions. The reason for this is that such models require fine meshes and very long compute times. Also for materials being processed at high temperatures and undergoing a phase change, there is currently a great lack of material property data available, especially between the solidus and liquidus temperatures of the material. The question of speed and materials properties could be answered by using parallel tools and linking up models across the respective length scales.

17.2   PHYSICS AND DISCRETISATION PROCEDURES

Computational mechanics has grown extensively over the last thirty years and is now a common tool both in academia and industry. Numerous camps have developed with the community based on specific physical phenomena of interest. Researchers in fluid dynamics have developed techniques originally based on finite difference procedures for modelling complex nonlinear flow phenomena based on Navier-Stokes equations; see [Pat80, Hir90]. The traditional discretisation technique adopted by the fluids community has been the finite volume method, traditionally using an Eulerian space frame co-ordinate system and structured type meshes. The structural and solid mechanics community have tended to use the finite element discretisation procedure based on a Lagrangian space frame co-ordinate system using unstructured type meshes, see [ZT85]. With regard to the actual discretisation procedure, it should be noted that both the finite element and finite volume methods can be viewed as different versions of the weighted residual method, see [OCZ92, BC95]. For the Galerkin finite element method the weighting functions are the shape functions which define the variation of a dependent variable over an element. The integration of the governing equations using this method results in volume integrals which can be approximated using quadrature methods. For the finite volume method the weighting function is taken as unity and the integration of the governing equations over defined control volumes results in both volume and surface integrals. Here the fluxes are approximated at the surfaces joining neighbouring control volumes. Lately attempts have been made by numerous groups at extending finite element methods to fluid mechanics and finite volume methods to solid mechanics. Also comparisons between FE and FV formulations for solid mechanics phenomena have shown close agreement between both methods, see [OCZ92, BC95].

Work on finite volume algorithms using unstructured meshes originated in the 1960's, see [Win]. Because of the large compute time needed to solve complex Computational Fluid Dynamics (CFD) problems the trend in the 1970's and 80's was to use structured type meshes. This allowed the use of iterative line-by-line solvers which required less processor memory than whole field solvers traditionally required for unstructured type meshes. The advantage of using finite volume methods, especially for highly nonlinear flows with phase changes, is the inherent local conservation property that this procedure allows. It may be argued that such FV formulations are potentially superior to the traditional FE methods because on this property. Also with structured meshes the advantage is that mesh generation can be accomplished very quickly. The drawback of using structured meshes is that the boundary of the computational domain needs to be modelled crudely. During the last decade coupled FV procedures have been developed for flow, heat transfer with phase changes and solid mechanics using unstructured meshes.

The finite volume method can either use cell centred or vertex centred control volumes, see Figure 1. For the cell centred variety the control volumes are defined as being the same as the mesh elements. For this control volume definition the dependent variables are stored at the mesh centre. Vertex based control volumes are established around the vertices of the mesh by joining mesh centres to mesh face centres. For the vertex based procedure the dependent variables are stored at the mesh vertices.

The governing equations for continuum physical phenomena such as fluid flow, heat transfer, electromagnetics and solid mechanics can be expressed in the following standard form, see [Cro96] :

$$\frac{\partial}{\partial t}(\rho A\phi) + \nabla.Q = \nabla.(\Gamma\nabla\phi) + S$$

This equation governs the conservation of mass, momentum, energy and force over any specified domain or control volume. Using the finite volume method we can integrate these equations over pre-defined control volumes and use the divergence theorem to transform volume integrals into surface integrals. From this procedure we obtain the following equation :

$$\frac{\partial}{\partial t}\int_v (\rho A\phi) + \int_v \nabla.Q = \int_v \nabla.(\Gamma\nabla\phi) + \int_v S$$

where the convection and diffusion terms are now approximated around the surface of the control volume. The transient and source contributions still being approximated over the whole control volume. The following terms can be substituted into the above generic formulation for each specific set of physics under consideration :

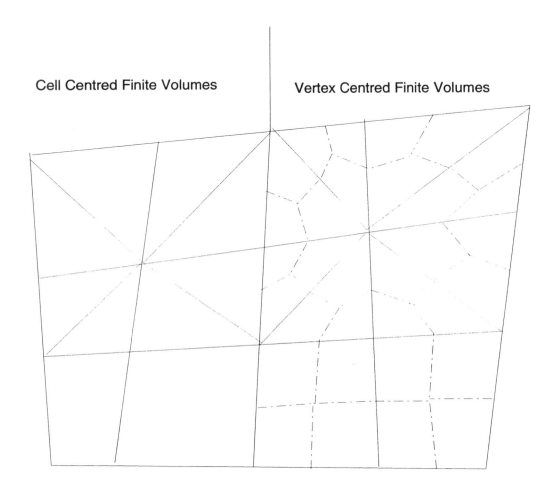

Finite Volume Meshes

Figure 1   Finite Volume Mesh Construction

.	ϕ	A	Γ_ϕ	S	Q
Mass	1	1	0	S_{mass}	$\rho\underline{v}$
Momentum	$\underline{v}$	1	Γ_v	$(S + \underline{JXB} - \nabla p)$	$\rho\underline{v}.\underline{v}$
Heat Transfer	h	1	k/c	S_h	$\rho\underline{v}h$
Electromagnetics	$\underline{B}$	1	η	$(\underline{B}\nabla)\underline{v}$	$\underline{u}.\underline{B}$
Elasticity	$\underline{u}$	$\frac{\partial}{\partial t}$	μ	$\rho\underline{f}_b$	$\mu(\nabla\underline{u})^T + \lambda[\nabla.\underline{u} - (2\mu + 3\lambda)\alpha T]\underline{I}$

17.3   PHYSICA : A MULTIPHYSICS SOFTWARE ENVIRONMENT

After many years of model and associated software development with regards to materials processing phenomena, it became apparent that from a software engineering point of view a great deal of computing tasks are generic in discretising and solving differential equations governing a physical process. For finite volume procedures the evaluation of fluxes across control volume faces, volume sources and coefficients of linear solvers in iterative procedures is generic, being based on the mesh geometry and material properties in a cell. As such these procedures are generally independent of the physics being addressed. PHYSICA is a software environment that has been developed in full recognition of the following :

- A software tool can be structured such that nodes, control volume faces and volumes can all be calculated automatically and treated as objects. Figure 2 shows the mapping between model space and solution space, which deals with distinct objects.
- Given that the mesh connectivity is described by a memory management system, it is straight forward to extend this system to incorporate all other run time information and variables.
- All equation solvers are generic and can be constructed so that they can be called interchangeably with consistent data structures.

Given that the memory manager manages all the mesh connectivity and hence control volume connectivity, the focus for the user is on the objects and the evaluation of the relevant terms making up the systems matrix (i.e diffusion, convection, sources, etc). Figure 3 shows the different hierarchical levels present in the PHYSICA environment. Here the most generic tools such as memory manager and input/output tools are present. The extent of user interaction with the software will increase the further we travel up the hierarchy. At the top level we have routines which are available to include new physical phenomena. It is emphasised that for general users this is the only level needed for user interaction.

At the highest level of the framework we have numerous models which address different physical phenomena. Each of these modules are coupled via the source terms, so that, for example, solidification has an effect on the fluid flow. It is emphasised here that users should be able to specify complex physical models as occurs in, for example, PHOENICS or CFX and other CFD based commercial codes.

In the first version of PHYSICA the following capabilities are present :

- Tetrahedral, wedge and hexahedral mesh shapes with data structures for full polyhedral meshes and control volumes.
- A full range of linear solvers including PCCG, SOR, Jacobi.

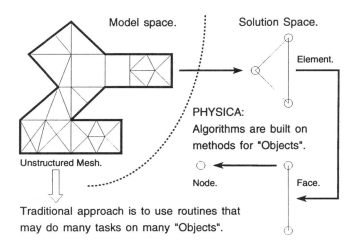

Figure 2   PHYSICA - Design principles

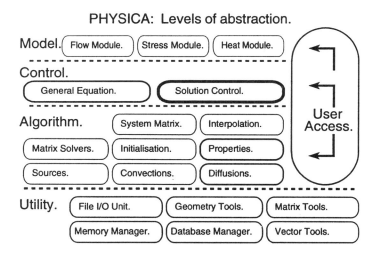

Figure 3   PHYSICA - Levels of code abstraction

- SIMPLE based solution procedure for fully compressible Navier-Stokes equations based on the cell centred finite volume procedure. A modified Rhie and Chow interpolation method is used to estimate velocities at cell faces, whilst a number of differencing techniques are present for the convection terms.
- Enthalpy based solidification/melting procedures coupled with the fluid flow solver.
- Elasto-viscoplastic solid mechanics solver based on small strains. Here the vertex based finite volume technique has been used where displacements are located at the mesh vertices and stress/strain values are located at the face integration points.
- Maxwell's equations and the electromagnetic effects on solidification/melting and fluid flow.
- Interfaces to numerous pre-post processing packages, including PATRAN, FAM and Femsys.

17.4   MULTIPHYSICS MODELLING APPLICATIONS

The shape casting process, see [Cam91, BCC$^+$96] is a classic example of a multiphysics process. During this process a cold mould is filled with molten metal which cools and solidifies. As the cooling progresses thermal gradients are set up within the cast material. In the liquid regions this will promote thermal convection which will redistribute heat around the component. Before the onset of solidification the cast material is in full thermal contact with its surrounding mould, but in the solidified state the cast material is able to move away from the mould and form a gap. The formation of this gap will affect the heat transfer coefficient at this point which in turn will affect the heat loss from the cast to the mould. To simulate the casting process, computational modellers must address the following phenomena :

- Fluid flow analysis of mould cavity filling.
- Solidification and evolution of latent heat.
- Thermal convection.
- Deformation, hence stress analysis as the cast shape develops.
- Porosity formation within the cast material.

In this manufacturing process the foundry engineer is attempting to minimise the amount of porosity and the magnitude of residual stress present in the cast component. This will require solving all the above phenomena using a specific mesh and specific material properties. The PHYSICA package was used for this analysis to predict both the final stress state of the cast component plus the likely areas of porosity. Here the equations governing the conservation of mass, momentum, heat energy and force are coupled and solved to predict changes in all the governing physics at the continuum level. For solidification modelling the Enthalpy procedure is used, see [VS91]. As the casting solidifies it is assumed to be governed by an Elasto-Viscoplastic constitutive law where the viscoplastic strains are assumed to follow a Perzyna relationship, see [Per66, TBC95]. A simple contact model has been used which assumes that the surrounding mould is rigid. Porosity is a phenomena which is governed at the micro scale (i.e dendrite level). At the macro scale the simulations use a Niyama empirical relationship, see [NUMS82], based on local cooling rates, to predict likely areas of porosity. Figure 4 shows predictions for thermal convection in the cast component at an early time before the onset of solidification. Here it can be seen that extensive flow is occurring from the cylindrical feeder, above the plate component, to the rest of the plate. The second diagram in this figure shows the final areas of likely porosity. As would be expected with such a shape, porosity is predicted in the step section as the feeding paths from the cylindrical feeder would be cut off resulting in solidification shrinkage that could not be fed with molten metal. Figure 5 shows both deformation across the centre section of the casting as well as the final stress profiles across the whole casting. It is interesting to see how the cast component has deformed within the mould. As noted above the formation of gaps between the casting and mould will effect the manner in which heat continues to be transmitted to the mould and the general rate of cooling. Including gap effects in the analysis gives predictions for cooling profiles which are 8% different from those calculated when ignoring deformation effects.

Another interesting area of modelling is semi-levitation melting, see [PHCC95] . Figure 6 shows the experimental set up for this analysis. Here a prototype version of the PHYSICA program has been used to solve the coupled electromagnetic, fluid flow and melting phenomena associated with this process. The idea is that electromagnetic fields can be used to levitate an ingot in an inert environment as it melts. This will greatly aid controlling defects in the material resulting from oxidation. This is a very complex problem which requires coupling the equations governing fluid flow, heat transfer and electromagnetic effects. Figure 7 shows results from this analysis for both the flow and solidification front. The flow vectors show how the magnetic field is driving the flow front in a circular fashion. Also the darker areas of the shaded plot show areas of early solidification. The next stage which is currently underway is to incorporate deformation/stress effects on this analysis.

17.5   EXPLOITATION OF PARALLEL COMPUTERS

Simulation of multiphysics processes such as castings can take extensive amounts of time in solving a large number of coupled nonlinear equations. The memory requirement to hold such problems runs into the hundreds of megabytes. Parallel processing offers a means of lowering run times and allowing the solution of larger problems. A parallel version of PHYSICA has been developed, see [MCJ95] using the method of geometric domain decomposition with message passing. Here the problem mesh is partitioned into P sub domains which are mapped onto P processors. This is a Single Program Multi Data paradigm in which each processor has a copy of the code which it uses to operate on a part of the problem. Data dependencies between different sub-domains are satisfied through the exchange of messages between processors. This provides a parallel code that offers portability, parallel efficiency, scalability to large P and scalability to large problems. Non-blocked or asynchronous communications, where communications are overlapped with calculation, are used where possible to overcome the communication overhead. Figure 8 shows the parallel efficiency for a multiphysics simulation involving coupled solidification and deformation phenomena. The graph shows the speedups observed using different number of elements. Using more processors for smaller mesh sizes shows that the interprocessor communication times begin to dominate the solution times.

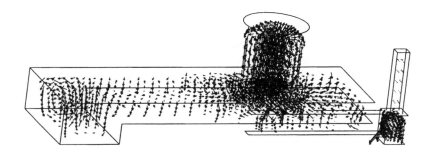

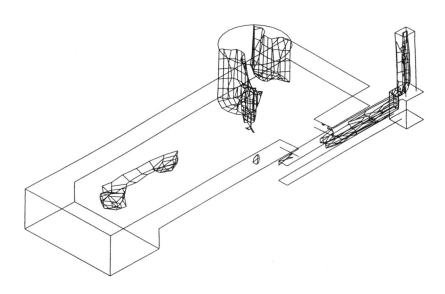

Figure 4 Fluid Flow and Porosity Predictions

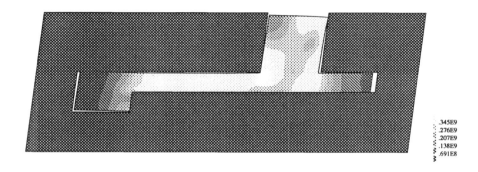

```
    .345E9
    .276E9
    .207E9
    .138E9
    .691E8
```

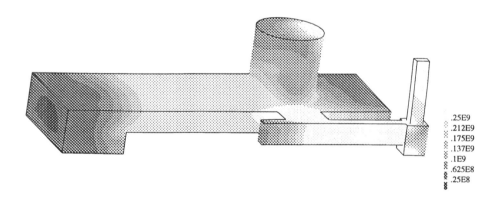

```
    .25E9
    .212E9
    .175E9
    .137E9
    .1E9
    .625E8
    .25E8
```

Figure 5   Deformation Predictions

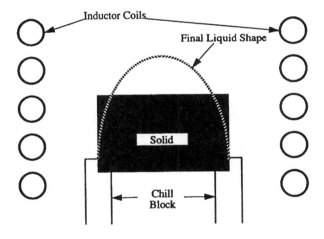

Figure 6   Idealised levitation casting device.

17.6   LINKING UP THE LENGTH SCALES

Continuum models (or macro models) are governed by what is happening at the micro
and atomistic levels for the material. Such effects as microstructure formation in a solid
component will govern the way in which the component will behave mechanically. Also
the interactions between atoms at the nanoscale will govern the thermodynamic and
mechanical properties of materials. Thermophysical properties such as specific heat
capacity and thermal conductivity as well as thermomechanical data such as Youngs
modulus have been calculated for binary alloys using Nano-scale simulations [RTT].
Work on extending such simulations to tertiary and higher order alloys typical of
casting materials is underway. Such expertise in modelling at the Nano-scale and the
resulting information obtained could be used within continuum level models to aid in
obtaining a greater representation of material properties behaviour. For example, in
the casting of metal components there is still a great deal of uncertainty in the variation
of the materials properties with regards temperature in the mushy zone. It is hoped
that atomistic models will be able to procide temperature dependent materials data
within the mushy zone for use in continuum based models.

17.7   CONCLUSIONS

Multiphysics modelling is now becoming a reality and software tools such as PHYSICA
are becoming available to enable such simulations. Hurdles still exist for modellers such
as the use of parallel architectures and communication across the length scales, but
as outlined above work in addressing these issues is now underway.

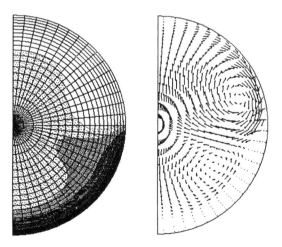

Figure 7   Levitation Casting Results.

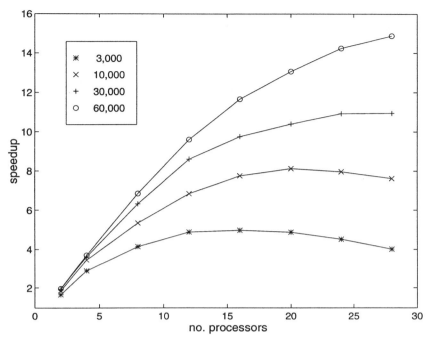

Figure 8 Computational Speedups for Different Mesh Sizes.

References

[BC95] Bailey C. and Cross M. (1995) A finite volume procedure to solve elastic solid mechanics problems in three dimensions on an unstructured mesh. *Int Jnl Num Meth Engng* 38: 1757–1776.

[BCC+96] Bailey C., Chow P., Cross M., Pericleous K., and Fryer Y. (1996) Multiphysics modelling of the metals casting process. *Proc Roy Soc London* 452: 459–486.

[Cam91] Campbell J. (1991) *Castings*. Butterworth-Heinemann, Oxford.

[Cro96] Cross M. (1996) Modelling of industrial multi-physics processes - a key role for computational mechanics. *IMA Jnl Math Appl. Bus Ind* 7: 3–21.

[Hir90] Hirsch C. (1990) *Numerical Computation of Internal and External Flows.* Wiley, Chichester.

[MCJ95] McManus K., Cross M., and Johnson S. (1995) Integrated flow and stress using an unstructured mesh on distributed memory parallel systems. In *Parallel Computational Fluid Dynamics - New Algorithms and Application.* North-Holland, Amsterdam.

[NUMS82] Niyama E., Uchida T., Morikawa M., and Saito S. (1982) Method of shrinkage prediction and its application to steel casting practice. *AFS Cat Metals Res* pages 52–63.

[OCZ92] Onate O., Cerrera E., and Zienkiewicz O. (1992) A Finite volume format for structural mechanics. *Int Jnl Num Meth Engng* 37: 181–201.

[Pat80] Patankar S. V. (1980) *Numerical Heat Transfer and Fluid Flow.* Hemisphere, New York.

[Per66] Perzyna P. (1966) Fundamental problems in viscoplasticity. *Adv. Appl. Mech.* pages 243–377.

[PHCC95] Pericleous K., Hughes M., Cook D., and Cross M. (1995) Mathematical Modelling of the solidification of tin with electromagnetic stirring. In *The 14th Riga Intl Conf on Magnetohydrodynamics.*

[RTT] Rafii-Tabar H. and Tambarajah A.Study of two dimensional nano-scale melting transitions in aluminium via computer based simulation. *Jnl of Nano Technology* .

[TBC95] Taylor G., Bailey C., and Cross M. (1995) Material non-linearity within a finite volume framework for the simulation of a metal casting process. In *In Computational Plasticity*, pages 1459–1470. Pinridge, Swansea.

[VS91] Voller V. and Swaminathan C. (1991) General source-based method for solidification phase change. *Numer. Heat Transfer* pages 175–189.

[Win] Winslow A. M.Numerical solution of quasilinear poisson equation in a nonuniform triangle mesh. *J. Comput. Phys* 1: 149–172.

[ZT85] Zienkiewicz O. C. and Taylor R. L. (1985) *The finite element method.* McGraw-Hill, New York.

18

Non-Newtonian flow modelling for the processing industry

I. Mutlu, P. Townsend and M. F. Webster

Institute of Non-Newtonian Fluid Mechanics,
Department of Computer Science, University of Wales,
Singleton Park, Swansea, U.K.

18.1   INTRODUCTION

Considerable effort has been devoted over the years to the development of constitutive equations capable of realistic modelling for the viscoelastic behaviour demonstrated by polymer solutions and melts (see e.g. [SP94, DC88, TS95, KL87]). Particular attention has been devoted more recently to the study of models for elongational flows, as extensional effects may play a significant role in many complex flows [KL87, Tan89, QMBA90, BTW94]. The need to study models that incorporate extension and shear behaviour is clear (see e.g. [SP94, DC88, TS95, KL87, SS90, Tan89, QMBA90, BTW94]). One such model is the nonlinear differential Phan-Thien/Tanner (PTT) model, which is capable of giving at least qualitative predictions for shear and elongational properties of both polymer solutions and melts [KL87, Tan89, CTW93]. The polymer melts of relevance in the present study exhibit shear-thinning and extension-thinning behaviour. In this paper a transient finite element algorithm is used to solve a rheologically complex flow of relevance to the polymer processing industry. This is substantiated by a practical example derived from an extrusion cable-coating process, that of wire-coating with polymer melts. This consists of a converging annular die flow followed by a drag flow of the coating onto the cable. Here, the former portion of the flow is addressed to provide some guidance on stress build-up within the die. Consideration is given to the material characterisation using an exponential PTT model, suitable for shear or extension dominated flows. We study the theoretical response of this model in two types of deformation: simple shear and uniaxial extension. This provides the background for the study of more complex flows, which display a combination of such deformation. A Taylor-Galerkin/Pressure correction scheme [CTW93] is employed that utilises Petrov-Galerkin streamline upwinding (SUPG) and a time-accurate procedure to achieve steady-state solutions.

The Mathematics of Finite Elements and Applications
Edited by J. R. Whiteman © 1997 John Wiley & Sons Ltd.

This introduces a fractional staged scheme per time step cycle within which both iterative and direct solvers are invoked. The scheme possesses semi-implicit properties and is element-based, which makes it practical to solve large complex systems of nonlinear partial differential equations of mixed parabolic-hyperbolic type. This is typical of the mathematical problems associated with the flow of viscoelastic fluids.

18.2 GOVERNING EQUATIONS

The non-dimensional system of equations for the incompressible, isothermal flow of a viscoelastic fluid may be given in terms of those for momentum transport,

$$Re\frac{\partial \mathbf{u}}{\partial t} = \nabla \cdot (2\mu_2 \mathbf{D} + \tau) - Re\,\mathbf{u} \cdot \nabla \mathbf{u} - \nabla p, \tag{18.1}$$

conservation of mass,

$$\nabla \cdot \mathbf{u} = 0, \tag{18.2}$$

and constitutive law, here taken as an exponential Phan-Thien/Tanner (PTT) model,

$$We\frac{\partial \tau}{\partial t} = 2\mu_1 \mathbf{D} - f\tau - We\left(\mathbf{u} \cdot \nabla \tau - \nabla \mathbf{u} \cdot \tau - (\nabla \mathbf{u} \cdot \tau)^T + \xi(\mathbf{D}\tau + (\mathbf{D}\tau)^T)\right). \tag{18.3}$$

In nomenclature, $\mathbf{u}$ is a velocity vector, p is a fluid pressure, τ is a polymeric extra-stress tensor, $\mathbf{D}$ is a rate of deformation tensor, t is the independent time variable, ∇ is the gradient operator, ^{T} denotes matrix transpose. Non-dimensionalisation is performed through scales of L/U for time and $\mu U/L$ for stress and pressure, where U and L are a characteristic velocity and length respectively. Here μ is a fluid viscosity given by $\mu = \mu_1 + \mu_2$, where μ_1 is a polymeric viscosity and μ_2 is a solvent viscosity, so that non-dimensional values $\mu_i^* = \mu_i/\mu$ are implied henceforth. Non-dimensional groups of Reynolds and Weissenberg number are defined as $Re = \rho U L/\mu$ and $We = \lambda_1 U/L$, governing levels of inertia and elasticity respectively, where ρ is the fluid density and λ_1 is a relaxation time. For the exponential PTT(ϵ, ξ) model, the material function f is defined as;

$$f = \exp(\epsilon\,We\,\mathrm{tr}(\tau)/\mu_1). \tag{18.4}$$

The parameters ϵ $(\epsilon \geq 0)$ and ξ $(0 \leq \xi \leq 2)$ in equations (18.3) and (18.4) are material parameters that govern the elongational and shear properties of the model.

The class of Phan-Thien/Tanner PTT(ϵ, ξ) constitutive equations are able to provide reasonable fits for shear and elongational material properties of many common polymer melts. A solvent viscosity is added in this model so that the constant viscosity Oldroyd-B model may be recovered as a special case of the PTT model, given by PTT(0,0). The solution of (18.1-18.3) is sought with an appropriate imposition of boundary and initial conditions, here under an axisymmetric frame of reference. A decoupled scheme is employed to obtain steady-state solutions. This permits Newtonian/Generalized Newtonian kinematics to be used as frozen coefficients for the constitutive equation. This choice is made for purely practical reasons, anticipating highly elastic settings and to stabilise the numerical computations for such nonlinear problems. For highly elastic flow conditions, often encountered with polymer melts, the constitutive equation is convection-dominated. Under such circumstances, consistent streamline upwind Petrov-Galerkin weighting [CTW93] is employed to suppress streamwise oscillations in the discrete solution. This is an effective mechanism to deal with highly elastic convective instabilities of a streamwise nature. High temporal accuracy through a Jacobi iteration is used to suppress cross-stream oscillations [GTW96]. Five Jacobi iterations per fractional equation stage are used for a converging annular flow considered below. A Courant number stability constraint is used to determine the time step. Details of the numerical scheme are given elsewhere [CTW93, GTW96, CTW94, HTJTW90].

18.3   THE PHAN-THIEN/TANNER MODEL

Within the complex flows of interest both shear and extension are present. It is instructive, therefore, to analyse first the PTT model response in pure shear and extension in isolation.

18.3.1   SIMPLE SHEAR FLOW

In a steady, fully developed, simple-shear flow, with velocity $\mathbf{u} = (u, v)$ defined as $u = 0$ and $v = \dot{\gamma}r$, stress components are determined via (18.3) as,

$$\tau_{zz} = \frac{(2 - \xi)\,We\,\dot{\gamma}}{f}\tau_{rz}, \tag{18.5}$$

$$\tau_{rr} = -\frac{\xi\,We\,\dot{\gamma}}{f}\tau_{rz}, \tag{18.6}$$

and

$$\tau_{rz} = \frac{\mu_1 f \dot{\gamma}}{f^2 + (We\dot{\gamma})^2\xi(2 - \xi)} = (\mu_s - \mu_2)\dot{\gamma}, \tag{18.7}$$

where $\dot{\gamma}$ is the non-dimensional shear rate. The first normal stress difference, N_1, and the shear viscosity, μ_s, are defined as;

$$n_1 = \tau_{zz} - \tau_{rr}, \tag{18.8}$$

and

$$\mu_s = \frac{\mu_1 f}{f^2 + (We\dot{\gamma})^2\xi(2 - \xi)} + \mu_2. \tag{18.9}$$

For this type of flow, the parameter f may be expressed recursively as,

$$\log f = \frac{2\epsilon(We\dot{\gamma})^2(1 - \xi)}{f^2 + (We\dot{\gamma})^2\xi(2 - \xi)}, \tag{18.10}$$

and is determined using an iterative (Newton-Raphson) technique.

18.3.2 UNIAXIAL EXTENSIONAL FLOW

In a steady uniaxial extensional flow with extension rate $\dot{\epsilon}$, and velocity components $v = \dot{\epsilon}z$ and $u = -\dot{\epsilon}r/2$, the stress response is given via equation (18.3) as:

$$\tau_{zz} = \frac{2\mu_1\dot{\epsilon}}{f - 2We\dot{\epsilon}(1-\xi)} = (\mu_E - 3\mu_2)\dot{\epsilon} + \tau_{rr}, \tag{18.11}$$

$$\tau_{rr} = \tau\theta\theta = \frac{-\mu_1\dot{\epsilon}}{f - 2We\dot{\epsilon}(1-\xi)}, \tag{18.12}$$

and

$$\tau_{rz} = 0, \tag{18.13}$$
$$\tag{18.14}$$

where μ_E is the elongational viscosity defined as;

$$\mu_E = \mu_1 \left(\frac{2}{f - 2We\dot{\epsilon}(1-\xi)} + \frac{1}{f + We\dot{\epsilon}(1-\xi)} \right) + 3\mu_2. \tag{18.15}$$

Under the above conditions, the parameter f in equations (18.11)-(18.15) is determined from the recursive relationship:

$$\log f = \frac{6\epsilon(We\dot{\epsilon})^2(1-\xi)}{(f - 2We\dot{\epsilon}(1-\xi))(f + We\dot{\epsilon}(1-\xi))}. \tag{18.16}$$

18.3.3 FIT TO EXPERIMENTAL DATA

A single representative relaxation time of $\lambda_1 = 5$ sec. is assumed from oscillatory shear rheometrical measurements for a low-density-polyethylene, corresponding to a value of $We = 28$ for the complex converging annular flow to follow. Using the above PTT model rheometrical response, a multi-variate sensitivity analysis for the material parameters of ϵ, ξ, and μ_1 has been conducted in both steady shear and pure extension experiments. A best fit is sought in steady shear to the experimental shear viscosity and first normal stress difference, and in pure extension to the elongational viscosity. The polymeric viscosity weighting, $\mu_1=0.9995$, is taken to constrain the tail of the shear viscosity (μ_s) curve, to attain the upper end of the shear rate experimental range, $\mathcal{O}(10^4)$, fixing the second Newtonian plateau. This implies a solvent viscosity contribution of $\mu_2 = 0.0005$, which renders a limiting Maxwellian form for (18.1). Then, the choice of ϵ parameter sets the departure of the μ_s-curve from the first Newtonian plateau at low shear rates, and the degree of shear-thinning prior to attaining the second Newtonian plateau already established. Finally, adjustment of the ξ parameter has only marginal influence and that solely on the shear viscosity fit; its selection sets the level of elevation of the μ_s-curve in the shear-thinning region, as ξ tends to zero. The parameter combination of $\epsilon = 0.1$, $\xi = 0.01$, and $\mu_1 = 0.9995$, is found to predict the experimental data to the precision shown in Figure 1. Figures 1(a) and 1(b) provide the dimensional experimental steady shear data and the corresponding fits of shear viscosity and first normal stress difference, respectively. The fit to experimental elongational viscosity-strain rate data, representing material response in pure extension, is provided in Figure 1(c). The optimality of fit is governed firstly by matching to the shear viscosity data, followed by that to the elongational viscosity. The first normal stress difference response is then inherited as a consequence. The shear viscosity data is best fit to shear rates of $\mathcal{O}(10^2)$, and elongational viscosity at extension rates of $\mathcal{O}(10^2)$. These are typical values encountered in the converging annular flow, dealt with below. An analytical test problem involving a fully developed annular flow is considered to study the effect of varying We on shear viscosity. Figure 2 shows the variation in the non-dimensional shear viscosity with We at various radial locations (corresponding to various shear rates) for a PTT(0.1,0.01) model. This reflects anticipated entry flow behaviour in the converging annular flow to follow. It can be seen from this figure that there is a sharp decrease in shear viscosity as We is increased. Figure 2 also shows that an increase in shear rate near solid walls ($\rho = 0.683$ and 0.767) causes a decrease in shear viscosity. The decrease in shear viscosity causes the shear stress, τ_{rz}, to decrease with increasing We, as apparent from (18.7). The normal stress component, τ_{zz}, is dependent upon τ_{rz} and thus μ_s, as observed in (18.5) and (18.7). As We is increased (at a constant $\dot{\epsilon}$), the shear viscosity decreases, causing τ_{zz} to decrease correspondingly.

18.4   CONVERGING ANNULAR FLOW

18.4.1   PROBLEM SPECIFICATION

A schematic diagram of the converging annular flow is presented in Figure 3(a). Boundary conditions at the shear flow inlet consist of analytical values of the PTT(0.1,0.01) stresses (τ_{ij}) determined assuming a fully developed Newtonian annular flow velocity profile. V_i is the inlet velocity, V_o is the outlet velocity, and R_i, R_o are the inner and outer radii respectively. Characteristic velocity and length scales are taken as mean inlet velocity$\times 10$ and hydraulic radius of $(R_o - R_i) \times 12$, with frame of reference to a complete extrusion-coating scenario. Simulations are conducted at a fixed level of inertia with $Re = 2.78 \times 10^{-4}$ for a standard flow rate, and 5.56×10^{-4} at a doubled flow rate. A uniform, structured, triangular mesh is employed, to capture the sharp Generalized Newtonian flow velocity gradients near die boundaries. In the streamwise direction, this mesh consists of 34 elements in the inlet region, 34 elements in the converging flow section, and 6 elements over the outlet zone. The cross-stream direction of flow is spanned with 10 elements. A Bird-Carreau model is employed to describe the inelastic kinematics. The shear viscosity, μ_s , for this model is defined as (cf. Binding et al. [BBG+96]):

$$\mu_s(\dot{\gamma}) = \mu \left(1 + K_1 \dot{\gamma}\right)^{(m-1)/2} \tag{18.17}$$

where K_1 is a natural time constant, $\dot{\gamma}$ is a power-law index, and is a generalised shear rate equal to $2\sqrt{II}$, II being the second invariant of the rate-of-strain tensor. With a power-law index of 0.372, values of the above parameters are taken to fit steady shear experimental data provided in Walters et al. [WBE94] and demonstrated in Figure 1(a). Corresponding kinematics are computed for this inelastic model satisfying (18.1) and (18.2), and used subsequently as frozen coefficients for the viscoelastic computations. This decoupled approach is founded on the assumption that there is a reasonably close correspondence between inelastic and viscoelastic kinematics.

18.4.2 RESULTS AND DISCUSSION

Results are presented in terms of stress and velocity gradients, with the former traced along characteristic sample lines streamwise to the flow. General flow trends indicate that fully developed stress and velocity fields are established along the inlet of the flow domain, followed by a stress build-up over the converging section, before finally reestablishing an almost fully developed flow at the outlet. The base flow kinematics for a Newtonian model at a doubled flow rate are presented in the form of shear and extension rate contour plots in Figure 3(b),3(c), over the latter half of the annular converging flow where maximum values are anticipated. From these fields, localised peak values may be discerned arising at the end of the converging section; reaching extension rates of 27 units (300 sec^{-1} dimensionally), and shear rates of 54 units (600 sec^{-1} dimensionally) that persist throughout the outlet zone. These values correlate reasonably well to the rheometrical data discussed earlier in section 18.3.3. The Generalised Newtonian kinematics adopt similar patterns, with elevation of peak extension and shear rates by a factor of two. As expected, shear thinning gives rise to flatter velocity profiles within the core flow, that are steeper near the boundary. At a standard flow rate, the maximum magnitude of velocity for the Generalized Newtonian fluid is 0.638, as opposed to 0.751 for the Newtonian fluid. Corresponding values at doubled flow rate are 1.503 and 1.275, respectively. This situation is reversed as one approaches the solid boundaries, with the velocity magnitude being smaller for the Newtonian than the Generalized Newtonian fluid.

Viscoelastic results are reported for $5 \leq We \leq 28$. At each value of We in the range, PTT(0.1,0.01) stresses for that particular value of We are employed as inlet boundary conditions. To commence each transient simulation, initial stress fields on the domain are assumed to be relaxed. Figure 4 depicts viscoelastic stresses for a standard flow rate setting, along two sample lines, R4 ($r = 0.188, z = 0$) within core flow, and R5 ($r = 177, z = 0$) at the boundary. Both Newtonian and Generalized Newtonian frozen kinematics yield higher stresses on the boundary, in contrast to the values on the inner core flow sample line, as expected. Along any sample line, the flow becomes fully developed within $\approx 12\%$ of the entry region length. The discrepancy in τ_{zz} at the inlet is due to the imposed Newtonian kinematics. Sharp oscillations are observed as the inlet section joins the converging section, followed by a stress build-up in the converging flow region. Along the inner sample line, R4, τ_{zz} stresses are marginally larger for the frozen Newtonian velocity field, than for the Generalized Newtonian field. This results as a direct consequence of shear-thinning. On the boundary, however, this situation is reversed, with stresses for the frozen Newtonian velocity field being lower than those for the corresponding Generalized Newtonian velocity field. These larger shear and normal stresses for the Generalized Newtonian model arise as a consequence of the domination of the large shear rate over the low viscosity level in the boundary layer (inferred by the viscometric equations (18.5)-(18.7) derived for a simple shear flow of a PTT(0.1,0.01) model).

The stress build-up in the converging flow section is followed by stress relaxation over the outlet. An increase in We results in a decrease in the level of τ_{zz} stress. The elastic τ_{rz} stresses are approximately nine to eleven times lower than the τ_{zz}

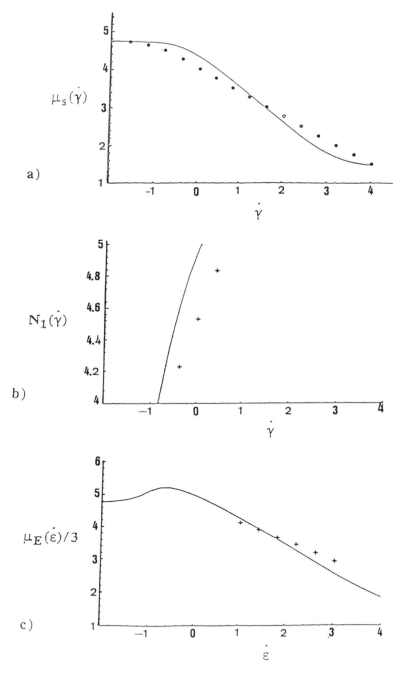

Figure 1   Fit to experimental data a) Shear velocity, b) First normal stress difference, c) Elongational viscosity. (——— PTT(0.1,0.01), $\mu_1 = 0.9995$, $\lambda_1 = 5$sec; +, o experimental data [WBE94])

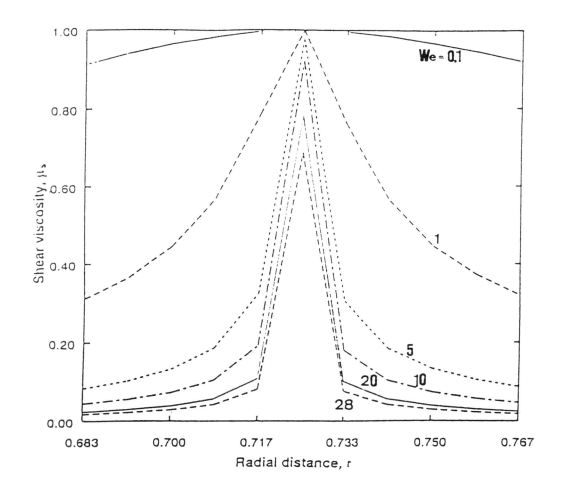

Figure 2   Variation in shear viscosity with Weissenberg number. (PTT(0.1,0.01),
$\mu_1 = 0.995$, $We = 0.1, 1, 5, 10, 20, 28$, annular flow)

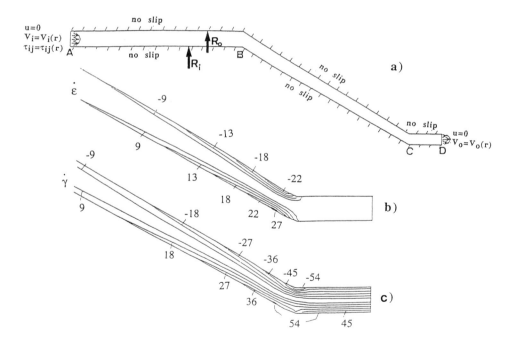

Figure 3 Converging annular flow. a) Computational geometry, b) Extension rates, c) Shear rates for Newtonian model.

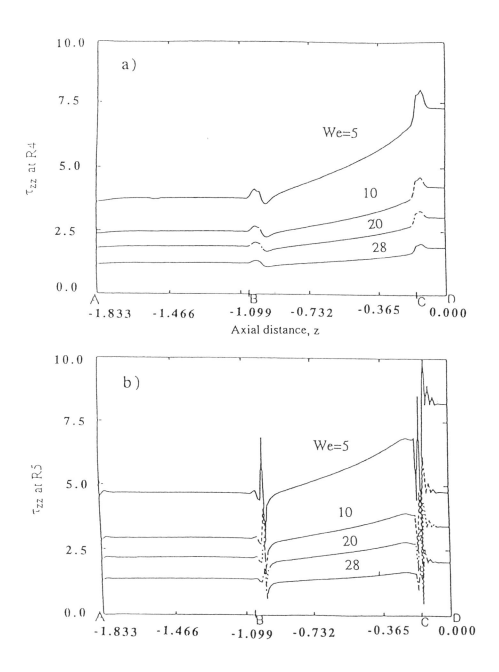

Figures 4(a) and 4(b) Converging annular flow. Comparison of τ_{zz} stresses along two samples lines for PTT(0.1,0.01) model with frozen Newtonian kinematics.

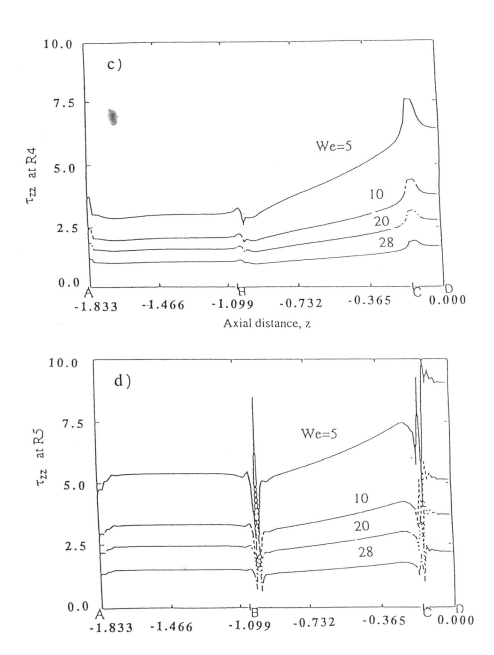

Figures 4(c) and 4(d) Converging annular flow. Comparison of τ_{zz} stresses along two samples lines for PTT(0.1,0.01) model with frozen Generalised Newtonian kinematics.

stresses, and follow similar trends throughout the flow to the normal stress. No significant difference is observed in τ_{rz} values between instances of frozen Newtonian or Generalized Newtonian kinematics.

18.5 CONCLUSIONS

This paper has described a viscoelastic simulation with an exponential PTT model using material characterisation for polymer melts. Fits to experimental data have been performed to determine the necessary constitutive model parameters. The use of a single set of material parameters, which provides a reasonably good fit to both steady shear and pure extension experimental data, has been investigated. The numerical procedure adopted has permitted a phased study of a complex flow encountered in the processing industry. For the converging annular flow, the decoupled scheme results show that elastic extra stresses with Newtonian and Generalized Newtonian kinematics are not significantly different. Under a single mode analysis, at the lower values of We considered, there is significant stress build-up within the converging die section. This build-up is gradually reduced as We is increased beyond twenty. To gain a more complete picture a multi-mode analysis is called for, that takes account of a number of relaxation times from the spectrum available, each with their associated partial zero shear-rate viscosity weightings. More recently, a fully coupled analysis has been conducted and found to yield good correspondence between elastic stresses for coupled and decoupled approaches. Hence, the pragmatic and stabilising decoupled procedure is vindicated.

18.6 ACKNOWLEDGEMENTS

The authors gratefully acknowledge the many helpful discussions relevant to this subject with Dr. A. R. Blythe, Dr. S. Gunter and Dr. B. Butler of BICC Cables Ltd., and Mr. A. A. Mosquera of Applied Computing and Engineering Ltd.

References

[BBG+96]   Binding D. M., Blythe A. R., Gunter S., Mosquera A. A., Townsend P., and Webster M. F. (1996) Modelling Polymer Melt Flows in Wirecoating Processes. To appear in J. Non-Newtonian Fluid Mech.

[BTW94]   Baloch A., Townsend P., and Webster M. F. (1994) Extensional Effects in Flows Through Contractions with Abrupt or Rounded Corners. *J. Non-Newtonian Fluid Mech.* 54: 285–302.

[CTW93]   Carew E. O. A., Townsend P., and Webster M. F. (1993) A Taylor-Petrov-Galerkin Algorithm for Viscoelastic Flow. *J. Non-Newtonian Fluid Mech.* 50: 253–287.

[CTW94]   Carew E. O. A., Townsend P., and Webster M. F. (1994) Taylor-Galerkin Algorithms for Viscoelastic Flow: Application to a Model Problem. *Num. Meth. Partial Diff. Equations* 10: 171–190.

[DC88]   Debaut B. and Crochet M. J. (1988) Extensional Effects in Complex Flows. *J. Non-Newtonian Fluid Mech.* 30: 169–184.

[GTW96]   Gunter S., Townsend P., and Webster M. F. (1996) The Simulation of Some Model Viscoelastic Extensional Flows. To appear in J. Numerical Meth. Fluids.

[HTJTW90] Hawken D. M., Tamaddon-Jahromi H. R., Townsend P., and Webster M. F. (1990) A Taylor-Galerkin Based Algorithm for Viscous Incompressible Flow. *Int. J. Numerical Meth. Fluids* 10: 327–351.

[KL87]   Khan S. A. and Larson R. G. (1987) Comparison of Simple Constitutive Equations for Polymer Melts in Shear and Biaxial and Uniaxial Extensions. *J. Rheology* 31: 207–234.

[QMBA90]   Quinzani L. M., McKinley G. H., Brown R. A., and Armstrong R. C. (1990) Modeling the Rheology of Polyisobutylene Solutions. *J. Rheology* 34: 705–748.

[SP94]   Saramito P. and Piau J. M. (1994) Flow Characteristics of Viscoelastic Fluids in an Abrupt Contraction by Using Numerical Modeling. *J. Non-Newtonian Fluid Mech.* 52: 263–288.

[SS90]   Schunk P. R. and Scriven L. E. (1990) Constitutive Equation for Modeling Mixed Extension and Shear in Polymer Solution Processing. *J. Rheology* 34: 1085–1119.

[Tan89]   Tanner R. I. (November 1989) Constitutive Model for Seventh Workshop on Numerical Computations in Viscoelastic Flows. Technical report, Department of Mechanical Engineering, University of Sydney.

[TS95]   Tirtaatmadja V. and Sridhar T. (1995) Comparison of Constitutive Equations for Polymer Solutions in Uniaxial Extension. *J. Rheology* 39: 1133–1160.

[WBE94]   Walters K., Binding D. M., and Evans R. E. (May 1994) Modelling the Rheometric Behaviour of 3 Polyethylene Melts. Technical report, Univ. Wales, Aberystwyth.

19

Global–Local Computational Procedures for the Modelling of Composite Structures

J. N. Reddy

Department of Mechanical Engineering,
Texas A&M University,
College Station,
Texas 77843–3123, U.S.A.

19.1 INTRODUCTION

Composite materials consist of two or more materials which together produce desirable properties that cannot be achieved with any of the constituents alone. Fiber–reinforced composite materials, for example, consist of high strength and high modulus *fibers* in a *matrix* material. Unlike their homogeneous isotropic counterparts, the heterogenous anisotropic constitution of fiber–reinforced laminated composite structures often results in the appearance of many unique phenomena that can occur on vastly different geometric scales, *i.e.*, at the global or laminate level, the ply level, or the fiber/matrix level. When the main emphasis of the analysis is to determine the overall global response of the laminated component, for example, gross deflections, critical buckling loads, fundamental vibration frequencies and associated mode shapes, such global behavior can often be accurately determined using relatively simple equivalent–single–layer laminate theories (ESL theories), especially for very thin laminates. Two commonly used examples of simple ESL theories are the classical and the first–order shear deformation theories [Red96].

As laminated composite materials undergo the transition from secondary structural components to primary critical structural components, the goals of analysis must be broadened to include a highly accurate assessment of localized regions where damage initiation is likely. The simple ESL laminate theories that often prove adequate for modeling secondary structures are of limited value in modeling primary structures for two reasons. First of all, most primary structural components are considerably thicker than secondary components, thus even the determination of the global response

The Mathematics of Finite Elements and Applications
Edited by J. R. Whiteman © 1997 John Wiley & Sons Ltd.

may require a refined laminate theory that accounts for thickness effects. Second, the assessment of localized regions of potential damage initiation begins with an accurate determination of the three–dimensional state of stress and strain at the ply level, regardless of whether damage prediction and assessment is desired at the ply level or at the fiber/matrix level. The simple ESL laminate theories are most often incapable of accurately determining the 3–D stress field at the ply level. Thus the analysis of primary composite structural components may require the use of 3–D elasticity theory or a layerwise laminate theory [Red87] that contains full 3–D kinematics and constitutive relations.

Here we present the displacement–based full layerwise theory of Reddy [Red87], and a variable kinematic model that incorporates both equivalent single–layer theories and layerwise theories for global–local analysis is described. Applications of the computational procedure to the free edge problem are discussed.

19.2   LAYERWISE THEORY

19.2.1   DISPLACEMENT FIELD

In the layerwise theory of Reddy [Red87, Rob93], the displacement field in the $k-$th layer of the laminate is written as

$$u^k(x,y,z) = \sum_{j=1}^{m} u_j^k(x,y)\phi_j^k(z) \tag{19.1}$$

$$v^k(x,y,z) = \sum_{j=1}^{m} v_j^k(x,y)\phi_j^k(z) \tag{19.2}$$

$$w^k(x,y,z) = \sum_{j=1}^{n} w_j^k(x,y)\psi_j^k(z) \tag{19.3}$$

where u^k, v^k, and w^k represent the total displacement components in the x, y and z directions, respectively, of a material point initially located at (x,y,z) in the undeformed laminate, and $\phi_j^k(z)$ and $\psi_j^k(z)$ are continuous functions of the thickness coordinate z. In general, $\psi^k \neq \phi^k$.

The functions $\phi_j^k(z)$ and $\psi_j^k(z)$ are selected to be layerwise continuous functions. For example, they can be chosen to be the one–dimensional Lagrange interpolation functions of the thickness coordinate, in which case, (u_j^k, v_j^k, w_j^k) denote the values of (u^k, v^k, w^k) at the $j-$th plane. The number of nodes, n, through the layer thickness define the polynomial degree $p = n+1$ of $\psi_j^k(z)$, which are defined only within the $k-$th numerical layer (see Figure 1). The functions $u_j^k(x,y)$, $v_j^k(x,y)$, and $w_j^k(x,y)$ represent the displacement components of all points located on the $j-$th plane (defined by $z = z_j$) in the undeformed laminate.

Since the thickness variation of the displacement components is defined in terms of piecewise Lagrangian interpolation functions, the displacement components will be continuous through the laminate thickness, but the transverse strains will be discontinuous across the interface between adjacent thickness subdivisions.

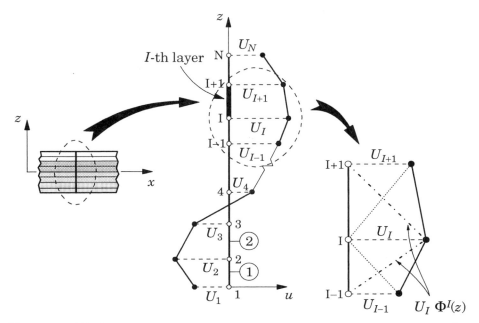

Figure 1 Displacement representation and the linear approximation functions $\Phi^I(z)$ used in the layerwise theory.

Any desired degree of displacement variation through the thickness is easily obtained by either adding more one–dimensional finite element subdivisions through the thickness (h–refinement) or using higher order Lagrangian interpolation polynomials (p–refinement) through the thickness. The layerwise concept introduced here is very general in that the number of subdivisions through the thickness can be greater than, equal to, or less than the number of material layers through the thickness and each layer can have linear, quadratic or higher–order polynomial variations of the displacements. When the number of numerical subdivisions used is less than the number of material layers, it amounts to the sublaminate concept where several material layers are represented as an equivalent, single, homogeneous layer.

The total displacement field of the laminate can be written as

$$u(x,y,z) \;=\; \sum_{I=1}^{N} U_I(x,y)\Phi^I(z) \tag{19.4}$$

$$v(x,y,z) \;=\; \sum_{I=1}^{N} V_I(x,y)\Phi^I(z) \tag{19.5}$$

$$w(x,y,z) \;=\; \sum_{I=1}^{M} W_I(x,y)\Psi^I(z) \tag{19.6}$$

where (U_I, V_I, W_I) denote the nodal values of (u,v,w), N is the number of nodes and Φ^I are the *global* interpolation functions (see Figure 1) for the discretization of the inplane displacements through thickness, and M is the number of nodes and Ψ^I are the global interpolation functions for discretization of the transverse displacement through thickness.

19.2.2 EQUATIONS OF MOTION

The governing equations of equilibrium for the layerwise theory can be derived using the principle of virtual displacements. We obtain

$$\frac{\partial N_{xx}^I}{\partial x} + \frac{\partial N_{xy}^I}{\partial y} - Q_x^I = 0 \tag{19.7}$$

$$\frac{\partial N_{xy}^I}{\partial x} + \frac{\partial N_{yy}^I}{\partial y} - Q_y^I = 0 \tag{19.8}$$

$$\frac{\partial \tilde{Q}_x^I}{\partial x} + \frac{\partial \tilde{Q}_y^I}{\partial y} - \tilde{Q}_z^I + q_b\delta_{I1} + q_t\delta_{IM} = 0 \tag{19.9}$$

$$\left\{ \begin{array}{c} N_{xx}^I \\ N_{yy}^I \\ N_{xy}^I \end{array} \right\} = \int_{-\frac{h}{2}}^{\frac{h}{2}} \left\{ \begin{array}{c} \sigma_{xx} \\ \sigma_{yy} \\ \sigma_{xy} \end{array} \right\} \Phi^I \, dz \tag{19.10}$$

$$\left\{ \begin{array}{c} Q_x^I \\ Q_y^I \end{array} \right\} = \int_{-\frac{h}{2}}^{\frac{h}{2}} \left\{ \begin{array}{c} \sigma_{xz} \\ \sigma_{yz} \end{array} \right\} \frac{d\Phi^I}{dz} \, dz \, , \quad \left\{ \begin{array}{c} \tilde{Q}_x^I \\ \tilde{Q}_y^I \\ \tilde{Q}_z^I \end{array} \right\} = \int_{-\frac{h}{2}}^{\frac{h}{2}} \left\{ \begin{array}{c} \sigma_{xz} \\ \sigma_{yz} \\ \sigma_{zz} \end{array} \right\} \Psi^I \, dz \tag{19.11}$$

where σ_{xx}, σ_{yy} and σ_{xy} represent the membrane stresses, and σ_{xz}, σ_{yz} and σ_{zz} denote the transverse shear and normal stresses.

19.2.3 FINITE ELEMENT MODEL

The displacement finite element model corresponding to the full layerwise theory is developed by substituting an assumed interpolation of the displacement field into the principle of virtual displacements for a representative finite element of the plate. Suppose that the displacement field is interpolated as

$$U_I(x,y,t) = \sum_{j=1}^p U_I^j \psi_j(x,y) \tag{19.12}$$

$$V_I(x,y,t) = \sum_{j=1}^p V_I^j \psi_j(x,y) \tag{19.13}$$

$$W_I(x,y,t) = \sum_{j=1}^q W_I^j \varphi_j(x,y) \tag{19.14}$$

where p and q are the number of nodes per 2–D element used to approximate the inplane and transverse deflections, respectively, and U_I^j, V_I^j, and W_I^j are the values of the displacements U_I, V_I, and W_I, respectively, at the j–th node of the 2–D finite element representing the j–th plane of the plate element. The functions $\psi_j(x,y)$ $(j = 1,2,\cdots,p)$ and $\varphi_j(x,y)$ $(j = 1,2,\cdots,q)$ are the two–dimensional Lagrangian interpolation polynomials associated with the j–th node of the two–dimensional finite element.

The semidiscrete finite element equations are obtained by substituting equations (19.12)-(19.14) into the principle of virtual work for a typical element. Here we use the same interpolation for all three displacements: $\varphi_j = \psi_j$ $(p = q)$ and $\Psi_I = \Phi_I$ $(M = N)$.

The following two–dimensional finite elements $(N = M)$ are used here with isoparametric formulations:

E4: Four–node Lagrange quadrilateral element.
E8: Eight–node serendipity quadrilateral element.
E9: Nine–node Lagrange quadrilateral element.

Each of these elements may be used in conjunction with one or more linear, quadratic, or cubic (denoted L,Q,C respectively) 1–D Lagrange elements through the thickness to create a wide variety of different layerwise finite elements. We use the notation E9–Q3 to denote a two–dimensional E9 element with three quadratic subdivisions through the thickness.

The full layerwise finite element model is the same as an associated conventional 3–D displacement finite element model in terms of interpolation capability and problem size for a 3–D body with parallel top and bottom surfaces. The layerwise element has some analysis advantages over the conventional 3–D elements. The layerwise format maintains a 2–D type data structure similar to finite element models of 2–D ESL theories. This provides several advantages over conventional 3–D finite element models. First the volume of the input data is reduced. Secondly, the inplane 2–D mesh and the transverse 1–D mesh can be refined independently of each other without having to reconstruct a 3–D finite element mesh. The 2–D type data structure also allows efficient formulation of the element stiffness matrices [Red96, Rob93].

19.3 VARIABLE KINEMATIC FORMULATIONS

19.3.1 INTRODUCTION

All multiple model methods represent an attempt to distribute limited computational resources in an optimal manner to achieve maximum solution accuracy with minimal solution cost, subject to certain problem specific constraints. This task often requires the joining of incompatible finite element meshes and/or incompatible mathematical models. Note that for the case of joining incompatible mathematical models, the numerical methods used to implement each of the mathematical models may be the same or different; often the finite element method is used to implement each of the models. The traditional difficulty with multiple model analyses is the maintenance of displacement continuity and force equilibrium along boundaries separating incompatible subregions.

The analysis of composite laminates has provided the incentive for the development of many of the reported multiple model methods [Tho90, Mao91, Ami92, Red92, Rob97], due mainly to the heterogeneous nature of composite materials and the wide range of scales of interest (*i.e.*, micromechanics level, lamina level, laminate level, structural component level). In general, the broad spectrum of multiple model methods can be divided into two categories: (1) the sequential or multi–step methods, and (2) the simultaneous methods. Most of the sequential multiple model methods reported to date are developed for global–local analysis. Typically the global region (*i.e.*, the entire computational domain) is analyzed with an economical, yet adequate model to determine the displacement or force boundary conditions for a subsequent analysis of the local region (*i.e.*, a small subregion of particular interest). The local region might be modeled with a highly refined mesh of the same ESL laminate elements or it might be modeled with 3–D finite elements. The sequential multiple methods attempt to iteratively establish force equilibrium along the global–local boundary, in addition to imposing displacement continuity. Each of the iterative sequential methods uses the same type of mathematical model for both the global region and the local region. A number of simultaneous multiple model methods have been reported in the literature. These methods are characterized by a simultaneous analysis of the entire computational domain where different subregions are modeled using different mathematical models and/or distinctly different levels of domain discretization. The simultaneous methods explicitly account for the full interaction of the different subregions and are thus directly extendable to nonlinear analysis. For a more detailed review of the multiple model methods, the reader may consult [Rob97].

19.3.2 *MULTIPLE ASSUMED DISPLACEMENT FIELDS*

While both full layerwise finite elements and conventional 3–D finite elements permit an accurate determination of 3–D ply level stress fields, they are computationally expensive to use; thus it is most often impractical to discretize an entire laminate with these types of elements. To accurately capture the localized 3–D stress fields in a tractable manner, it is usually necessary to resort to a simultaneous multiple model approach in which different subregions of the laminate are described with different types of mathematical models. The objective of such a simultaneous multiple model analysis is to match the most appropriate mathematical model with each subregion based on the physical characteristics, applied loading, expected behavior, and level of solution accuracy desired of each subregion. In the previously mentioned works that use some form of hierarchical, multiple assumed displacement fields, both displacement fields are based on the same mathematical model, hence the subregions differed only in the level of refinement of the interpolated solution. Reddy and Robbins [Red92, Rob97] are the first ones to employ hierarchical multiple assumed displacement fields to model different subregions with different mathematical models (*e.g.*, FSDT and LWPT).

Although simultaneous multiple model methods are simple in concept, the actual implementation of such techniques is complicated and cumbersome due mainly to the need for maintaining displacement continuity across subregion boundaries separating incompatible subdomains. To avoid such difficulties, a new hierarchical, variable kinematic finite element that provides the framework for a very general, robust, simultaneous multiple model methodology for laminated composite plates, is developed. The variable kinematic finite elements possess the following attributes:

1. The kinematics and constitutive relations of the element can be conveniently changed, thus allowing the element to represent a variety of different mathematical models from the very simple to the very complex.
2. Different types of elements can be conveniently connected together in the same computational domain, thus permitting different subregions to be described by

different mathematical models. One might also think of the variable kinematic finite element as a very sophisticated, adaptable, transition element that circumvents the need for more than one type of transition element.

The hierarchical, variable kinematic finite element is developed using a multiple assumed displacement field approach, *i.e.*, by superimposing two or more different types of assumed displacement fields in the same finite element domain. In general, the multiple assumed displacement field can be expressed as

$$u_i(x, y, z) = u_i^{ESL}(x, y, z) + u_i^{LWT}(x, y, z) \tag{19.15}$$

where i=1,2,3, and $u_1 = u, u_2 = v$, and $u_3 = w$ are the displacement components in the x, y, and z directions, respectively. The reference plane of the plate coincides with the xy−plane. The underlying foundation of the displacement field is provided by u_i^{ESL}, which represents the assumed displacement field for any desired 'equivalent single–layer' theory. The second term u_i^{LWT} represents the assumed displacement field for any desired full layerwise theory. The layerwise displacement field is included as an optional, incremental enhancement to the basic ESL displacement field, so that the element can have full 3–D modeling capability when needed. Depending on the desired level of accuracy, the element may use none, part, or all of the layerwise field to create a series of different elements having a wide range of kinematic complexity. For example, discrete layer transverse shear effects can be added to the element by including u_1^{LWT} and u_2^{LWT}. Discrete layer transverse normal effects can be added to the element by including u_3^{LWT}. Displacement continuity is maintained between these different types of elements by simply enforcing homogeneous essential boundary conditions on the incremental layerwise variables, thus eliminating the need for multi–point constraints, penalty function methods or special transition elements.

To illustrate the usefulness of a finite element based on the assumed displacement field of (19.15), consider a specific case where the individual displacement fields are defined as follows:

u_i^{ESL}: First Order Shear Deformation Field

$$
\begin{aligned}
u_1^{ESL}(x, y, z) &= u_0(x, y) + z\phi_x(x, y) & (19.16) \\
u_2^{ESL}(x, y, z) &= v_0(x, y) + z\phi_y(x, y) & (19.17) \\
u_3^{ESL}(x, y, z) &= w_0(x, y) & (19.18)
\end{aligned}
$$

u_i^{LWT}: Layerwise Field of Reddy [Red87]

$$
u_1^{LWT}(x, y, z) = \sum_{I=1}^{N} U_I(x, y)\Phi^I(z) \tag{19.19}
$$

$$
u_2^{LWT}(x, y, z) = \sum_{I=1}^{N} V_I(x, y)\Phi^I(z) \tag{19.20}
$$

$$
u_3^{LWT}(x, y, z) = \sum_{I=1}^{M} W_I(x, y)\Psi^I(z) \tag{19.21}
$$

where (U_I, V_I, W_I) denote the nodal values of $(u_1, u_2, u_3)=(u, v, w)$, N is the number of nodes (or $N - 1$ is the number of subdivisions) through thickness and Φ^I are the

1–D (global) interpolation functions for the discretization of the inplane displacements through thickness; and M and Ψ^I have similar meaning for the discretization of the transverse displacement through thickness. It should be noted that the layerwise field given by (19.19)-(19.21) is sufficiently general to model any of the deformation modes that can be modeled by the ESL field given in (19.16)-(19.18); thus for elements that use all the variables shown in (19.16)-(19.18) and (19.19)-(19.21), there will be five redundant variables that must be set to zero (or ignored) to permit a unique solution for the remaining variables. The ESL variables are essential for connecting different types of elements. Therefore, the following five of the layerwise variables should be set to zero:

$$U_1 = U_N = 0, \quad V_1 = V_N = 0, \quad W_1 = 0 \tag{19.22}$$

The remaining layerwise nodal variables have the meaning of incremental enhancement to the ELS displacement field.

19.3.3 FINITE ELEMENT MODEL

A hierarchy of three distinct types of plate elements can be obtained from the composite displacement field of equation (19.15), where the individual displacement expansions are defined by (19.16)-(19.18) and (19.19)-(19.21). The first and simplest type of element is the first order shear deformation element (or FSDT element). This element is formed using equation (19.16)-(19.18) while ignoring (19.19)-(19.21). The second type of element is the Type I layerwise element (or LWT1 element). The LWT1 element is formed using (19.16)-(19.18),(19.19) and (19.20), while ignoring (19.21); thus, it is a partial layerwise element. Four of the layerwise variables should be set to zero (or simply ignored) to remove the redundancy from the composite displacement field (e.g., $U_1 = U_N = V_1 = V_N = 0$). Like the FSDT element, the LWT1 element assumes a state of zero transverse normal stress and thus does not explicitly account for transverse normal strain. This is implicitly achieved by using a reduced constitutive matrix similar to the FSDT element. The inclusion of (19.19) and (19.20) in the LWT1 element provides discrete layer transverse shear effects, unlike the simple gross transverse shear effect included in the FSDT element. Thus the LWT1 element is applicable to thick laminates and often yields results comparable to 3–D finite elements while using approximately two thirds the number of degrees of freedom. The third and most complex element is the Type II layerwise element (or LWT2 element). This element is formed using both (19.16)-(19.18) and (19.19)-(19.21), thus it is a full layerwise element. The composite displacement field contains five redundant variables. Thus five layerwise variables are chosen (e.g., U_1, U_N, V_1, V_N, W_1) and set to zero or simply ignored. The LWT2 element explicitly accounts for all six strain components in a kinematically correct manner, i.e., the in–plane strains are C^1–continuous through the laminate thickness while the transverse strains are C^0–continuous through the laminate thickness. The inclusion of the full layerwise field provides the LWT2 element with both discrete layer transverse shear effects and discrete layer transverse normal effects. The LWT2 element uses a full constitutive matrix $(\bar{Q}_{ij} = \bar{C}_{ij})$, and it is equivalent in accuracy and cost to a stack of conventional 3–D finite elements.

Note that the various element types are created by hierarchically adding variables to the basic first order shear deformation field. The matrix form of the finite element equations that result from the hierarchical use of (19.16)-(19.18), (19.19)-(19.21), and (19.11) within a single element domain is given by

$$\begin{bmatrix} [K^{EE}] & [K^{EL}] \\ [K^{LE}] & [K^{LL}] \end{bmatrix} \begin{Bmatrix} \{U^E\} \\ \{U^L\} \end{Bmatrix} = \begin{Bmatrix} \{F^E\} \\ \{F^L\} \end{Bmatrix} \tag{19.23}$$

where $[K^{EE}]$ represents the element stiffness matrix for an equivalent single–layer FSDT element and $[K^{LL}]$ represents the element stiffness matrix for a full layerwise element. The remaining submatrices represent coupling stiffnesses between the three different displacement fields. Based on the particular type of element desired, the appropriate terms in the composite stiffness matrix are identified and computed. Since all three element types possess the first order shear deformation variables of (19.16)-(19.18), these different types of elements can easily be simultaneously connected together in the same computational domain by simply setting certain layerwise variables of (19.19)-(19.21) to zero along the incompatible boundary.

When maintaining strict subregion compatibility (SSC), all three displacement components are continuous across all types of subregion boundaries (FSDT/LWT1, LWT1/LWT2, LWT2/FSDT). In contrast, relaxed subregion compatibility (RSC) maintains total continuity of the in–plane displacement components across all types of subregion boundaries, but does not maintain total continuity of the transverse displacement component across FSDT/LWT2 or LWT1/LWT2 boundaries. This relaxation is often useful for obtaining accurate transverse normal stresses within LWT2 subregions, near FSDT/LWT2 or LWT1/LWT2 boundaries, since it eliminates the transverse pinching or stretching of the laminate near these boundaries, and effectively allows the transverse normal strain to react to the local dominant inplane strains. Within those portions of LWT2 subregions sufficiently removed from FSDT/LWT2 or LWT1/LWT2 boundaries, both strict and relaxed conditions yield the same stress distributions.

19.4   ILLUSTRATIVE EXAMPLES

The problem of determining the free edge stress fields in laminates subjected to inplane extension or bending is used to illustrate the variable kinematic finite element (VKFE) model methodology. The free edge problem is ideally suited for global–local analysis, because the 3–D stress field exists only in a boundary region (*i.e.*, free edge) of the laminate and elsewhere only a 2–D stress state exists. Thus, the LWT2 elements can be used in the free edge (local) region and ESL elements can be used everywhere else (global region) to capture the stress fields accurately.

To demonstrate the accuracy and economy afforded by the variable kinematic finite elements, a global–local analysis is performed to determine the nature of the free edge stress field in three different laminates subjected to axial extension: $(45/-45)_s$, $(45/0-45/90)_s$, and $(45/0-45/90/90/-45/0/45)_s$. The three laminates have length $2a$, width $2b$, and thickness $2h$. Each of the three laminates has a length–to–width ratio of 10 (*i.e.*, $a/b = 10$). The material plies in each laminate are of equal thickness h_k. The following geometric differences exist among the three laminates:

$$(45/-45)_s \text{ laminate:} \qquad \tfrac{b}{h} = 4, \ h = 2h_k, \ \tfrac{b}{h_k} = 8 \tag{19.24}$$

$(45/0/-45/90)_s$ laminate: $\frac{b}{h} = 15$, $h = 4h_k$, $\frac{b}{h_k} = 60$ (19.25)

$(45/0/-45/90)_{2s}$ laminate: $\frac{b}{h} = 15$, $h = 8h_k$, $\frac{b}{h_k} = 120$ (19.26)

Each of the material plies in the three laminates is idealized as a homogeneous, orthotropic material; the material properties (expressed in the principal material coordinate system) are defined below.
Material plies in the four–layer laminate:

$$E_1 = 20 \times 10^6 \text{ psi}, \ E_2 = E_3 = 2.1 \times 10^6 \text{ psi} \tag{19.27}$$

$$G_{12} = G_{23} = G_{13} = 0.85 \times 10^6 \text{ psi}, \ \nu_{12} = \nu_{23} = \nu_{13} = 0.21 \tag{19.28}$$

Material plies in the eight– and sixteen–layer laminates:

$$E_1 = 19.5 \times 10^6 \text{ psi}, \ E_2 = E_3 = 1.48 \times 10^6 \text{ psi} \tag{19.29}$$

$$G_{12} = G_{23} = G_{13} = 0.8 \times 10^6 \text{ psi}, \ \nu_{12} = \nu_{23} = \nu_{13} = 0.3 \tag{19.30}$$

The origin of the global coordinate system coincides with the centroid of each of the 3–D composite laminates. The x–coordinate is taken along the length, the y–coordinate is taken across the width, and the z–coordinate is taken through the thickness of the laminate. Since the laminate is symmetric about the xy–plane, only the upper half of each laminate is modeled. Thus the computational domain is defined by $(-a < x < a, -b < y < b, 0 < z < h)$. The displacement boundary conditions for all three laminates are:

$$u_1(a, y, z) = u_0 \ , \ \ u_1(-a, y, z) = u_2(-a, 0, 0) = u_2(a, 0, 0) = u_3(x, y, 0) = 0 \quad (19.31)$$

The slight differences in geometry and material properties among the three laminates allow comparison with solutions published in the literature. For the $(45/-45)_s$ laminate, Wang and Choi [Wan82a, Wan82b] have developed a quasi–3D elasticity solution, while Whitcomb et al. [Whi82] produced a solution from a highly refined, quasi–3D finite element model. For the $(45/0/-45/90)_s$ and $(45/0/-45/90/90/-45/0/45)_s$ laminates, Whitcomb and Raju [Whi85] obtained quasi–3D finite element solutions using a highly refined mesh.

The variable kinematic finite elements are used in a simultaneous multiple model analysis (global–local analysis) of these three laminates in order to accurately yet efficiently determine the free edge stresses near the middle of one of the two free edges. The global region is modeled using first order shear deformable elements (FSDT); the local region, where accurate 3–D stresses are desired, is modeled with LWT2 elements.

First the $(45/-45)_s$ laminate will be used to assess the effects of subregion compatibility type (SSC or RSC) and size of the local LWT2 subregion on the accuracy of the computed transverse stresses near the free edge. For this purpose, five different finite element meshes are created. The 2–D, in–plane discretization for all five meshes is exactly the same, consisting of a 5×11 mesh of eight–node, quadratic, 2–D, quadrilateral finite elements (see Figure 2). All elements have the same length $(2a/5)$; however, the width of the elements decreases as the free edge at (x, b, z) is approached. The widths of the eleven rows of elements, as one moves away from the refined free edge, are $h_k/16, h_k/16, h_k/8, h_k/4, h_k/2, h_k, h_k, 2h_k, 3h_k, 3h_k$, and $5h_k$

Table 1 Description of global–local meshes for the $(45/-45)_s$ laminate under axial extension. h_k = thickness of a single material ply. All five VKFE meshes have the exact same inplane discretization (5×11).

Remarks	Mesh 1	Mesh 2	Mesh 3	Mesh 4	Mesh 5
Number of Elements in Local LWT2 Region	3×4	3×5	3×6	3×7	5×11
Width of Local Region	$\frac{1}{2}h_k$	h_k	$2h_k$	$3h_k$	$16h_k$
Length of Local Region	$\frac{6}{5}a$	$\frac{6}{5}a$	$\frac{6}{5}a$	$\frac{6}{5}a$	$2a$
Total Number of Active D.O.F. in VKFE Mesh (Strict Compatibility)	1,986	2,400	2,814	3,228	9,116
Total Number of Active D.O.F. in VKFE Mesh (Relaxed Compatibility)	2,354	2,800	3,246	3,690	9,116

(h_k=ply thickness). The five meshes differ only in the width of the local region where LWT2 elements are used. The LWT2 elements used in the local region employ eight quadratic layers through the laminate thickness (four per material layer) as shown in Figure 3. The thickness of the numerical layers decreases as the $+45/-45$ interface is approached. From bottom to top, the layer thicknesses are $0.533h_k$, $0.267h_k$, $0.133h_k$, $0.083h_k$, $0.083h_k$, $0.133h_k$, $0.267h_k$, and $0.533h_k$.

Table 1 summarizes the five meshes used for the $(45/-45)_s$ laminate. Note that mesh 5 is not a global–local mesh. Mesh 5 uses LWT2 elements throughout the entire computational domain, thus serving as a control mesh for judging the accuracy of the four global–local meshes. In meshes 1 through 4, the local region (LWT2 elements) is adjacent to the free edge (x, b, z) and is centered about the plane $(0, y, z)$. In meshes 1 through 4, the length of the local region spans three fifths of the total length of the laminate; however, the width of the local region differs in each mesh ranging from $h_k/2$ to $3h_k$. Two runs are made with each of the four global–local meshes, the first using strict subregion compatibility along the FSDT–LWT2 boundary, and the second using relaxed subregion compatibility along the FSDT–LWT2 boundary.

The stresses are computed via the constitutive relations at the reduced Gauss points within the individual layers of each LWT2 element. All stresses are nondimensionalized as follows:

$$\bar{\sigma}_{ij} = \frac{20\sigma_{ij}}{\varepsilon_0 E_1} \tag{19.32}$$

where ε_0 is the nominal applied axial strain of $u_0/(2a)$. The stress distributions shown

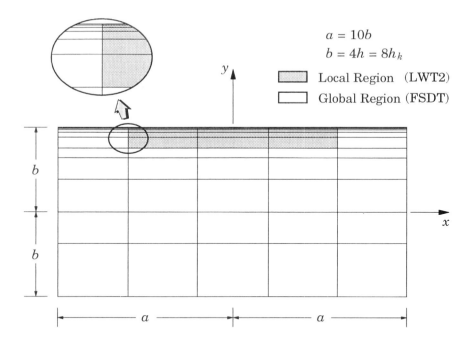

$a = 10b$
$b = 4h = 8h_k$

Local Region (LWT2)
Global Region (FSDT)

Figure 2   The 2–D mesh of variable kinematic finite elements used to model a $(45/-45)_s$ laminate under axial extension. All elements are eight–node quadrilaterals.

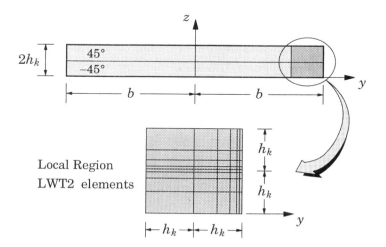

Figure 3   Discretization within the local LWT2 region, on the yz–plane. The eight–node LWT2 elements in the local region use quadratic layers.

in Figures 4 and 6 are generated by computing the nondimensionalized stresses at a series of adjacent reduced Gauss points, and then connecting these points with straight lines.

Figures 4 and 5 show the distribution of the interlaminar stresses σ_{zz} and σ_{xz} near the free edge. The results in these two figures were obtained using relaxed subregion compatibility conditions. The stresses presented in Figure 4 are computed at the reduced Gauss points near the middle of the refined free edge, $i.e.$, along the line $(-0.115a, 0.998b, z)$. This is also the reduced Gauss point located furthest from the FSDT/LWT2 boundary. The stresses presented in Figure 4 are computed at the reduced Gauss points closest to the line $(0, y, h_k)$, $i.e.$, along the line $(-0.115a, y, 1.014h_k)$. All four of the global local meshes are successful in identifying the spikes in σ_{zz} and σ_{xz} that occur at the $45/-45$ interface. The results of meshes 3 and 4 are graphically indistinguishable from the results of the control mesh, Mesh 5. While meshes 1 and 2 exhibit some error, they do capture the qualitative nature of the transverse stress distributions near the free edge. In meshes 1 and 2, the transverse shear stress is predicted more accurately than the transverse normal stress. The results also indicate that the boundary layer thickness (or the width of the local region) should be at least $2h_k$ to capture both interlaminar stresses accurately.

To illustrate the accuracy of the variable kinematic elements in determining the free edge stress field for more complex laminates, a simultaneous multiple model analysis (global–local analysis) is performed on an eight–ply $(45/0/-45/90)_s$ laminate, and a sixteen–ply $(45/0/-45/90/90/-45/0/45)_s$ laminate. Both of these laminates are subjected to axial extension similar to the previously examined $(45/-45)_s$ laminate. The in–plane discretization consists of a 5×11 2–D mesh of eight–node quadrilateral elements as shown previously in Figure 2. The local region is discretized with a 3×6 mesh of LWT2 elements. For the $(45/0/-45/90)_s$ laminate, the LWT2 elements contain 12 quadratic layers (three per material ply). Within each material ply the three quadratic layers have thicknesses of $0.25h_k, 0.5h_k$, and $0.25h_k$ from bottom to top. For the $(45/0/-45/90/90/-45/0/45)_s$ laminate, the LWT2 elements contain 16 quadratic layers (two per material ply). Within each material ply both of the quadratic layers have thicknesses of $0.5h_k$. The $(45/0/-45/90)_s$ model contains 4,382 active degrees of freedom while the $(45/0/-45/90/90/-45/0/45)_s$ model contains 5,638 active degrees of freedom.

The computed transverse shear stress and transverse normal stress distributions for these two laminates are shown in Figures 6 and 7. The present results show excellent agreement with the quasi–3D finite element solutions of Whitcomb and Raju [Whi85] (not included in the figure). For both laminates the maximum transverse normal stress occurs at the intersection of the 90/90 interface and the free edge, while the maximum transverse shear stress occurs at the intersection of the 45/0 and $0/-45$ interfaces with the free edge.

Both of these laminates have enough distinct material plies to make a full 3–D analysis prohibitively expensive, thus a sequential or simultaneous multiple model analysis is the only reasonable alternative. Many laminates have a very large number of distinct material plies, thus even with a multiple model analysis, the investigator may have to resort to using the sublaminate approach ($i.e.$, ply grouping) within the local LWT2 region. In this case the investigator would identify a target group of adjacent material plies that would receive one or more numerical layers each, while the remaining plies are grouped into one or more numerical layers and effectively homogenized. By performing several of these analyses, the investigator can gradually piece together a picture of the 3–D stress state through the laminate thickness within the local LWT2 region.

19.5 SUMMARY AND CONCLUSIONS

In this paper a generalized layerwise theory proposed and advanced by the author is described, and a hierarchical, displacement–based, global–local finite element model that permits an accurate, efficient, and convenient analysis of localized three–dimensional effects in laminated composite plates is also presented. By superimposing a hierarchy of assumed displacement fields in the same finite element domain, a variable kinematic finite element model is developed. All displacement fields in the hierarchy share the same assumed inplane variation, but differ in their assumed transverse variation.

The underlying foundation of the variable kinematic element's composite displacement field is provided by a two–dimensional 'equivalent single-layer' plate theory (e.g., the first order shear deformation theory). The layerwise displacement field of Reddy [Red87] is included as an optional, incremental enhancement to the displacement field of the two–dimensional plate theory, so that the element can have full three–dimensional modeling capability when needed. Depending on the desired level of accuracy, an element can use none, part, or all of the layerwise field to create a hierarchy of different elements having a wide range of kinematic complexity and representing a number of different mathematical models. Discrete layer transverse shear effects and discrete layer transverse normal effects can be independently added to the element by including appropriate terms from the layerwise field.

In a 2–D mesh of variable kinematic finite elements, each one of the elements is capable of simulating any of the element types in the hierarchy. Due to the hierarchical nature of the multiple assumed displacement fields, displacement continuity can be maintained between different types of elements in the hierarchy (i.e., elements based on different mathematical models) by simply enforcing homogeneous essential boundary conditions on certain terms in the composite displacement field along the incompatible boundary. This simple process can easily be automated and subsequently removed form the concern of the user/analyst. Thus, in a single 2–D mesh of variable kinematic finite elements, it is possible to designate several different subregions that are described by elements that are based on different mathematical models. The variable kinematic elements circumvent the inconvenience and problems associated with the traditional methods of maintaining displacement continuity across incompatible subdomains (e.g., multipoint constraints, special transition elements, and penalty methods).

19.5.1 ACKNOWLEDGEMENT

The support of this research by the Army Research Office (ARO) through Grant No. DAAH04–96–1–0080 is gratefully acknowledged.

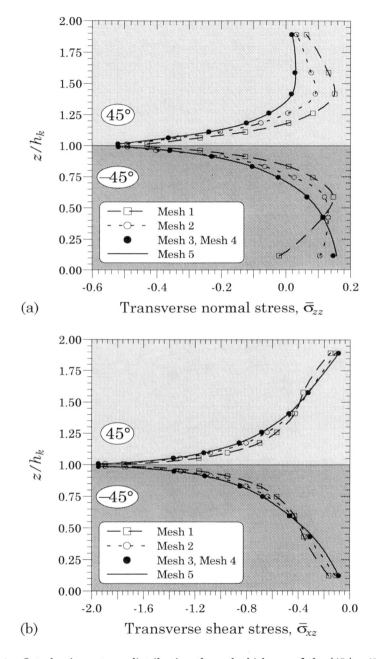

(a)

(b)

Figure 4   Interlaminar stress distribution through thickness of the $(45/-45)_s$ laminate near free edge. Results computed for meshes 1 through 5 with relaxed subregion compatibility.

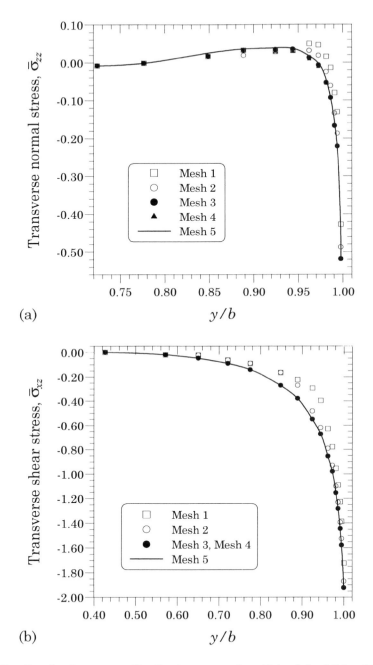

(a)

(b)

Figure 5   Interlaminar stress distribution across the width of the $(45/-45)_s$ laminate near the upper $45/-45$ interface $(z = 1.014 h_k)$. Results computed for meshes 1 through 5 with relaxed subregion compatibility.

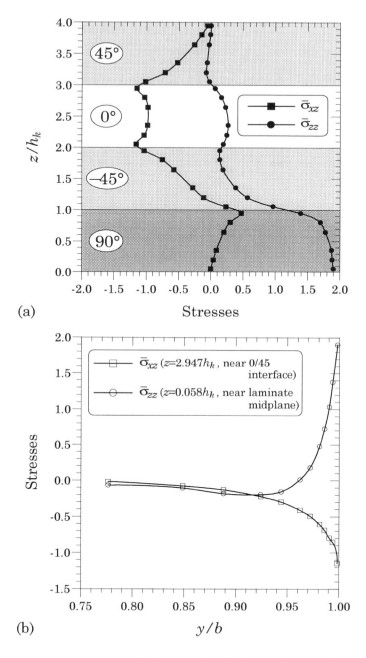

(a)

(b)

Figure 6   Interlaminar stress distributions in $(45/0/-45/90)_s$ laminate under axial extension. (a) Through the thickness near the free edge ($x = -0.115a, y = 0.998b$). (b) Across the width ($x = -0.115a$).

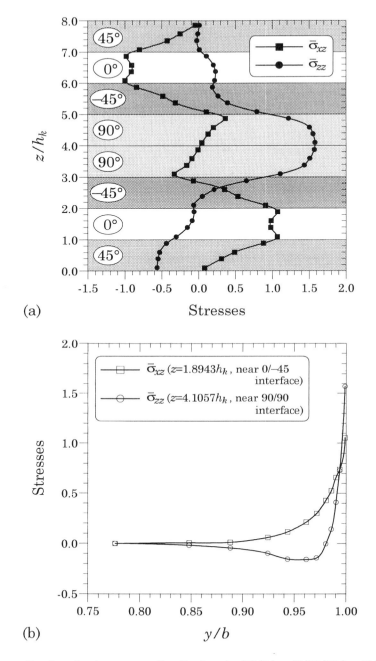

(a)

(b)

Figure 7   Interlaminar stress distributions in $(45/0/-45/90/90/-45/0/45)_s$ laminate under axial extension. (a) Through the thickness near the free edge $(x = -0.115a, y = 0.998b)$. (b) Across the width $(x = -0.115a)$.

References

[Ami92]   Aminpour, A. A. and McCleary, S.L. and Ransom, J. B., and Housner, J. M. (1992) A Global/Local Analysis Method for Treating Details in Structural Design. In Noor A. K. (ed) *Adaptive, Multilevel, and Hierarchical Computational Strategies*, volume 157, pages 119–137. ASME, AMD.

[Mao91]   Mao, K. M., and Sun, C. T. (1991) A Refined Global–Local Finite Element Analysis Method. *International Journal for Numerical Methods in Engineering* 32: 29–43.

[Red87]   Reddy, J. N. (1987) A Generalization of Two–Dimensional Theories of Laminated Composite Plates. *Communications in Applied Numerical Methods* 3: 173–180.

[Red92]   Reddy, J. N., and Robbins, D. H. (1992) Analysis of Composite Laminates Using Variable Kinematic Finite Elements. *RBCM–Journal of the Brazilian Society of Mechanical Sciences* 14(4): 299–326.

[Red96]   Reddy, J. N. (1996) *Mechanics of Laminated Plates: Theory and Analysis*. CRC Press, Boca Raton, FL.

[Rob93]   Robbins, D. H., Jr., and Reddy, J. N. (1993) Modeling of Thick Composites Using a Layerwise Laminate Theory. *International Journal for Numerical Methods In Engineering* 36: 655–677.

[Rob97]   Robbins, D. H., Jr., and Reddy, J. N. (1996/1997) Variable Kinematic Modeling of Laminated Composite Plates. *International Journal for Numerical Methods in Engineering* in press.

[Tho90]   Thompson, D. M., and Griffin, Jr., O. H. (1990) 2–D to 3–D Global/Local Finite Element Analysis of Cross–Ply Composite Laminates. *Journal of Reinforced Plastics and Composites* 9: 492–502.

[Wan82a]   Wang, S. S. and Choi, I. (1982) Boundary–Layer Effects in Composite Laminates, Part I: Free Edge Stress Singularities. *Journal of Applied Mechanics* 49: 541–548.

[Wan82b]   Wang, S. S., and Choi, I. (1982) Boundary–Layer Effects in Composite Laminates, Part II: Free Edge Stress Solutions and Basic Characteristics. *Journal of Applied Mechanics* 49: 549–560.

[Whi82]   Whitcomb, J. D., Raju, I. S., and Goree, J. G. (1982) Reliability ofthe Finite Element Method for Calculating Free Edge Stresses in Composite Laminates. *Computers and Structures* 15(1): 23–37.

[Whi85]   Whitcomb, J. D., and Raju, I. S. (1985) Analysis of Interlaminar Stresses in Thick Composite Laminates With and Without Edge Delamination. In Johnson W. S. (ed) *Delamination and Debonding of Materials*, number 876 in ASTM STP, pages 69–94. American Society for Testing and Materials, Philedelphia.

20

A Viscoelastic Higher-Order Beam Finite Element

A. R. Johnson[1] and A. Tessler[2]

[1]Vehicle Structures Directorate, MS 240
Army Research Laboratory
NASA Langley Research Center
Hampton, VA 23681-0001
[2]Computational Structures Branch, MS 240
NASA Langley Research Center
Hampton, VA 23681-0001

20.1   INTRODUCTION

Advanced composites technology has led to the use of highly viscous, low-modulus materials that are combined with the traditional high-modulus, load carrying materials to produce stiff, highly damped structures. The quasi-static and dynamic analyses of such structures require improvements in the material damping representation over the standard proportional damping schemes. Halpin and Pagano [HP68] demonstrated that the relaxation moduli for anisotropic solids produce symmetric matrices that can be expanded in a Prony series form (i.e., a series of exponentially decaying terms). Early viscoelastic models for small deformations of composites focused on computing the complex moduli for anisotropic solids from the elastic properties of the fibers and the complex modulus properties of the matrix [Has70a, Has70b]. Recently, various classical constitutive models have been used including generalized Maxwell and Kelvin-Voigt solids [ADdS91, SWW94]. These constitutive models have practical value since they provide adequate approximations for the dynamic softening and hysteresis effects—the phenomena that are not directly proportional to strain rates.

   In this paper, a brief review of the history integral form of the Maxwell solid is presented to provide background for the differential constitutive law. The interested reader is referred to Coleman and Noll [CN61] and Schapery [Sch74] for more comprehensive discussions on the subject. A new differential constitutive law for the Maxwell solid is then derived. This form is a special case of the model developed by Johnson and Stacer [JS93] for large strain viscoelastic deformations of rubber. It has

The Mathematics of Finite Elements and Applications
Edited by J. R. Whiteman © 1997 John Wiley & Sons Ltd.

also been used by Johnson et al. [JTM95a, JTM95b] to formulate a viscoelastic, large-displacement shell finite element. The differential constitutive law is then combined with the higher-order beam theory and finite element formulation of Tessler [Tes91] providing viscoelastic capability for thick beams. The additional strain variables required in the constitutive law are replaced with element nodal variables that are conjugate to the elastic nodal variables. This results in only minor modifications to the elastic finite element code. Finally, several numerical examples are presented demonstrating the viscoelastic, thick-beam response under quasi-static loading.

20.2   MAXWELL SOLID IN HISTORY INTEGRAL FORM

The stress-strain relation for a linear elastic material can be written in the tensor form as

$$\sigma_{ij} = C_{ijkl}\varepsilon_{kl} \tag{20.1}$$

where σ_{ij} are the stress components, C_{ijkl} are the elastic stiffness coefficients, and ε_{kl} are the strains. When a linear viscoelastic material is subjected to an instantaneous incremental strain, $\Delta\varepsilon_{kl}$, the time dependent stress takes the form

$$\sigma_{ij}(t) = C_{ijkl}\Delta\varepsilon_{kl} + \sigma_{ij}^v(t) \tag{20.2}$$

where the viscous stresses, $\sigma_{ij}^v(t)$, are monotonically decreasing functions of time. The Boltzmann superposition method is often used to approximate (20.2) as follows. The viscous stresses, $\sigma_{ij}^v(t)$, are factored such that

$$\sigma_{ij}^v(t) = C_{ijkl}^v(t)\Delta\varepsilon_{kl} \tag{20.3}$$

The functions $C_{ijkl}^v(t)$ are referred to as time dependent moduli. These monotonically decreasing functions are approximated in time using a Prony series, i.e.,

$$C_{ijkl}^v(t) = C_{ijkl}^* \sum_{n=1}^{N} e^{-\frac{t}{\tau_n}} \tag{20.4}$$

where $\tau_n, C_{ijkl}^* \geq 0$. The stresses then become

$$\sigma_{ij}(t) = C_{ijkl}\Delta\varepsilon_{kl} + \sum_{n=1}^{N} C_{ijkl}^* e^{-\frac{t}{\tau_n}} \Delta\varepsilon_{kl} \tag{20.5}$$

The above approximation is extended to the case of a continuously deforming solid by associating the continuous time dependent strain with an incremental strain history and convoluting (20.5) in time. The approximation to the time dependent stresses becomes

$$\sigma_{ij}(t) = C_{ijkl} \sum_{m=1}^{M} \Delta_m\varepsilon_{kl} + \sum_{n=1}^{N} C_{ijkl}^* \sum_{m=1}^{M} H(t-t_m)e^{-\frac{(t-t_m)}{\tau_n}} \Delta_m\varepsilon_{kl} \tag{20.6}$$

where the strain increments are set at times t_m for $m = 1, \ldots, M$, and use is made of the Heaviside unit step function, $H(t-t_m)$. At this juncture, it is often customary to define the viscous moduli as functions of the relative time, $t - t_m$, i.e.,

$$^*C_{ijkl}(t - t_m) = \sum_{n=1}^{N} C^*_{ijkl} H(t - t_m) e^{-\frac{(t-t_m)}{\tau_n}} \tag{20.7}$$

The constitutive model in (20.6) then takes the form

$$\sigma_{ij}(t) = C_{ijkl} \sum_{m=1}^{M} \Delta_m \varepsilon_{kl} + \sum_{m=1}^{M} {}^*C_{ijkl}(t - t_m)\Delta_m \varepsilon_{kl} \tag{20.8}$$

Assuming that strains are smooth functions of time, and taking the limit as $\Delta_m t = t_{m+1} - t_m \to 0$ for all m, gives rise to

$$\sigma_{ij}(t) = C_{ijkl}\varepsilon_{kl}(t) + \int_{\tau=-\infty}^{t} {}^*C_{ijkl}(t - \tau)\frac{d\varepsilon_{kl}(\tau)}{d\tau}d\tau \tag{20.9}$$

where it is noted once again that the viscous moduli, $^*C_{ijkl}(t - \tau)$, are monotonically decreasing in time. Equation (20.9) is often referred to as the history integral for a linear Maxwell solid.

In many practical applications, adequate time-dependent stress predictions can be obtained with only a small number of terms in the Prony series, (20.4). The constitutive model as presented in (20.9) requires that the history of the measurable kinematics, $\varepsilon_{kl}(\tau)$, be known in addition to the Prony series. This leads to computational algorithms that must determine how much of the history to retain in order to update the viscous stress approximation as time evolves.

20.3 MAXWELL SOLID IN DIFFERENTIAL FORM

Following Johnson et al., [JS93, JTM95a, JTM95b] new constitutive equations for a linear viscous solid are derived. The new constitutive equations are in differential form and they are equivalent to the history integral form just described. Departing from the history integral formulation at (20.6), defining internal strain variables, $^*_n\varepsilon_{kl}$, which relate to the strains as,

$$\Delta_m {}^*_n\varepsilon_{kl}(t - t_m) = H(t - t_m)e^{-\frac{(t-t_m)}{\tau_n}}\Delta_m \varepsilon_{kl} \tag{20.10}$$

introducing (20.10) into (20.6) and factoring out C^*_{ijkl}, the stresses appear as

$$\sigma_{ij}(t) = C_{ijkl} \sum_{m=1}^{M} \Delta_m \varepsilon_{kl} + C^*_{ijkl} \sum_{n=1}^{N} \sum_{m=1}^{M} \Delta_m {}^*_n\varepsilon_{kl}(t - t_m) \tag{20.11}$$

Following the Maxwell solid formulation, it is assumed that the strains are smooth functions of time. In the limit as $\Delta_m t = t_{m+1} - t_m \to 0$ for all m, (20.11) becomes

$$\sigma_{ij}(t) = C_{ijkl}\varepsilon_{kl}(t) + C^*_{ijkl} \sum_{n=1}^{N} \int_{\tau=-\infty}^{t} d^*_n\varepsilon_{kl}(t - \tau) \tag{20.12}$$

It is desirable to derive a differential equation for the time dependent strain variables. In the limit as $\Delta_m t \to 0$, (20.10) becomes

$$d\,_n^*\varepsilon_{kl}(t-\tau) = H(t-\tau)e^{-\frac{(t-\tau)}{\tau_n}}d\varepsilon_{kl}(\tau) \tag{20.13}$$

Integrating (20.13) with respect to the history, τ, yields

$$_n^*\varepsilon_{kl}(t) = \int_{\tau=-\infty}^{t} \frac{d\varepsilon_{kl}(\tau)}{d\tau}e^{-\frac{(t-\tau)}{\tau_n}}d\tau \tag{20.14}$$

Differentiating (20.14) with respect to the current time, t, yields

$$\frac{d\,_n^*\varepsilon_{kl}(t)}{dt} = -\frac{1}{\tau_n}\left[\int_{\tau=-\infty}^{t} \frac{d\varepsilon_{kl}(\tau)}{d\tau}e^{-\frac{(t-\tau)}{\tau_n}}d\tau\right] + \frac{d\varepsilon_{kl}(t)}{dt} \tag{20.15}$$

Substituting (20.14) into (20.15) yields the differential equations for the internal strain variables in the form

$$\frac{d\,_n^*\varepsilon_{kl}}{dt} + \frac{_n^*\varepsilon_{kl}}{\tau_n} = \frac{d\varepsilon_{kl}}{dt} \qquad \text{for each } n. \tag{20.16}$$

Introducing (20.14) into (20.12) results in the stress equation given by

$$\sigma_{ij}(t) = C_{ijkl}\varepsilon_{kl}(t) + C_{ijkl}^* \sum_{n=1}^{N} {}_n^*\varepsilon_{kl}(t) \tag{20.17}$$

Equations (20.16) and (20.17) represent the constitutive model in differential form. Note that this particular form is for the case of a material with a relaxation modulus given by (20.4). Also, for a material whose modulus is expressed by (20.4), the constitutive model of equations (20.16) and (20.17) is equivalent to the history integral model given by (20.9). In what follows, the differential form of the constitutive model is explored in the context of a higher-order beam theory and its associated finite element.

20.4   VISCOELASTIC HIGHER-ORDER BEAM

In this section a quasi-static Maxwell solid version of Tessler's [Tes91] higher-order beam theory is formulated and a simple beam finite element is derived. The beam dimensions and sign convention are shown in Figure 1. The viscoelastic constitutive model for the beam that is consistent with equations (20.16) and (20.17) can be written in matrix form as

$$\mathbf{s}(t) = \mathbf{C}\,\mathbf{e} + {}^*\mathbf{C}\sum_{n=1}^{N} {}_n^*\mathbf{e}(t) \tag{20.18}$$

and

$$\frac{d\,_n^*\mathbf{e}}{dt} + \frac{_n^*\mathbf{e}}{\tau_n} = \frac{d\mathbf{e}}{dt} \qquad \text{for each } n. \tag{20.19}$$

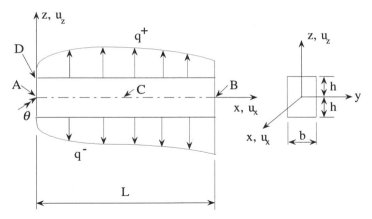

Boundary conditions applied at points A & B.

Stresses and displacements computed at C & D.

Figure 1   Thick-beam geometry, kinematics, and loading.

where

$$\mathbf{s}^T = (\sigma_{xx}, \sigma_{zz}, \tau_{xz})$$

$$\mathbf{e}^T = (\varepsilon_{xx}, \varepsilon_{zz}, \gamma_{xz})$$

$${}_n^*\mathbf{e}^T = ({}_n^*\varepsilon_{xx}, {}_n^*\varepsilon_{zz}, {}_n^*\gamma_{xz})$$

$$\mathbf{C} = \begin{bmatrix} C_{11} & C_{13} & 0 \\ C_{13} & C_{33} & 0 \\ 0 & 0 & C_{55} \end{bmatrix} \quad \text{and} \quad {}^*\mathbf{C} = \begin{bmatrix} C_{11}^* & C_{13}^* & 0 \\ C_{13}^* & C_{33}^* & 0 \\ 0 & 0 & C_{55}^* \end{bmatrix}$$

In this higher-order theory, the components of the displacement vector are approximated through the beam thickness by way of five kinematic variables, i.e.,

$$u_x(x, z, t) = u(x, t) + h\zeta\theta(x, t) \tag{20.20}$$

$$u_z(x, z, t) = w(x, t) + \zeta w_1(x, t) + \left(\zeta^2 - \frac{1}{5}\right) w_2(x, t) \tag{20.21}$$

where $\zeta = z/h$ denotes a nondimensional thickness coordinate and $2h$ is the total thickness. Note that $u(x, t)$ represents the midplane (i.e. reference plane) axial displacement, $\theta(x, t)$ is the bending rotation of the cross-section of the beam, $w(x, t)$ is the weighted-average deflection [Tes91], and $w_1(x, t)$ and $w_2(x, t)$ are the higher-order transverse displacement variables enabling a parabolic distribution of $u_z(x, z, t)$ through the thickness. The above displacement assumptions give rise to the following thickness distributions for the strains: a linear axial strain, a cubic transverse normal strain, and a quadratic transverse shear strain. These strain components have the following form, [Tes91],

$$\varepsilon_{xx} = u(x, t)_{,x} + h\zeta\theta(x, t)_{,x} \tag{20.22}$$

$$\varepsilon_{zz} = \frac{w_1(x, t)}{h} + \phi_z(\zeta)\frac{w_2(x, t)}{h^2} + \phi_x(\zeta)\theta(x, t)_{,x} \tag{20.23}$$

$$\gamma_{xz} = \phi_{xz}(\zeta)(w(x,t),_x + \theta(x,t)) \tag{20.24}$$

where

$$\phi_x(\zeta) = \frac{h\nu_{13}\zeta(4 - 7\zeta^2)}{17}, \quad \phi_z(\zeta) = \frac{14\zeta h(3 - \zeta^2)}{17}, \quad \phi_{xz}(\zeta) = \frac{5(1 - \zeta^2)}{4}$$

The simplest finite element approximation of this beam theory, as explored in [Tes91], involves a three-node configuration (see Figure 2) which is achieved by the following interpolations

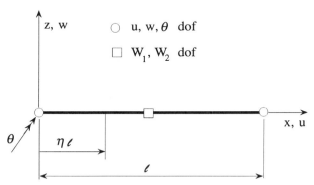

Figure 2   A three-node, higher-order-theory thick-beam element.

$$u(\eta,t) = (1 - \eta)u_0(t) + \eta u_1(t) \tag{20.25}$$

$$\theta(\eta,t) = (1 - \eta)\theta_0(t) + \eta\theta_1(t) \tag{20.26}$$

$$w(\eta,t) = (1 - \eta)w_0(t) + \eta w_1(t) - \frac{\ell}{2}\eta(1 - \eta)(\theta_0(t) - \theta_1(t)) \tag{20.27}$$

$$w_1(\eta,t) = W_1(t) \tag{20.28}$$

$$w_2(\eta,t) = W_2(t) \tag{20.29}$$

where $\eta = x/\ell$ is the nondimensional element length coordinate. Note that the nodal degrees-of-freedom at the two ends of the element are subscripted with indices 0 and 1. Since the strains do not possess derivatives of the $w_1(\eta,t)$ and $w_2(\eta,t)$ variables, these variables need not be continuous at the element nodes and, hence, their simplest approximation is constant for each element. Their corresponding degrees-of-freedom are attributed to a node at the element midspan.

For a quasi-static loading, the virtual work statement for an element of volume V with the differential form of the Maxwell constitutive law included can be written as

$$\int \mathbf{e}^T \mathbf{C}\delta e\, dV + \sum_{n=1}^N \int \overset{*}{_n}\mathbf{e}^T \, {}^*\mathbf{C}\delta \overset{*}{_n}\mathbf{e}\, dV - \delta W = 0 \tag{20.30}$$

where the first integral represents the internal virtual work done by the elastic stresses, the second is the internal virtual work done by the viscous stresses, and δW is the virtual work done by the external forces. Introducing equations (20.25)–(20.29) into equations (20.20)–(20.21) and substituting the results into equations (20.22)–(20.24) yields finite element approximations of the strains in terms of the nodal variables, i.e.,

$$e = \mathbf{B}\mathbf{u} \tag{20.31}$$

where

$$\mathbf{B} = \begin{bmatrix} -\frac{1}{\ell} & 0 & -\frac{z}{\ell} & 0 & 0 & \frac{1}{\ell} & 0 & \frac{z}{\ell} \\ 0 & 0 & -\frac{\phi_x}{\ell} & \frac{1}{h} & \frac{\phi_z}{h^2} & 0 & 0 & \frac{\phi_x}{\ell} \\ 0 & -\frac{\phi_{xz}}{\ell} & \frac{\phi_{xz}}{2} & 0 & 0 & 0 & \frac{\phi_{xz}}{\ell} & \frac{\phi_{xz}}{2} \end{bmatrix}$$

and $\mathbf{u}^T = (u_0, w_0, \theta_0, W_1, W_2, u_1, w_1, \theta_1)$ denotes the element nodal displacement vector.

A set of analogous nodal variables, $\overset{*}{n}\mathbf{u}$, and corresponding viscous strains, $\overset{*}{n}\mathbf{e}$, are introduced. These are related by

$$\overset{*}{n}\mathbf{e} = \mathbf{B}\,\overset{*}{n}\mathbf{u} \tag{20.32}$$

The virtual work statement for an element then becomes

$$\mathbf{u}^T \int \mathbf{B}^T \mathbf{C} \mathbf{B}\, dV\, \delta\mathbf{u} + \sum_{n=1}^{N} \overset{*}{n}\mathbf{u}^T \int \mathbf{B}^T {}^*\mathbf{C}\mathbf{B}\, dV\, \delta\,\overset{*}{n}\mathbf{u} - \delta W = 0 \tag{20.33}$$

By defining the integrals in (20.33) as stiffness matrices, there results

$$\mathbf{u}^T \mathbf{k}\, \delta\mathbf{u} + \sum_{n=1}^{N} \overset{*}{n}\mathbf{u}^T\, {}^*\mathbf{k}\, \delta\,\overset{*}{n}\mathbf{u} - \delta W = 0 \tag{20.34}$$

Since (20.19) implies $\delta\mathbf{e} = \delta\,\overset{*}{n}\mathbf{e}$ when t is constant, the virtual work takes on a simpler form

$$\left[\mathbf{u}^T \mathbf{k} + \sum_{n=1}^{N} \overset{*}{n}\mathbf{u}^T\, {}^*\mathbf{k} \right] \delta\mathbf{u} - \delta W = 0 \tag{20.35}$$

This implies that at any given time the element equilibrium equations are

$$\mathbf{k}\mathbf{u} + \sum_{n=1}^{N} {}^*\mathbf{k}\,\overset{*}{n}\mathbf{u} = \mathbf{f} \qquad \text{for each element} \tag{20.36}$$

where $\mathbf{f}$ denotes the element consistent load vector due to the external loading. Introducing equations (20.31) and (20.32) into the differential equations for the strain variables in (20.19) yields

$$\frac{d\,\overset{*}{n}\mathbf{u}}{dt} + \frac{\overset{*}{n}\mathbf{u}}{\tau_n} = \frac{d\mathbf{u}}{dt} \qquad \text{for each } n \tag{20.37}$$

The global equilibrium equations are determined by the standard assembly of the element equations, (20.36). Note, there is no assembly for (20.37). The variables $\overset{*}{n}\mathbf{u}$ are independent from element to element (recall, these variables carry the time dependent information for the material within the element). The global equilibrium equations at a given time are

$$\mathbf{K}\mathbf{u}_g = \mathbf{F}_{mech} - \mathbf{F}_{visc} \tag{20.38}$$

where $\mathbf{u}_g$ denotes the global nodal variable vector, $\mathbf{K}$ is the elastic stiffness matrix, $\mathbf{F}_{mech}$ is the global force vector due to mechanical loads, and $\mathbf{F}_{visc}$ is the assembled vector for $\sum_{n=1}^{N} {}^*\mathbf{k}_n^* \mathbf{u}$. The viscoelastic problem is solved by simultaneously integrating the differential and algebraic equations expressed by equations (20.37) and (20.38), where the latter is subject to the appropriate boundary restraints.

As far as the finite element implementation is concerned, a conventional linear elastic code can be readily adapted to perform the viscoelastic analysis for a Maxwell material, i.e., for a material whose relaxation stiffness coefficients can be modeled with a Prony series. First, the instantaneous stiffness coefficients, C_{ijkl}^*, are used in place of the elastic values to compute the element viscous stiffness matrices, ${}^*\mathbf{k}$, which are stored for repeated use. The internal nodal variables for each element, ${}_n^*\mathbf{u}$, are set equal to their initial values and stored. A predictor-corrector algorithm is then used to integrate equations (20.37) and (20.38) in time. The predictor-corrector integration algorithm used in this effort is described in the Appendix.

20.5   APPLICATIONS

Numerical solutions representative of stress relaxation, creep, and cyclic creep for a thick orthotropic beam are presented. The beam elastic stiffness coefficients ($\mathbf{C}$ matrix in (20.18)) for the state of plane stress can be written in terms of engineering material constants as

$$C_{11} = \frac{E_x}{1 - \nu_{xz}\nu_{zx}}, C_{33} = \frac{E_z}{1 - \nu_{xz}\nu_{zx}}, C_{13} = \nu_{xz}C_{33}, C_{55} = G_{xz}$$

A unidirectional E-glass/epoxy laminate is considered for which the material constants are: $E_x = 38.6$ GPa, $E_z = 8.27$ GPa, $G_{xz} = 4.14$ GPa, $\nu_{xz} = 0.26$, and $\nu_{zx} = \nu_{xz}E_x/E_z$. The viscous relaxation properties were computed from complex modulus vs. frequency data for the E-glass/epoxy reported in [GP76]. The equations for the real and imaginary parts of the modulus of a Maxwell series were least-squares fit to the data in a frequency range of 45 Hz–145 Hz. The least squares fit was performed with the constraint that the moduli in the Maxwell series each be positive. The series was defined with ten time constants; $\tau_n = 1.0\text{E}{-}4, 3.162\text{E}{-}4, 1.0\text{E}{-}3, 3.162\text{E}{-}3, 1.0\text{E}{-}2,$ ..., $1.0\text{E}{+}1$, and infinity. The Prony series was scaled so that its equivalent static value (at t equal to infinity) was unity. The resulting series is

$$P(t) = 1.0 + 0.01755e^{-0.0001t} + 0.000257e^{-0.01t} + 0.072014e^{-0.3162t}$$

The time dependent stiffness values for the E-glass/epoxy are given by $\mathbf{C}^\nu = \mathbf{C}P(t)$. The beam dimensions are as follows: $L = 0.1m$, $2h = 0.02m$, and $b = 0.01m$ (refer to Figure 1).

Example 1. A cantilever beam with w, u, θ fixed at point A has a prescribed deflection w at point B that is ramped from 0 to -1 cm in 0.05 sec and then held constant at -1 cm. Figure 3 depicts the value of the maximum axial stress computed at point D as a function of time. Also shown are the elastic and viscous stress components comprising the total stress. The decay of the total viscoelastic stress to its elastic value as time is increased demonstrates the expected step-strain relaxation behavior.

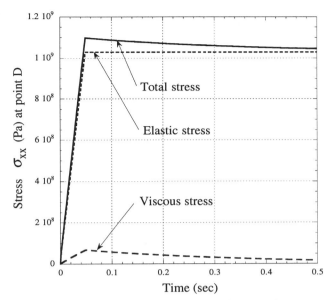

Figure 3   Cantilevered beam under prescribed tip deflection. Stress in top surface at
support.

Example 2. A simply-supported beam with w and u fixed at point A and w fixed
at point B is subject to a uniform, top-surface pressure, $q^+(t)$. The time-dependent
value of the pressure is ramped from 0.0 to 1.0 MPa in 0.05 sec and then held constant.
Figure 4 shows the value of the w deflection at the midspan of the beam. The viscous
solution shows the expected creep response.

Example 3. The simply-supported beam in the preceding example is subject to a
harmonic pressure loading given by

$$q^+(t) = 0, \quad t \leq 0 \text{ and } q^+(t) = 0.05*(1 - \cos(50\pi t))\text{MPa}, \quad t > 0$$

Figure 5 depicts the transverse normal stress and strain values at the midspan (point
C) versus time. The upward drift of the maximum and minimum values of the strain
demonstrates cyclic creep.

20.6   CONCLUSION

A differential form of the Maxwell viscous solid constitutive theory has been derived
and implemented within a higher-order-theory beam finite element. The finite element
formulation is attractive for several reasons: (1) The constitutive constants are the
same as those needed in the classical history-integral model, and they are also readily
available from step-strain relaxation tests, (2) The state variables are conjugate to
the elastic strain measures; hence, they are consistent with the kinematic assumptions
of the elastic formulation, (3) The update of the state variables can be performed in
a parallel computing environment, allowing the viscous force vector in the equations

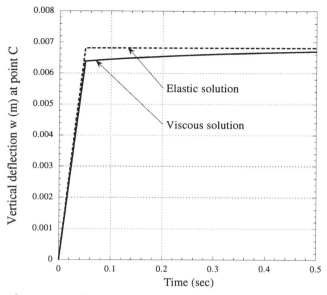

Figure 4   Simply-supported beam under prescribed pressure loading. Weighted-average
deflection, w, at center of beam.

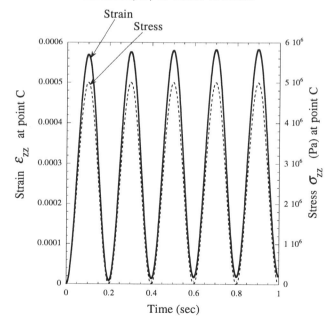

Figure 5   Simply-supported beam under harmonic pressure loading. Transverse normal
strain and stress at beam center versus time.

of motion to be determined efficiently within the predictor-corrector algorithm, (4) Applications of time-dependent displacements and loads are performed within the same finite element algorithm, and (5) The higher-order beam theory accounts for both transverse shear and transverse normal deformations—the effects that need to be accounted for in thick and highly orthotropic beams and high-frequency dynamics.

The computational examples for the problems of relaxation, creep, and cyclic creep clearly demonstrated the predictive capabilities of the finite element formulation.

20.7 ACKNOWLEDGMENT

The first author is grateful to the United States Army Research Standardization Group, Europe for supporting his participation in MAFELAP96 and to the Army Research Laboratory at the NASA Langley Research Center for supporting this research.

References

[ADdS91] Argyris J., Doltsinis I. S., and da Silva V. D. (1991) Constitutive modelling and computation of non-linear viscoelastic solids, Part I. Rheological models and numerical integration techniques. *Comput. Methods Appl. Mech. Engrg.* 88: 135–163.

[CN61] Coleman B. D. and Noll W. (1961) Foundations of linear viscoelasticity. *Reviews of Modern Physics* 33(2): 239–249.

[GP76] Gibson R. F. and Plunkett R. (1976) Dynamic mechanical behavior of fiber-reinforced composites: Measurement and analysis. *J. Composite Materials* 10: 325–341.

[Has70a] Hashin Z. (1970) Complex moduli of viscoelastic composites—I. General theory and applications to particulate composites. *Int. J. Solids Structures* 6: 539–552.

[Has70b] Hashin Z. (1970) Complex moduli of viscoelastic composites—II. Fiber reinforced materials. *Int. J. Solids Structures* 6: 797–807.

[HP68] Halpin J. C. and Pagano N. J. (1968) Observations on linear anisotropic viscoelasticity. *J. Composite Materials* 2(1): 68–80.

[JS93] Johnson A. R. and Stacer R. G. (1993) Rubber viscoelasticity using the physically constrained system's stretches as internal variables. *Rubber Chemistry and Technology* 66(4): 567–577.

[JTM95a] Johnson A. R., Tanner J. A., and Mason A. J. (1995) A kinematically driven anisotropic viscoelastic constitutive model applied to tires. In *Computational Modeling of Tires*. NASA. compiled by A. K. Noor and J. A. Tanner.

[JTM95b] Johnson A. R., Tanner J. A., and Mason J. A. (1995) A viscoelastic model for tires analyzed with nonlinear shell elements. University of Akron. presented at the Fourteenth Annual Meeting and Conference on Tire Science and Technology.

[Sch74]   Schapery R. A. (1974) Viscoelastic behavior and analysis of composite materials. In Sendeckyj G. P. (ed) *Composite Materials*, volume 2. Academic Press.

[SWW94] Shaw S., Warby M. K., and Whiteman J. R. (1994) Numerical techniques for problems of quasistatic and dynamic viscoelasticity. In *The Mathematics of Finite Elements and Applications*. John Wiley & Sons. edited by J. R. Whiteman.

[Tes91]   Tessler A. (1991) A two-node beam element including transverse shear and transverse normal deformations. *Int. J. for Numer. Methods Eng.* 32: 1027–1039.

20.8   APPENDIX

The predictor-corrector integration algorithm is described below. The storage requirements, beyond the requirements for the elastic problem, involves two sets of vectors, ${}^*_n\mathbf{u}$, for each element, two global vectors, $\mathbf{u}_g$, a global force vector, $\mathbf{F}_{visc}$, a full set of element viscous stiffness matrices, ${}^*\mathbf{k}$, and a matrix which carries the Prony series information for each element. The time integration algorithm used in this effort provides a trapezoidal solution to the differential equations while simultaneously solving the algebraic equilibrium equations. The elastic (static) finite element code already contains the assembly and solver subroutines needed. Building the viscous (quasi-static) finite element code only requires that additional storage be allotted and a few modified call statements be added. The algorithm is outlined below.

Time Integration Algorithm

A. Initialize variables $(1 \Rightarrow t, 2 \Rightarrow t + \Delta t)$

${}^*_n\mathbf{u}_2 = \mathbf{0} \; \forall$ elements.

$\mathbf{u}_{g,2} = \mathbf{0}$

$\beta_{1n} = \left(1 - \frac{\Delta t}{2\tau_n}\right) / \left(1 + \frac{\Delta t}{2\tau_n}\right)$ integration constant.

$\beta_{2n} = 1/ \left(1 + \frac{\Delta t}{2\tau_n}\right)$ integration constant.

B. Move data to next time step.

${}^*_n\mathbf{u}_1 = {}^*_n\mathbf{u}_2 \; \forall$ elements.

$\mathbf{F}_{mech,1} = \mathbf{F}_{mech,2}$

$\mathbf{F}_{visc,1} = \mathbf{F}_{visc,2}$

$\mathbf{u}_{g,1} = \mathbf{u}_{g,2}$

$t = t + \Delta t$

$\mathbf{F}_{mech,2} = \mathbf{F}_{mech}(t)$

$$\mathbf{u}_{g,2} = \mathbf{K}^{-1}[\mathbf{F}_{mech,2} - \mathbf{F}_{visc,1}] \text{ seed for Step C.}$$

C. Update internal variables (element components of global vectors implied).
$$\overset{*}{_n}\mathbf{u}_2 = \beta_{1n}\overset{*}{_n}\mathbf{u}_1 + \beta_{2n}(\mathbf{u}_{g,2} - \mathbf{u}_{g,1})$$

D. Update viscous forces.
$$\mathbf{F}_{visc,2} \text{ assemble } \sum_{n=1}^{N} {}^{*}\mathbf{k}\,\overset{*}{_n}\mathbf{u}_2$$

E. Update global displacements.
$$\mathbf{u}_{g,2} = \mathbf{K}^{-1}[\mathbf{F}_{mech,2} - \mathbf{F}_{visc,2}]$$

F. If changes to $\mathbf{u}_{g,2}$ are small then go to Step B.
Else go to Step C.

21

A Block Red-Black SOR Method for a Two-Dimensional Parabolic Equation Using Hermite Collocation

Stephen H. Brill[1] and George F. Pinder[2]

[1]Department of Mathematics and Statistics
University of Vermont
Burlington, Vermont 05405
U. S. A.

[2]Department of Civil and Environmental Engineering
University of Vermont
Burlington, Vermont 05405
U. S. A.

21.1 INTRODUCTION

In [LHHR95], Lai *et al.* study a block Jacobi method for solving the two-dimensional Poisson equation

$$\nabla^2 u = \frac{\partial^2 u}{\partial x^2} + \frac{\partial^2 u}{\partial y^2} = H(x, y) \tag{21.1}$$

defined on the interior of the unit square $\mathcal{S} = [0, 1] \times [0, 1]$, discretized by the collocation method with a uniform mesh, given Dirichlet boundary conditions

$$u(x, y) = C(x, y), \qquad (x, y) \in \partial \mathcal{S}. \tag{21.2}$$

They determine eigenvalues for the iteration matrix of their block Jacobi method and then use the theory in [You71] to determine a formula for ω_{opt}, the optimal relaxation factor ω for the block SOR method associated with their block Jacobi scheme. In this paper, we explain how to extend their work to ensure that the optimal SOR method is parallelizable by using a red-black ordering scheme. We then use these ideas to

The Mathematics of Finite Elements and Applications
Edited by J. R. Whiteman © 1997 John Wiley & Sons Ltd.

efficiently solve the two-dimensional parabolic equation

$$\frac{\partial u}{\partial t} - \frac{\partial^2 u}{\partial x^2} - \frac{\partial^2 u}{\partial y^2} = -H(x, y, t)$$

with Dirichlet boundary conditions and suitable initial conditions.

21.2   HERMITE CUBIC POLYNOMIALS

21.2.1   ONE-DIMENSIONAL FORMULATION

Let $u(x)$ be a function defined on the interval $[0, 1]$. Partition the interval using n equally spaced nodes $0 = x_0, x_1, \ldots, x_m = 1$, where $m = n - 1$ is the number of elements. Let $h = 1/m$.

Consider the functions (cf. [Pic94])

$$f_j(x) = \begin{cases} \dfrac{(x - x_{j-1})^2}{h^3}[2(x_j - x) + h], & x_{j-1} \leq x \leq x_j \\ \dfrac{(x_{j+1} - x)^2}{h^3}[3h - 2(x_{j+1} - x)], & x_j \leq x \leq x_{j+1} \\ 0, & \text{otherwise} \end{cases}$$

and

$$g_j(x) = \begin{cases} \dfrac{(x - x_{j-1})^2(x - x_j)}{h^2}, & x_{j-1} \leq x \leq x_j \\ \dfrac{(x_{j+1} - x)^2(x - x_j)}{h^2}, & x_j \leq x \leq x_{j+1} \\ 0, & \text{otherwise} \end{cases}$$

These are the Hermite cubic polynomials that we use as basis functions in our collocation approach. Notice that

$$f_j(x_i) = \delta_{i,j}$$

$$\frac{df_j}{dx}(x_i) = f'_j(x_i) = 0 \quad \forall i, j$$

$$g_j(x_i) = 0 \quad \forall i, j$$

$$\frac{dg_j}{dx}(x_i) = g'_j(x_i) = \delta_{i,j},$$

where $\delta_{i,j}$ is the Kronecker symbol.

Let $u_j = u(x_j)$ and let $u'_j = u'(x_j) = \frac{du}{dx}(x_j)$ for $j = 0, 1, \ldots, m$. Then the cubic polynomial interpolating the u_j's and the u'_j's is

$$\hat{u}(x) = \sum_{j=0}^{m} (u_j f_j(x) + u'_j g_j(x)). \tag{21.3}$$

In [Pap83], Papatheodorou uses $g^\star_j(x) = \frac{g_j(x)}{h}$ (in place of $g_j(x)$) when forming (21.3). He makes this choice (also used in [LHHR95]) because eigenvalue analysis is much easier using $g^\star_j(x)$ instead of $g_j(x)$. It is easily seen that the iteration matrices studied herein that one obtains using $g_j(x)$ and $g^\star_j(x)$ are identical. In this paper, we use $g_j(x)$ in the computer code that generates the numerical results and employ $g^\star_j(x)$ for our analysis.

21.2.2 TWO-DIMENSIONAL FORMULATION

Let $u(x, y)$ be a function defined on S. Partition ∂S by using n equally spaced nodes in both the x- and y-directions. Letting $m = n - 1$ and $h = 1/m$, we partition S into m^2 square elements, where the dimensions of each element are $h \times h$. If we consider two-dimensional bi-cubic Hermite basis polynomials, we obtain, by analogy to (21.3):

$$\hat{u}(x, y) = \sum_{q=0}^{m} \sum_{r=0}^{m} [u_{qr} f_q(x) f_r(y) + u^x_{qr} g_q(x) f_r(y) + u^y_{qr} f_q(x) g_r(y) + u^{xy}_{qr} g_q(x) g_r(y)],$$

$$(21.4)$$

where

$$u_{qr} = u(x_q, y_r)$$

$$u^x_{qr} = \frac{\partial u}{\partial x}(x_q, y_r)$$

$$u^y_{qr} = \frac{\partial u}{\partial y}(x_q, y_r)$$

$$u^{xy}_{qr} = \frac{\partial^2 u}{\partial x \partial y}(x_q, y_r).$$

We see that $\hat{u}(x, y)$ interpolates the functions u, $\frac{\partial u}{\partial x}$, $\frac{\partial u}{\partial y}$, and $\frac{\partial^2 u}{\partial x \partial y}$ at the grid points (x_q, y_r), for $q, r = 0, \ldots, m$.

21.3 COLLOCATION DISCRETIZATION OF THE PDE

If the interpolating polynomial (21.4) is introduced into the governing equation (21.1), we obtain

$$\frac{\partial^2 \hat{u}}{\partial x^2} + \frac{\partial^2 \hat{u}}{\partial y^2} - H(x, y) = E(x, y),$$

where $E(x, y)$ is an error function.

We see that at each of the n^2 grid points (x_q, y_r), we have four degrees of freedom, namely u_{qr}, u^x_{qr}, u^y_{qr}, and u^{xy}_{qr}. However, on the boundary ∂S, many of these values are known. In particular, we know (from (21.2))

$$u_{qr} = u(x_q, y_r) = C(x_q, y_r)$$

for all nodes (grid points) on ∂S. In addition, we can calculate

$$u^x_{qr} = \frac{\partial u}{\partial x}(x_q, y_r) = \frac{\partial C}{\partial x}(x_q, y_r)$$

on the north and south boundaries and

$$u^y_{qr} = \frac{\partial u}{\partial y}(x_q, y_r) = \frac{\partial C}{\partial y}(x_q, y_r)$$

on the east and west boundaries. We therefore know the values of a total of $8n - 4$ degrees of freedom and do not know the values of $4m^2$ degrees of freedom. Therefore, to uniquely determine these $4m^2$ degrees of freedom we require $4m^2$ equations, or 4 equations per element. To achieve this, we choose four points (x_k, y_ℓ) in the interior of each element and enforce $E(x_k, y_\ell) = 0$ at each of these $4m^2$ "collocation points". It is known (from [Cel83]) that the optimal choices for the collocation points for the symmetric differential operator given in (21.1) are the so-called "Gauss points". On the interval $[-1, 1]$, the Gauss points are $\pm z$, where $z = 3^{-1/2}$. On the square element $[-1, 1] \times [-1, 1]$, the Gauss points are $(-z, -z)$, $(-z, z)$, $(z, -z)$, and (z, z). Transforming these four Gauss points into each of the m^2 elements of our mesh defines the full set of $4m^2$ "collocation" equations. These can be written

$$\sum_{q=0}^{m} \sum_{r=0}^{m} \{[f_q''(x_k)f_r(y_\ell) + f_q(x_k)f_r''(y_\ell)]u_{qr} + [g_q''(x_k)f_r(y_\ell) + g_q(x_k)f_r''(y_\ell)]u_{qr}^x$$

$$+[f_q''(x_k)g_r(y_\ell) + f_q(x_k)g_r''(y_\ell)]u_{qr}^y + [g_q''(x_k)g_r(y_\ell) + g_q(x_k)g_r''(y_\ell)]u_{qr}^{xy}\}$$

$$= H(x_k, y_\ell), \quad (21.5)$$

where (x_k, y_ℓ) varies over all $4m^2$ collocation points.

There are many ways in which to number the unknowns and equations. Each numbering system will define a different structure for the matrix arising from the system of linear equations (21.5) that we must solve. We use a numbering proposed by [Cel83] and by [LHHR95], which is depicted pictorially in Figure 1 for the case of $n = 5$. In the figure, h_{ij} indicates the approximate location of collocation point (x_j, y_i). It is seen that the matrix equation that arises from this numbering for $n = 5$ is

$$\widetilde{M}\widetilde{\mathbf{v}} = \widetilde{\mathbf{k}}, \quad (21.6)$$

where

$$\widetilde{M} = \begin{bmatrix} A_2 & A_3 & -A_4 & & & & & \\ A_4 & A_1 & -A_2 & & & & & \\ & & A_1 & A_2 & A_3 & -A_4 & & \\ & & A_3 & A_4 & A_1 & -A_2 & & \\ & & & & A_1 & A_2 & A_3 & -A_4 \\ & & & & A_3 & A_4 & A_1 & -A_2 \\ & & & & & & A_1 & A_2 & -A_4 \\ & & & & & & A_3 & A_4 & -A_2 \end{bmatrix},$$

$$\widetilde{\mathbf{v}} = \begin{bmatrix} \mathbf{v}_0^T & \mathbf{v}_1^T & \mathbf{v}_2^T & \mathbf{v}_3^T & \mathbf{v}_4^T & \mathbf{v}_5^T & \mathbf{v}_6^T & \mathbf{v}_7^T \end{bmatrix}^T,$$

$$\widetilde{\mathbf{k}} = \begin{bmatrix} \mathbf{k}_0^T & \mathbf{k}_1^T & \mathbf{k}_2^T & \mathbf{k}_3^T & \mathbf{k}_4^T & \mathbf{k}_5^T & \mathbf{k}_6^T & \mathbf{k}_7^T \end{bmatrix}^T.$$

The vectors $\mathbf{v}_i$ and $\mathbf{k}_i$ are given by

$$\mathbf{v}_i = \begin{bmatrix} v_{i0} & v_{i1} & v_{i2} & v_{i3} & v_{i4} & v_{i5} & v_{i6} & v_{i7} \end{bmatrix}^T,$$

$$\mathbf{k}_i = \begin{bmatrix} k_{i0} & k_{i1} & k_{i2} & k_{i3} & k_{i4} & k_{i5} & k_{i6} & k_{i7} \end{bmatrix}^T,$$

where $k_{ij} = H(x_j, y_i) - (BV_{ij})$. Here BV_{ij} indicates any known boundary value information that appears on the left side of (21.5) that is pertinent to the equation defined at collocation point (x_j, y_i). It is clear that BV_{ij} may be non-zero only when (x_j, y_i) is in a boundary element.

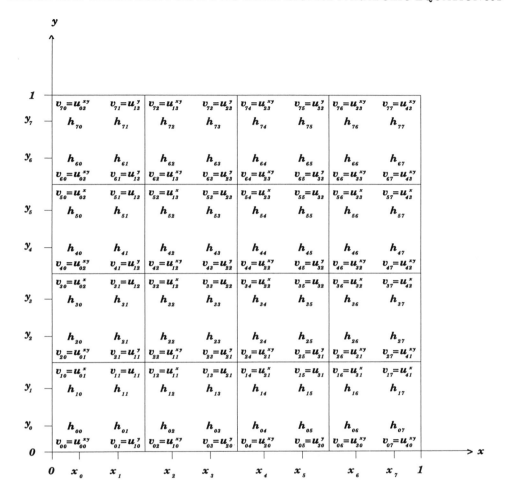

Figure 1 numbering of equations and unknowns

The submatrices A_i, $i = 1, 2, 3, 4$, all have the structure:

$$
A_i =
\begin{bmatrix}
a_{i2} & a_{i3} & -a_{i4} & & & & & \\
a_{i4} & a_{i1} & -a_{i2} & & & & & \\
a_{i1} & a_{i2} & a_{i3} & -a_{i4} & & & & \\
a_{i3} & a_{i4} & a_{i1} & -a_{i2} & & & & \\
& & a_{i1} & a_{i2} & a_{i3} & -a_{i4} & & \\
& & a_{i3} & a_{i4} & a_{i1} & -a_{i2} & & \\
& & & & a_{i1} & a_{i2} & -a_{i4} & \\
& & & & a_{i3} & a_{i4} & -a_{i2} &
\end{bmatrix}.
$$

Although the above example is for $n = 5$, it should be clear how the corresponding matrices and vectors would appear for different values of n.

It is seen in [LHHR95] and [Pap83] that $a_{ij} = \frac{\bar{a}_{ij}}{9h^2}$, where

$$
\begin{array}{llll}
\bar{a}_{11} = -24 - 18\sqrt{3}, & \bar{a}_{12} = -12 - 8\sqrt{3}, & \bar{a}_{13} = 24, & \bar{a}_{14} = 3 + \sqrt{3}, \\
\bar{a}_{21} = -12 - 8\sqrt{3}, & \bar{a}_{22} = -3 - 2\sqrt{3}, & \bar{a}_{23} = 3 - \sqrt{3}, & \bar{a}_{24} = 0, \\
\bar{a}_{31} = 24, & \bar{a}_{32} = 3 - \sqrt{3}, & \bar{a}_{33} = -24 + 18\sqrt{3}, & \bar{a}_{34} = -12 + 8\sqrt{3}, \\
\bar{a}_{41} = 3 + \sqrt{3}, & \bar{a}_{42} = 0, & \bar{a}_{43} = -12 + 8\sqrt{3}, & \bar{a}_{44} = -3 + 2\sqrt{3}.
\end{array}
$$

$$(21.7)$$

21.4 BLOCK JACOBI METHOD FOR POISSON'S EQUATION

To begin, the matrix $\widetilde{M}$ is partitioned into

$$
\widetilde{M} =
\left[
\begin{array}{cc|cc|cc|cc}
A_2 & A_3 & -A_4 & & & & & \\
A_4 & A_1 & -A_2 & & & & & \\
& A_1 & A_2 & A_3 & -A_4 & & & \\
& A_3 & A_4 & A_1 & -A_2 & & & \\
& & & A_1 & A_2 & A_3 & -A_4 & \\
& & & A_3 & A_4 & A_1 & -A_2 & \\
& & & & & A_1 & A_2 & -A_4 \\
& & & & & A_3 & A_4 & -A_2
\end{array}
\right],
$$

which we write more concisely as

$$
\widetilde{M} =
\left[
\begin{array}{cccc}
A_F & B_F & & \\
C_F & A & B & \\
& C & A & B \\
& & C & A & B_L \\
& & & C_L & A_L
\end{array}
\right].
\tag{21.8}
$$

The block Jacobi method is then defined by

$$
\widetilde{D}\widetilde{\mathbf{v}}^{(p+1)} = (\widetilde{L} + \widetilde{U})\widetilde{\mathbf{v}}^{(p)} + \widetilde{\mathbf{k}},
\tag{21.9}
$$

where $\widetilde{\mathbf{v}}^{(p)}$ is the approximation to $\widetilde{\mathbf{v}}$ after p iterations and where $\widetilde{M}$ is split into $\widetilde{M} = \widetilde{D} - \widetilde{L} - \widetilde{U}$, where

$$
\widetilde{D} =
\left[
\begin{array}{ccccc}
A_F & & & & \\
& A & & & \\
& & A & & \\
& & & A & \\
& & & & A_L
\end{array}
\right],
$$

$$
-\widetilde{L} =
\left[
\begin{array}{ccccc}
C_F & & & & \\
& C & & & \\
& & C & & \\
& & & C_L &
\end{array}
\right],
$$

and

$$-\tilde{U} = \begin{bmatrix} B_F & & & \\ & B & & \\ & & B & \\ & & & B_L \end{bmatrix}.$$

We solve (21.9) for $\tilde{\mathbf{v}}^{(p+1)}$ as follows. First, we note that each of the $m+1$ rows of $\tilde{D}$ in (21.9) defines a matrix equation, which is entirely decoupled from the rest. Hence, each of these $m+1$ matrix equations may be solved simultaneously in parallel. We see that each of these equations is of the form

$$\check{A}\check{\mathbf{v}} = \check{\mathbf{k}},$$

where $\check{A} = A_F, A_L$ or A; $\check{\mathbf{v}} =$ the corresponding vector of unknowns; and $\check{\mathbf{k}} =$ the corresponding right-hand-side vector of known values. For the case where $\check{A} = A_F$ or A_L, it is clear that $\check{A}$ is block tridiagonal, with the blocks being 2×2 matrices. For example, consider

$$\check{A} = A_F = A_2 = \begin{bmatrix} a_{22} & a_{23} & -a_{24} & & & & & \\ a_{24} & a_{21} & -a_{22} & & & & & \\ & a_{21} & a_{22} & a_{23} & -a_{24} & & & \\ & a_{23} & a_{24} & a_{21} & -a_{22} & & & \\ & & & a_{21} & a_{22} & a_{23} & -a_{24} & \\ & & & a_{23} & a_{24} & a_{21} & -a_{22} & \\ & & & & & a_{21} & a_{22} & -a_{24} \\ & & & & & a_{23} & a_{24} & -a_{22} \end{bmatrix}.$$

We employ a direct block tridiagonal solver to obtain $\check{\mathbf{v}}_k$.

The case where $\check{A} = A$ is just slightly more complicated. Here we see that

$$\check{A} = \begin{bmatrix} A_1 & -A_2 \\ A_1 & A_2 \end{bmatrix},$$

which has the structure

$$\check{A} = \begin{bmatrix} \times & \times & \times & & & & & & \times & \times & \times & & & & & \\ \times & \times & \times & & & & & & \times & \times & \times & & & & & \\ & \times & \times & \times & \times & & & & & \times & \times & \times & \times & & & \\ & \times & \times & \times & \times & & & & & \times & \times & \times & \times & & & \\ & & & \times & \times & \times & \times & & & & & \times & \times & \times & \times & \\ & & & \times & \times & \times & \times & & & & & \times & \times & \times & \times & \\ & & & & & \times & \times & \times & & & & & & \times & \times & \times \\ & & & & & \times & \times & \times & & & & & & \times & \times & \times \\ \times & \times & \times & & & & & & \times & \times & \times & & & & & \\ \times & \times & \times & & & & & & \times & \times & \times & & & & & \\ & \times & \times & \times & \times & & & & & \times & \times & \times & \times & & & \\ & \times & \times & \times & \times & & & & & \times & \times & \times & \times & & & \\ & & & \times & \times & \times & \times & & & & & \times & \times & \times & \times & \\ & & & \times & \times & \times & \times & & & & & \times & \times & \times & \times & \\ & & & & & \times & \times & \times & & & & & & \times & \times & \times \\ & & & & & \times & \times & \times & & & & & & \times & \times & \times \end{bmatrix}.$$

Permuting the rows and columns of $\check{A}$ via a similarity transformation (see [LHHR95]) gives

$$
A' = \left[
\begin{array}{cccc|cccc|cccc|cccc}
\times & \times & \times & \times & \times & \times & & & & & & & & & & \\
\times & \times & \times & \times & \times & \times & & & & & & & & & & \\
\times & \times & \times & \times & \times & \times & & & & & & & & & & \\
\times & \times & \times & \times & \times & \times & & & & & & & & & & \\
\hline
 & & \times & \times & \times & \times & \times & \times & \times & \times & & & & & & \\
 & & \times & \times & \times & \times & \times & \times & \times & \times & & & & & & \\
 & & \times & \times & \times & \times & \times & \times & \times & \times & & & & & & \\
 & & \times & \times & \times & \times & \times & \times & \times & \times & & & & & & \\
\hline
 & & & & & & \times & \times & \times & \times & \times & \times & \times & \times & & \\
 & & & & & & \times & \times & \times & \times & \times & \times & \times & \times & & \\
 & & & & & & \times & \times & \times & \times & \times & \times & \times & \times & & \\
 & & & & & & \times & \times & \times & \times & \times & \times & \times & \times & & \\
\hline
 & & & & & & & & & & \times & \times & \times & \times & \times & \times \\
 & & & & & & & & & & \times & \times & \times & \times & \times & \times \\
 & & & & & & & & & & \times & \times & \times & \times & \times & \times \\
 & & & & & & & & & & \times & \times & \times & \times & \times & \times \\
\end{array}
\right],
$$

which is clearly block tridiagonal, with the blocks being 4×4 matrices. Obviously, we must also permute correspondingly the entries of $\check{\mathbf{v}}$ (giving $\mathbf{v}'$) and those of $\check{\mathbf{k}}$ (giving $\mathbf{k}'$). We then employ a direct block tridiagonal solver on $A'\mathbf{v}' = \mathbf{k}'$ to obtain $\mathbf{v}'$.

21.5   RED-BLACK SOR FOR POISSON'S EQUATION

While the equations in (21.9) may be solved simultaneously in parallel, we find that the rate at which the sequence $\{\tilde{\mathbf{v}}^{(p)}\}$ converges to $\tilde{\mathbf{v}}$ is unacceptably slow. This motivates us to seek a method with a faster convergence rate that can still take advantage of parallelism.

We recall (21.8):

$$
\widetilde{M} = \left[
\begin{array}{ccccc}
A_F & B_F & & & \\
C_F & A & B & & \\
 & C & A & B & \\
 & & C & A & B_L \\
 & & & C_L & A_L
\end{array}
\right]
\begin{array}{c}
0 \\ 1 \\ 2 \\ 3 \\ 4
\end{array},
\tag{21.10}
$$

where the last column gives the block row number of $\widetilde{M}$.

Via a similarity transformation, we permute the rows and columns of $\widetilde{M}$ (and correspondingly the entries of $\tilde{\mathbf{v}}$ and $\tilde{\mathbf{k}}$) in (21.6) and (21.10) to obtain

$$
M\mathbf{v} = \mathbf{k},
\tag{21.11}
$$

where

$$
M = \left[
\begin{array}{cc|cc}
A_F & & B_F & \\
 & A & C & B \\
 & & A_L & C_L \\
\hline
C_F & B & A & \\
 & C & B_L & A
\end{array}
\right.
\left|
\begin{array}{c}
0 \\
2 \\
4 \\
1 \\
3
\end{array}
\right] .
$$

More precisely, M is obtained from $\widetilde{M}$ by writing from top to bottom all the even numbered block rows of $\widetilde{M}$ (in ascending order), followed by all the odd numbered block rows of $\widetilde{M}$ (in ascending order). We abbreviate M as

$$
M = \left[
\begin{array}{cc}
D_R & M_U \\
M_L & D_B
\end{array}
\right] .
$$

Correspondingly, we write

$$
\mathbf{v} = \left[
\begin{array}{c}
\mathbf{v}_R \\
\mathbf{v}_B
\end{array}
\right] \text{ and } \mathbf{k} = \left[
\begin{array}{c}
\mathbf{k}_R \\
\mathbf{k}_B
\end{array}
\right] .
$$

Analogously to (21.9), we split M into $M = D - L - U$, where

$$
D = \left[
\begin{array}{cc}
D_R & \\
 & D_B
\end{array}
\right], \quad -L = \left[
\begin{array}{cc}
 & \\
M_L &
\end{array}
\right], \quad \text{and } -U = \left[
\begin{array}{cc}
 & M_U \\
 &
\end{array}
\right] .
$$

Then the standard block SOR formulation is

$$
(D - \omega L)\mathbf{v}^{(p+1)} = [(1 - \omega)D + \omega U]\mathbf{v}^{(p)} + \omega \mathbf{k}, \tag{21.12}
$$

where the *relaxation factor* ω is chosen such that $1 < \omega < 2$. Dividing (21.12) into its red (top) and black (bottom) parts, we obtain

$$
D_R \mathbf{v}_R^{(p+1)} = (1 - \omega)D_R \mathbf{v}_R^{(p)} - \omega M_U \mathbf{v}_B^{(p)} + \omega \mathbf{k}_R \tag{21.13}
$$

and

$$
\omega M_L \mathbf{v}_R^{(p+1)} + D_B \mathbf{v}_B^{(p+1)} = (1 - \omega)D_B \mathbf{v}_B^{(p)} + \omega \mathbf{k}_B. \tag{21.14}
$$

We now introduce the vectors

$$
\mathbf{z}_c^{(p+1)} = \mathbf{v}_c^{(p+1)} - \mathbf{v}_c^{(p)},
$$

where the color subscript $c = R$ or B. We also introduce the *color dependent residual vectors* $\mathbf{r}_c^{(a,b)}$, defined as

$$
\mathbf{r}_R^{(a,b)} = \mathbf{k}_R - \left(D_R \mathbf{v}_R^{(a)} + M_U \mathbf{v}_B^{(b)} \right)
$$

and

$$
\mathbf{r}_B^{(a,b)} = \mathbf{k}_B - \left(M_L \mathbf{v}_R^{(a)} + D_B \mathbf{v}_B^{(b)} \right),
$$

where the superscripts (a) and (b) denote iteration level. By considering (21.11), it is clear that these residual vectors measure how close the approximants $\mathbf{v}_R^{(a)}$ and $\mathbf{v}_B^{(b)}$ are to $\mathbf{v}_R$ and $\mathbf{v}_B$, components of the true solution of (21.11). Algebraically manipulating (21.13) and (21.14) and using the notation introduced above yields

$$D_R \mathbf{z}_R^{(p+1)} = \omega \mathbf{r}_R^{(p,p)} \tag{21.15}$$

and

$$D_B \mathbf{z}_B^{(p+1)} = \omega \mathbf{r}_B^{(p+1,p)}, \tag{21.16}$$

which are of a form and structure very similar to that of (21.9). In the SOR algorithm, we compute $\mathbf{v}^{(p+1)}$ using (21.15) and (21.16).

It is clear that we have still maintained a high degree of parallelism by using this red-black SOR scheme. Evidently, all of the red equations in (21.13) may be solved simultaneously in parallel. Once we have obtained $\mathbf{v}_R^{(p+1)}$ from (21.13), we may solve all the black equations in (21.14) simultaneously in parallel, obtaining $\mathbf{v}_B^{(p+1)}$.

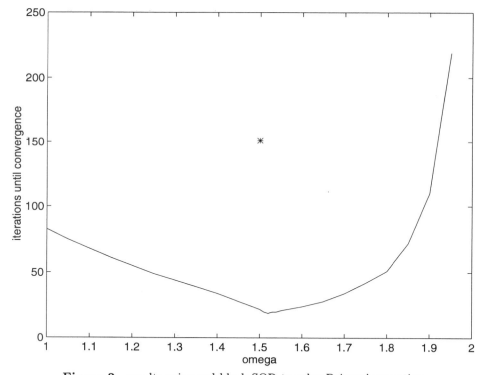

Figure 2   results using red-black SOR to solve Poisson's equation

Numerical results are illustrated in Figure 2. We ran our version of the algorithm for both the Jacobi method and the red-black SOR method using various values of ω. We chose $m = 10$ and the boundary conditions and the function $H(x, y)$ such that $u = x^3 \sin \pi y$. Letting $\mathbf{r}^{(p,p)} = \left[\left(\mathbf{r}_R^{(p,p)} \right)^T \quad \left(\mathbf{r}_B^{(p,p)} \right)^T \right]^T$, our convergence criterion was that $\|\mathbf{r}^{(p,p)}\|_\infty < 0.001$. The Jacobi method needed 151 iterations to converge, which is indicated by an asterisk in the middle of the graph. By comparison, for $\omega = 1.52$, the SOR method required only 19 iterations to converge. Indeed, according to the theory in [LHHR95] and [You71], the optimal ω for this problem is $\omega_{\mathrm{opt}} \approx 1.5068$, which agrees well with our numerical investigation.

21.6 THE PARABOLIC EQUATION

We now seek to solve the parabolic equation

$$\frac{\partial u}{\partial t} = \frac{\partial^2 u}{\partial x^2} + \frac{\partial^2 u}{\partial y^2} - H(x, y, t), \tag{21.17}$$

defined on the interior of $\mathcal{S}$, discretized by the collocation method with a uniform mesh, given Dirichlet boundary conditions and initial condition

$$u(x, y, t) = C(x, y, t), \qquad (x, y) \in \partial \mathcal{S}, \quad u(x, y, 0) = W(x, y).$$

We approximate the time derivative by

$$\frac{\partial u}{\partial t} = \frac{u^{(q+1)} - u^{(q)}}{\Delta t}, \tag{21.18}$$

where the superscript (q) indicates the value of u after q time steps.

Recalling (21.6), we see that matrix $\widetilde{M}$ was formed by evaluating $\frac{\partial^2 \hat{u}}{\partial x^2} + \frac{\partial^2 \hat{u}}{\partial y^2}$ at the collocation points. Correspondingly, we form matrix $\widetilde{P}$ by evaluating $\hat{u}$ at the collocation points. Clearly, $\widetilde{P}$ has precisely the same structure as that of $\widetilde{M}$. Letting p_{ij} be the non-trivial entries of $\widetilde{P}$ (just as the a_{ij}'s in (21.7) are the non-trivial entries of $\widetilde{M}$), we see that the numbers p_{ij} are given by

$$
\begin{array}{llll}
\bar{p}_{11} = 86 + 48\sqrt{3} & \bar{p}_{12} = 13 + 7\sqrt{3} & \bar{p}_{13} = 22 & \bar{p}_{14} = 5 + \sqrt{3} \\
\bar{p}_{21} = 13 + 7\sqrt{3} & \bar{p}_{22} = 2 + \sqrt{3} & \bar{p}_{23} = 5 - \sqrt{3} & \bar{p}_{24} = 1 \\
\bar{p}_{31} = 22 & \bar{p}_{32} = 5 - \sqrt{3} & \bar{p}_{33} = 86 - 48\sqrt{3} & \bar{p}_{34} = 13 - 7\sqrt{3} \\
\bar{p}_{41} = 5 + \sqrt{3} & \bar{p}_{42} = 1 & \bar{p}_{43} = 13 - 7\sqrt{3} & \bar{p}_{44} = 2 - \sqrt{3}
\end{array}
$$

If we now introduce (21.18) and the interpolating polynomial (21.4), where the interpolating polynomial (21.4) and forcing function now have time dependence, i.e. u_{qr}, u_{qr}^x, u_{qr}^y, u_{qr}^{xy}, and H are now functions also of t, into (21.17) and evaluate the right side of (21.17) at the collocation (Gauss) points at time $\theta\, t^{(q+1)} + (1 - \theta)\, t^{(q)}$, where $0 \le \theta \le 1$, then we obtain the matrix form of the collocation discretization of the parabolic PDE:

$$\frac{\widetilde{P}\widetilde{\mathbf{v}}^{(q+1)} - \widetilde{P}\widetilde{\mathbf{v}}^{(q)}}{\Delta t} = \theta\, [\widetilde{M}\widetilde{\mathbf{v}}^{(q+1)} - \widetilde{\mathbf{k}}^{(q+1)}] + (1 - \theta)\, [\widetilde{M}\widetilde{\mathbf{v}}^{(q)} - \widetilde{\mathbf{k}}^{(q)}]. \tag{21.19}$$

Letting $\tau = \theta\, \Delta t$ and $\bar{\tau} = (1 - \theta)\, \Delta t$, we may express (21.19) as

$$(\widetilde{P} - \tau\widetilde{M})\widetilde{\mathbf{v}}^{(q+1)} = (\widetilde{P} + \bar{\tau}\widetilde{M})\widetilde{\mathbf{v}}^{(q)} - (\tau\, \widetilde{\mathbf{k}}^{(q+1)} + \bar{\tau}\widetilde{\mathbf{k}}^{(q)}). \tag{21.20}$$

In examining (21.20), we see that this equation defines how we may move from time step (q) to time step $(q+1)$. In particular, at time step (q), all the vectors on the right side of (21.20) contain known values. Letting $\widetilde{Q} = (\widetilde{P} - \tau\widetilde{M})$ and $\widetilde{\mathbf{b}}^{(q)} = $ the right side of (21.20), we may write (21.20) as

$$\widetilde{Q}\widetilde{\mathbf{v}}^{(q+1)} = \widetilde{\mathbf{b}}^{(q)}, \qquad (21.21)$$

which is of a form and structure identical to those of (21.6). We may therefore apply to (21.21) the block red-black SOR algorithm that we developed for (21.6). That is, at each time step in (21.21) we iterate to convergence using block red-black SOR.

21.7   EIGENVALUES AND RESULTS

Using the work in [LHHR95] as a guide, we determined the eigenvalues of the block Jacobi matrix one would use to solve (21.21). These eigenvalues may be computed using the following recipe:

$$\theta_k = \frac{k\pi}{m}$$

$$c_k = \cos\theta_k$$

$$r_k = \sqrt{43 + 40c_k - 2c_k^2}$$

$$\alpha_k^\pm = \frac{(3-\sqrt{3})[(7-c_k)\gamma^2 + 24(8+c_k)\gamma - 288\sqrt{3}(3\sqrt{3}\pm r_k)]}{(3+\sqrt{3})(7-c_k)\gamma^2 + 24(45 + 29\sqrt{3} - 2\sqrt{3}c_k)\gamma + 1728(10 + 6\sqrt{3} - c_k)}$$

$$\beta_k^\pm = \frac{(19-9\sqrt{3})[11(7-c_k)\gamma^2 + 24(4+23c_k)\gamma + 864(-37 - 8c_k \pm 3\sqrt{3}r_k)]}{(169 + 69\sqrt{3})(7-c_k)\gamma^2 + 24(1247 + 993\sqrt{3} + 184c_k + 6\sqrt{3}c_k)\gamma + 1728(122 + 54\sqrt{3} - 59c_k)}$$

where $k = 1, \ldots, m-1$ and $\gamma = \frac{h^2}{\tau}$. Then form the sets

$$\{\lambda_1, \lambda_2, \ldots, \lambda_{2m}\} =$$
$$\left\{ \frac{(3-\sqrt{3})\gamma + 24(3 - 2\sqrt{3})}{(3+\sqrt{3})\gamma + 24(3 + 2\sqrt{3})}, \frac{(\sqrt{3} - 1)\gamma + 24(2\sqrt{3} - 3)}{(\sqrt{3} + 1)\gamma + 24(2\sqrt{3} + 3)}, \alpha_1^+, \alpha_1^-, \ldots, \alpha_{m-1}^+, \alpha_{m-1}^- \right\}$$

and

$$\{\bar{\lambda}_1, \bar{\lambda}_2, \ldots, \bar{\lambda}_{2m}\} =$$
$$\left\{ \frac{(9-4\sqrt{3})\gamma + 12(9 - 7\sqrt{3})}{(9+4\sqrt{3})\gamma + 12(9 + 7\sqrt{3})}, \frac{(-4+3\sqrt{3})\gamma + 36(-5 + 3\sqrt{3})}{(4+3\sqrt{3})\gamma + 36(5 + 3\sqrt{3})}, \beta_1^-, \beta_1^+, \ldots, \beta_{m-1}^-, \beta_{m-1}^+ \right\}.$$

Now let $\phi_{jk} = (\lambda_j - \bar{\lambda}_j)\, c_k$ for $k = 1, \ldots, m-1$ and $j = 1, \ldots, 2m$. Then $\sigma(J)$, the set of eigenvalues of the block Jacobi matrix, is (cf. [LHHR95])

$$\sigma(J) = \{\mu : \mu = \pm\lambda_j, \ j = 1, \ldots, 2m\}$$

$$\cup \left\{ \mu : \mu = \frac{1}{2}\left(\phi_{jk} \pm \sqrt{\phi_{jk}^2 + 4\lambda_j\bar{\lambda}_j}\right), \ j = 1, \ldots, 2m, \ k = 1, \ldots, m-1 \right\}.$$

Given this recipe for the computation of eigenvalues of the Jacobi matrix, one can use the theory in [LHHR95] and [You71] to compute ω_{opt} for the optimal block SOR method. It can also be shown that all these eigenvalues must have modulus less than unity, irrespective of the value of γ. Thus, the Jacobi method for the model parabolic problem must converge for any γ.

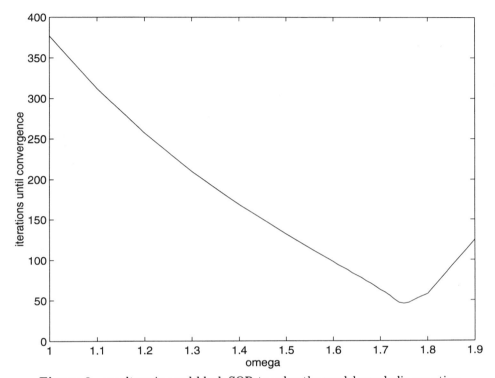

Figure 3 results using red-black SOR to solve the model parabolic equation

For an example, we ran both the block Jacobi method and block SOR method for various values of ω on the parabolic problem. The boundary and initial conditions and function $H(x, y, t)$ were chosen such that $u = x^2 y^3 (1 + e^{-t})$. We chose $m = 32$, $\theta = \frac{1}{2}$ and let the code run over one time step, from $t = 0$ to $t = \Delta t = 0.1$. The convergence criterion was that the infinity norm of the residual vector had to be less than 10^{-5}. For the Jacobi method, 747 iterations were required for convergence. The number of iterations needed for convergence of the SOR method is illustrated in Figure 3 for various values of ω. The value of ω that gave us the fewest number of iterations (namely 46 iterations) was $\omega = 1.75$. This agrees well with the value of ω_{opt} given by the theory, namely $\omega_{\text{opt}} \approx 1.7384$.

21.8   SUMMARY

Given the work of [LHHR95] Lai *et al.*, we developed herein a fast and parallelizable SOR method for the numerical solution of Poisson's equation on the unit square with uniform mesh and Dirichlet boundary conditions. We then extended these techniques to the numerical solution of a model parabolic equation. Our numerical results agree with our analytic results, showing that using our block red-block SOR method on the parabolic equation with appropriately chosen relaxation factor ω gives much faster results than does the block Jacobi method.

References

[Cel83]   Celia M. A. (1983) *Collocation on Deformed Finite Elements and Alternating Direction Collocation Methods.* PhD thesis, Princeton University.

[LHHR95] Lai Y.-L., Hadjidimos A., Houstis E. N., and Rice J. R. (1995) On the Iterative Solution of Hermite Collocation Equations. *SIAM J. Matrix Anal. Appl.* 16: 254–277. (Also Technical Report, Purdue University, 1992).

[Pap83]   Papatheodorou T. S. (1983) Block AOR Iteration for Nonsymmetric Matrices. *Math. Comp* 41: 511–525.

[Pic94]   Piccirilli D. T. (1994) Using the Collocation Method with Splines under Tension and Upstream Weighting to Solve the One-Dimensional Convection-Diffusion Equation. Master's thesis, University of Vermont.

[You71]   Young D. M. (1971) *Iterative Solution of Large Linear Systems.* Academic Press, New York.

22

A FEM Approach for the Analysis of Optimal Control of Distributed Parameter Systems using a Maximum Principle

A. Bazezew[1], J. C. Bruch, Jr.[1] and J. M. Sloss[2],

[1]Department of Mechanical and Environmental Engineering,
University of California, Santa Barbara, U.S.A.

[2]Department of Mathematics,
University of California, Santa Barbara, U.S.A.

22.1   INTRODUCTION

Vibrations of structures that are subjected to dynamic loads arise in a wide variety of engineering applications. The need to suppress such unwanted vibrations has been the focus of structural control. The progress in research and development of structural control has been greatly influenced by the need to improve the performance of mechanical systems such as flexible attachments for spacecrafts, flexible robot links, slender components, [Bal82].

Control of such systems,whose mathematical formulations are usually described by partial differential equations, can be achieved by distributed control. In distributed control, the control is distributed over the entire or part of the spatial interior of the system. Mathematical tools for solving for the optimal distributed control have been formulated [Sad83], [Sad89], [Slo95], and [Bru95]. In particular, a maximum principle has been formulated that establishes a relationship between the optimal control of the system and an adjoint variable.

Construction of numerical methods for solving optimal control problems has been one of the areas of on-going research [Orl93], [Baz96]. Herein a finite element algorithm is presented to solve the optimal distributed control problem for beams with constant and variable cross-sections. This problem, by means of a maximum principle, is reduced to a problem of solving an initial-terminal-boundary-value problem for

The Mathematics of Finite Elements and Applications
Edited by J. R. Whiteman © 1997 John Wiley & Sons Ltd.

a system of equations involving the state variable and an adjoint variable. These variables are coupled by means of the optimal control, as well as by means of the terminal conditions. A recursive algorithm that is efficient, stable and accurate is developed that solves the resulting initial-terminal-boundary-value problem. An index of performance is used that consists of the sum of quadratic functionals of displacement and velocity as well as a penalty functional involving the control force. Numerical results are presented which compare the behaviour of the controlled and uncontrolled systems. Numerical results are also compared to analytical results for the constant cross-section case. The approach presented herein is very useful, in particular, for structural vibration problems that are non-separable.

22.2   FORMULATION OF THE OPTIMAL CONTROL PROBLEM

Consider the transverse vibration of a damped cantilevered beam of length l described by the following equation of motion:

$$\rho A(x)w_{tt} + cw_t + [EI(x)w_{xx}]_{xx} = f(x,t) \quad 0 < x < l, \ 0 < t < T \quad (22.1)$$

where $w(x,t)$ represents the transverse displacement of the beam, ρ is the density of beam material, $A(x)$ is the cross-sectional area of the beam, E is Young's modulus, $I(x)$ is beam's area moment of inertia, c is the damping coefficient, T is a fixed finite time, and $f(x,t) \in U_{ad}$ is the control force with the admissible set U_{ad} defined as:

$$U_{ad} = \{f(x,t) \,|\, f \in L^2(0,T; L^2(0,l))\}$$

The subscripts t and x refer to derivatives with respect to that particular variable.
    The state variable $w(x,t)$ is subject to the initial conditions

$$w(x,0) = \phi(x), \quad w_t(x,0) = \psi(x) \quad (22.2)$$

where $\phi(x) \in H^2(0,l)$ and $\psi(x) \in L^2(0,l)$. The boundary conditions are:

$$w(0,t) = 0, \quad EI(0)w_x(0,t) = 0 \quad (22.3)$$
$$EI(l)w_{xx}(l,t) = 0, \quad [EI(x)w_{xx}]_x \,|_{x=l} = 0 \quad (22.4)$$

The objective of the control problem is to find $f^*(x,t) \in U_{ad}$ which minimizes a measure of displacement and velocity at time T. Using the concept of Pareto optimality [Lei73], the multi-objective problem is: Minimize an index of performance $J(f)$ involving multiple objectives given by the weighted sum

$$J(f) = \sum_{k=1}^{3} \mu_k J_k(f) \quad (22.5)$$

where

$$J_1(f) = \int_0^l w^2(x,T;f)dx, \quad J_2(f) = \int_0^l w_t^2(x,T;f)dx, \quad J_3(f) = \int_0^T \int_0^l f^2(x,t)dxdt$$

in which μ_1, μ_2, μ_3 are weighting constants satisfying the conditions $\mu_1, \mu_2 \geq 0$, $\mu_3 > 0$ and $\mu_1 + \mu_2 > 0$.

The control problem is now expressed as: Determine an optimal control function $f^*(x,t) \in U_{ad}$ such that

$$J(f^*) = \min_{f \in U_{ad}} J(f) \tag{22.6}$$

with $w(x,t)$ subject to equations (22.1)-(22.4).

The solution to the formulated control problem is obtained by using a maximum principle. Let $v(x,t)$ be an adjoint variable satisfying the differential equation:

$$\rho A(x)v_{tt} - cv_t + [EI(x)v_{xx}]_{xx} = 0 \quad 0 < x < l, \ 0 < t < T \tag{22.7}$$

subject to boundary conditions similar to (22.3)-(22.4) and terminal conditions:

$$v(x,T) = -2\mu_2(\rho A(x))^{-1}w_t(x,T) \tag{22.8}$$

$$v_t(x,T) = 2\mu_1(\rho A(x))^{-1}w(x,T) - 2c\mu_2(\rho A(x))^{-2}w_t(x,T) \tag{22.9}$$

Let $f^*(x,t) \in U_{ad}$ be the unique optimal control function of problem (22.6). Let $f^0(x,t) \in U_{ad}$ and $w^0 = w(x,t; f^0)$ where w^0 satisfies equations (22.1)-(22.4) with $f(x,t) = f^0(x,t)$. Further, let $v^0 = v(x,t; f^0)$ be the solution to the adjoint problem given by equations (22.7)-(22.9) with $f = f^0$. If f^0 satisfies

$$\max_{f \in U_{ad}} H\left[x,t; v^0(x,t), f(x,t)\right] = H\left[x,t; v^0(x,t), f^0(x,t)\right] \tag{22.10}$$

almost everywhere in $(0,l) \times (0,T)$, and the functional H is given by

$$H\left[x,t; v(x,t), f(x,t)\right] = v(x,t)f(x,t) - \mu_3 f^2(x,t), \tag{22.11}$$

then the maximum principle states that

$$f^*(x,t) = f^0(x,t).$$

Assuming the optimal control exists and f^0 satisfies condition (22.10), then f^0 is indeed the optimal control. From (22.10) and (22.11) it follows that

$$f^*(x,t) = v^0(x,t)/(2\mu_3) \tag{22.12}$$

Thus to determine the optimal control function $f^*(x,t)$ given by (22.12), one needs to evaluate $v^0(x,t)$ which is coupled to the state variable $w(x,t)$ through the terminal conditions. Figure 1 schematically shows the problem formulation which will be solved numerically by using the finite element method in the space-time domain.

22.3   FINITE ELEMENT SOLUTION

The FEM equations for the approximate solution of the state variable $w(x,t)$ together with its adjoint variable $v(x,t)$, using Hermitian polynomials for the shape functions, are given by:

$$M\ddot{\hat{\mathbf{w}}} + C\dot{\hat{\mathbf{w}}} + K\hat{\mathbf{w}} = \hat{\mathbf{f}} \tag{22.13}$$

$$M\ddot{\hat{\mathbf{v}}} - C\dot{\hat{\mathbf{v}}} + K\hat{\mathbf{v}} = 0 \tag{22.14}$$

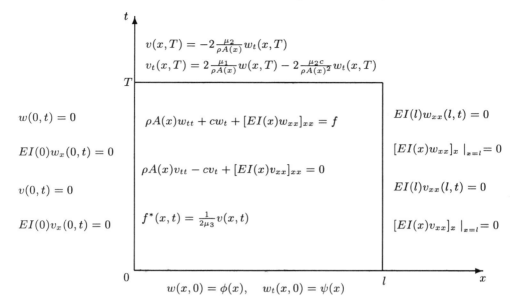

Figure 1 Optimal distributed control problem formulation for the vibrating beam in the space-time domain

with appropriate initial and boundary conditions, where M, C, K are the mass, damping and stiffness matrices, respectively; $\hat{\mathbf{w}}(t) = [\hat{w}_1(t) \ \ldots \ \hat{w}_n(t)]^T$ are nodal displacements $(1,\ldots,n/2)$ and slopes $(n/2+1,\ldots,n)$; $\hat{\mathbf{v}}(t) = [\hat{v}_1(t) \ \ldots \ \hat{v}_n(t)]^T$; $\hat{\mathbf{f}}(t) = \left[\hat{f}_1(t) \ \ldots \ \hat{f}_n(t)\right]^T$ is the equivalent nodal controlling force vector; and n is the number of unknown degrees of freedom at time t.

The mass, stiffness, damping matrices and force vector for an element of length l_e are given, respectively, by

$$M_e = \int_0^{l_e} \mathbf{H}^T(x)\rho A(x)\mathbf{H}(x)dx$$

$$K_e = \int_0^{l_e} \left(\frac{\partial^2 \mathbf{H}(x)}{\partial x^2}\right)^T EI(x) \left(\frac{\partial^2 \mathbf{H}(x)}{\partial x^2}\right) dx$$

$$C_e = \int_0^{l_e} \mathbf{H}^T(x)c\mathbf{H}(x)dx$$

$$\hat{\mathbf{f}}_e = \left[\int_0^{l_e} \mathbf{H}^T(x)\mathbf{H}(x)dx/(2\mu_3)\right]\hat{\mathbf{v}}_e = m_e\hat{\mathbf{v}}_e$$

where $\mathbf{H}(x) = [H_1(x) \ H_2(x) \ H_3(x) \ H_4(x)]^T$ in which H_i are the Hermitian polynomials, $i = 1,\ldots,4$, and $\hat{\mathbf{v}}_e = [\hat{v}_{1_e}(t) \ \hat{v}_{2_e}(t) \ \hat{v}_{3_e}(t) \ \hat{v}_{4_e}(t)]^T$.

Equations (22.13) and (22.14) are solved by using the Newmark algorithm for step-by-step integration [Bat96]. Equations for approximating the velocity, displacement and their x-derivatives at time $t + \Delta t$ are assumed

$$\dot{\hat{\mathbf{w}}}(t + \Delta t) = \dot{\hat{\mathbf{w}}}(t) + \left[(1 - \delta)\ddot{\hat{\mathbf{w}}}(t) + \delta\ddot{\hat{\mathbf{w}}}(t + \Delta t)\right]\Delta t \tag{22.15}$$

$$\hat{\mathbf{w}}(t + \Delta t) = \hat{\mathbf{w}}(t) + \dot{\hat{\mathbf{w}}}(t)\Delta t + \left[(0.5 - \alpha)\ddot{\hat{\mathbf{w}}}(t) + \alpha\ddot{\hat{\mathbf{w}}}(t + \Delta t)\right]\Delta t^2 \tag{22.16}$$

where δ and α are parameters that can be determined to obtain integration accuracy and stability. For the problems studied herein $\delta = 0.5$ and $\alpha = 0.25$. In addition the equilibrium equation (22.13) is considered at time $t + \Delta t$.

$$M\ddot{\hat{\mathbf{w}}}(t + \Delta t) + C\dot{\hat{\mathbf{w}}}(t + \Delta t) + K\hat{\mathbf{w}}(t + \Delta t) = \hat{\mathbf{f}}(t + \Delta t) \tag{22.17}$$

Equation (22.16) can be used to solve for $\ddot{\hat{\mathbf{w}}}(t + \Delta t)$ in terms of $\hat{\mathbf{w}}(t + \Delta t)$ and values of $\hat{\mathbf{w}}$, $\dot{\hat{\mathbf{w}}}$ and $\ddot{\hat{\mathbf{w}}}$ at t. Having done this, equation (22.15) can be solved for $\dot{\hat{\mathbf{w}}}(t + \Delta t)$ in terms of $\hat{\mathbf{w}}(t + \Delta t)$ and values of $\hat{\mathbf{w}}$, $\dot{\hat{\mathbf{w}}}$ and $\ddot{\hat{\mathbf{w}}}$ at t. These relations are then substituted into equation (22.17) which is solved for $\hat{\mathbf{w}}(t + \Delta t)$ which in turn yields $\dot{\hat{\mathbf{w}}}(t + \Delta t)$ and $\ddot{\hat{\mathbf{w}}}(t + \Delta t)$. Similar equations are generated for $\hat{\mathbf{v}}$, except at time $t - \Delta t$ since a backward recursive scheme will be used. The problem is solved recursively and the algorithm used is presented below.

22.4 NUMERICAL ALGORITHM

- **STEP I.**   Set
  - *   k=1   (k is the iteration counter)
  - * $\hat{v}^{(k)}(i, j) = 0$ where $\hat{v}^{(k)}(i, j) = v_i^{(k)}[(j - 1)\Delta t]$; $i = 1, 2, \ldots, n$;
    $j = 1, 2, \ldots, m + 1$, $m = (T/\Delta t)$ and $t = (j - 1)\Delta t$
  - * $\hat{f}^{(k)}(i, j) = 0$
  - * $\hat{w}_c(i, j) = 0$ ($\hat{w}_c$ is a test matrix)
  - * ε (ε is the desired error limit)
- **STEP II.**   Solve equation (22.13) for $\hat{w}^{(k)}(i, j)$ using the Newmark algorithm.

  — (A) Determine various parameters :
  - *   Form stiffness, mass, and damping matrices K, M, and C, respectively
  - *   Initialize $\hat{\mathbf{w}}(0)$, $\dot{\hat{\mathbf{w}}}(0)$, $\ddot{\hat{\mathbf{w}}}(0)$
    $\hat{w}_i(0) = \hat{w}(i, 1) = \phi(i\Delta x)$; $\dot{\hat{w}}_i(0) = \hat{w}(i, 1) = \psi(i\Delta x)$; $i = 1, \ldots, n/2$;
    $\hat{w}_i(0) = \hat{w}(i, 1) = \phi_x(i\Delta x)$; $\dot{\hat{w}}_i(0) = \hat{w}(i, 1) = \psi_x(i\Delta x)$; $i = n/2+1, \ldots, n$;
    $\ddot{\hat{\mathbf{w}}}(0) = M^{-1}[-C\dot{\hat{\mathbf{w}}}(0) - K\hat{\mathbf{w}}(0) + \hat{\mathbf{f}}(0)]$
  - *   Set time-step Δt, and the integration constants α and δ
  - *   Evaluate:
    $a_0 = 1/(\alpha\Delta t^2)$, $a_1 = \delta/(\alpha\Delta t)$, $a_2 = 1/(\alpha\Delta t)$, $a_3 = 1/(2\alpha) - 1$,
    $a_4 = \delta/\alpha - 1$, $a_5 = \Delta t(\delta/\alpha - 2)/2$, $a_6 = \Delta t(1 - \delta)$, $a_7 = \delta\Delta t$

  - *   Form effective stiffness matrix K^{ef}

    $$K^{ef} = K + a_0 M + a_1 C$$

  - *   Triangularize K^{ef}

    $$K^{ef} = LDL^T$$

— (B) For each time-step:

  * Calculate effective $\hat{\mathbf{f}}^{ef}$ load at time $t + \Delta t$

$$\hat{\mathbf{f}}^{ef}(t + \Delta t) = \hat{\mathbf{f}}(t + \Delta t) + M[a_0\hat{\mathbf{w}}(t) + a_2\dot{\hat{\mathbf{w}}}(t) + a_3\ddot{\hat{\mathbf{w}}}(t)] + \\ C[a_1\hat{\mathbf{w}}(t) + a_4\dot{\hat{\mathbf{w}}}(t) + a_5\ddot{\hat{\mathbf{w}}}(t)]$$

  * Solve for $\hat{\mathbf{w}}(t + \Delta t)$

$$LDL^T\hat{\mathbf{w}}(t + \Delta t) = \hat{\mathbf{f}}^{ef}(t + \Delta t)$$

  * Calculate $\ddot{\hat{\mathbf{w}}}(t + \Delta t)$ and $\dot{\hat{\mathbf{w}}}(t + \Delta t)$

$$\ddot{\hat{\mathbf{w}}}(t + \Delta t) = a_0[\hat{\mathbf{w}}(t + \Delta t) - \hat{\mathbf{w}}(t)] - a_2\dot{\hat{\mathbf{w}}}(t) - a_3\ddot{\hat{\mathbf{w}}}(t)$$
$$\dot{\hat{\mathbf{w}}}(t + \Delta t) = \dot{\hat{\mathbf{w}}}(t) + a_6\ddot{\hat{\mathbf{w}}}(t) + a_7\ddot{\hat{\mathbf{w}}}(t + \Delta t)$$

- **STEP III**.

  — Solve equation (22.14) for $\hat{v}^{k+1}(i, j)$ using the Newmark algorithm presented in **STEP II** with time-step $-\Delta t$ starting at t=T with terminal conditions

$$\ddot{\hat{v}}^{(k+1)}(i, m + 1) = -2\mu_2[\rho A(i\Delta x)]^{-1}\dot{\hat{w}}^{(k)}(i, m + 1),$$
$$\dot{\hat{v}}^{(k+1)}(i, m+1) = 2\mu_1[\rho A(i\Delta x)]^{-1}\dot{\hat{w}}^{(k)}(i, m+1) - 2c\mu_2[\rho A(i\Delta x)]^{-2}\dot{\hat{w}}^{(k)}(i, m+1),$$
$$i = 1,\ldots n/2;$$

  and similar forms for $\hat{v}$ and $\dot{\hat{v}}$ with $i = n/2 + 1, \ldots, n$ for the slopes,
$$\ddot{\hat{\mathbf{v}}}(T) = M^{-1}[C\dot{\hat{\mathbf{v}}}(T) - K\hat{\mathbf{v}}(T)].$$

  — Calculate $\hat{f}^{(k+1)}(i, j)$ as given by
$$\hat{f}^{(k+1)}(i, j) = \frac{(1-\theta)}{(2\mu_3)}\sum_{e=1}^{N} m_e\hat{v}_e^{(k)}[(j - 1)\Delta t] + \frac{\theta}{(2\mu_3)}\sum_{e=1}^{N} m_e\hat{v}_e^{(k+1)}[(j - 1)\Delta t]$$

  where N is the number of elements that contribute to $\hat{f}^{(k+1)}$ at node (i,j)

- **STEP IV**. Check for convergence:

  — If $| \hat{w}^{(k)}(i, j) - \hat{w}_c(i, j) | < \varepsilon$, stop
  — Else, set
    * $\hat{w}_c(i, j) = \hat{w}^{(k)}(i, j)$
    * $k = k + 1$
    * Go to **STEP II**(B)

Note: In **STEP II** a forward recurrence scheme in time is used, while a backward recurrence scheme in time is used in **STEP III**.

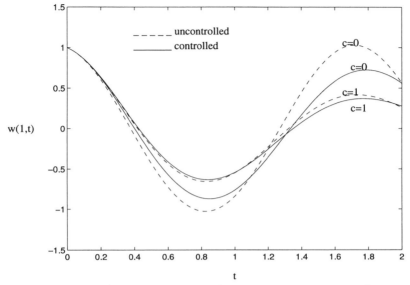

Figure 2   Displacement of the tip of the constant cross-section beam

22.5   NUMERICAL RESULTS

Results of a study of beam motion for the optimal distributed control problem are given for specific examples. A cantilever beam of constant cross-section and a stepped cantilever beam are considered. The beams are modelled with eight elements (n=16) and CAL80 [Wil88] is used in the numerical solution of the FEM model.

For the constant cross-section beam the following properties are assumed: $\rho A = 1$, $EI = 1$, and $l = 1$ (i.e., essentially equations (22.1) and (22.7) are similar to being non-dimensionalized with the appropriate definitions for x and t).

For the stepped beam, rectangular cross-sections are considered with the ratio of successive areas taken as 0.707. Moreover, $\rho = 1$, $E = 1$, and $l = 1$ (non-dimensionalized in each section). In applying the terminal conditions, equations (22.8) and (22.9), at each node point the average of the constant areas to the left and right of this point was used.

For both cases weighting factors $\mu_1 = \mu_2 = \mu_3 = 1$ are used. The initial conditions assumed are: $\phi(x)$ =the eigenfunction corresponding to the fundamental frequency and $\psi(x) = -\phi(x)$. Other necessary parameters are: $c = 0$ and 1; $\varepsilon = 10^{-6}$; $\theta = 0.05$; $T = 2.0$; and $\Delta t = 0.005$. Results are presented in the accompanying plots. The maximum number of iterations to obtain convergence of the solutions was $k_{max} = 47$.

For the constant cross-section beam, Figure 2 shows a comparison of the controlled and uncontrolled tip displacements for $c = 0$ and $c = 1$ of the FEM solution. The controlled numerical results are compared to the analytical solution for the constant cross-section beam in Figure 3. Figure 4 shows the controlling force required for $c = 0$ and $c = 1$ for this case. Figures 5 and 6 show similar results for the stepped cross-section beam.

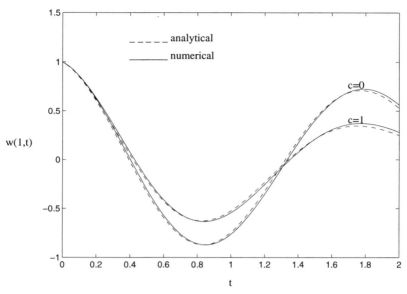

Figure 3   Comparison of the displacement of the tip of the constant cross-section beam

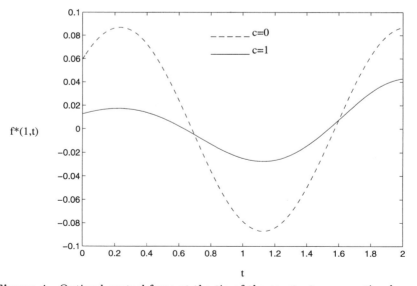

Figure 4   Optimal control force at the tip of the constant cross-section beam

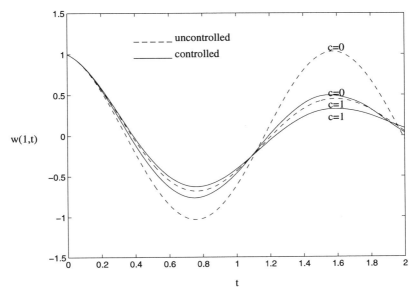

Figure 5 Displacement of the tip of the stepped cross-section beam

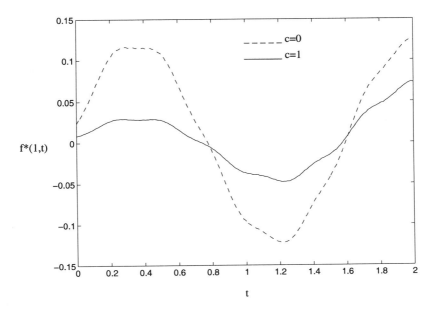

Figure 6 Optimal control force at the tip of the stepped cross-section beam

22.6 ACKNOWLEDGEMENT

The first author would like to thank the Higher Education Commission of the Ministry of Education, Addis Ababa, Ethiopia for support during the early stage of this research.

References

[Bal82] Balas, M.J. (1982) Trends in Large Space Structure Control: Fondest Hopes, Wildest Dreams. *IEEE Trans. on Automatic Control* 27: 522–532.

[Bat96] Bathe, K.J. (1996) *Finite Element Procedures in Engineering Analysis.* Prentice-Hall, Inc., Englewood Cliffs, New Jersey.

[Baz96] Bazezew, A., Bruch, Jr., J.C. and Sloss, J.M. (1996) Optimal (Distributed or Boundary) Control of Continuous Systems Solved Numerically in a Space-Time Domain. (Submitted). Submitted.

[Bru95] Bruch, Jr., J.C., Adali, S., Sloss, J.M. and Sadek, I.S. (1995) Maximum Principle for the Optimal Control of a Hyperbolic Equation in One Dimension: II. Application. *J. of Optimization Theory and Appl.* 87: 287–300.

[Lei73] Leitmann, G. and Schmitendorf, W.E. (1973) Some Sufficiency Conditions for Pareto Optimal Control. *ASME J. Dyn. Sys. Meas. Cont.* 95: 356–362.

[Orl93] Orlov, Y.V. and Razumovskii, D.D. (1993) Numerical Methods for Solving Problems of Optimal Generalized Control. *J. Automation and Remote Control* 54: 740–746.

[Sad83] Sadek, I.S. (1983) Theory and Application of the Maximum Principle for Optimal Control of Systems with Distributed Parameters. Ph.D. Dissertation, Department of Mathematics,, University of California, Santa Barbara.

[Sad89] Sadek, I.S., Bruch, Jr., J.C., Sloss, J.M. and Adali, S. (1989) Control of a Variable Cross-section Beam by Distributed Forces. *J. of Mech. Struct.* 16: 313–333.

[Slo95] Sloss, J.M., Sadek, I.S., Bruch, Jr.,J.C. and Adali, S. (1995) Maximum Principle for the Optimal Control of a Hyperbolic Equation in One Dimension: I.Theory. *J. of Optimization Theory and Appl.* 87: 33–45.

[Wil88] Wilson, E.L. and Bayo, E. (1988) CAL80: Computer Assisted Learning of Structural Analysis. Computer Aided Learning, University of California, Santa Barbara.

23

On some applications of $H\sigma$–stable wavelet–like hierarchical finite element space decompositions

Panayot S. Vassilevski

Center of Informatics and Computing Technology/CLPIP,
Bulgarian Academy of Sciences,
"Acad. G. Bontchev" street, Block 25 A,
1113 Sofia, Bulgaria

23.1   INTRODUCTION

In this paper we are concerned with the construction of efficient numerical methods for matrix problems arising from finite element methods for elliptic partial differential equations. In practical computation, the standard nodal basis for the finite element space is often chosen as the computational basis and the resulting matrices are ill-conditioned. In Vassilevski and Wang [VW97] the objective was to seek a substitution for the standard nodal basis so that the stiffness matrix arising from the new basis is well–conditioned, preserving the two major properties required for a computationally feasible basis: (a) the basis functions must be computable and (b) they must also have local support, hence the resulting stiffness matrix is sparse. This paper will first review the construction of local projection operators by the wavelet-like method proposed in [VW96a]. The latter wavelet–like projection operators have a main application in the construction of a stable Riesz basis with the above mentioned features for the finite element application to elliptic problems.

Attempts in the search for a stable Riesz basis with some restrictions, either on the mesh or on the analysis, have been made in Griebel and Oswald [GO94], Kotyczka and Oswald [KO95], and Stevenson [Ste95a], [Ste95b]. For a comparative study on the construction of economical Riesz bases for Sobolev spaces we refer to Lorentz and Oswald [LO96]. The method from Vassilevski and Wang [VW96a], [VW97] is general and provides a satisfactory answer for most elliptic equations. It is based on modifying the existing (unstable) hierarchical basis by using operators which are approximations of the L^2–projections onto coarse finite element spaces. For more details we refer

The Mathematics of Finite Elements and Applications
Edited by J. R. Whiteman © 1997 John Wiley & Sons Ltd.

to [VW97]. The construction and the main properties of the stable Riesz basis are reviewed in Section 23.2.

In the present paper we stress two applications of the stable wavelet–like local projection operators (denoted further by π_k) which are the main ingredient in the construction of stable Riesz bases for finite element spaces.

The first application, presented in Section 23.3, is in the construction of AMLI (Algebraic Multi–Level Iteration) preconditioners for matrices in normal form (i.e., $A^T A$) in the case of convection–diffusion (non–symmetric) finite element elliptic equations. We prove, under an H^2–regularity assumption (commonly imposed when studying convergence of multilevel methods in the L^2–norm, cf. e.g. Bank and Dupont [BD81] and Goldstein, Manteuffel and Parter [GMP93]), that in three space dimensions the AMLI method of hybrid type (see further Definition 23.3) is both of optimal order and spectrally equivalent to $B = A^T A$. This method may be an alternative to the classical W–cycle multigrid with sufficiently many smoothing iterations.

The second main application is based on the H^σ–boundedness of the local projection operators π_k. In Section 23.4, we use this fact for the construction of approximate harmonic extension operators. This technique has already been used in Bramble and Vassilevski [BV96] for constructing preconditioners in the interface domain decomposition (or DD) technique that allow for inexact subdomain solvers. Here, we present another application in the dual to the DD method; namely, the domain embedding technique. For the latter we refer to Nepomnyaschikh [Nep91b], [Nep91a], see also Proskurowski and Vassilevski [PV94] and the references given therein. The application of the H^σ–boundedness of the local projection operators in the domain embedding context was possible due to an algebraic fact, that a strengthened Cauchy–Schwarz inequality for the matrix implies the same strengthened Cauchy–Schwarz inequality for the inverse matrix. The detailed presentation is given in Section 23.4.

For practical aspects of the wavelet–like modification of the classical HB (hierarchical basis) or equivalently, of the bounded local projection operators π_k, we refer to Vassilevski and Wang [VW96b], [VW97] and Bramble and Vassilevski [BV96].

23.2 A STABLE RIESZ BASIS BY WAVELET METHOD

In this section we review the construction of local projections π_k which are H^σ–stable, $\sigma \in (0, 1]$, and provide computationally feasible Riesz bases for the finite element space $V = V_J$. The bilinear form of main interest is the one from the second-order elliptic problems. The method to be presented here was proposed by Vassilevski and Wang in [VW96a]. It relies on the fundamental estimate due to Oswald [Osw94] which characterizes the Sobolev space norms $\|.\|_\sigma$, $\sigma \in [0, 1]$ for finite element spaces:

$$\sum_{j=1}^{k} h_j^{-2\sigma} \|(Q_j - Q_{j-1})v\|_0^2 \leq \sigma_N \|v\|_\sigma^2 \quad \text{for all } v \in V_k. \tag{23.1}$$

Here, $V_0 \subset V_1 \subset \cdots \subset V_J$ is a sequence of nested conforming finite element spaces contained in $H_0^1 = H_0^1(\Omega)$ obtained by $J \geq 1$ successive steps of uniform refinement of an initial coarse triangulation $\mathcal{T}_0$ of the polygonal domain $\Omega \subset \mathbb{R}^d$, $d = 2, 3$. The kth level meshsize is denoted by h_k and we assume that $h_k = \frac{1}{2}h_{k-1} = 2^{-k}h_0$. Also, the kth level triangulation (the set of elements at level k) will be denoted by $\mathcal{T}_k$. Finally, $Q_k : L^2(\Omega) \to V_k$ is the standard L^2–projection operator. Here and in what follows by $\|\cdot\|_\sigma$, $\sigma \in [0,1]$ we will denote the norm in the Sobolev space $H_0^\sigma(\Omega)$. (The space H_0^σ is obtained by interpolation between the spaces H_0^1 and L^2, see, e.g., Bramble [Bra95].)

23.2.1   ON THE BASIS CONSTRUCTION

Define the L^2–projection operators $Q_k : L^2(\Omega) \to V_k$ by the identity,

$$(Q_k v, \psi)_0 = (v, \psi)_0 \quad \text{for all } \psi \in V_k.$$

Also, assume that there are computationally feasible approximations $Q_k^a : L^2(\Omega) \to V_k$ to Q_k such that for some small tolerance $\tau > 0$ the following estimate holds:

$$\|(Q_k - Q_k^a)v\|_0 \leq \tau \|Q_k v\| \quad \text{for all } v \in L^2(\Omega). \tag{23.2}$$

We also need the nodal interpolation operators $I_k : C(\Omega) \to V_k$ defined for any continuous function ψ by $I_k \psi = \sum_{i=1}^{n_k} \psi(x_i)\phi_i^{(k)}$. Here, $\{\phi_i^{(k)},\ i = 1,\dots,n_k\}$ is the nodal basis of V_k. That is, $\phi_i^{(k)}(x_j) = \delta_{i,j}$–the Kronecker symbol, when x_j runs over all the nodal degrees of freedom $\mathcal{N}_k$ of (the node set at kth discretization level) of V_k. Note that $\{\phi_i^{(k-1)},\ i = 1,\dots,n_{k-1}\} \cup \{\phi_i^{(k)},\ i = n_{k-1}+1,\dots,n_k\}$ also forms a basis, called the two–level hierarchical basis of V_k.

Definition 23.1 (WAVELET–LIKE LOCAL PROJECTION OPERATORS). *The projection operators of major interest are defined as follows:*

$$\pi_k = \prod_{j=k}^{J-1}(I_j + Q_j^a(I_{j+1} - I_j)), \tag{23.3}$$

with $\pi_J = I$.

It is clear that $\pi_k \psi = \psi$ if $\psi \in V_k$ since $I_j\psi + Q_j^a(I_{j+1} - I_j)\psi = I_j\psi = \psi$ for $j \geq k$ based on $(I_{j+1} - I_j)\psi = 0$ and $I_j\psi = \psi$. This also implies that $\pi_k^2 = \pi_k$.

Note also that $\pi_{k-1}(I_k - I_{k-1})\phi = Q_{k-1}^a(I_k - I_{k-1})\phi$ and $\pi_k - \pi_{k-1} = (I - Q_{k-1}^a)(I_k - I_{k-1})\pi_k$. Then, the components in the definition for the wavelet–like multilevel hierarchical basis read as follows:

$$\{\phi_i^{(0)}, i = 1,\dots,n_0\} \bigcup_{j=1}^{k}\{(I - Q_{j-1}^a)\phi_i^{(j)},\ i = n_{j-1}+1,\dots,n_j\}. \tag{23.4}$$

The above components $\{(I - Q_{j-1}^a)\phi_i^{(j)},\ i = n_{j-1}+1,\dots,n_j\}$ can be seen as a modification of the classical hierarchical basis components based on the interpolation operator I_k since $(I - Q_{j-1}^a)\phi_i^{(j)} = (I - Q_{j-1}^a)(I_j - I_{j-1})\phi_i^{(j)}$; the modification of the classical hierarchical basis components $\{(I_j - I_{j-1})\phi_i^{(j)},\ i = n_{j-1}+1,\dots,n_j\}$ comes from the additional term $Q^a_{j-1}(I_j - I_{j-1})\phi_i^{(j)}$. In other words, the modification was made by subtracting from each nodal hierarchical basis function $\phi_i^{(j)}$ its approximate L^2–projection $Q^a_{k-1}\phi_i^{(j)}$ onto the coarse level $k - 1$. The modified hierarchical basis functions are close relatives of the known Battle–Lemarié wavelets [Dau92].

Observe that in the limit case when $Q^a_k = Q_k$ (i.e., $\tau = 0$ in (23.2)), we get

$$\pi_k v = Q_k I_{k+1} Q_{k+1} I_{k+2} \ldots Q_{J-1} I_J v = Q_k Q_{k+1} \ldots Q_{J-1} v = Q_k v.$$

Therefore, $\pi_k v = Q_k v$; i.e., π_k reduces to the exact L^2–projection Q_k. As is well–known, the L^2–projection operators are bounded both in $H^1_0(\Omega)$ and $L^2(\Omega)$. Note finally that in this case we also have $\pi_k - \pi_{k-1} = Q_k - Q_{k-1}$. The decomposition of V based on the operators $Q_k - Q_{k-1}$ was a starting point in the construction of the BPX–method in Bramble, Pasciak and Xu [BPX90] (see also the survey Xu [Xu92]), and its development later in Zhang [Zha92] and Oswald [Osw94]. This gives us a hope that the hierarchical multilevel basis corresponding to the above choice of the operators π_k may yield a stable Riesz basis if τ is sufficiently small.

23.2.2   PRELIMINARY ESTIMATES

For an analysis of the multilevel basis (23.4) we need some auxiliary estimates already presented in Vassilevski and Wang [VW96a].

The following result on estimating the error $e_j = (\pi_j - Q_j)v$ will play an important role in our analysis:

Lemma 23.1. *Let C_R be a mesh-independent upper bound of the L^2-norm for the operator $I_s - I_{s-1} : V_s \to V_s$. Then,*

$$\|R_{s-1}v\|_0 \le C_R \tau \|v\|_0 \quad \text{for all } v \in V_s. \tag{23.5}$$

For a given $\sigma \in (0,1]$, assume that $\tau > 0$ is sufficiently small such that,

$$(1 + C_R\tau)\frac{1}{2^\sigma} \le q = Const < 1. \tag{23.6}$$

Then, there exits an absolute constant C such that for $e_j = (\pi_j - Q_j)v$, $v \in V_k$ there holds:

$$\sum_{j=1}^k h_j^{-2\sigma}\|e_j\|_0^2 \le C\tau^2 \sum_{j=1}^k h_j^{-2\sigma}\|(Q_j - Q_{j-1})v\|_0^2 \le C\tau^2 \sigma_N \|v\|_\sigma^2 \quad \forall v \in V_k. \tag{23.7}$$

Remark 23.1. *Note that in order to have L^2–stability of the deviations one has to assume that the tolerance τ is level dependent, i.e., one needs a tolerance $\tau = \mathcal{O}(J^{-1})$. Then*

$$\sum_{s=1}^k \|e_{s-1}\|_0^2 \le C\|v\|_0^2, \quad \text{for all } v \in V_k. \tag{23.8}$$

Lemma 23.2. *Let $V_k^1 = (I - M_{k-1})V_k^{(1)}$, with $V_k^{(1)} \equiv (I_k - I_{k-1})V_k$, be the modified hierarchical subspace of level k for any given L^2–bounded operators M_j. Then, there are* *positive constants c_1 and c_2 independent of k such that for any $\psi_1 = (I - M_{k-1})\phi_1 1 \in V_k^1$, with $\phi_1 \in V_k^{(1)}$,*

$$c_1 \|\phi_1\|_r^2 \leq \|\psi_1\|_r^2 \leq c_2 \|\phi_1\|_r^2, \qquad r = 0, 1. \tag{23.9}$$

We recall, that $\|.\|_1$ stands for the norm in the Sobolev space $H_0^1(\Omega)$ and $\|.\|_0$ denotes the $L^2(\Omega)$–norm.

Lemma 23.3. *Given v and let $v_1^{(k)} = (\pi_k - \pi_{k-1})v$. There exists a sufficiently small constant $\tau_0 > 0$ such that if the approximate projections Q_k^a satisfy (23.2) with $\tau \in (0, \tau_0)$ (see (23.6)), then*

$$\|v\|_1^2 \simeq \sum_{k=0}^{J} h_k^{-2} \|v_1^{(k)}\|_0^2.$$

For proofs of Lemma 23.1, Lemma 23.2, and Lemma 23.3, see Vassilevski and Wang [VW97].

23.2.3 STABILITY ANALYSIS

Here we study the Riesz property of the wavelet–like multilevel hierarchical basis defined in (23.4); that is, we show the H^1-stability of the approximate wavelet basis defined in (23.4).

For any $v \in V$ let

$$v = \sum_{x_i \in \mathcal{N}_0} c_{0,i} \phi_i^{(0)} + \sum_{k=1}^{J} \sum_{x_i \in \mathcal{N}_k^{(1)}} c_{k,i}(I - Q_{k-1}^a)\phi_i^{(k)} \tag{23.10}$$

be its representation with respect to the approximate wavelet basis. The corresponding coefficient norm of v is given by

$$\|v\| = \left(h_0^{d-2} \sum_{x_i \in \mathcal{N}_0} c_{0,i}^2 + \sum_{k=1}^{J} h_k^{d-2} \sum_{x_i \in \mathcal{N}_k^{(1)}} c_{k,i}^2 \right)^{1/2}, \qquad (d = 2, 3). \tag{23.11}$$

Our main result in this section is the following norm equivalence:

Theorem 23.1. *There exists a small (but fixed) $\tau_0 > 0$ such that if the approximate projections Q_k^a satisfy (23.2) with $\tau \in (0, \tau_0)$, then there are positive constants c_1 and c_2 satisfying*

$$c_1 \|v\|^2 \leq \|v\|_1^2 \leq c_2 \|v\|^2 \qquad \forall v \in V. \tag{23.12}$$

In other words, the modified hierarchical basis is a stable Riesz basis for the second order elliptic and Stokes problems. The equivalence relation (23.12) shall be abbreviated as $\|v\|^2 \simeq \|v\|_1^2$.

Proof. We first rewrite (23.10) as follows:

$$v = \sum_{k=0}^{J} v_1^{(k)}, \tag{23.13}$$

where, with $Q_{-1}^a = 0$,

$$v_1^{(k)} = \sum_{x_i \in \mathcal{N}_k^{(1)}} c_{k,i}(I - Q_{k-1}^a)\phi_i^{(k)} \in V_k^1. \tag{23.14}$$

Furthermore, by letting $\phi^{(k)} = \sum_{x_i \in \mathcal{N}_k^{(1)}} c_{k,i}\phi_i^{(k)} \in V_k^{(1)}$ we see that $v_1^{(k)} = (I - Q_{k-1}^a)\phi^{(k)}$. Thus, by using (23.9) in Lemma 23.2 (with $r = 0$ and $M_{k-1} = Q_{k-1}^a$) we obtain

$$\|\phi^{(k)}\|_0^2 \simeq \|v_1^{(k)}\|_0^2. \tag{23.15}$$

Since $\phi^{(k)} \in V_k^{(1)}$, then

$$\|\phi^{(k)}\|_0^2 \simeq h_k^d \sum_{x_i \in \mathcal{N}_k^{(1)}} c_{k,i}^2.$$

Combining the above with (23.15) yields

$$\|v\|^2 \simeq \sum_{k=0}^{J} h_k^{-2}\|v_1^{(k)}\|_0^2.$$

This, together with Lemma 23.3, completes the proof of the theorem.   □

Corollary 23.1. *For any fixed $\sigma \in (0,1]$, denote*

$$\|v\|_\sigma^2 = \sum_{k=0}^{J} h_k^{-2\sigma}\|v_1^{(k)}\|_0^2, \quad v_1^{(k)} = (\pi_k - \pi_{k-1})v.$$

Then,

$$\|v\|_\sigma \simeq \|v\|_\sigma,$$

for all $v \in H_0^\sigma(\Omega)$ (the latter space is defined by interpolation between $H_0^1(\Omega)$ and $L^2(\Omega)$, see e.g. Bramble [Bra95]) restricted to the finite element space $V = V_J$. The constants in the norm equivalence depend on σ as indicated in (23.6). For $\sigma = 0$ as mentioned in Remark 23.1, we have to assume a tolerance $\tau = \mathcal{O}(J^{-1})$.

23.3 ALGEBRAIC MULTI–LEVEL ITERATION (AMLI) PRECONDITIONERS FOR $A^T A$

In this section we construct optimal order preconditioners for solving systems of linear algebraic equations $A\mathbf{x} = \mathbf{b}$ transformed in the normal form $A^T A\mathbf{x} = A^T \mathbf{b}$ which may be useful for solving convection–diffusion finite element elliptic equations and also in certain domain embedding methods (when A is symmetric).

Consider the homogeneous Dirichlet boundary value problem for the following second-order elliptic equation:

$$\mathcal{L}(u) \equiv -\nabla \cdot (a(x)\nabla u) + \mathbf{b}(x) \cdot \nabla u + c(x)u = f(x), \quad x \in \Omega,$$
$$(23.16)$$

where $a = a(x)$ is a symmetric and positive definite matrix with bounded and measurable entries; $\mathbf{b} = \mathbf{b}(x)$ and $c = c(x)$ are given bounded functions; $f = f(x)$ is a function in $H^{-1}(\Omega)$.

Note that we do not have here in mind the singularly perturbed case of convection dominated problems of the form (23.16).

A weak form for the Dirichlet problem of (23.16) seeks $u \in H_0^1(\Omega)$ satisfying

$$a(u, v) = f(v) \qquad \forall v \in H_0^1(\Omega), \tag{23.17}$$

where

$$a(u, v) = \int_\Omega (a(x)\nabla u \cdot \nabla v + \mathbf{b}(x) \cdot \nabla u\, v + c(x)uv)\, dx$$

and $f(v)$ is the action of the linear functional f on v.

Let us approximate (23.17) by using the Galerkin method with, say, continuous piecewise linear polynomials. If $V = V_h$ denotes the finite element space associated with a prescribed triangulation of Ω with mesh size h, then the Galerkin approximation is given as the solution of the following problem: *Find $u_h \in V_h$ satisfying*

$$a(u_h, \phi) = f(\phi) \qquad \forall \phi \in V_h. \tag{23.18}$$

It has been shown that the discrete problem (23.18) has a unique solution when the mesh size h is sufficiently small. Details can be found in [Sch74], [SW96].

To shorten the exposition, we assume that $a(\cdot, \cdot)$ is H_0^1–coercive.

We summarize the assumptions on $a(., .)$:

- $H_0^1 \times H_0^1$–*boundedness*, i.e., there exists a positive constant γ_2 such that

$$a(v, w) \leq \gamma_2 (a_0(v, v))^{\frac{1}{2}} (a_0(w, w))^{\frac{1}{2}}, \quad \text{for all } v, w \in H_0^1(\Omega);$$
$$(23.19)$$

- H_0^1–*coercivity*, i.e., there exists a constant γ_1 such that

$$a(v, v) \geq \gamma_1^2 a_0(v, v) \quad \text{for all } v \in H_0^1(\Omega); \tag{23.20}$$

Here, $a_0(., .)$ is, for example, the principal symmetric and positive definite part of $a(., .)$.

23.3.1 BOUNDS OF LOCAL PROJECTION OPERATORS

We consider any local projection operator $\pi_k : \mathcal{C} \to V_k$ where $\mathcal{C}$ is for example the space of continuous functions that in particular contains $V = V_J$ the finite element space under consideration.

Let

$$a_0(\pi_k v, \pi_k v) \leq \eta^2(k_0) a_0(\pi_{k+k_0} v, \pi_{k+k_0} v) \quad \text{for all } v \in \mathcal{C}. \tag{23.21}$$

We assume that the norm–bound η of π_k in energy–norm may only depend on the level difference $k_0 \geq 0$. Our main application, though will be when $\eta(k_0)$ is independent of k_0, i.e., for stable local projection operators π_k.

We now state the following regularity assumption:

Let P_k be the, associated with $a(.,.)$, (non–symmetric) elliptic projection operator $P_k : H_0^1 \to V_k$, i.e., $a(P_k v, \psi) = a(v, \psi)$ for all $\psi \in V_k$. Similarly, define $P_k^\star : H_0^1 \to V_K$ by $a(\psi, P_k^\star v) = a(\psi, v)$ for all $\psi \in V_k$.

Let also λ_k be the largest eigenvalue of the symmetric positive definite operator associated with $a_0(.,.)$ restricted to V_k, i.e, $\lambda_k = \sup\limits_{v \in V_k} \dfrac{a_0(v,v)}{\|v\|_0^2}$.

We assume:

- *Assumption I:* (FULL ELLIPTIC REGULARITY) The following optimal L^2–error estimates hold:

$$\begin{aligned} \lambda_k \|v - P_k v\|_0^2 &\leq \sigma_R a_0(v - P_k v, v - P_k v), \quad \text{for all } v \in H_0^1, \\ \lambda_k \|v - P_k^\star v\|_0^2 &\leq \sigma_R a_0(v - P_k^\star v, v - P_k^\star v), \quad \text{for all } v \in H_0^1. \end{aligned} \tag{23.22}$$

The first error estimate in (23.22) implies:

$$\lambda_k a_0(v - P_k v, v - P_k v) \leq \sigma_A^2 \|Av\|_0^2, \quad \text{for all } v \in V. \tag{23.23}$$

We also assume the following standard inverse estimate:
- *Assumption II:* (INVERSE ESTIMATE)

$$\lambda_m \simeq h_m^{-2} \simeq 2^{2m} h_0^{-1}. \tag{23.24}$$

A main result of this section is the following estimate:

$$\|A_k \pi_k v\|_0 \leq \sigma_E \eta(k_0) \|A_{k+k_0} v\|_0 \quad \text{for all } v \in V_{k+k_0}. \tag{23.25}$$

Based on the boundedness estimate for $a(.,.)$ one has:

$$\begin{aligned} \|A_k \psi\|_0^2 &= a(\psi, A_k \psi) \\ &\leq \gamma_2 \sqrt{a_0(\psi, \psi)} \sqrt{a_0(A_k \psi, A_k \psi)} \\ &\leq \gamma_2 \sqrt{\lambda_k} \sqrt{a_0(\psi, \psi)} \|A_k \psi\|_0, \end{aligned}$$

which shows

$$\begin{aligned} \|A_k \psi\|_0 &\leq \gamma_2 \sqrt{\lambda_k} \sqrt{a_0(\psi, \psi)} \\ &\leq \tfrac{\gamma_2}{\gamma_1} \sqrt{\lambda_k} \sqrt{a(\psi, \psi)} \quad \text{for all } \psi \in V_k. \end{aligned} \tag{23.26}$$

Consider the estimates:

$$\begin{aligned} \|A_k \pi_k v\|_0 &\leq \|A_k \pi_k (v - P_k v)\|_0 + \|A_k P_k v\|_0 \\ &\leq \gamma_2 \lambda_k^{\frac{1}{2}} (a_0(\pi_k(v - P_k v), \pi_k(v - P_k v)))^{\frac{1}{2}} + \|A_k P_k v\|_0 \\ &\leq \gamma_2 \eta(k_0) (\lambda_k a_0(v - P_k v, v - P_k v))^{\frac{1}{2}} + \|A_k P_k v\|_0 \\ &\leq \gamma_2 \eta(k_0) \sigma_A \|Av\|_0 + \|A_k P_k v\|_0 \\ &\leq \gamma_2 \eta(k_0) \sigma_A \|Av\|_0 + \|A_k P_k v\|_0. \end{aligned} \tag{23.27}$$

Now, noting that

$$\|A_k P_k v\|_0^2 = a(P_k v, A_k P_k v) = a(v, A_k P_k v) = (Av, A_k P_k v) \leq \|Av\|_0 \|A_k P_k v\|_0,$$

we get

$$\|A_k P_k v\|_0 \leq \|Av\|_0. \tag{23.28}$$

Combining estimates (23.27–23.28), estimate (23.25) then follows letting $\sigma_E = \gamma_2 \sigma_A + 1$ and $A = A_{k+k_0}$.

23.3.2 SPECTRAL EQUIVALENCE ESTIMATES FOR THE TWO–LEVEL PRECONDITIONER

In this susbsection we derive some spectral relations and define a two–level preconditioner based on the direct space decomposition $V_{k+k_0} = (I - \pi_k)V_{k+k_0} \oplus V_k$.

One then immediately gets the following spectral equivalence result: Let $(Sv_2, v_2) \equiv \inf\limits_{v \in V_{k+k_0}:\, \pi_k v = v_2} \|A_{k+k_0} v\|_0^2$, $v \in V_{k+k_0}$, define the Schur complement of $B_{k+k_0} \equiv A_{k+k_0}^T A_{k+k_0}$ with respect to the direct decomposition $V_{k+k_0} = (I - \pi_k)V_{k+k_0} \oplus V_k$; i.e., let

$$A_{k+k_0} = \begin{bmatrix} A_{11} & A_{12} \\ A_{21} & A_{22} \end{bmatrix} \begin{matrix} \} (I - \pi_k)V_{k+k_0} \\ \} V_k \end{matrix}. \tag{23.29}$$

Note that $A_{22} = A_k$.

Then $S = B_{22} - B_{21} B_{11}^{-1} B_{12}$ where $B_{k+k_0} = \{B_{rs}\}_{r,s=1}^2 \ (= A_{k+k_0}^T A_{k+k_0})$. That is,

$$
\begin{aligned}
B_{11} &= A_{11}^T A_{11} + A_{21}^T A_{21} \\
B_{12} &= A_{11}^T A_{12} + A_{21}^T A_{22} \\
B_{21} &= A_{12}^T A_{11} + A_{22}^T A_{21} \\
B_{22} &= A_{22}^T A_{22} + A_{12}^T A_{12} = B_k + A_{12}^T A_{12}.
\end{aligned}
$$

Noting then that $(A_{12}v_2, A_{12}v_2) = a(v_2, A_{12}v_2) = a(v_2, P_k^\star A_{12}v_2) = (A_k v_2, P_k^\star A_{12}v_2) \leq \|P_k^\star\|_0 \|A_{12}v_2\|_0 \|A_k v_2\|_0 \leq \sigma(k_0) \|A_{12}v_2\|_0 \|A_k v_2\|_0$, i.e.,

$$\|A_{12}v_2\|_0 \leq \sigma(k_0) \|A_k v_2\|_0, \quad \text{where } v_2 = \pi_k v, \ v_1 = (I - \pi_k)v,$$

one gets the estimates, from (23.25) and $(Sv_2, v_2) \leq (B_k v_2, v_2) + \|A_{12}v_2\|_0^2$,

$$(B_k v_2, v_2) \leq \sigma_E^2 \eta^2(k_0)(Sv_2, v_2), \quad (Sv_2, v_2) \leq (1 + \sigma^2(k_0))(B_k v_2, v_2)$$

for all $v_2 \in V_k$.

Therefore, we have proved the following result.

Lemma 23.4. *The Schur complement* $S : V_k \to V_k$ *of* $B_{k+k_0} = A_{k+k_0}^T A_{k+k_0}$ *with respect to the direct decomposition* $V_{k+k_0} = (I - \pi_k)V_{k+k_0} \oplus V_k$ *and the operator* $B_k = A_k^T A_k$ *satisfy the spectral estimates:*

$$\frac{1}{1 + \sigma^2(k_0)}(Sv_2, v_2) \leq (B_k v_2, v_2) \leq \sigma_E^2 \eta^2(k_0)\,(Sv_2, v_2) \quad \text{for all } v_2 \in V_k.$$

The constant $\sigma(k_0)$ is defined from the following norm bound:

$$\|P_k^\star v\|_0 \leq \sigma(k_0)\|v\|_0, \quad \text{for all } v \in V_{k+k_0},$$

and admits the following asymptotic behavior with respect to $k_0 \to \infty$:

$$\sigma(k_0) \leq Ch_k/h_{k+k_0} \simeq 2^{k_0}. \tag{23.30}$$

Proof. Estimate (23.30) is seen by duality, i.e., from *Assumption I* and the behavior of $\lambda_m \simeq h_m^{-2}$, i.e., *Assumption II* (which is actually a standard inverse estimate). To see that, consider the estimates:

$$\begin{aligned}
a_0(v - P_k^\star v, v - P_k^\star v) &\leq \gamma_1^{-2} a(v - P_k^\star v, v - P_k^\star v) \\
&= \gamma_1^{-2} a(v, v - P_k^\star v) \\
&\leq \gamma_2 \gamma_1^{-2} [a_0(v,v)]^{\frac{1}{2}} [a_0(v - P_k^\star v, v - P_k^\star v)]^{\frac{1}{2}},
\end{aligned}$$

which imply,

$$a_0(v - P_k^\star v, v - P_k^\star v) \leq \gamma_2^2/\gamma_1^4 \, a_0(v,v) \quad \text{for all } v. \tag{23.31}$$

Consider then the estimate (based on *Assumption I* and (23.31)):

$$\begin{aligned}
\|v - P_k^\star v\|_0 &\leq Ch_k \, (a_0(v - P_k^\star v, v - P_k^\star v))^{\frac{1}{2}} \\
&\leq Ch_k \gamma_2/\gamma_1^2 \, (a_0(v,v))^{\frac{1}{2}} \\
&\leq Ch_k \sqrt{\lambda_{k+k_0}} \, \|v\|_0 \\
&\leq Ch_k/h_{k+k_0} \, \|v\|_0,
\end{aligned}$$

which shows the desired asymptotic behavior of $\sigma(k_0)$ since $\|P_k^\star v\|_0 \leq \|v\|_0 + \|v - P_k^\star v\|_0 \simeq \frac{h_k}{h_{k+k_0}} \|v\|_0$. $\square$

As a corollary, one may formulate the following two–level method for B_{k+k_0}.

Definition 23.2. *The operator*

$$M_{TL} = \begin{bmatrix} \widehat{B}_{11} & 0 \\ B_{21} & \widehat{B}_k \end{bmatrix} \begin{bmatrix} I & \widehat{B}_{11}^{-1} B_{12} \\ 0 & I \end{bmatrix}$$

defines a two–level preconditioner to B_{k+k_0}. Here $\widehat{B}_{11}$ is symmetric positive definite approximation to B_{11} and similarly, $\widehat{B}_k$ is symmetric positive definite approximation to B_k or to $S_D \equiv B_{22} - B_{21}\widehat{B}_{11}^{-1} B_{12}$.

The analysis of the two–level preconditioner is based on Lemma 23.4 and in the case of $\widehat{B} = S_D$, on the estimate

$$\frac{1}{1 + \sigma^2(k_0)}(S_D v_2, v_2) \leq (B_k v_2, v_2) \leq \sigma_E^2 \eta^2(k_0) \, (S_D v_2, v_2) \quad \text{for all } v_2 \in V_k, \tag{23.32}$$

which is proved in the same way as Lemma 23.4 using the inequalities:

$$\begin{aligned}
(S_D v_2, v_2) &\leq (B_{22} v_2, v_2) \\
&\leq (1 + \sigma^2(k_0))(B_k v_2, v_2) \\
&\leq (1 + \sigma^2(k_0))\sigma_E^2 \eta^2(k_0)(S v_2, v_2) \\
&\leq (1 + \sigma^2(k_0))\sigma_E^2 \eta^2(k_0)(S_D v_2, v_2).
\end{aligned}$$

Here, we have also assumed that $\widehat{B}_{11}$ is properly scaled, namely, that $(B_{11}v_1, v_1) \leq (\widehat{B}_{11}v_1, v_1)$ for all v_1, which implies the inequality $(Sv_2, v_2) \leq (S_D v_2, v_2)$, used in the last line above. One can also show that $(B_{11}v_1, v_1) \leq (1 + \sigma^2(k_0))\sigma_E^2 \eta^2(k_0)(Bv, v)$, based on Lemma 23.4 and inequality $(B_{22}v_2, v_2) \leq (1 + \sigma^2(k_0))\sigma_E^2 \eta^2(k_0)(Sv_2, v_2)$. Thus, B_{11} and B_{22} allow for approximations, and the relative condition number of the two–level preconditioner M_{TL} with respect to B can be estimated in terms of $\mathcal{H}(k_0) \equiv (1 + \sigma^2(k_0))\sigma_E^2 \eta^2(k_0)$ and the constants involved in the spectral equivalence relations between $\widehat{B}_{11}$ and B_{11} and between $\widehat{B}_k$ and B_k. More details are found from Vassilevski [Vas97], see also Bank [Ban96] and the analysis in the next subsection.

23.3.3 THE AMLI PRECONDITIONER

In this section we define the AMLI preconditioner and present the analysis of its relative condition number with respect to $B = B_{k+k_0}$.

One can generalize the preconditioner from Definition 23.2 to the multi–level case in a standard way, cf. Axelsson and Vassilevski [AV90], Vassilevski [Vas92] or the survey paper by Vassilevski [Vas97].

Definition 23.3. *Let* $M_0 = A_0$. *For* $s = 1, 2, \ldots, \left[\frac{J}{k_0}\right]$ *and* $m = min\{J, sk_0\}$, $k = (s-1)k_0$ *consider* $B_m = \{B_{r,s}\}_{r,s=1}^2$. *The operator*

$$M_m = \begin{bmatrix} \widehat{B}_{11} & 0 \\ B_{21} & \widetilde{S}_k \end{bmatrix} \begin{bmatrix} I & \widehat{B}_{11}^{-1}B_{12} \\ 0 & I \end{bmatrix}$$

defines the Algebraic Multi–Level Iteration (AMLI) preconditioner to B_m. *Here,* $\widehat{B}_{11}$ *is symmetric positive definite approximation to* B_{11}, *and*

$$\widetilde{S}_k^{-1} \equiv [I - p_\nu(M_k^{-1}S_D)] S_D^{-1}, \quad S_D \equiv B_{22} - B_{21}\widehat{B}_{11}^{-1}B_{12} \tag{23.33}$$

where

$$p_\nu(t) = \frac{1 + T_\nu\left(\frac{\alpha_1 + \alpha_2 - 2t}{\alpha_2 - \alpha_1}\right)}{1 + T_\nu\left(\frac{\alpha_1 + \alpha_2}{\alpha_2 - \alpha_1}\right)}, \tag{23.34}$$

and $T_\nu(t)$ *is the Chebyshev polynomial of first kind and degree* $\nu = \nu_{k_0} > 1$. *Also,* $\alpha_2 = 1 + \sigma^2(k_0)$ *and* $\alpha \equiv \alpha_1\sigma_E^2\eta^2(k_0)$ *is a lower bound of the minimum eigenvalue of* $M_k^{-1}B_k$, *which has to be estimated.*

Remark 23.2. *To implement one action of* M_m^{-1}, *as readily seen, one performs two inverse actions of* $\widehat{B}_{11}$, *matrix–vector products with the blocks* B_{21} *and* B_{12} *of* B *as well as* ν *inverse actions of* M_k *(already defined by induction) and also* $\nu - 1$ *actions of* S_D *which is based on the actions of the blocks of* B, B_{21}, B_{22} *and* B_{12} *and one more action of* $\widehat{B}_{11}^{-1}$ *for each action of* S_D.

For the analysis of the AMLI preconditioner M_m the following estimate will be useful:

$$(Sv_2, v_2) \leq (S_D v_2, v_2) \text{ all } v_2, \quad \text{all } v, \tag{23.35}$$

provided $\widehat{B}_{11}$ is scaled such that

$$(B_{11}v_1, v_1) \le (\widehat{B}_{11}v_1, v_1) \le (1 + \beta(k_0)) (B_{11}v_1, v_1) \quad \text{for all } v_1.$$

$$(23.36)$$

We also have:

$$(B_{22}v_2, v_2) \le (1 + \sigma^2(k_0))(B_k v_2, v_2) \le (1 + \sigma^2(k_0))\sigma_E^2\eta^2(k_0) (Sv_2, v_2) \quad \text{for all } v_2.$$

This estimate, letting

$$\mathcal{H}(k_0) = (1 + \sigma^2(k_0))\sigma_E^2\eta^2(k_0),$$

$$(23.37)$$

implies the strengthened Cauchy–Schwarz inequality

$$(B_{21}v_1, v_2) \le \gamma(k_0) [(B_{11}v_1, v_1)]^{\frac{1}{2}} [(B_{22}v_2, v_2)]^{\frac{1}{2}} \quad \text{for all } v_1, v_2,$$

$$(23.38)$$

where $\gamma(k_0) = \sqrt{1 - \frac{1}{\mathcal{H}(k_0)}}$.

The latter inequality in turn implies the estimate,

$$(B_{11}v_1, v_1) \le \mathcal{H}(k_0) (B_m v, v), \quad \text{for all } v = v_1 + v_2, \ v_2 = \pi_k v.$$

$$(23.39)$$

Choose now $\nu > \mathcal{H}(k_0)$. Then there exists a sufficiently small $\alpha > 0$ such that the following inequality holds:

$$1 + \beta(k_0)\mathcal{H}(k_0) + \mathcal{H}^2(k_0) \frac{(1 - \widetilde{\alpha})^\nu}{\alpha \left[\sum_{l=1}^{\nu} (1 + \sqrt{\widetilde{\alpha}})^{\nu-l}(1 - \sqrt{\widetilde{\alpha}})^{l-1} \right]^2} \le \frac{1}{\alpha}, \ \widetilde{\alpha} = \frac{\alpha}{\mathcal{H}(k_0)}.$$

$$(23.40)$$

We note that (23.40) (after multiplying it by α) reduces to $\frac{1}{\nu^2}\mathcal{H}^2(k_0) \le 1$ by letting $\alpha \to 0$.

We are now in a position to prove the main result concerning the spectral equivalence relations between the AMLI–preconditioner M_m and B_m.

Theorem 23.2. *Assume that $\nu > \mathcal{H}(k_0)$ and let α satisfy inequality (23.40). Then, the following estimates hold:*

$$(M_m v, v) \le (B_m v, v) \le \frac{1}{\alpha} (M_m v, v) \quad \text{for all } v \in V_m.$$

$$(23.41)$$

Proof. The proof follows from a standard induction argument. We have $M_0 = B_0$, hence (23.41) holds for any $\alpha \le 1$. Assume now, that for some $s \ge 1$ and $k = (s-1)k_0$ (23.41) is valid. Then for $((M_m - B_m)v, v) = ((\widehat{B}_{11} - B_{11})v_1, v_1) + ((\widetilde{S}_k - S_D)v_2, v_2)$ first the following estimate holds:

$$0 \le ((M_m - B_m)v, v), \quad \text{for all } v,$$

due to the choice of the polynomial p_ν in (23.34) (and α as in (23.40)), the induction assumption, and (23.32).

We will use next the following technical fact,

$$\sup\left\{\frac{p_\nu(t)}{1-p_\nu(t)} : t \in \left[\alpha\frac{1}{\sigma_E^2\eta^2(k_0)}, 1+\sigma^2(k_0)\right]\right\} = \frac{\sup\limits_{t\in[\alpha_1,\,\alpha_2]} p_\nu(t)}{1-\sup\limits_{t\in[\alpha_1,\,\alpha_2]} p_\nu(t)}$$

and that $\sup\limits_{t\in[\alpha_1,\,\alpha_2]} p_\nu(t) = \frac{2}{1+T_\nu\left(\frac{\alpha_2+\alpha_1}{\alpha_2-\alpha_1}\right)}$, where $\alpha_1 = \frac{\alpha}{\sigma_E^2\eta^2(k_0)}$ and $\alpha_2 = 1+\sigma^2(k_0)$.

Therefore, for $\widetilde{\alpha} = \frac{\alpha_1}{\alpha_2} = \frac{\alpha}{\mathcal{H}(k_0)}$, $\sup\limits_{t\in[\alpha_1,\,\alpha_2]} p_\nu(t) = \frac{4q^\nu}{(1+q^\nu)^2}$, $q = \frac{1-\sqrt{\widetilde{\alpha}}}{1+\sqrt{\widetilde{\alpha}}}$, which in

particular implies that

$$\frac{\sup\limits_{t\in[\alpha_1,\,\alpha_2]} p_\nu(t)}{1-\sup\limits_{t\in[\alpha_1,\,\alpha_2]} p_\nu(t)} = \frac{4q^\nu}{(1-q^\nu)^2} = \frac{4(1-\sqrt{\widetilde{\alpha}})^\nu(1+\sqrt{\widetilde{\alpha}})^\nu}{\left[(1+\sqrt{\widetilde{\alpha}})^\nu-(1-\sqrt{\widetilde{\alpha}})^\nu\right]^2}$$

$$= \frac{(1-\widetilde{\alpha})^\nu}{\widetilde{\alpha}\left[\sum\limits_{l=1}^{\nu}(1+\sqrt{\widetilde{\alpha}})^{\nu-l}(1-\sqrt{\widetilde{\alpha}})^{l-1}\right]^2} \simeq \frac{\mathcal{H}(k_0)}{\alpha\nu^2}, \quad \alpha \to 0.$$

Also, by the choice of $\widehat{B}_{11}$ (see (23.36)), the spectral relations (23.32) and inequality (23.40), we have:

$$((M_m - B_m)v, v) \le \beta(k_0)(B_{11}v_1, v_1)$$
$$+(S_D(I - p_\nu(M_k^{-1}S_D))^{-1}v_2, v_2) - (S_Dv_2, v_2)$$
$$\le \beta(k_0)\mathcal{H}(k_0)(B_mv, v)$$
$$+(S_Dv_2, v_2)\sup\left\{\frac{p_\nu(t)}{1-p_\nu(t)} : t \in [\alpha\lambda_{\min}(B_k^{-1}S_D), \lambda_{\max}(B_k^{-1}S_D)]\right\}$$
$$\le \beta(k_0)\mathcal{H}(k_0)(B_mv, v)$$
$$+(B_{22}v_2, v_2)\sup\left\{\frac{p_\nu(t)}{1-p_\nu(t)} : t \in [\alpha\lambda_{\min}(B_k^{-1}S_D), \lambda_{\max}(B_k^{-1}S_D)]\right\}$$
$$\le \beta(k_0)\mathcal{H}(k_0)(B_mv, v)$$
$$+(Sv_2, v_2)\mathcal{H}(k_0)\sup\left\{\frac{p_\nu(t)}{1-p_\nu(t)} : t \in \left[\alpha\frac{1}{\sigma_E^2\eta^2(k_0)}, 1+\sigma^2(k_0)\right]\right\}$$
$$\le \left\{\beta(k_0)\mathcal{H}(k_0) + \frac{\mathcal{H}^2(k_0)}{\alpha}\frac{(1-\widetilde{\alpha})^\nu}{\left[\sum\limits_{l=1}^{\nu}(1+\sqrt{\widetilde{\alpha}})^{\nu-l}(1-\sqrt{\widetilde{\alpha}})^{l-1}\right]^2}\right\}(B_mv, v)$$
$$\le (\tfrac{1}{\alpha} - 1)(B_mv, v).$$

The latter inequality completes the proof. $\qquad\qquad\square$

Now, we emphasize the asymptotic behavior of $\mathcal{H}(k_0)$. Assume at this point that the local projection operator $\pi_k : V_{k+k_0} \to V_k$ is H_0^1–bounded, i.e., $\eta(k_0)$ is independent of k_0. Such operators are available based on approximate wavelet modification of the classical HB (hierarchical basis) as described in the previous section. Then, one has:

$$\mathcal{H}(k_0) \simeq \sigma^2(k_0) \simeq 2^{2k_0}, \quad k_0 \to \infty. \qquad\qquad (23.42)$$

From the complexity requirement we also have the restriction on ν,

$$2^{dk_0} > \nu.$$

The latter inequality together with $\nu > \mathcal{H}(k_0) \simeq 2^{2k_0}$ show the following main result:

Theorem 23.3. *In three space dimensions, $d = 3$, the AMLI preconditioner M_m is both, of optimal complexity, and spectrally equivalent to B_m if k_0 is sufficiently large and the polynomial degree ν satisfies $2^{dk_0} > \nu > \mathcal{H}(k_0) \simeq 2^{2k_0}$.*

23.3.4   ESTIMATION OF THE CONDITION NUMBER OF B_{11}

In this section we show that the condition number of the major block B_{11} on the diagonal of B_{k+k_0} grows at most like $\left(\frac{h_k}{h_{k+k_0}}\right)^3$, and hence for k_0 fixed B_{11} is well–conditioned.

We begin, for any $v_1 = (I - \pi_k)v$, with the standard approximation property of the L^2–projection operator Q_k:

$$\|Q_k v_1 - v_1\|_0 \le Ch_k \sqrt{a_0(v_1, v_1)} \le Ch_k \gamma_1^{-1} \sqrt{(A_{11}v_1, v_1)}.$$

Then, based on the deviation estimate (23.7) (recall that π_k is close to the exact L^2–projection operator Q_k) and on the coercivity estimate (note that $\pi_k v_1 = 0$):

$$\|Q_k v_1\|_0 = \|(Q_k - \pi_k)v_1\|_0 \le Ch_k \sqrt{a_0(v_1, v_1)} \le Ch_k \gamma_1^{-1} \sqrt{(A_{11}v_1, v_1)},$$

one arrives at the following approximation estimate:

$$\|v_1\|_0^2 \le (\|Q_k v_1\|_0 + \|v_1 - Q_k v_1\|_0)^2 \le c_1^{-1}\lambda_k^{-1}(A_{11}v_1, v_1) \le c_1^{-1}\lambda_k^{-1}\|A_{11}v_1\|_0\|v_1\|_0.$$

Therefore, one gets

$$\lambda_k \|v_1\|_0 \le c_1^{-1}\|A_{11}v_1\|_0. \tag{23.43}$$

We also have, using estimate (23.26) and the proven L^2–coercivity of A_{11}, (23.43),

$$
\begin{aligned}
\|A_{21}v_1\|_0 \;&\le\; \|Av_1\|_0 \\
&\le\; \sqrt{\lambda_{k+k_0}}\,\tfrac{\gamma_2}{\gamma_1}(Av_1, v_1)^{\frac{1}{2}} \\
&=\; \sqrt{\lambda_{k+k_0}}\,\tfrac{\gamma_2}{\gamma_1}(A_{11}v_1, v_1)^{\frac{1}{2}} \\
&\le\; \sqrt{\lambda_{k+k_0}}\,\tfrac{\gamma_2}{\gamma_1}\left(\|A_{11}v_1\|_0\|v_1\|_0\right)^{\frac{1}{2}} \\
&\le\; \sqrt{\tfrac{\lambda_{k+k_0}}{c_1\lambda_k}}\,\tfrac{\gamma_2}{\gamma_1}\|A_{11}v_1\|_0.
\end{aligned}
$$

Then, since $B_{11} = A_{11}^T A_{11} + A_{21}^T A_{21}$, the latter estimate and the coercivity one (23.43), show the required eigenvalue bounds stated in the next theorem.

Theorem 23.4. *The following estimates hold:*

$$c_1^2\lambda_k^2 \le \frac{\|A_{11}v_1\|_0^2}{\|v_1\|_0^2} \le \frac{(B_{11}v_1, v_1)}{\|v_1\|_0^2} \le \left[1 + \left(\tfrac{\gamma_2}{\gamma_1}\right)^2 \tfrac{\lambda_{k+k_0}}{c_1\lambda_k}\right] \frac{\|A_{11}v_1\|_0^2}{\|v_1\|_0^2}$$

$$\le \left[1 + \left(\tfrac{\gamma_2}{\gamma_1}\right)^2 \tfrac{\lambda_{k+k_0}}{c_1\lambda_k}\right] \lambda_{k+k_0}^2.$$

That is, the condition number of B_{11} grows at most like $\left(\frac{\lambda_{k+k_0}}{\lambda_k}\right)^3 \simeq 2^{6k_0}$. Therefore, for bounded k_0 (as required in our application), the block B_{11} will be well–conditioned and hence relatively easy to approximate (such as in (23.36)).

23.4   BOUNDED EXTENSION OPERATORS

In this section we review one more application of the bounded local projection operators π_k to construct H^1–bounded extension operators, first used in Bramble and Vassilevski [BV96]. Here we emphasize the application of such bounded extension operators in the case of constructing preconditioners for finite element elliptic equations by the domain embedding technique.

The main fact used in the presented domain embedding technique is that one can transform the given problem based on bounded extension operators to a more stable form that allows for approximate solutions (preconditioners) in the embedding (more regular) domain Ω. We also take advantage of an algebraic fact (perhaps first proven here) that a strengthened Cauchy–Schwarz inequality for the matrix implies a strengthened Cauchy–Schwarz inequality for the inverse matrix with the same constant.

23.4.1   WAVELET–LIKE EXTENSION OPERATORS

In this section we describe a general framework for constructing computationally feasible H^1–bounded extension operators E that extend data defined on a interface boundary Γ into the interior of subdomains $\{\Omega_i\}_{i=1}^p$ that compose the original domain Ω. In the domain embedding application (to be considered in the following section) we will have $p = 2$ and Ω_1 will be a given in some sense irregular domain. Ω will be the embedding, computationally more regular, domain (e.g., parallelepiped) which will contain Ω_1. Then $\Omega_2 = \Omega \setminus \partial\Omega_1$. Finally, the interface Γ will be (a part) of $\partial\Omega_1$ across which Ω_1 is embedded in Ω, i.e., $\Gamma = \partial\Omega_1 \cap \partial\Omega_2$,

Consider the bilinear form $a(u, \varphi) = \int_\Omega a\nabla u \cdot \nabla\varphi \, dx$, $u, \varphi \in H_0^1(\Omega)$ and let $V = V_J$ be the finite element space of continuous piecewise linear functions corresponding to a triangulation $\mathcal{T}_h$ which we assume is obtained by $J \geq 1$ successive steps of uniform refinement of an initial coarse triangulation $\mathcal{T}_0$ of Ω. We also assume that the non–overlapping domain partitioning of Ω

$$\Omega = \Gamma \cup \Omega_1 \cup \Omega_2 \cup \cdots \cup \Omega_p, \tag{23.44}$$

consists of subdomains Ω_i that are coarse–grid domains (that is each Ω_i is completely covered by elements from $\mathcal{T}_0$). This implies that the interface Γ is also completely covered by boundaries of elements from $\mathcal{T}_0$. Let h_0 be the meshsize corresponding to $\mathcal{T}_0$, then $h_k = 2^{-k}h_0$ will be the meshsize corresponding to $\mathcal{T}_k$ if we divide each edge of the elements of $\mathcal{T}_{k-1}$ in two equal parts to create the elements of $\mathcal{T}_k$.

The coefficient $a = a(x)$, $x \in \Omega$ which can be a $d \times d$ symmetric positive definite matrix, bounded uniformly in $\overline{\Omega}$. It is assumed that a varies smoothly in each Ω_i but may have large jumps across Γ. For the purpose of constructing preconditioners for $a(.,.)$ it is sufficient to assume that a is piecewise constant with respect to the partition $\{\Omega_i\}$ of Ω. Let

$$
\begin{aligned}
\rho_{i,\text{max}} &= \sup_{x\in\Omega_i} \max_{\xi\in\mathbb{R}^d} \frac{\xi^T a(x)\xi}{\xi^T\xi} \\
\rho_{i,\text{min}} &= \inf_{x\in\Omega_i} \min_{\xi\in\mathbb{R}^d} \frac{\xi^T a(x)\xi}{\xi^T\xi},
\end{aligned} \tag{23.45}
$$

be bounds of the variation of the coefficient matrix $a(x)$ in each subdomain Ω_i, $i = 1, 2, \cdots, p$.

Definition 23.4 (EXTENSION OPERATOR). *Consider the following extension operator $\mathcal{E}_0 : V|_\Gamma \to V$, defined by*

$$\mathcal{E}_0 \varphi|_{\Omega_i} = \sum_{k=1}^{J} E_k^0 (q_k^i - q_{k-1}^i)\varphi_i.$$

Here E_k^0 is the trivial extension of any function $\psi_i \in V_k|_{\partial\Omega_i}$ by zero at the interior kth level nodes in Ω_i and $q_k^i : L^2(\partial\Omega_i) \to V_k|_{\partial\Omega_i}$ are the L^2–projection operators restricted to the boundaries of Ω_i.

It is clear then that

$$a(\mathcal{E}_0\varphi, \mathcal{E}_0\varphi) \leq \sum_{i=1}^{p} \rho_{i,\max} \int_{\Omega_i} \nabla \mathcal{E}_0\varphi_i \cdot \nabla \mathcal{E}_0\varphi_i dx. \tag{23.46}$$

Using the following $|.|_1$–seminorm characterization of space $H^1(\Omega_i) \cap V_k|_{\Omega_i}$

$$|v|_{1,\,\Omega_i}^2 \simeq \inf_{v:\, v = \sum_{k=1}^{J} v_k,\, v_k \in V_k|_{\Omega_i}} \sum_{k=1}^{J} h_k^{-2}\|v_k\|_{0,\,\Omega_i}^2,$$

used for $v := \mathcal{E}_0\varphi_i$ and replacing v_k with $E_k^0\varphi_i$ in (23.46), we end up with the following upper bound,

$$\begin{aligned} a(\mathcal{E}_0\varphi, \mathcal{E}_0\varphi) &\leq \sum_{i=1}^{p} \rho_{i,\max} \sum_{k=1}^{J} h_k^{-2}\|E_k^0\varphi\|_{0,\,\Omega_i}^2 \\ &\leq \tau_1^{-1} \sum_{i=1}^{p} \rho_{i,\max} \sum_{k=1}^{J} h_k^{-1}\|(q_k^i - q_{k-1}^i)\varphi_i\|_{0,\,\partial\Omega_i}^2. \end{aligned} \tag{23.47}$$

Next, use the quasi–optimality of the $L^2(\partial\Omega_i)$ projections $\{q_k^i\}_{k=0}^{J}$; namely:

$$\sum_{k=1}^{J} h_k^{-1}\|(q_k^i - q_{k-1}^i)\varphi_i\|_{0,\,\partial\Omega_i}^2 \simeq \inf_{\varphi_i = \sum_{k=1}^{J} \varphi_i^{(k)},\, \varphi_i^{(k)} \in V_k|_{\partial\Omega_i}} \sum_{k=1}^{J} h_k^{-1}\|\varphi_i^{(k)}\|_{0,\,\partial\Omega_i}^2. \tag{23.48}$$

Given an arbitrary $v \in V$ such that $v|_{\partial\Omega_i} = \varphi_i$, consider the decomposition,

$$v|_{\Omega_i} = Q_0^i(v|_{\Omega_i}) + \sum_{k=1}^{J}(Q_k^i - Q_{k-1}^i)(v|_{\Omega_i}).$$

Here Q_k^i are the $L^2(\Omega_i)$ projections onto the spaces $V_k|_{\Omega_i}$.

Letting $\varphi_i^{(k)} = (Q_k^i - Q_{k-1}^i)(v|_{\Omega_i})|_{\partial\Omega_i}$ we get $\varphi_i = Q_0^i(v|_{\Omega_i})|_{\partial\Omega_i} + \sum_{k=1}^{J} \varphi_i^{(k)}$.

Therefore, we can use this decomposition in (23.48) to get:

$$
\begin{aligned}
\sum_{k=1}^{J} h_k^{-1} \|(q_k^i - q_{k-1}^i)\varphi_i\|_{0,\,\partial\Omega_i}^2 &\leq C \sum_{k=1}^{J} h_k^{-1} \|(Q_k^i - Q_{k-1}^i)(v|_{\Omega_i})|_{\partial\Omega_i}\|_{0,\,\partial\Omega_i}^2 \\
&\leq C \sum_{k=1}^{J} h_k^{-2} \|(Q_k^i - Q_{k-1}^i)(v|_{\Omega_i})\|_{0,\,\Omega_i}^2 \\
&\leq C \|\nabla v\|_{0,\,\Omega_i}^2 \\
&\leq C \rho_{i,\min}^{-1} \int_{\Omega_i} a(x)\nabla v \cdot \nabla v \, dx \, \cdot
\end{aligned}
\tag{23.49}
$$

Here, we used a standard inverse inequality to bound boundary integrals $\int_{\partial\Omega_i}$ by domain integrals $\int_{\Omega_i}$.

Combining (23.47) and (23.49), by summation over i, and taking infimum over $v \in V$ such that $v|_{\partial\Omega_i} = \varphi_i$, the following main norm–bound is proved:

Theorem 23.5. *The extension operator $\mathcal{E}_0$ defined in Definition 23.4 is H^1–bounded and satisfies an estimate of the form:*

$$
a(\mathcal{E}_0\phi, \mathcal{E}_0\phi) \leq \eta_0 \inf_{v\in V:\ v|_\Gamma=\phi} a(v,v) \quad \text{for any } \phi \in V|_\Gamma \, \cdot
$$

Here, η_0 depends only on the local variation of the coefficient matrix $a(x)$ defined in (23.45), i.e., on $\max_{1\leq i\leq p} \frac{\rho_{i,max}}{\rho_{i,min}}$.

For practical computations, however we have to replace the exact $L^2(\partial\Omega_i)$–projection operators q_k^i by their computationally feasible approximations π_k^i as introduced in Section 23.2, Definition 23.1 (defined respectively for the finite element spaces restricted to the boundaries $\partial\Omega_i$).

Definition 23.5 (WAVELET–LIKE EXTENSION OPERATOR). *Consider the following extension operator $\mathcal{E} : V|_\Gamma \to V$, defined by*

$$
\mathcal{E}\varphi|_{\Omega_i} = \sum_{k=1}^{J} E_k^0(\pi_k^i - \pi_{k-1}^i)\varphi_i, \quad \varphi_i = \varphi|_{\partial\Omega_i} \, \cdot
\tag{23.50}
$$

Then using Corollary 23.1 (based on Lemma 23.1 for $\sigma = \frac{1}{2}$ and the norm characterization of $H^{\frac{1}{2}}(\partial\Omega_i)$) one has the following modification of Theorem 23.5.

Theorem 23.6. *The extension operator $\mathcal{E}$ defined in Definition 23.5 is H^1–bounded and satisfies an estimate of the form:*

$$
a(\mathcal{E}\phi, \mathcal{E}\phi) \leq \eta \inf_{v\in V:\ v|_\Gamma=\phi} a(v,v) \quad \text{for any } \phi \in V|_\Gamma \, \cdot
\tag{23.51}
$$

Here, η depends only on the local variation of the coefficient matrix $a(x)$ defined in (23.45), i.e., on $\max_{1\leq i\leq p} \frac{\rho_{i,max}}{\rho_{i,min}}$.

23.4.2 APPROXIMATE HARMONIC EXTENSION OPERATOR AND RELATED STRENGTHENED CAUCHY INEQUALITY

Let A be a stiffness matrix coming from a finite element discretization of a second order elliptic problem on a domain Ω with Dirichlet boundary conditions. Let now Ω_1 be a subdomain of Ω which is exactly covered by elements of the given triangulation of Ω. In practice we are interested in a problem formulated in Ω_1 which for a reason of simplicity we then embed in a more standard computational domain Ω such as a rectangle or a cube in three dimensions. Let the interface boundary across which Ω_1 is embedded in Ω be Γ. We consider, for the original problem formulated in Ω_1, Dirichlet boundary conditions. In such a way, if we use domain decomposition ordering of the degrees of freedom in Ω, with respect to the partitioning $\Omega = \Omega_1 \cup \Gamma \cup \Omega_2$, $\Omega_2 \equiv \Omega \setminus \overline{\Omega}_1$, the stiffness matrix A_1 associated with the original problem will be a principal submatrix of A, the matrix associated with the problem in the extended domain Ω, i.e., we will have,

$$
A = \begin{bmatrix} A_1 & A_{10} & 0 \\ A_{01} & A_0 & A_{02} \\ 0 & A_{20} & A_2 \end{bmatrix} \begin{matrix} \} \ \Omega_1 \\ \} \ \Gamma \\ \} \ \Omega_2 \equiv \Omega \setminus \overline{\Omega}_1 \end{matrix} \tag{23.52}
$$

The methods that we will be interested in for solving the problem

$$
A_1 \mathbf{x}_1 = \mathbf{b}_1, \tag{23.53}
$$

will be based on the existence of efficient preconditioners for the matrix A, i.e., for problems in the standard computational domain Ω.

For ease of presentation we will reorder the block matrix A into the following two–by–two block form:

$$
A = \begin{bmatrix} A_{11} & A_{12} \\ A_{21} & A_{22} \end{bmatrix} \begin{matrix} \} \ \Omega_1 \cup \Omega_2 \\ \} \ \Gamma \end{matrix}. \tag{23.54}
$$

Note that A_{11} is block diagonal with blocks on its main diagonal A_1 and A_2. Hence if we derive a preconditioner $\hat{B}_{11}$ for A_{11}, it will induce in a natural way a preconditioner B_1 for the original matrix A_1 whose inverse actions can be computed via $B_1^{-1} \mathbf{v}_1 = \left(\hat{B}_{11}^{-1} \begin{bmatrix} \mathbf{v}_1 \\ 0 \end{bmatrix} \begin{matrix} \} \ \Omega_1 \\ \} \ \Omega_2 \end{matrix} \right)\Big|_{\Omega_1}$.

For the construction of the preconditioners for A_1 we will need a mapping $\mathcal{E}$ that transforms data given on the interface boundary Γ to data in the interior of $\Omega \setminus \Gamma = \Omega_1 \cup \Omega_2$. The requirement that we impose on this mapping is to be bounded in the energy norm defined by the bilinear form $a(.,.)$ from which the stiffness matrix A is computed. In vector–matrix notation, we have a (rectangular) matrix $\begin{bmatrix} E_{12} \\ I \end{bmatrix} \begin{matrix} \} \ \Omega \setminus \Gamma = \Omega_1 \cup \Omega_2 \\ \} \ \Gamma \end{matrix}$, such that the following estimate holds:

$$
\left(\begin{bmatrix} E_{12} \\ I \end{bmatrix} \mathbf{v}_2 \right)^T A \left(\begin{bmatrix} E_{12} \\ I \end{bmatrix} \mathbf{v}_2 \right) \leq \eta \inf_{\mathbf{v}_1} \begin{bmatrix} \mathbf{v}_1 \\ \mathbf{v}_2 \end{bmatrix}^T A \begin{bmatrix} \mathbf{v}_1 \\ \mathbf{v}_2 \end{bmatrix}. \tag{23.55}
$$

Here η is a positive constant (≥ 1) independent of the mesh size. In other words, the mapping $\mathcal{E}$ that extends given data on Γ into the interior of the subdomains Ω_1 and

Ω_2 is approximate "harmonic" with respect to the given bilinear form $a(.,.)$. Using the function–bilinear form notation we can write (23.55) as

$$a(\mathcal{E}v_2, \mathcal{E}v_2) \leq \eta \inf_{v|_\Gamma = v_2} a(v, v),$$

i.e., the extension provided by the mapping $\mathcal{E}$ is quasi–optimal. The case $\eta = 1$ corresponds to the exact "harmonic" extension, i.e., $E_{12} = -A_{11}^{-1} A_{12}$. The latter mapping is impractical since it would require exact solutions in the subregions and this was our original intent (i.e., to solve a problem posed in Ω_1 with a coefficient matrix A_1; recall that $A_{11} = \begin{bmatrix} A_1 & 0 \\ 0 & A_2 \end{bmatrix} \begin{matrix} \} \, \Omega_1 \\ \} \, \Omega_2 \end{matrix}$). A possible choice of computationally feasible approximate harmonic extension mappings $\mathcal{E}$ is given in Definition 23.5.

Consider now the (square) transformation matrix,

$$E = \begin{bmatrix} I & E_{12} \\ 0 & I \end{bmatrix} \begin{matrix} \} & \Omega_1 \cup \Omega_2 \\ \} & \Gamma \end{matrix}$$

and then consider the following transformed matrix (for this approach we refer to Vassilevski and Axelsson [VA94] and Bramble and Vassilevski [BV96]),

$$\hat{A} = E^T A E = \begin{bmatrix} \hat{A}_{11} & \hat{A}_{12} \\ \hat{A}_{21} & \hat{A}_{22} \end{bmatrix}. \tag{23.56}$$

One easily derives the relations:

$$\begin{aligned} \hat{A}_{11} &= A_{11} \\ \hat{A}_{12} &= A_{12} + A_{11} E_{12} \\ \hat{A}_{21} &= A_{21} + E_{12}^T A_{11} \\ \hat{A}_{22} &= \begin{bmatrix} E_{12} \\ I \end{bmatrix}^T A \begin{bmatrix} E_{12} \\ I \end{bmatrix}. \end{aligned} \tag{23.57}$$

The important observation is that the first block has not been changed after transforming A to $\hat{A}$. Also notice the form of the block $\hat{A}_{22}$ of $\hat{A}$. On its basis the estimate (23.55) takes the equivalent form:

$$\mathbf{v}_2^T \hat{A}_{22} \mathbf{v}_2 \leq \eta \inf_{\hat{\mathbf{v}}_1} \begin{bmatrix} \hat{\mathbf{v}}_1 \\ \mathbf{v}_2 \end{bmatrix}^T \hat{A} \begin{bmatrix} \hat{\mathbf{v}}_1 \\ \mathbf{v}_2 \end{bmatrix} = \eta \inf_{\mathbf{v}_1} \begin{bmatrix} \mathbf{v}_1 \\ \mathbf{v}_2 \end{bmatrix}^T A \begin{bmatrix} \mathbf{v}_1 \\ \mathbf{v}_2 \end{bmatrix} \quad (\hat{\mathbf{v}}_1 = \mathbf{v}_1 - E_{12} \mathbf{v}_2). \tag{23.58}$$

It will be demonstrated in what follows that the transformed form of A, $\hat{A}$ is more stable in the sense that it allows for certain approximations of the blocks on the main diagonal of $\hat{A}$ and also allows for approximations to its corresponding Schur complements. The same applies for the block–entries and respective Schur complements of the inverse of $\hat{A}$. The main tool in proving these facts is the following strengthened Cauchy inequality valid for the two–by–two block structure of $\hat{A}$:

$$\mathbf{v}_2^T \hat{A}_{21} \hat{\mathbf{v}}_1 \leq \gamma \sqrt{\hat{\mathbf{v}}_1^T A_{11} \hat{\mathbf{v}}_1} \sqrt{\mathbf{v}_2^T \hat{A}_{22} \mathbf{v}_2}, \quad \text{for all } \hat{\mathbf{v}}_1, \, \mathbf{v}_2, \tag{23.59}$$

where $\gamma = \sqrt{1 - \frac{1}{\eta}}$, i.e., $\gamma \in [0, 1)$ (strictly less than one). This follows from inequality (23.58) as shown in Vassilevski [Vas92] (see also the proof of Theorem 23.7).

We will be needing the following corollaries of the strengthened Cauchy inequality (found already in Axelsson and Gustafsson [AG83]).

Lemma 23.5. *Let A be a symmetric positive definite, two–by–two block matrix,*

$$A = \begin{bmatrix} A_{11} & A_{12} \\ A_{21} & A_{22} \end{bmatrix},$$

that satisfies the strengthened Cauchy (or Cauchy–Bunyakowsky–Schwarz) inequality,

$$\mathbf{v}_2^T A_{21} \mathbf{v}_1 \leq \gamma \left[\mathbf{v}_1^T A_{11} \mathbf{v}_1 \right]^{\frac{1}{2}} \left[\mathbf{v}_2^T A_{22} \mathbf{v}_2 \right]^{\frac{1}{2}} \quad \text{for all } \mathbf{v}_1, \ \mathbf{v}_2,$$

for some positive constant γ strictly less than one. Consider now the following Schur complements of A:

$$\begin{aligned} S_1 &= A_{11} - A_{12} A_{22}^{-1} A_{21} \text{ and} \\ S_2 &= A_{22} - A_{21} A_{11}^{-1} A_{12}. \end{aligned}$$

Then the following inequalities hold:

$$\begin{aligned} \mathbf{v}_1^T S_1 \mathbf{v}_1 &\geq (1-\gamma^2) \mathbf{v}_1^T A_{11} \mathbf{v}_1 \quad \text{for all } \mathbf{v}_1 \\ \mathbf{v}_2^T S_2 \mathbf{v}_2 &\geq (1-\gamma^2) \mathbf{v}_2^T A_{22} \mathbf{v}_2 \quad \text{for all } \mathbf{v}_2. \end{aligned} \tag{23.60}$$

23.4.3   THE STRENGTHENED CAUCHY INEQUALITY FOR THE INVERSE MATRIX

In this section we prove the main result on which our construction of the domain embedding preconditioners is based. Namely, we show that given a two–by–two block matrix A, symmetric positive definite, for which a strengthened Cauchy inequality holds with a constant $\gamma \in [0, 1)$ (as in Lemma 23.5), then for the induced two–by–two block structure of A^{-1} the same strengthened Cauchy inequality, with the same constant γ holds.

Theorem 23.7. *Let A be a symmetric positive definite, two–by–two block matrix,*

$$A = \begin{bmatrix} A_{11} & A_{12} \\ A_{21} & A_{22} \end{bmatrix}. \tag{23.61}$$

We assume that there exists a positive constant γ, strictly less than one, such that the following strengthened Cauchy (or Cauchy–Bunyakowsky–Schwarz) inequality holds:

$$\mathbf{v}_2^T A_{21} \mathbf{v}_1 \leq \gamma \left[\mathbf{v}_1^T A_{11} \mathbf{v}_1 \right]^{\frac{1}{2}} \left[\mathbf{v}_2^T A_{22} \mathbf{v}_2 \right]^{\frac{1}{2}} \quad \text{for all } \mathbf{v}_1, \ \mathbf{v}_2. \tag{23.62}$$

Consider then $B = A^{-1} = \begin{bmatrix} B_{11} & B_{12} \\ B_{21} & B_{22} \end{bmatrix}$ partitioned according to the block partitioning of A as in (23.61). Then for the same constant $\gamma \in (0, 1)$, the following strengthened Cauchy inequality holds:

$$\mathbf{v}_2^T B_{21} \mathbf{v}_1 \leq \gamma \left[\mathbf{v}_1^T B_{11} \mathbf{v}_1 \right]^{\frac{1}{2}} \left[\mathbf{v}_2^T B_{22} \mathbf{v}_2 \right]^{\frac{1}{2}} \quad \text{for all } \mathbf{v}_1, \ \mathbf{v}_2. \tag{23.63}$$

Or equivalently, we have:

$$\mathbf{v}_1^T B_{11} \mathbf{v}_1 \leq \frac{1}{1-\gamma^2} \begin{bmatrix} \mathbf{v}_1 \\ \mathbf{v}_2 \end{bmatrix}^T B \begin{bmatrix} \mathbf{v}_1 \\ \mathbf{v}_2 \end{bmatrix}, \quad \text{for all } \mathbf{v}_1, \ \mathbf{v}_2. \tag{23.64}$$

Proof. Consider the following block–factorization form of A,

$$A = \begin{bmatrix} I & 0 \\ A_{21}A_{11}^{-1} & I \end{bmatrix} \begin{bmatrix} A_{11} & 0 \\ 0 & S \end{bmatrix} \begin{bmatrix} I & A_{11}^{-1}A_{12} \\ 0 & I \end{bmatrix},$$

where $S = A_{22} - A_{21}A_{11}^{-1}A_{12}$ is the corresponding Schur complement. Note that S is symmetric and positive definite. This block–factorization form of A implies the following block factorization of $B = A^{-1}$:

$$B = \begin{bmatrix} I & -A_{11}^{-1}A_{12} \\ 0 & I \end{bmatrix} \begin{bmatrix} A_{11}^{-1} & 0 \\ 0 & S^{-1} \end{bmatrix} \begin{bmatrix} I & 0 \\ -A_{21}A_{11}^{-1} & I \end{bmatrix}.$$

This factorization in turn shows the following exact representation of $B = A^{-1}$:

$$B = \begin{bmatrix} A_{11}^{-1} + A_{11}^{-1}A_{12}S^{-1}A_{21}A_{11}^{-1} & -A_{11}^{-1}A_{12}S^{-1} \\ -S^{-1}A_{21}A_{11}^{-1} & S^{-1} \end{bmatrix},$$

i.e., we have,

$$\begin{aligned} B_{11} &= A_{11}^{-1} + A_{11}^{-1}A_{12}S^{-1}A_{21}A_{11}^{-1}, \\ B_{12} &= -A_{11}^{-1}A_{12}S^{-1}, \\ B_{21} &= -S^{-1}A_{21}A_{11}^{-1}, \text{ and} \\ B_{22} &= S^{-1}. \end{aligned} \qquad (23.65)$$

As it is shown in Vassilevski [Vas97] the strengthened Cauchy inequality (23.63) and the inequality (23.64) are equivalent. This is seen by looking at the nonnegative quadratic form

$$(\mathbf{v} + t\mathbf{w})^T B(\mathbf{v} + t\mathbf{w}) - (1 - \gamma^2)\mathbf{v}_1^T B_{11}\mathbf{v}_1 \geq 0,$$

for any real t and any vectors $\mathbf{v}$, $\mathbf{w}$ of the form $\mathbf{v} = \begin{bmatrix} \mathbf{v}_1 \\ 0 \end{bmatrix}$ and $\mathbf{w} = \begin{bmatrix} 0 \\ \mathbf{v}_2 \end{bmatrix}$. I.e.,

$$\gamma^2 \mathbf{v}_1^T B_{11}\mathbf{v}_1 + 2\mathbf{v}_2^T B_{21}\mathbf{v}_1 t + \mathbf{v}_2^T B_{22}\mathbf{v}_2 t^2 \geq 0.$$

Its discriminant $D = (\mathbf{v}_2^T B_{21}\mathbf{v}_1)^2 - \gamma^2\mathbf{v}_1^T B_{11}\mathbf{v}_1\mathbf{v}_2^T B_{22}\mathbf{v}_2$ must be non–positive, and this is inequality (23.63). The converse is also true – the non–positivity of the discriminant D (which is the strengthened Cauchy inequality) implies that the above quadratic form is nonnegative, and this is (for $t = 1$) inequality (23.64).

To show (23.64), we use the formulas for the block entries of B given in (23.65).

The given strengthened Cauchy inequality (23.7) for A implies the following relations between matrix blocks of A (due to Lemma 23.5):

$$\mathbf{v}_2^T S\mathbf{v}_2 \geq (1 - \gamma^2)\mathbf{v}_2^T A_{22}\mathbf{v}_2 \quad \text{for all } \mathbf{v}_2,$$

which implies

$$\mathbf{v}_2^T A_{22}^{-1}\mathbf{v}_2 \geq (1 - \gamma^2)\mathbf{v}_2^T S^{-1}\mathbf{v}_2 \quad \text{for all } \mathbf{v}_2.$$

The latter inequality used for $\mathbf{v}_2 = A_{21}\mathbf{v}_1$ reads,

$$\mathbf{v}_1^T A_{12}A_{22}^{-1}A_{21}\mathbf{v}_1 \geq (1 - \gamma^2)\mathbf{v}_1^T A_{12}S^{-1}A_{21}\mathbf{v}_1 \quad \text{for all } \mathbf{v}_1. \qquad (23.66)$$

Similarly, for the other Schur complement $S_1 \equiv A_{11} - A_{12}A_{22}^{-1}A_{21}$, the original inequality also implies (due to Lemma 23.5),

$$\mathbf{v}_1^T(A_{11} - A_{12}A_{22}^{-1}A_{21})\mathbf{v}_1 = \mathbf{v}_1^T S_1 \mathbf{v}_1 \geq (1 - \gamma^2)\mathbf{v}_1^T A_{11}\mathbf{v}_1 \quad \text{for all } \mathbf{v}_1,$$

or which is the same,

$$\gamma^2 \mathbf{v}_1^T A_{11}\mathbf{v}_1 \geq \mathbf{v}_1^T A_{12}A_{22}^{-1}A_{21}\mathbf{v}_1 \quad \text{for all } \mathbf{v}_1.$$

The last inequality, together with (23.66) imply

$$\frac{\gamma^2}{1 - \gamma^2}\mathbf{v}_1^T A_{11}\mathbf{v}_1 \geq \mathbf{v}_1^T A_{12}S^{-1}A_{21}\mathbf{v}_1 \quad \text{for all } \mathbf{v}_1.$$

The latter inequality is equivalent to (by letting $\mathbf{v}_1 := A_{11}^{-1}\mathbf{v}_1$)

$$\frac{\gamma^2}{1 - \gamma^2}\mathbf{v}_1^T A_{11}^{-1}\mathbf{v}_1 \geq \mathbf{v}_1^T A_{11}^{-1}A_{12}S^{-1}A_{21}A_{11}^{-1}\mathbf{v}_1 \quad \text{for all } \mathbf{v}_1.$$

Then we have,

$$
\begin{aligned}
\frac{1}{1-\gamma^2}\mathbf{v}_1^T A_{11}^{-1}\mathbf{v}_1 &= (1 + \frac{\gamma^2}{1-\gamma^2})\mathbf{v}_1^T A_{11}^{-1}\mathbf{v}_1 \\
&\geq \mathbf{v}_1^T(A_{11}^{-1} + A_{11}^{-1}A_{12}S^{-1}A_{21}A_{11}^{-1})\mathbf{v}_1 \\
&= \mathbf{v}_1^T B_{11}\mathbf{v}_1, \quad \text{for all } \mathbf{v}_1.
\end{aligned}
$$

The latter inequality shows, since $A_{11}^{-1} = B_{11} - B_{12}B_{22}^{-1}B_{21}$ is a Schur complement of the symmetric positive definite matrix B, that

$$\mathbf{v}_1^T B_{11}\mathbf{v}_1 \leq \frac{1}{1 - \gamma^2} \inf_{\mathbf{v}_2} \begin{bmatrix}\mathbf{v}_1\\\mathbf{v}_2\end{bmatrix}^T B \begin{bmatrix}\mathbf{v}_1\\\mathbf{v}_2\end{bmatrix}.$$

Thus the proof is complete.   □

23.4.4 THE CONSTRUCTION OF PRECONDITIONERS FOR A_{11} ON THE BASIS OF AVAILABLE PRECONDITIONERS FOR A

In this section we present our approach for deriving preconditioners for the principle submatrix A_{11} of A (where A is partitioned in a two–by–two block form as in (23.61)) on the basis of any available preconditioner for A itself. Here we assume that either A itself satisfies a strengthened Cauchy inequality or that it can be transformed to $\hat{A} = E^T A E$, with $E = \begin{bmatrix} I & E_{12} \\ 0 & I \end{bmatrix}$, for which a strengthened Cauchy inequality holds. The crucial part of the construction of the preconditioners is that the actions of E_{12} and E_{12}^T on vectors $\mathbf{v}_2$ and $\mathbf{v}_1$, respectively, be computable. For the domain embedding application E_{12} comes from the approximate harmonic extension mapping $\mathcal{E}$ (see (23.55)).

Let now M be a given symmetric positive definite preconditioner to A and let the following spectral equivalence relations hold:

$$\gamma_1 \mathbf{v}^T A\mathbf{v} \leq \mathbf{v}^T M\mathbf{v} \leq \gamma_2\mathbf{v}^T A\mathbf{v}, \quad \text{for all } \mathbf{v}, \tag{23.67}$$

for some positive constants γ_1, γ_2. Hence M^{-1} will be spectrally equivalent to $B \equiv A^{-1}$. Then $E^{-1}M^{-1}E^{-T}$ will be spectrally equivalent to $\hat{B} \equiv \hat{A}^{-1} = E^{-1}BE^{-T}$. As a corollary, we obtain that $\hat{B}_{11} = [I, \ 0]\hat{B}\begin{bmatrix}I\\0\end{bmatrix}$ will be spectrally equivalent to $[I, \ 0]E^{-1}M^{-1}E^{-T}\begin{bmatrix}I\\0\end{bmatrix} = [I, \ -E_{12}]M^{-1}\begin{bmatrix}I\\-E_{12}^T\end{bmatrix}$. By Theorem 23.7, $\hat{B}_{11}$ is spectrally equivalent to A_{11}^{-1} $(= \hat{B}_{11} - \hat{B}_{12}\hat{B}_{22}^{-1}\hat{B}_{21})$, hence $[I, \ 0]E^{-1}M^{-1}E^{-T}\begin{bmatrix}I\\0\end{bmatrix} = [I, \ -E_{12}]M^{-1}\begin{bmatrix}I\\-E_{12}^T\end{bmatrix}$ provides the inverse actions of a spectrally equivalent preconditioner to A_{11}.

More specifically, from (23.67) used in the following equivalent form

$$\gamma_2^{-1}\mathbf{v}^T\hat{B}\mathbf{v} \leq \mathbf{v}^TE^{-1}M^{-1}E^{-T}\mathbf{v} \leq \gamma_1^{-1}\mathbf{v}^T\hat{B}\mathbf{v}, \quad \text{for all } \mathbf{v},$$

we get as a corollary, letting $\mathbf{v} = \begin{bmatrix}\mathbf{v}_1\\0\end{bmatrix}$,

$$\gamma_2^{-1}\mathbf{v}_1^T\hat{B}_{11}^{-1}\mathbf{v}_1 \leq \mathbf{v}_1^T[I, \ -E_{12}]M^{-1}\begin{bmatrix}I\\-E_{12}^T\end{bmatrix}\mathbf{v}_1 \leq \gamma_1^{-1}\mathbf{v}_1^T\hat{B}_{11}\mathbf{v}_1.$$

The latter inequality combined with the inequality from Theorem 23.7,

$$\mathbf{v}_1^TA_{11}^{-1}\mathbf{v}_1 \leq \mathbf{v}_1^T\hat{B}_{11}^{-1}\mathbf{v}_1 \leq \frac{1}{1-\gamma^2}\mathbf{v}_1^TA_{11}^{-1}\mathbf{v}_1, \quad \text{for all } \mathbf{v}_1,$$

imply the desired spectral equivalence relation:

$$\gamma_2^{-1}\mathbf{v}_1^TA_{11}^{-1}\mathbf{v}_1 \leq \mathbf{v}_1^T[I, \ -E_{12}]M^{-1}\begin{bmatrix}I\\-E_{12}^T\end{bmatrix}\mathbf{v}_1 \leq \gamma_1^{-1}\frac{1}{1-\gamma^2}\mathbf{v}_1^TA_{11}^{-1}\mathbf{v}_1 \quad \text{for all } \mathbf{v}_1. \tag{23.68}$$

That is, we have:

Theorem 23.8. *Let M be a spectrally equivalent preconditioner for A and let $E = \begin{bmatrix}I & E_{12}\\0 & I\end{bmatrix}$ with computable actions of E_{12} and E_{12}^T, transform A to $\hat{A} = E^TAE$, for which a strengthened Cauchy inequality with a constant $\gamma \in [0,1)$ holds. Then the mapping*

$$[I, \ -E_{12}]M^{-1}\begin{bmatrix}I\\-E_{12}^T\end{bmatrix}, \tag{23.69}$$

provides the inverse actions of a spectrally equivalent preconditioner for the principal submatrix A_{11} of A. More specifically, we have

$$Cond\left([I, \ -E_{12}]M^{-1}\begin{bmatrix}I\\-E_{12}^T\end{bmatrix}A_{11}\right) \leq \frac{1}{1-\gamma^2}Cond\,(M^{-1}A) \leq \frac{1}{1-\gamma^2}\frac{\gamma_2}{\gamma_1}, \tag{23.70}$$

where the constants γ_2 and γ_1 are from the spectral equivalence relation (23.67).

ACKNOWLEDGMENT

This work of the author was partially supported by the Bulgarian Ministry for Education, Science and Technology under grant I–504, 1995.

References

[AG83]   Axelsson O. and Gustafsson I. (1983) Preconditioning and two–level multigrid methods of arbitrary degree of approximations. *Math. Comp.* 40: 219–242.

[AV90]   Axelsson O. and Vassilevski P. (1990) Algebraic multilevel preconditioning methods, II. *SIAM J. Numer. Anal.* 27: 1569–1590.

[Ban96] Bank R. E. (1996) Hierarchical bases and the finite element method. *Acta Numerica* .

[BD81]   Bank R. E. and Dupont T. (1981) An optimal order process for solving finite element equations. *Math. Comp.* 36: 35–51.

[BPX90] Bramble J. H., Pasciak J. E., and Xu J. (1990) Parallel multilevel preconditioners. *Math. Comp.* 55: 1–22.

[Bra95] Bramble J. H. (1995) *Multigrid methods.* Pitman Research Notes in Mathematical Series, # 234, Longman Scientific & Technical, (second edition), UK.

[BV96]   Bramble J. H. and Vassilevski P. S. (1996) Wavelet–like extension operators in non–overlapping domain decomposition algorithms. Institute for Scientific Computation, Texas A & M University, College Station, TX (in preparation).

[Dau92] Daubechies I. (1992) *Ten Lectures on Wavelets.* SIAM, Philadelphia, PA.

[GMP93] Goldstein C. I., Manteuffel T. A., and Parter S. V. (1993) Preconditioning and boundary conditions without H^2 estimates: L^2 condition numbers and the distribution of the singular values. *Math. Comp.* 30: 343–376.

[GO94]   Griebel M. and Oswald P. (1994) Tensor product type subspace splittings and multilevel iterative methods for anisotropic problems. *Advances in Computational Mathematics* 4: 171–206.

[KO95]   Kotyczka U. and Oswald P. (1995) Piecewise linear prewavelets of small support. In Chui C. and Schumaker L. L. (eds) *Approximation Theory VIII, vol. 2 (wavelets and multilevel approximation)*, pages 235–242. World Scientific, Singapore.

[LO96]   Lorentz R. and Oswald P. (1996) Constructing economical riesz bases for sobolev spaces. talk, presented at the Domain Decomposition Conference held in Bergen, Norway, June 3–8, 1996, GMD, Bonn.

[Nep91a] Nepomnyaschikh S. V. (1991) Mesh theorems of traces, normalization of function traces and their inversion. *Soviet J. Numer. Anal. and Math. Modelling* 6: 223–242.

[Nep91b] Nepomnyaschikh S. V. (1991) Method of splitting into subspaces for solving boundary value problems in complex form domain. *Soviet J. Numer. Anal. and Math. Modelling* 6: 1–26.

[Osw94] Oswald P. (1994) *Multilevel Finite Element Approximation. Theory and Applications.* Teubner Skripten zur Numerik, Teubner, Stuttgart.

[PV94] Proskurowski W. and Vassilevski P. S. (1994) Preconditioning capacitance matrix problems in domain imbedding. *SIAM J. Sci. Comput.* 15: 77–88.

[Sch74] Schatz A. H. (1974) An observation concerning Ritz–Galerkin methods with indefinite bilinear forms. *Math. Comp.* 28: 959–962.

[Ste95a] Stevenson R. (1995) Robustness of the additive and multiplicative frequency decomposition multilevel method. *Computing* 54: 331–346.

[Ste95b] Stevenson R. (1995) A robust hierarchical basis preconditioner on general meshes. Report # 9533, Department of Mathematics, University of Nijmegen, Nijmegen, The Netherlands.

[SW96] Schatz A. H. and Wang J. (1996) Some new error estimates for Ritz-Galerkin methods with minimal regularity assumptions. *Math. Comp.* 65: 19–27.

[VA94] Vassilevski P. S. and Axelsson O. (1994) A two–level stabilizing framework for interface domain decomposition preconditioners. In Dimov I. T., Sendov B., and Vassilevski P. S. (eds) *Proceedings of the Third International Conference O(h^3), Advances in Numerical Methods and Applications*, pages 196–202. World Scientific, Singapore, New Jersey, London, Hong Kong.

[Vas92] Vassilevski P. S. (1992) Hybrid V–cycle algebraic multilevel preconditioners. *Math. Comp.* 58: 489–512.

[Vas97] Vassilevski P. S. (1997) On two ways of stabilizing the hierarchical basis multilevel methods. *SIAM Review (to appear)* .

[VW96a] Vassilevski P. S. and Wang J. (1996) Stabilizing the hierarchical basis by approximate wavelets, I: Theory. *Numer. Linear Alg. Appl. (submitted)* .

[VW96b] Vassilevski P. S. and Wang J. (1996) Stabilizing the hierarchical basis by approximate wavelets, II: Implementation and Numerical Experiments. *SIAM J. Sci. Comput. (submitted)* .

[VW97] Vassilevski P. S. and Wang J. (1997) Wavelet–like methods in the design of efficient multilevel preconditioners for elliptic pdes. In Dahmen W., Kurdila A., and Oswald P. (eds) *Multiscale Wavelet Methods for PDEs, (Wavelet Analysis and its Applications Series).* Academic Press (to appear).

[Xu92] Xu J. (1992) Iterative methods by space decomposition and subspace correction. *SIAM Review* 34: 581–613.

[Zha92] Zhang X. (1992) Multilevel Schwarz methods. *Numer. Math.* 63: 521–539.

24

Covariant mixed finite element method; A first approach

J.-M. Thomas

Laboratoire de Mathématiques Appliquées
UPRES A 5033, Université de Pau & CNRS,
Avenue de l'Université, 64000 Pau (France).

24.1 INTRODUCTION

We extend the Raviart-Thomas mixed finite element method for a second order elliptic problem defined on a plane domain to problems defined on a parameterized surface of $\mathbb{R}^3$. We will analyse mixed finite element methods for a second order elliptic problem defined on a surface Ω in the Euclidean 3D-space. A first generalization of such methods to thin shell problems can be found in [TT96]. Another important generalization is supplied by electromagnetism problems.

Let Ω be a parametrized smooth surface in $\mathbb{R}^3$. More precisely, we assume the existence of a smooth injective mapping Φ from the Euclidean 2-dimensional space $\mathbb{R}^2$ into the Euclidean 3-dimensional space $\mathbb{R}^3$ such that

$$\Omega = \Phi(\Theta)$$

where the parameter domain Θ is for instance the unit square of $\mathbb{R}^2$; for computing we shall assume the effective knowledge of this mapping . Let f be a given numerical function defined on Ω. We consider the problem associated with the Beltrami equation on this surface Ω:

$$- \operatorname{Div}(\mathbf{grad}\, u) = f \quad \text{on } \Omega$$

with the Dirichlet condition : $u = 0$ on $\partial\Omega$. In the above equation, the operator Div is the covariant divergence of the tangent vector field $\mathbf{grad}\, u$, the gradient of the function u. If we consider u as a function of the parameters, u is solution of an equation of the type :

$$-\frac{1}{\sqrt{a}} \operatorname{div}\left(\sqrt{a}\,\mathbf{grad}\, u\right) = f \quad \text{on } \Theta$$

where now the operator div is the classical divergence operator of a vector field defined on a Euclidean space and where $\sqrt{a}$ is the rate of the area measure on Ω per the area measure on Θ.

The Mathematics of Finite Elements and Applications
Edited by J. R. Whiteman © 1997 John Wiley & Sons Ltd.

In Section 24.3, we will give a mixed formulation of this problem; in mixed formulations of second order problems (see [RT77] and for a more general presentation see [BF91], [RT91]), the main unknown is the vector field $\mathbf{s} = \mathbf{grad}\, u$ while the numerical function appears only as a Lagrangian multiplier. We prove the existence and the uniqueness of the solution of the covariant mixed formulation. In Section 24.4, we will analyse an abstract covariant mixed finite element method; a priori errors bounds are given. A generalization of the lowest order Raviart-Thomas mixed finite element is then presented. However, we must first describe the difficulty specifically due to a Riemannian (non Euclidean) metric. In particular we must be precise as to which unknowns are retained for the vector field : its contravariant components or its covariant components ? This choice is sensible in the construction of the finite element method. So in Section 24.2, some fundamental results of differential geometry and of tensor calculus are recalled.

24.2   TECHNICAL PRELIMINARIES

The following elementary notions of differential geometry and tensorial calculus may be found in numerous treatises on the subject; e.g. in [Lic50]; our presentation follows essentially that of [Cia96] with slight differences in the notations.
Let Θ be for example the unit square $]0,1[\,\times\,]0,1[$ of $\mathbb{R}^2$ and let $\Phi : \overline{\Theta} \to \mathbb{R}^3$ be an injective -mapping of class C^2. The surface Ω of $\mathbb{R}^3$ is defined by

$$\Omega = \Phi\,(\Theta)\,.$$

We use the Einstein summation convention, according to which a suffix which appears twice in a formula is automatically summed over; moreover we will use Latin indices and exponents belonging to the set $\{1,2,3\}$ and Greek indices and exponents belonging to the set $\{1,2\}$. The only exception to these rules will concern the index h which is kept as a discretization parameter : as usual, h will tend to zero and there is no summation convention for repeated h !
Let $(\mathbf{e}_i)$ be the canonical basis of the Euclidean space $\mathbb{R}^3$ and (x^i) be the coordinates of a point $\mathbf{x} \in \mathbb{R}^3$; with the above conventions, we write: $\mathbf{x} = x^i \mathbf{e}_i$. If (Φ^i) are the components of the mapping Φ, a generic point of the surface Ω is defined by

$$x^i = \Phi^i\,(t^\alpha)$$

where (t^α) are the coordinates of a generic point $\mathbf{t}$ of the square Θ in the Euclidean space $\mathbb{R}^2$. In other words, (t^α) are the two parameters which define a point $\mathbf{x}$ of the surface Ω. We write ∂_α in place of $\partial/\partial t^\alpha$, the partial derivative with respect to the α-th variable of the parameters space.
With each point $\mathbf{x}$ of the surface Ω, we associate the two vectors

$$\mathbf{a}_\alpha = (\partial_\alpha \mathbf{\Phi}) \circ \Phi^{-1}$$

and assume that these two vectors are linearly independent ; so they constitute a basis of the tangent plane to the surface at the point $\mathbf{x}$; this basis is called the covariant basis. The two vectors $\mathbf{a}^\alpha$ of the same tangent plane defined by the relations

$$\mathbf{a}^\alpha \cdot \mathbf{a}_\beta = \delta^\alpha_\beta$$

constitute the contravariant basis. The $\left(\delta_\beta^\alpha\right)$ are the Kronecker symbols: $\delta_\beta^\alpha = 1$ if $\alpha = \beta$, $= 0$ if not.

One then defines the metric tensor of the surface, also called the fundamental tensor, whose covariant components $a_{\alpha\beta}$ are

$$a_{\alpha\beta} := \mathbf{a}_\alpha \cdot \mathbf{a}_\beta$$

while the contravariant components $a^{\alpha\beta}$ are

$$a^{\alpha\beta} := \mathbf{a}^\alpha \cdot \mathbf{a}^\beta$$

and the Christoffel symbols $\Gamma_{\alpha\beta}^\gamma$ are given by

$$\Gamma_{\alpha\beta}^\gamma := \partial_\alpha \mathbf{a}_\beta \cdot \mathbf{a}^\gamma.$$

Note the symmetries:

$$a_{\alpha\beta} = a_{\beta\alpha},, \quad a^{\alpha\beta} = a^{\beta\alpha}, \quad \Gamma_{\alpha\beta}^\gamma = \Gamma_{\beta\alpha}^\gamma$$

and the Ricci identities:

$$\partial_\gamma a_{\alpha\beta} = \Gamma_{\gamma\alpha}^\sigma a_{\sigma\beta} + \Gamma_{\gamma\beta}^\sigma a_{\sigma\alpha}.$$

The area element along the surface Ω is $d\Omega = \sqrt{a}\,d\Theta$, where $d\Theta = dt^1 dt^2$ is the Lebesgue measure in $\mathbb{R}^2$ and

$$a := \det\left(a_{\alpha\beta}\right) = 1/\det\left(a^{\alpha\beta}\right).$$

The following result will be useful for the mixed formulation:

Lemma 24.1 *We have for $\alpha = 1, 2$*

$$\Gamma_{\alpha\beta}^\beta = \frac{1}{2a}\partial_\alpha a.$$

Proof. On the one hand, from the Ricci identities, we deduce

$$(\partial_\sigma a_{\alpha\beta}) a^{\alpha\beta} = 2\,\Gamma_{\sigma\alpha}^\alpha;$$

on the other hand, from the explicit relations between the $a^{\alpha\beta}$ and the $a_{\alpha\beta}$, it is easy to verify

$$\partial_\sigma a = a\,(\partial_\sigma a_{\alpha\beta})\,a^{\alpha\beta}.$$

Lemma 24.1 is a consequence of these two results. ☐

Now if $u : \Omega \to \mathbb{R}$ is a numerical function defined on the surface Ω, we designate by $(\partial_\alpha u)$ the covariant components of the vector field $\mathbf{grad}\,u$ defined as functions on Ω; more precisely : $\partial_\alpha u = (\partial (u \circ \Phi)/\partial t^\alpha) \circ \Phi^{-1}$.

Let $\mathbf{s} : \Omega \to T\Omega$ be a vector field defined on the surface Ω in the tangent bundle of Ω: $\mathbf{s}(\mathbf{x})$ is a vector of the tangent plane to Ω at the point $\mathbf{x} \in \Omega$; we designate by $(s^\alpha(\mathbf{x}))$ the contravariant components of this vector and $(s_\alpha(\mathbf{x}))$ the contravariant components. The correspondences between contravariant and covariant components are

$$s^\alpha = a^{\alpha\beta} s_\beta, \quad s_\alpha = a_{\alpha\beta} s^\beta.$$

The covariant derivative of the contravariant component s^β with respect to the α-th variable of the parameter space is by definition

$$D_\alpha s^\beta := \partial_\alpha s^\beta + \Gamma^\beta_{\sigma\alpha} s^\sigma.$$

The covariant derivative of the vector field is then defined by

$$\text{Div}\, \mathbf{s} := D_\alpha s^\alpha.$$

Lemma 24.2 *We have the identity*

$$\text{Div}\, \mathbf{s} = \frac{1}{\sqrt{a}} \partial_\alpha \left(\sqrt{a}\, s^\alpha \right).$$

Proof. By the definition of the covariant derivatives, we have

$$\text{Div}\, \mathbf{s} = D_\alpha s^\alpha = \partial_\alpha s^\alpha + \Gamma^\alpha_{\beta\alpha} s^\beta.$$

With the identity obtained in Lemma 24.1 one gets

$$\text{Div}\, \mathbf{s} = \partial_\alpha s^\alpha + \left(\frac{1}{2a} \partial_\beta a \right) s^\beta = \partial_\alpha s^\alpha + \left(\frac{1}{2a} \partial_\alpha a \right) s^\alpha$$

and so the above identity is obtained. $\square$

24.3   COVARIANT MIXED FORMULATION.

We consider the model problem: find a function $u : \Omega \to \mathbb{R}$ and a vector field $\mathbf{s} : \Omega \to T\Omega$ which are solutions of the first order partial differential system

$$\begin{cases} \mathbf{s} = \mathbf{grad}\, u & \text{on } \Omega \\ \text{Div}\, \mathbf{s} + f = 0 & \text{on } \Omega \end{cases} \qquad (\mathcal{P})$$

with for instance the Dirichlet boundary condition

$$u = 0 \quad \text{on } \partial\Omega.$$

One can use either the covariant components of $\mathbf{s}$ or the contravariant components; one notes than it seems more natural to use the covariant components for writing the first equation of $(\mathcal{P})$, which is in physical problems associated with a constitutive law, whereas it seems more natural to use the contravariant components for writing the second equation of $(\mathcal{P})$, which is associated with the equilibrium principle. With the use of these covariant and contravariant components, the system $(\mathcal{P})$ becomes

$$\begin{cases} s_\alpha = \partial_\alpha u & \text{on } \Omega \\ D_\alpha s^\alpha + f = 0 & \text{on } \Omega. \end{cases}$$

In the mixed formulation, equilibrium preservation is fundamental; so we choose to keep as the main unknowns u and (s^α) the contravariant components of the vector field $\mathbf{s}$; as a consequence the first equation of $(\mathcal{P})$ is now written:

$$s^\alpha = a^{\alpha\beta}\,\partial_\beta u \quad \text{on } \Omega.$$

We introduce the Sobolev spaces

$$W = H(\mathrm{Div};\Omega) := \left\{\mathbf{r} = (r^\alpha) \in \left(L^2(\Omega)\right)^2 ; \ \mathrm{Div}\,\mathbf{r} = D_\alpha r^\alpha \in L^2(\Omega)\right\}$$

and

$$M = L^2(\Omega).$$

Equipped with the norm

$$\|\mathbf{r}\|_W = \left\{\|a_{\alpha\beta} r^\alpha r^\beta\|_{0,\Omega} + \|D_\alpha r^\alpha\|_{0,\Omega}^2\right\}^{\frac{1}{2}}$$

$$\|v\|_M = \|v\|_{0,\Omega}$$

where

$$\|v\|_{0,\Omega}^2 := \int_\Omega |v|^2\,d\Omega = \int_\Theta |v|^2\,\sqrt{a}\,d\Theta$$

the spaces W and M are Hilbert spaces. Using the identity obtained in Lemma 24.2, one can remark that this Hilbert space $W = H(\mathrm{Div};\Omega)$ is isomorphic to the Hilbert space $H(\mathrm{div};\Omega)$ equipped with the norm

$$\|\mathbf{r}\|_{H(div;\Omega)} = \left\{\|r^1\|_{0,\Omega}^2 + \|r^2\|_{0,\Omega}^2 + \|\partial_\alpha r^\alpha\|_{0,\Omega}^2\right\}^{\frac{1}{2}};$$

however this norm has no link with the physics of the problem.

The mixed formulation of problem $(\mathcal{P})$ is :

$$\begin{cases} \mathbf{s} \in W; \ u \in M; \\[2mm] \forall \mathbf{r} \in W, \quad \displaystyle\int_\Omega a_{\alpha\beta}\,s^\beta\,r^\alpha\,d\Omega + \int_\Omega u\,D_\alpha r^\alpha\,d\Omega = 0 \\[2mm] \forall v \in M, \quad \displaystyle\int_\Omega v\,D_\alpha s^\alpha\,d\Omega = -\int_\Omega f v\,d\Omega . \end{cases} \qquad (24.1)$$

Proposition 24.1 For all $f \in L^2(\Omega)$, the formulation (24.1) has a unique solution $(\mathbf{s}, u) \in W \times M$. Moreover u is the solution of the variational problem

$$\begin{cases} u \in H_0^1(\Omega); \\[2mm] \forall v \in H_0^1(\Omega), \quad \displaystyle\int_\Omega a^{\alpha\beta}\,\partial_\beta u\,\partial_\alpha v\,d\Omega = \int_\Omega f v\,d\Omega \end{cases} \qquad (24.2)$$

and the pair (s,u) is solution of $(\mathcal{S})$.

Before giving the proof of this Proposition, let us state an essential tool in the analysis of problems with covariant derivatives.

Lemma 24.3 *For all $v \in H_0^1(\Omega)$ and all $\mathbf{r} \in H(\mathrm{Div}; \Omega)$ we have the Green's formula*

$$\int_\Omega r^\alpha \, \partial_\alpha v \, d\Omega = -\int_\Omega v \, D_\alpha r^\alpha \, d\Omega.$$

Proof. Using a classical Green's formula on the Euclidean domain Θ, we get for $v \in H_0^1(\Omega)$ and $\mathbf{r} \in H(\mathrm{Div}; \Omega)$

$$\int_\Omega r^\alpha \, \partial_\alpha v \, d\Omega = -\int_\Omega v \, \frac{1}{\sqrt{a}} \, \partial_\alpha \left(\sqrt{a} r^\alpha\right) d\Omega.$$

Using Lemma 24.3, we obtain the above Green's formula on Ω. $\square$

Proof of the Proposition 24.1. For the uniqueness of the solution of Problem (24.1), it is sufficient by linearity to show that the homogeneous (case $f = 0$) problem (24.1) admits only the trivial solution $\mathbf{s} = \mathbf{0}$, $u = 0$.

For the existence of the solution of Problem (24.1), let u be the solution of (24.2) and $\mathbf{s}$ be the vector field defined by $\mathbf{s} = \mathbf{grad}\, u$. We have $u \in H_0^1(\Omega)$, subspace of M, and using Green's formula one proves that $\mathbf{s}$ belongs to W. Finally one can verify that this pair $(\mathbf{s}, u)$ is the solution of (24.1). $\square$

Remark 24.1 *A more instructive proof of this Proposition consists of using the Babuška-Brezzi theory framework for mixed methods (see [BF91], [RT91]). In particular, one proves the LBB-condition :*

$$\inf_{v \in M} \sup_{\mathbf{r} \in W} \frac{\displaystyle\int_\Omega v \, Div\, \mathbf{r} \, d\Omega}{\|\mathbf{r}\|_W \, \|v\|_M} > 0;$$

For this proof, see [Tho96].

24.4   COVARIANT MIXED FINITE ELEMENT

Let W_h be a finite dimensional subspace of M and M_h be a finite dimensional subspace of M. We consider the discrete problem : find a pair $(\mathbf{s}_h, u_h)$ which is the solution of

$$\begin{cases} \mathbf{s}_h \in W_h \,;\; u_h \in M_h; \\[2mm] \forall \mathbf{r}_h \in W_h, \quad \displaystyle\int_\Omega a_{\alpha\beta}\, s_h^\beta \, r_h^\alpha \, d\Omega + \int_\Omega u_h \, D_\alpha r_h^\alpha \, d\Omega = 0 \\[2mm] \forall v_h \in M_h, \quad \displaystyle\int_\Omega v_h \, D_\alpha s_h^\alpha \, d\Omega = -\int_\Omega f \, v_h \, d\Omega. \end{cases} \qquad (24.3)$$

Proposition 24.2 *With the assumption :*

$$\forall v_h \in M_h, \; \exists \mathbf{r}_h \in W_h, \; \int_\Omega v_h \, D_\alpha r_h^\alpha \, d\Omega = \|v_h\|_M^2$$

we have the existence and the uniqueness of the solution of Problem (3). Moreover, if we assume more precisely that there exists a constant c, independent of h, such that

$$\forall v_h \in M_h, \ \exists \mathbf{r}_h \in W_h, \ \begin{cases} \| \mathbf{r}_h \|_W \leq c \, \| v_h \|_M \\ \text{and} \\ \displaystyle\int_\Omega v_h \, D_\alpha r_h^\alpha \, d\Omega = \| v_h \|_M^2 \end{cases}$$

then we have the a priori error bound between $(\mathbf{s}, u)$ the solution of (24.1), and $(\mathbf{s}_h, u_h)$, the solution of (24.3), :

$$\begin{cases} \| \mathbf{s} - \mathbf{s}_h \|_W + \| u - u_h \|_M \leq \\ \qquad C \, (\inf_{\mathbf{r}_h \in W_h} \| \mathbf{s} - \mathbf{r}_h \|_W + \inf_{v_h \in M_h} \| u - v_h \|_M) \end{cases}$$

with C a constant which is independent of h.

The proof of these abstract results of Proposition 24.2 are straightforward.

As a first example of this theory, we consider now the following finite element approximation. Let $\mathcal{T}_h$ be a regular triangulation of the square Θ :

$$\overline{\Theta} = \bigcup_{T \in \mathcal{T}_h} T$$

where all T are triangles (e.g. isosceles right angled triangles) of diameter no greater than h. Then

$$\overline{\Omega} = \bigcup_{T \in \mathcal{T}_h} \Phi(T)$$

is a triangulation of the surface Ω by the curvilinear triangles $K = \Phi(T)$. We designate by $P_0(T)$ the space of functions which are constant on T; we designate by $D_1(T)$ the lowest order Raviart-Thomas space of vector functions defined on T :

$$D_1(T) = (P_0(T))^2 + \mathbf{t} \, P_0(T).$$

In the literature this space D_1 is often denoted RT_0, see [WY95], [Whe96]. For $K = \Phi(T)$, $D_1(K)$ will be the space

$$D_1(K) = \left\{ \mathbf{r} \in (L^2(K))^2 \, ; \, \mathbf{r} \circ \Phi \in D_1(T) \right\}$$

while $P_0(K) = \{v \in L^2(K) \, ; \, v \circ \Phi \in P_0(T)\}$ is nothing other than the space of functions which are constant on K.

Now we make the choices :

$$W_h = \{\mathbf{r}_h \in H(\mathrm{Div}; \Omega); \, \sqrt{a} r_h \in D_1(K) \quad \text{for all } K\} \qquad (24.4)$$

and

$$M_h = \{v_h \in L^2(\Omega) \, ; \, v \in P_0(K) \text{ for all } K\}. \qquad (24.5)$$

In this case, one can prove :

Proposition 24.3 *With the choices (24.4) for W_h and (24.5) for M_h, Problem 24.3 has a unique solution and we have the error bound*

$$\|\mathbf{s} - \mathbf{s}_h\|_{H(\mathrm{Div};\Omega)} + \|u - u_h\|_{L^2(\Omega)} \leq C\,h\,\|u\|_{H^2(\Omega)}\,.$$

Remark 24.2 *For v_h given in M_h, we cannot find $\mathbf{r}_h$ in W_h such that $\mathrm{Div}\,\mathbf{r}_h = v_h$ since for $\mathbf{r}_h \in W_h$ the function $\mathrm{Div}\,\mathbf{r}_h$ is not constant over the finite element K. However the discrete LBB-condition holds under the current assumption.*

References

[BF91]   Brezzi F. and Fortin M. (1991) *Mixed and Hybrid Finite Element Methods.* Springer Verlag, Amsterdam.

[Cia96] Ciarlet P. G. (1996) *Mathematical Elasticity*, volume II: Plates and Shells. North-Holland, Amsterdam.

[Lic50] Lichnerowicz A. (1950) Eléments de calcul tensoriel. In *Collection Armand Colin, N 259 Section de Mathématiques.* Armand Colin, Paris.

[RT77]   Raviart P. A. and Thomas J.-M. (1977) A mixed finite element methods for second order elliptic problems. In Galligani I. and Magenes E. (eds) *Mathematical Aspects of Finite Element Methods*, volume 606, Lecture Notes in Mathematics, pages 292–315. Springer Verlag, Berlin.

[RT91]   Roberts J. E. and Thomas J.-M. (1991) Mixed and hybrid methods. In Ciarlet P. and Lions J. (eds) *Handbook of Numerical Analysis*, volume II, Finite Element Methods (Part 1), pages 527–637. Elsevier Science (North-Holland), Amsterdam.

[Tho96] Thomas J.-M. (1996) Méthodes d'éléments finis mixtes covariants. (to appear).

[TT96]   Tardea G. and Thomas J.-M. (1996) Dualisation du cisaillement transverse dans le modéle de Naghdi pour la flexion des coques minces. (to appear).

[Whe96] Wheeler M. F. (1996) Mixed finite element methods for modelling multiphase flow in porous media. In *MAFELAP 1996, (These Proceedings)*, pages 223–234.

[WY95]   Wheeler M. F. and Yotov I. (1995) Mixed finite element methods for modeling flow and transport in porous media. In Bourgeat A. and Others (eds) *Mathematical Modelling of Flow Trough Porous Media*, pages 337–357. World Scientific Publishing, Singapore.

25

Stress Singularities in Bonded Elastic Materials

Andreas Rössle and Anna-Margarete Sändig

Mathematisches Institut A/6,
Universität Stuttgart,
Pfaffenwaldring 57,
D-70511 Stuttgart, Germany.

25.1 INTRODUCTION

The structure of the asymptotic expansions for solutions of two-dimensional linear elastic boundary-transmission problems near corner points is investigated with respect to the multiplicities of the eigenvalues of a corresponding generalised eigenvalue problem. The coefficients in the expansions can increase in neighbourhoods of critical angles and of critical values of material parameters, where the type of the asymptotic expansions change. We analyse this situation for the Neumann-transmission problem for dissimilar isotropic materials under mechanical and thermal loadings. Programs, which find the critical angles and material parameters and calculate the singular terms, have been developed. We compare our results with FEM-results. We also remark that stress singularities of elastic fields caused by peculiarities of the domain, discontinuity of boundary conditions or inhomogeneous materials are of great significance in the mechanics of brittle fracture and in the mechanics of composite media.

We consider two-dimensional linear elastic problems in domains consisting of two subdomains Ω_1 and Ω_2 connected (bonded) along an interface (see Figure 1). We formulate the corresponding boundary-transmission problem and analyse the behaviour of the solutions near the corner point O described by an asymptotic expansion of the solution with respect to the distance r to the corner point. The structure of the expansion is strongly influenced by the mechanical and thermal loads, the geometry and the dissimilar materials.

We have to distinguish between two cases: First, the loads vanish in a neighbourhood of the corner points which means that we can use the theory of Kondrat'ev [Kon67], Maz'ya and Plamenevskij [MP84a], [MP84b] for elliptic boundary value problems in domains with conical points and the results of Nicaise and Sändig [NS94] for

The Mathematics of Finite Elements and Applications
Edited by J. R. Whiteman © 1997 John Wiley & Sons Ltd.

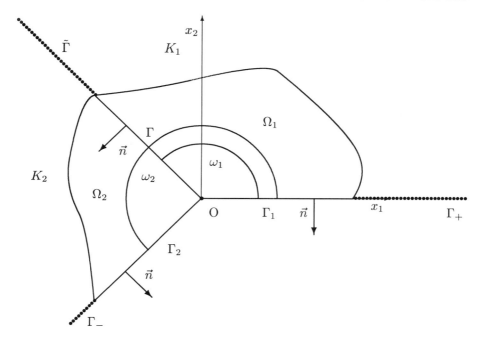

Figure 1   *Bonded dissimilar elastic materials with a notch*

interface problems both developed in appropriate weighted Sobolev spaces. Secondly, the presence of loads in the vicinity of corner points means that the right-hand sides of the governing differential equations are from standard Sobolev spaces. In this case additional singular terms can occur in the asymptotic expansions [Gri85], [MR92]. In particular, temperature loads do not vanish in neighbourhoods of corner points and therefore this situation is of special interest in mechanics [FMY93], [MY93], [Yos96]. In both cases above the form of the asymptotic expansion can change with respect to a perturbation of the geometry [more precisely of the opening angles ω_i (see Figure 1)] and with respect to a perturbation of the elastic material parameters in Ω_i. The reason for this is that a generalised boundary-eigenvalue problem for an operator bundle (resulting from a Mellin-transformation) generates an asymptotic expansion, and the different algebraic and geometrical multiplicities of the corresponding eigenvalues yield different asymptotic expansions. The coefficients in the asymptotic expansions are very sensitive (unstable) with respect to changes of the form of the asymptotics; they can increase and yield unwieldy asymptotic terms.

In this paper we analyse this situation for a boundary-transmission problem for a joint consisting of two dissimilar homogeneous, isotropic materials. At the exterior boundary segments the stress-vectors are given, and at the interface we have stress-jumps. The eigenvalue "one" with the rotational rigid motion as eigensolution is significant in this case, especially for different constant temperature fields.

The paper is organised as follows: First we formulate the problem and describe briefly the mathematical framework which yields the asymptotic expansions. Then we derive a method for the calculation of the asymptotic terms when the right-hand sides of the differential equations are from standard Sobolev spaces. This method yields a numerical algorithm which allows us to indicate the "critical" angles and those material parameters where the asymptotics changes, thus influencing the calculation of the singular terms. Finally, we give a numerical example and compare our results with those produced with the FEM [FMY93], [Leg88].

25.2 THE PROBLEM

Let Ω be a two-dimensional bounded domain consisting of two domains Ω_1 and Ω_2 of different isotropic or anisotropic elastic materials. The boundaries $\partial\Omega_i$, $i = 1, 2$, are divided into two parts; one part is the common boundary segment Γ (the interface), where transmission conditions are considered; the other part Γ_i consists of boundary segments, where boundary conditions are given (see Figure 1). There can be corner points; cracks ($\omega_1 = 2\pi$), or interface cracks ($\omega_2 = 2\pi$) are included as limit cases. The equilibrium equations and Hooke's law for small deformations yield the boundary value problem for the displacement fields u_k under mechanical loading

$$
\begin{aligned}
L_k\,u_k &= D^T C_k D\,u_k &= -f_k &\quad \text{in } \Omega_k \\
\sigma_k[u_k]\,n &= N^T C_k D\,u_k &= g_k &\quad \text{on } \Gamma_k
\end{aligned}
\;,
$$

where

$$
D^T = \begin{pmatrix} \partial_1 & 0 & \partial_2 \\ 0 & \partial_2 & \partial_1 \end{pmatrix}, \; C_k = \begin{pmatrix} C^{(k)}_{1111} & C^{(k)}_{1122} & C^{(k)}_{1112} \\ C^{(k)}_{1122} & C^{(k)}_{2222} & C^{(k)}_{2212} \\ C^{(k)}_{1112} & C^{(k)}_{2212} & C^{(k)}_{1212} \end{pmatrix}, \, N^T = \begin{pmatrix} n_1 & 0 & n_2 \\ 0 & n_2 & n_1 \end{pmatrix};
$$

$n = \begin{pmatrix} n_1 \\ n_2 \end{pmatrix}$ is the exterior unit normal vector and the C_k are symmetric matrices satisfying a strong ellipticity condition:

$$
\sum_{i,j,m,n=1}^{2} C^{(k)}_{ijmn}\, \xi_{ij}\, \overline{\xi}_{mn} \ge M \sum_{i,j=1}^{2} |\xi_{ij}|^2 \quad \forall \xi_{ij},\, \xi_{mn} \in \mathbb{C}.
$$

The transmission conditions on Γ read:

$$
\left.
\begin{aligned}
u_1 - u_2 &= 0 \\
\sigma_1[u_1]\,n - \sigma_2[u_2]\,n &= 0
\end{aligned}
\right\} \text{on } \Gamma ,
$$

where n is the exterior unit normal vector with respect to Ω_1. If additionally thermal loadings act in Ω_k then we have

$$
L_k\,u_k = -f_k + A_k\,\text{grad}\,T_k \quad \text{in } \Omega_k
$$

$$
\sigma_k[u_k]\,n = g_k + A_k\,T_k\,n \quad \text{on } \Gamma_k \tag{25.1}
$$

$$
\left.
\begin{aligned}
u_1 - u_2 &= 0 \\
\sigma_1[u_1]\,n - \sigma_2[u_2]\,n &= [A_1\,T_1 - A_2\,T_2]\,n
\end{aligned}
\right\} \text{on } \Gamma ,
$$

where $A_k = \left(a_{ij}^{(k)}\right)_{i,j=1,2}$ is the thermal expansion matrix with $a_{ij}^{(k)} =$
$\sum\limits_{m,l=1}^{2} C_{ijml}^{(k)} \alpha_{ml}^{(k)} = const$, $a_{ij}^{(k)} = a_{ji}^{(k)}$. T_k denotes the steady state temperature field in Ω_k and satisfies a boundary-transmission problem

$$a_{11}^{(k)}\frac{\partial^2 T_k}{\partial x_1^2} + 2a_{12}^{(k)}\frac{\partial^2 T_k}{\partial x_1 \partial x_2} + a_{22}^{(k)}\frac{\partial^2 T_k}{\partial x_2^2} = F_k \quad \text{in } \Omega_k$$

$$N_k(T_k) := \sum_{i,j=1}^{2} a_{ji}^{(k)}\frac{\partial T_k}{\partial x_j}\cos(n,x_i) = G_k \quad \text{on } \Gamma_k \tag{25.2}$$

$$\left.\begin{array}{rcl} T_1 - T_2 & = & H_1 \\ N_1(T_1) - N_2(T_2) & = & H_2 \end{array}\right\} \text{on } \Gamma ,$$

$k = 1, 2$. We consider solutions $T_k \in W_2^1(\Omega_k)$ of (25.2), which are uniquely determined except for constants. From regularity results [Kon67], [NS94], it follows that $T_k \in W_p^1(\Omega_k)$ for $p > 2$ for appropriate F_k, G_k, H_1 and H_2. Therefore the right-hand sides of (25.1) are from $L_p(\Omega_k)$, $W_p^{1-\frac{1}{p}}(\Gamma_k)$ and $W_p^{1-\frac{1}{p}}(\Gamma)$, respectively. Thus, there is an asymptotic expansion of the solutions $u_k \in W_2^1(\Omega_k)$ of (25.1) [Kon67], [MP84a], [MP84b], [NS94], [Sän96] in a neighbourhood of the corner point O (see Figure 1) with respect to the distance r:

$$\begin{pmatrix} u_1(r,\omega) \\ u_2(r,\omega) \end{pmatrix} = \eta\left[\sum_{\substack{0 < \operatorname{Re}\alpha_i < 2 - \frac{2}{p} \\ \alpha_i \neq 1}} c_i\, r^{\alpha_i} \begin{pmatrix} \phi_1(\ln r, \omega) \\ \phi_2(\ln r, \omega) \end{pmatrix}\right.$$

$$\left.+r\begin{pmatrix} v_1(\ln r, \omega) \\ v_2(\ln r, \omega) \end{pmatrix}\right] + \begin{pmatrix} w_1(r,\omega) \\ w_2(r,\omega) \end{pmatrix} , \tag{25.3}$$

where $w_k(r,\omega) \in W_p^2(\Omega_k)$, $k = 1, 2$ and η is a cut-off-function. Usually, the terms with $\operatorname{Re}\alpha_i < 1$ are called singular terms, since they yield unbounded stress singularities and the corresponding factors are called generalised stress intensity factors. There are integral formulae [MP84b], [BMS96] for the factors c_i ($\alpha_i \neq 0$ or 1), in which the entire right-hand sides of (25.1) occur, independently, whether they describe mechanical or thermal loadings. The term $r\begin{pmatrix} v_1(\ln r, \omega) \\ v_2(\ln r, \omega) \end{pmatrix}$, which also implies stress singularities in general, is not easy to handle, especially if we have loadings which do not vanish at the corner point O; e.g. constant temperature fields in Ω_k, $k = 1, 2$.

We analyse in this paper the situation for $\alpha = 1$ restricted to isotropic, homogeneous, elastic materials under the influence of constant temperature fields:

$$\begin{array}{rcl} L_k u_k & = & \mu_k \Delta u_k + (\lambda_k + \mu_k)\operatorname{grad} \operatorname{div} u_k = 0 \quad \text{in } \Omega_k \\ \sigma_k[u_k]\,n & = & \alpha_k[3\lambda_k + 2\mu_k]T_k\,n \quad \text{on } \Gamma_k \end{array} \tag{25.4}$$

$$\left.\begin{array}{rcl} u_1 - u_2 & = & 0 \\ (\sigma_1[u_1] - \sigma_2[u_2])\,n & = & [\alpha_1(3\lambda_1 - 2\mu_1)T_1 - \alpha_2(3\lambda_2 + 2\mu_2)T_2]\,n \end{array}\right\} \text{on } \Gamma ,$$

$\alpha_1 = const$, $\alpha_2 = const$, $T_1 = const$, $T_2 = const$. We study the following problem:
*Investigate the instabilities of the asymptotics (and the coefficients c_i) if α_i lies in a
neighbourhood of the "eigenvalue" $\alpha = 1$ and derive an algorithm for the calculation
of the displacement fields v_1 and v_2.*

25.3 CALCULATION OF THE ASYMPTOTICS

The asymptotic expansion can be calculated by considering only the local behaviour
of the solutions near corner points, having in mind that only the coefficients c_i depend
on the complete loads. Therefore we localise our problems, multiplying the solution u
by a cut-off function $\eta(x) = \eta(|\underline{x}|) = \eta(r) \in C^\infty(\overline{\mathbb{R}}_+)$, where

$$\eta(r) = \begin{cases} 1 & \text{for } r < \frac{\delta}{2} \\ 0 & \text{for } r > \delta \end{cases} .$$

Here δ is a small positive number such that the domain $\Omega_1 \cup \Omega_2$ coincides in a δ-
neighbourhood of the corner point with a double wedge $K_1 \cup K_2$ (see Figure 1). For
the calculation of the terms $r^{\alpha_i} \phi_i(\ln r, \varphi)$ we have to find non-trivial solutions of the
problem

$$L_k u_k = 0 \quad \text{in } K_k$$

$$\sigma_k[u_k] n = 0 \quad \text{on } \Gamma^\pm \tag{25.5}$$

$$\left. \begin{aligned} u_1 - u_2 &= 0 \\ \sigma_1[u_1] n - \sigma_2[u_2] n &= 0 \end{aligned} \right\} \text{on } \tilde{\Gamma} \quad,$$

where $\Gamma^\pm$ and $\tilde{\Gamma}$ are the edges of the corresponding infinite cones (see Figure 1).
Using polar coordinates and the Mellin-transformation with respect to r ($r\partial_r \to \alpha$)
or special ansatz-functions $s(\alpha, \varphi)$ we get a boundary-transmission problem

$$\mathcal{A}(\alpha)\, \hat{u} = \mathcal{A}(\alpha, \partial_\omega, \omega)\, \hat{u}(\alpha, \omega) = 0 \tag{25.6}$$

with ordinary differential operators, which depend on the complex parameter α. The
distribution of the generalised eigenvalues α_i of $\mathcal{A}(\alpha)$ in the strip $0 < \operatorname{Re}\alpha_i < 2 - \frac{2}{p}$
together with the Jordan chains of the corresponding eigenfunctions and associated
eigenfunctions lead, for the first sum in the asymptotic expansion (25.3), to:

$$\sum_{0<\operatorname{Re}\alpha_i<2-\frac{2}{p}} \sum_{\ell=1}^{I_i} \sum_{j=0}^{m_{i\ell}-1} c_{i\ell j}\, r^{\alpha_i} \left(\sum_{k=0}^{j} \frac{1}{k!} (\ln r)^k \begin{pmatrix} e^{(1)}_{i,\ell,j-k}(\alpha_i,\omega) \\ e^{(2)}_{i,\ell,j-k}(\alpha_i,\omega) \end{pmatrix} \right) , \tag{25.7}$$

where I_i is the geometrical multiplicity of α_i and $m_{i\ell}$ denotes the length of the Jordan
chains; that means

$$\sum_{q=0}^{m} \frac{1}{q!} \left(\frac{\partial}{\partial\alpha} \right)^q \mathcal{A}(\alpha)\, e_{i,\ell,m-q}(\alpha) \Big|_{\alpha=\alpha_i} = 0 \tag{25.8}$$

for $m = 0, 1, \ldots, m_{i\ell} - 1$, where $e_{i,\ell,m-q}(\alpha) = \begin{pmatrix} e^{(1)}_{i,\ell,m-q}(\alpha, \omega) \\ e^{(2)}_{i,\ell,m-q}(\alpha, \omega) \end{pmatrix}$. $\mathcal{A}(\alpha)$ consists

of the Mellin-transformed differential operators $\mathcal{L}(\alpha) = \begin{pmatrix} \mathcal{L}_1(\alpha) \\ \mathcal{L}_2(\alpha) \end{pmatrix}$ and the Mellin-

transformed boundary and transmission operators. The Riesz-index is defined by

$$R(\alpha_i) = \max_{\ell=1,\ldots,I_i} \{m_{i\ell}\}.$$

The calculation of the eigenvalues α_i of the Jordan chains

$$\begin{pmatrix} e^{(1)}_{i,\ell,k} \\ e^{(2)}_{i,\ell,k} \end{pmatrix} \qquad \begin{aligned} i &= 1, \ldots, N_0 \\ \ell &= 1, \ldots, I_i \\ k &= 0, \ldots, m_{i\ell} - 1 \end{aligned}$$

in (25.7) and of the field $\begin{pmatrix} v_1 \\ v_2 \end{pmatrix}$ in (25.3) is relatively easy, starting with a known

fundamental system of solutions $\left\{ f_1^{(k)}, \ldots, f_4^{(k)} \right\}_{k=1,2} = \{z_1, \ldots, z_8\}$ of the Mellin-

transformed Lamé equation systems:

$$\mathcal{L}_k(\alpha, \omega)\, \hat{u}_k(\alpha, \omega) = 0, \quad k = 1, 2.$$

Inserting $\sum_{i=1}^{8} d_i(\alpha) z_i(\alpha, \omega)$ into the Mellin-transformed boundary and interface

conditions we get a linear system of equations for the coefficient vector $d(\alpha) = (d_1(\alpha), \ldots, d_8(\alpha))^T$:

$$\mathcal{M}(\alpha)\, d(\alpha) = 0.$$

There are non-trivial solutions, if $\det \mathcal{M}(\alpha) = 0$, which yield the eigenvalues α_i. Furthermore, we get the Jordan chains for α_i by setting

$$e_{i,\ell,0}(\alpha_i, \omega) = \left. \sum_{j=1}^{8} d_j(\alpha) z_j(\alpha, \omega) \right|_{\alpha=\alpha_i}$$

$$e_{i,\ell,1}(\alpha_i, \omega) = \left. \frac{1}{1!} \frac{\partial}{\partial \alpha} \left[\sum_{j=1}^{8} d_j(\alpha) z_j(\alpha, \omega) \right] \right|_{\alpha=\alpha_i}$$

$$= \left. \sum_{j=1}^{8} \left[\tfrac{\partial}{\partial \alpha} d_j(\alpha) z_j(\alpha, \omega) + d_j(\alpha) \tfrac{\partial}{\partial \alpha} z_j(\alpha, \omega) \right] \right|_{\alpha=\alpha_i}$$

$$\vdots$$

$$e_{i,\ell,R(\alpha_i)-1}(\alpha_i, \omega) = \left. \frac{1}{(R(\alpha_i)-1)!} \frac{\partial^{R(\alpha_i)-1}}{\partial \alpha^{R(\alpha_i)-1}} \left[\sum_{j=1}^{8} d_j(\alpha) z_j(\alpha, \omega) \right] \right|_{\alpha=\alpha_i}$$

$$= \left. \frac{1}{(R(\alpha_i)-1)!} \sum_{j=1}^{8} \left[\sum_{k=0}^{R(\alpha_i)-1} \binom{R(\alpha_i)-1}{k} \frac{\partial^{R(\alpha_i)-1-k}}{\partial \alpha^{R(\alpha_i)-1-k}} d_j(\alpha) \frac{\partial^k}{\partial \alpha^k} z_j(\alpha, \omega) \right] \right|_{\alpha=\alpha_i}.$$

$$(25.9)$$

The coefficients $d_j(\alpha)\big|_{\alpha=\alpha_i}$, $\frac{\partial}{\partial\alpha}d_j(\alpha)\big|_{\alpha=\alpha_i}$, $\ldots$, $\frac{\partial^{R(\alpha_i)-1}}{\partial\alpha^{R(\alpha_i)-1}}d_j(\alpha)\big|_{\alpha=\alpha_i}$ are calculated from the relations

$$\sum_{\nu=0}^{\rho}\frac{1}{\nu!}\frac{\partial^\nu}{\partial\alpha^\nu}[\mathcal{M}(\alpha)]\frac{1}{(\rho-\nu)!}\frac{d^{\rho-\nu}}{d\alpha^{\rho-\nu}}d(\alpha)\bigg|_{\alpha=\alpha_i}=0 \quad , \quad \rho=0,\ldots,R(\alpha_i)-1. \quad (25.10)$$

The computation of the eigenvalues via the determinant method is done by a Newton-bisection-method; their distribution is shown in Figure 2.

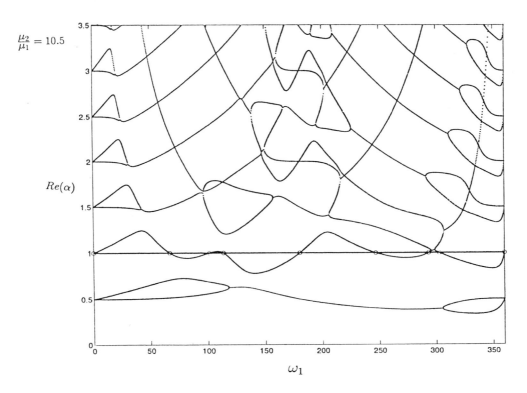

$\frac{\mu_2}{\mu_1}=10.5$

$Re(\alpha)$

ω_1

Figure 2 *Distribution of the eigenvalues for the interface crack in a joint* ($\nu_1=0.17$, $\nu_2=0.29$ *(Poisson's ratios)*, $\frac{\mu_1}{\mu_2}=10.5$, $\omega_2=2\pi$)

We now calculate the term $r\begin{pmatrix}v_1(\ln r,\omega)\\v_2(\ln r,\omega)\end{pmatrix}=r\,v(\ln r,\omega)$ in the expansion (25.3) for the problem (25.4) with constant right-hand sides. From the general theory [Kon67] it follows that problem (25.4) has a particular solution rv with

$$v=v(\ln r,\omega)=\sum_{i=0}^{R(1)}\frac{1}{i!}(\ln r)^i\,\psi_{R(1)-i}(\omega)\,.$$

Lemma 25.1 *The vector functions $\psi_{R(1)-i}(\omega)$ can be calculated by the following algorithm:*

$$\psi_0(\alpha,\omega)\Big|_{\alpha=1} \quad = \quad \psi_0(\omega) \quad = \quad \sum_{i=1}^{8} c_{0,i}\, z_i(\alpha,\omega)\Big|_{\alpha=1}$$

$$\psi_1(\alpha,\omega)\Big|_{\alpha=1} \quad = \quad \psi_1(\omega) \quad = \quad \sum_{i=1}^{8} \left[c_{1,i}\, z_i(\alpha,\omega) + c_{0,i}\tfrac{\partial}{\partial\alpha} z_i(\alpha,\omega)\right]\Big|_{\alpha=1}$$

$$\vdots$$

$$\psi_{R(1)}(\alpha,\omega)\Big|_{\alpha=1} \;=\; \psi_{R(1)}(\omega) \;=\; \tfrac{1}{R(1)!} \sum_{i=1}^{8}\left[\sum_{k=0}^{R(1)} \binom{R(1)}{k} c_{R(1)-k,i}\tfrac{\partial^k}{\partial\alpha^k} z_i(\alpha,\omega)\right]\Big|_{\alpha=1},$$

where the $(8\,[R(1)+1])$ unknown coefficients are solutions of the following equation system:

$$\sum_{\nu=0}^{\rho} \frac{1}{\nu!}\frac{\partial^\nu}{\partial\alpha^\nu}\left[\mathcal{M}(\alpha)\right]\Big|_{\alpha=1}\frac{1}{(\rho-\nu)!}c_{\rho-\nu} = 0 \quad for\ \rho=0,\ldots,R(1)-1 \qquad (25.11)$$

$$\sum_{\nu=0}^{R(1)} \frac{1}{\nu!}\frac{\partial^\nu}{\partial\alpha^\nu}\left[\mathcal{M}(\alpha)\right]\Big|_{\alpha=1}\frac{1}{(R(1)-\nu)!}c_{R(1)-\nu} = g \quad , \qquad (25.12)$$

where

$$g = \begin{pmatrix} \alpha^{(1)}\left(3\lambda^{(1)}+2\mu^{(1)}\right)T^{(1)}\,n_1^{(1)} \\ \alpha^{(1)}\left(3\lambda^{(1)}+2\mu^{(1)}\right)T^{(1)}\,n_2^{(1)} \\ \alpha^{(2)}\left(3\lambda^{(2)}+2\mu^{(2)}\right)T^{(2)}\,n_1^{(2)} \\ \alpha^{(2)}\left(3\lambda^{(2)}+2\mu^{(2)}\right)T^{(2)}\,n_2^{(2)} \\ 0 \\ 0 \\ \left[\alpha^{(1)}\left(3\lambda^{(1)}+2\mu^{(1)}\right)T^{(1)} - \alpha^{(2)}\left(3\lambda^{(2)}+2\mu^{(2)}\right)T^{(2)}\right]n_1^{(1)} \\ \left[\alpha^{(1)}\left(3\lambda^{(1)}+2\mu^{(1)}\right)T^{(1)} - \alpha^{(2)}\left(3\lambda^{(2)}+2\mu^{(2)}\right)T^{(2)}\right]n_2^{(1)} \end{pmatrix}$$

and $c_{\rho-\nu}$ is the vector with the components $c_{\rho-\nu,i}$, $i=1,\ldots,8$.

Sketch of the proof:

First we have to show that $rv = \sum_{i=0}^{R(1)} \frac{1}{i!}(\ln r)^i\, \psi_{R(1)-i}(\omega) = \begin{pmatrix} rv_1 \\ rv_2 \end{pmatrix}$ satisfies the equations

$$\mathcal{L}(r\partial_r,\omega,\partial_\omega)\, rv = \begin{pmatrix} \mathcal{L}_1(r\partial_r,\omega,\partial_\omega)\, rv_1 \\ \mathcal{L}_2(r\partial_r,\omega,\partial_\omega)\, rv_2 \end{pmatrix} = 0 \quad in\ \Omega = \Omega_1 \cup \Omega_2$$

using ideas in [MP84b]. Furthermore we have to guarantee that the system (25.11), (25.12) has a solution. The approach is the following: Take an appropriate (non-trivial) solution of (25.11), which always exists, and insert it into the system (25.12). Then we get

$$[\mathcal{M}(\alpha)]\Big|_{\alpha=1} c_{R(1)} = g - \sum_{\nu=1}^{R(1)} \frac{1}{\nu!} \frac{\partial^\nu}{\partial \alpha^\nu} \mathcal{M}(\alpha) \frac{1}{(R(1)-\nu)!} c_{R(1)-\nu} \quad .$$

The new right-hand side satisfies the solvability condition, which will be discussed for various cases.

25.4   INSTABILITIES AND NUMERICAL RESULTS

Some of the coefficients c_i in the asymptotic expansion (25.3) [or more precisely $c_{i\ell j}$ in (25.7)] and some of the coefficients of v can increase if we are near "critical" angles and critical values of the material parameters, that means near such geometrical or structural peculiarities, where the type of asymptotics changes. This situation was handled in [CD93a], [CD93b], [CD94], [MR92], [MR94] and [Sän96], where stabilisation procedures are also given.

Here we use the algorithm (25.11), (25.12) to calculate the coefficients $c_{\nu,i}$ of the functions v. The unboundedness of some of these coefficients allows a change of the Riesz-index, and therefore a change in the asymptotic expansion to be indicated. Let us illustrate this by an example: In most cases the geometrical multiplicity and the Riesz-index equal one for the eigenvalue $\alpha = 1$ (see Figure 2). This leads to the function

$$\text{v} = \sum_{i=1}^{8} \left[c_{1,i} z_i(\alpha, \omega) + c_{0,i} \frac{\partial}{\partial \alpha} z_i(\alpha, \omega) \right]\Big|_{\alpha=1} + \ln r \sum_{i=1}^{8} c_{0,i} z_i(\alpha, \omega)\Big|_{\alpha=1}$$

with the coefficients satisfying

$$\mathcal{M}(1)\, c_0 = 0$$

$$\mathcal{M}(1)\, c_1 + \tfrac{\partial \mathcal{M}}{\partial \alpha}(1)\, c_0 = g \quad .$$

(25.13)

In some exceptional cases (see Figure 2) we have geometrical multiplicity 1, but $R(1) = 2$. In the numerical example below four exceptional angles occur. In this case we have in the asymptotics a function

$$\text{v} = \sum_{i=1}^{8} \left[\tilde{c}_{2,i} z_i(\alpha, \omega) + \tilde{c}_{1,i} \frac{\partial}{\partial \alpha} z_i(\alpha, \omega) + \tfrac{1}{2!} \tilde{c}_{0,i} \frac{\partial^2}{\partial \alpha^2} z_i(\alpha, \omega) \right]\Big|_{\alpha=1}$$

$$+ \ln r \sum_{i=1}^{8} \left[\tilde{c}_{1,i} z_i(\alpha, \omega) + \tilde{c}_{0,i} \frac{\partial}{\partial \alpha} z_i(\alpha, \omega) \right]\Big|_{\alpha=1} + (\ln r)^2 \sum_{i=1}^{8} \tilde{c}_{0,i} z_i(\alpha, \omega)\Big|_{\alpha=1} \quad ,$$

where

$$\mathcal{M}(1)\, \tilde{c}_0 = 0$$

$$\mathcal{M}(1)\, \tilde{c}_1 + \tfrac{\partial \mathcal{M}}{\partial \alpha}(1)\, \tilde{c}_0 = 0$$

(25.14)

$$\tfrac{1}{2!}\mathcal{M}(1)\, \tilde{c}_2 + \tfrac{\partial \mathcal{M}}{\partial \alpha}(1)\, \tilde{c}_1 + \tfrac{1}{2!} \tfrac{\partial^2 \mathcal{M}}{\partial \alpha^2}(1)\tilde{c}_0 = g \quad .$$

A careful analysis (see [Rös96]) of the properties of the equation systems (25.13), (25.14) and a representation of the solutions by the well known Cramer formula show that the coefficient vectors c_0 and c_1 can have unbounded terms for certain right-hand sides g.

We have developed MATLAB-programs which enable us to identify critical angles and material parameters where the form of the asymptotic expansion changes. We have compared our calculations of critical angles with FEM-results [FMY93], [Leg88] and have seen good agreement (maximal difference to [FMY93]: 0.06°, maximal difference to [Leg88]: 0.002π). We are also able to compute both the coefficients $c_{k,i}$ in the integer terms by solving (25.11), (25.12) and the Jordan-chains by solving (25.10) in the cases $R(1) = 1, 2$. Higher Riesz-indices occur very seldom. In the vicinity of critical angles we have illustrated the expected increase of the integer-terms-coefficients described above. As an example which demonstrates these instabilities we consider a steel-concrete-joint. The given data are:

$\omega_2 = 270°$, 1. material: concrete, 2. Material: steel
$\mu_1 = 0.124 \cdot 10^{11} \frac{N}{m^2}$, $\lambda_1 = 0.064 \cdot 10^{11} \frac{N}{m^2}$, $\nu_1 = 0.17$, $\alpha_1 = 6.2 \cdot 10^{-6} \frac{1}{K}$, $T_1 = 10°C$,
$\mu_2 = 0.839 \cdot 10^{11} \frac{N}{m^2}$, $\lambda_2 = 1.159 \cdot 10^{11} \frac{N}{m^2}$, $\nu_2 = 0.29$, $\alpha_2 = 11.5 \cdot 10^{-6} \frac{1}{K}$, $T_2 = 50°C$

critical angles

$\omega_1^* = 14.67881546721049°$
$\omega_1^* = 61.71829227373731°$
$\omega_1^* = 91.41071877594439°$
$\omega_1^* = 113.6484475010777°$

1.) $\omega_1 \approx \omega_1^*$, $\omega_1 = 61.71829°$ $R(1) = 1$

We get
$c_1 \approx (12320.6, -481.7, 8131.6, 3505.8, 1991.6, 563.1, 1392.8, 0)^T$,
$c_0 \approx (0, 0, 0, 635.9, 0, 0, 0, 473.1)^T$.

2.) $\omega_1 = \omega_1^* = 61.71829227373731°$ (critical angle) $R(1) = 2$

The computation yields
$\tilde{c}_2 \approx 10^{-3} \cdot (-0.04, 0.10, 0.46, 0.42, 0.24, -0.34, 1.38, 0)^T$,
$\tilde{c}_1 \approx 10^{-3} \cdot (-0.2, 0.008, -0.13, -0.06, -0.03, -0.009, -0.02, 0)^T$
$\tilde{c}_0 \approx 10^{-3} \cdot (0, 0, 0, -0.01, 0, 0, 0, -0.008)^T$.

The generalised stress intensity factors [see (25.3)] also grow in an unbounded manner in the neighbourhood of critical angles or material parameters. This fact was stated by Munz/Yang [MY93] in several numerical experiments using FEM-methods.

References

[BMS96] Bochniak M., Maschaik M., and Sändig A.-M. (1996) Computation of stress intensity factors for 2-dim harmonic and elastic fields. To appear.

[CD93a] Costabel M. and Dauge M. (1993) General edge asymptotics of solutions of second order elliptic boundary value problems I. *Proc. Royal Soc. Edinburgh* 123A: 109–155.

[CD93b] Costabel M. and Dauge M. (1993) General edge asymptotics of solutions of second order elliptic boundary value problems II. *Proc. Royal Soc. Edinburgh* 123A: 157–184.

[CD94] Costabel M. and Dauge M. (1994) Stable asymptotics for elliptic systems on plane domains with corners. *Comm. Partial Differential equations* 19: 1677–1726.

[FMY93] Fett T., Munz D., and Yang Y. Y. (1993) The regular stress term in bonded dissimilar materials after a change in temperature. *Engineering Fracture Mechanics* 44: 185–194.

[Gri85] Grisvard P. (1985) *Elliptic problems in nonsmooth domains.* Pitman, Boston – London – Melbourne.

[Kon67] Kondrat'ev V. A. (1967) Boundary value problems for elliptic equations in domains with conical or angular points. *Transactions of the Moscow Mathematical Society* 16: 227–313.

[Leg88] Leguillon D. (1988) Sur le moment ponctuel appliqué à un secteur: le paradoxe de Sternberg-Koiter. *C. R. Academie de Sciences Paris* 307, série II: 1741–1746.

[MP84a] Maz'ya V. G. and Plamenevskij V. A. (1984) Estimates in L_p and in Hölder classes and the Miranda-Agmon maximum principle for solutions of elliptic boundary value problems in domains with singular points on the boundary. *American Mathematical Society, Translation* 123: 1–56.

[MP84b] Maz'ya V. G. and Plamenevskij V. A. (1984) On the coefficients in the a-symptotics of solutions of elliptic boundary value problems in domains with conical points. *American Mathematical Society, Translation* 123: 57–88.

[MR92] Maz'ya V. G. and Roßmann J. (1992) On a problem of Babushka (stable asymptotics of the solution to the Dirichlet problem for elliptic equations of second order in domains with angular points). *Mathematische Nachrichten* 155: 199–220.

[MR94] Maz'ya V. G. and Roßmann J. (1994) On the behaviour of solutions to the Dirichlet problem for second order elliptic equations near edges and polyhedral vertices with critical angles. *Zeitschrift für Analysis und ihre Anwendungen* 13: 19–47.

[MY93]   Munz D. and Yang Y. Y. (1993) Stresses near the edge of bonded dissimilar materials described by two stress intensity factors. *International Journal of Fracture* 60: 169–177.

[NS94]   Nicaise S. and Sändig A.-M. (1994) General interface problems, Part I and II. *Mathematical Methods in the Applied Sciences* 17: 395–430.

[Rös96]   Rössle A. (1996) Spannungssingularitäten für gekoppelte Strukturen in der Festkörpermechanik unter mechanischer und thermischer Belastung. Master's thesis, Universität Stuttgart, Mathematisches Institut A/6.

[Sän96]   Sändig A.-M. (1996) Asymptotics and its stabilization for solutions of elliptic boundary value problems in domains with conical points. Zeitschrift für Analysis und ihre Anwendungen, to appear.

[Yos96]   Yosibash Z. (1996) Numerical thermo-elastic analysis of singularities in two-dimensions. *International Journal of Fracture* 74: 341–361.

26

A 2D Finite Element Method for Magnetic Field Computations in Electrical Machines using an Advanced Material Characterisation

Luc R. Dupré[1] and Roger Van Keer[2]

[1]Department of Electrical Power Engineering,
University of Gent,
Sint-Pietersnieuwstraat 41, B-9000 Gent, Belgium

[2]Department of Mathematical Analysis,
University of Gent,
Galglaan 2, B-9000 Gent, Belgium

26.1 INTRODUCTION

In this paper we present a 2D finite element method for the evaluation of magnetic field patterns in a one tooth region of an electrical machine, taking into account a vector Preisach model for the material characterisation. In a 2D domain, such a hysteresis model is needed due to the local rotating magnetic flux excitations. These excitations in electrical machines result from the complexity of the magnetic circuit and of the magnetic motoric force distributions.

An outline of this paper is as follows. In Section 26.1 we derive the mathematical model of the physical problem, in particular paying attention to the special boundary conditions. Section 26.3 deals with the nonstandard variational formulation. First, the governing Maxwell equations are rewritten in a suitable way in terms of the scalar magnetic potential φ and its time derivative. This allows us to take into account the correct magnetic material parameters, viz the differential permeabilities. These parameters are unique and can be obtained from the hysteresis model used. Next, two possible types of source conditions are introduced. Finally, the resulting initial-boundary value problem for the potential φ is given an appropriate variational formulation, taking into account the nonlocal boundary conditions. In Section 26.4, the variational problem is solved numerically by a conforming FEM, using a properly constructed approximation space (for a quadratic mesh), followed by a modified Crank-Nicholson algorithm for the time discretisation. The latter includes a suitable iteration procedure to deal with the nonlinear hysteresis behaviour. In Section 26.5 we present a few numerical results for the local field patterns, from which the ion losses in electrical machines may be evaluated.

26.2   MATHEMATICAL MODEL

The magnetic behaviour of the material can be described in terms of the macroscopic fields, taking into account the hysteresis phenomena.

We consider a (cross section of a) single tooth region D, see Figure 1 (as usual, assuming uniformity in the third direction). The electrical conductivity σ is assumed to be zero.

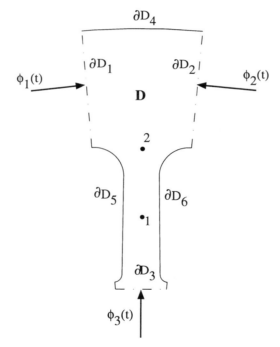

Figure 1   Model of one tooth region

The corresponding Maxwell equations for the magnetic field $\bar{H}=H_x\bar{1}_x + H_y\bar{1}_y$ and the magnetic induction $\bar{B} = B_x\bar{1}_x + B_y\bar{1}_y$, in the 2D domain D are, see e.g [HR81],

$$rot\bar{H} = 0, \qquad (26.1)$$
$$div\bar{B} = 0. \qquad (26.2)$$

The boundary ∂D of the region D is divided into six parts ∂D_1, ∂D_2, ..., ∂D_6. Enforcing a total flux $\phi_s(t)$ through the parts ∂D_s, s=1,2,3, the corresponding boundary conditions (BCs) are found to be

$$\phi_s(t) = \int_{\partial D_s} \bar{B} \cdot \bar{n} dl, \ t > 0, \ s = 1, 2, 3, \qquad (26.3)$$

$$\bar{H} \mathrm{x} \bar{n} = 0 \text{ on } \partial D_s, \ t > 0, \ s = 1, 2, 3, \qquad (26.4)$$

where $\bar{n}$ is the unit outward normal vector to the boundary part ∂D_s.
On the other hand, a zero flux leakage through ∂D_4, ∂D_5 and ∂D_6 results in the complementary BCs:

$$\bar{B} \cdot \bar{n} = 0 \text{ on } \partial D_s, t > 0, s = 4, 5, 6. \qquad (26.5)$$

The demagnetized state of the material at $t = 0$ is expressed by the initial condition (IC)

$$\bar{H}(x, y; t = 0) = 0, \forall (x, y) \in D. \qquad (26.6)$$

The BH-relation, showing the dependency of $\bar{B}$ on $\bar{H}(t)$ and its history $\bar{H}_{past}(t)$, is obtained from a vector Preisach model, see [May91], based upon the principle of elementary dipoles. Explicitely,

$$\bar{B}(\bar{H}, \bar{H}_{past}) = \frac{1}{\pi} \int_{-\frac{\pi}{2}}^{\frac{\pi}{2}} \bar{1}_\theta d\theta \int_{-\infty}^{\infty} d\alpha \int_{-\infty}^{\alpha} d\beta \ \eta_r(\theta, \alpha, \beta, t) P_r(\theta, \alpha, \beta). \qquad (26.7)$$

Here $\eta_r(\theta, \alpha, \beta, t)$ takes the time dependent value of the magnetisation of the dipole with parameters α and β on the axis which encloses an angle θ with the x-axis. The dependency of the *Preisach function* P_r on θ reflects the anisotropy of the material.

26.3   A NONSTANDARD VARIATIONAL FORMULATION

26.3.1   SCALAR POTENTIAL AND ITS TIME DERIVATIVE

First, we rewrite the Maxwell equations (26.1)-(26.2) in a suitable form. From (26.1) a scalar potential $\varphi(x, y; t)$ may be introduced such that $\bar{H} = -grad\varphi$ (of course, φ can only be determined up to a constant, the choice of which will be specified below). The usual reformulation of the Maxwell equations in terms of this scalar potential doesn't allow the material characteristics from the vector Preisach model, to be taken properly into account .

To overcome this difficulty, notice that the differential permeabilities $\mu_{xx} = \partial B_x/\partial H_x$, $\mu_{xy} = \partial B_x/\partial H_y$, $\mu_{yx} = \partial B_y/\partial H_x$ and $\mu_{yy} = \partial B_y/\partial H_y$ are *uniquely* defined by the vector Preisach model. Hence, the material characteristics should be incorporated by means of these permeabilities. Therefore, we pass to the auxiliarly unknown u, defined as

$$u(x, y; t) = \frac{\partial \varphi}{\partial t}. \tag{26.8}$$

Considering the time derivative of (26.2), we may arrive at the 2nd order elliptic DE:

$$\frac{\partial}{\partial x} \left(\mu_{xx} \frac{\partial u}{\partial x} + \mu_{xy} \frac{\partial u}{\partial y} \right) + \frac{\partial}{\partial y} \left(\mu_{yx} \frac{\partial u}{\partial x} + \mu_{yy} \frac{\partial u}{\partial y} \right) = 0. \tag{26.9}$$

On the other hand, the boundary conditions (26.3)-(26.5) lead to

$$\frac{d\phi_s(t)}{dt} = \int_{\partial D_s} \frac{d\bar{B}}{dt} \cdot \bar{n} dl, \ t > 0, \ s = 1, 2, 3, \tag{26.10}$$

$$\varphi = C_s(t) \text{ on } \partial D_s, \ t > 0, \ s = 1, 2, 3, \tag{26.11}$$

and

$$\frac{d\bar{B}}{dt} \cdot \bar{n} = 0 \text{ on } \partial D_s, \ t > 0, \ s = 4, 5, 6. \tag{26.12}$$

Here, to remove the degree of freedom involved in the scalar potential φ, we *choose*

$$\varphi = 0 \text{ on } \partial D_3, \ t > 0. \tag{26.13}$$

To this problem we must add the IC resulting from (26.6) and (26.13), viz

$$\varphi = 0, \ \forall(x, y) \in D, \ t = 0. \tag{26.14}$$

26.3.2 SOURCE CONDITIONS

We must account for two types of source conditions, similarly as in [VDM96].

(a) Under ϕ-*type excitation*, the total flux $\phi_s(t)$ through ∂D_s, s=1,2,3 is enforced (under the restriction that $\sum_{s=1}^{3} \phi_s(t) = 0, t > 0$, due to (26.2) and (26.3)).
Then, the uniform but time dependent value of the scalar potential φ on ∂D_s, denoted by $C_s(t)$, $s = 1$ or 2, is not given a priori, but must be determined as part of the problem.

(b) Under φ-*excitation*, the uniform value $\varphi(t) = C_s(t)$, $t > 0$, at ∂D_s, $s = 1$ and 2, is enforced, (recall (26.13)).
From the boundary value problem (26.9), (26.11)-(26.12), (26.14), we may obtain the magnetic induction $\bar{B}$. The total flux $\phi_s(t)$, $s = 1, 2$ or 3, then follows from (26.10) and (26.14).

26.3.3 VARIATIONAL PROBLEM

To derive a suitable variational form of this problem for the two types of source conditions mentioned, we introduce the function space

$$V = \{v \in W_2^1(D); v|_{\partial D_s} \text{ is a constant depending on s}, s = 1, 2, 3\} \qquad (26.15)$$

Here $W_2^1(D)$ is the usual first order Sobolov space on D and the condition " $v|_{\partial D_s}$ is constant" must be understood in the sense of traces, as defined e.g. in [Cia78]. Multiplying both sides of (26.9) with a test function $v(x, y) \in V$, integrating over D, applying Green's theorem and invoking (26.10), the problem (26.9)-(26.14) is found to be (formally) equivalent with the following variational problem:

Find a function $u(x, y; t)$, with $u(x, y; t) = \frac{\partial \varphi}{\partial t}$,
obeying $u \in W_2^1(D)$ and $\varphi \in V$ for every $t > 0$, such that

$$\int_D \left[(\mu_{xx} \frac{\partial u}{\partial x} + \mu_{xy} \frac{\partial u}{\partial y}) \frac{\partial v}{\partial x} + (\mu_{yx} \frac{\partial u}{\partial x} + \mu_{yy} \frac{\partial u}{\partial y}) \frac{\partial v}{\partial y} \right] dxdy = \sum_{s=1}^{3} \frac{d\phi_s(t)}{dt} v|_{\partial D_s}, \quad (26.16)$$

$$\forall \text{ v} \in \text{V}, t > 0,$$

and

$$\varphi = 0, \text{ for } t = 0. \qquad (26.17)$$

By the requirement $\varphi \in V$ for every $t > 0$, (26.11) is automatically taken into account.

26.4 A CONTINUOUS TIME CONFORMING FEM

Let τ_h be a regular triangulation of the tooth region D with mesh parameter h, as shown in Figure 2. We consider quadratic triangular finite elements.

By $N_j(x, y)$, $(j = 1, ..., J)$, we denote the *standard cardinal* basis functions, associated with all vertices and side midpoints (x_j, y_j), $(j = 1, ..., J)$, J being the total number of these nodes. Here, the nodes are numbered such that the first I, $I < J$, belong to the domain D or to the boundaries ∂D_4, ∂D_5 and ∂D_6. On the boundaries ∂D_1, ∂D_2 and ∂D_3 there are J_1, J_2 and J_3 nodes respectively, $(J - I = J_1 + J_2 + J_3)$. Let $C^0(\bar{D})$ be the space of continuous functions on $\bar{D}$ and let $P_2(T)$ be the space of polynomials of degree ≤ 2 on T, then one has, see e.g. [Cia78],

$$X_h := \{v \in C^0(\bar{D}) \mid v|_T \in P_2(T), \forall T \in \tau_h\} = span(N_j)_{j=1}^{J} \qquad (26.18)$$

and

$$X_{0h} := \{v \in X_h \mid v = 0 \text{ on } \partial D_s, s = 1, 2, 3\} = span(N_j)_{j=1}^{I}. \qquad (26.19)$$

We construct 3 special functions from the space X_h, viz

$$\psi_{I+1}(x, y) = \sum_{j=I+1}^{I+J_1} N_j(x, y), \qquad (26.20)$$

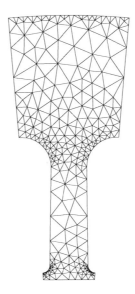

Figure 2   Triangulation τ_h of the domain D

$$\psi_{I+2}(x,y) = \sum_{j=I+J_1+1}^{I+J_1+J_2} N_j(x,y), \qquad (26.21)$$

$$\psi_{I+3}(x,y) = \sum_{j=I+J_1+J_2+1}^{J} N_j(x,y), \qquad (26.22)$$

showing the properties

$$\psi_{I+s} \equiv 1 \text{ on } \partial D_s, \psi_{I+s} \equiv 0 \text{ on } D\backslash \text{ triangles adjacent to } \partial D_s, \ s = 1,2,3. \quad (26.23)$$

Writing, for convenience, $\psi_j = N_j$, $1 \le j \le I$, we finally define $V_h \subset V$ by:

$$V_h = span(\psi_j)_{j=1}^{I+3}. \qquad (26.24)$$

The finite element approximation $\varphi_h(x,y;t) \in V_h$ of $\varphi(x,y;t)$ is defined by a problem similar to (26.16)-(26.17), now with V replaced by V_h. Here, we approximate the space dependency of μ_{kl}, by passing to $\hat{\mu}_{kl} \simeq \mu_{kl}$,

$$\hat{\mu}_{kl}(x,y;t,\varphi_h(x,y;t),\varphi_h^{(past)}(x,y;t)) =$$

$$\mu_{kl}(x_T^c,y_T^c,t,\varphi_h(x_T^c,y_T^c,t),\varphi_h^{(past)}(x_T^c,y_T^c,t)), \qquad (26.25)$$

$$(x,y) \in T, \ \forall T \in \tau_h, \ t > 0,$$

where (x_T^c,y_T^c) is the center of gravity of T. Thus the nonlinear and hysteresis effects, leading to the complicated form of the differential permeability μ_{kl}, can be taken properly into account. Here, μ_{kl} now depends upon the finite element approximation of the magnetic field $H(x, y; t)$ and its history $H^{(past)}(x, y; t)$, through φ_h and $\varphi_h^{(past)}$ respectively.

Recalling (26.13) and decomposing φ_h as

$$\varphi_h(x, y; t) = \sum_{j=1}^{I+2} \varphi_j(t)\psi_j(x, y), \quad t > 0, \tag{26.26}$$

we have $\varphi_j(t) = \varphi_h(x_j, y_j; t)$, $1 \leq j \leq I$, and $\varphi_{I+s}(t) = \varphi_h(x, y; t)|_{\partial D_s}$, $t > 0$, $s = 1, 2$.

Notice that in the case of ϕ-excitation (case (a) in Section 26.3), all coefficient functions $\varphi_j(t)$, $1 \leq j \leq I + 2$, are unknown, while in the case of φ-excitation (case (b) in Section 26.3), the coefficient functions $\varphi_{I+1}(t)$ and $\varphi_{I+2}(t)$ are given.

For the respective sets of unknown coefficient functions we are led to the following system of 1st order ODEs,

$$M(t, C(t), C^{(past)}(t)) \cdot \frac{dC}{dt} = F, \quad t > 0, \tag{26.27}$$

along with the ICs, cf. (26.14),

$$C(0) = 0, \tag{26.28}$$

where, the matrices involved read as follows.

case (a)

C and $C^{(past)}$ are the column matrices,

$$\begin{aligned} C(t) &= [\varphi_1(t), \varphi_2(t), ..., \varphi_{I+2}(t)]^T, \\ C^{(past)}(t) &= [\varphi_1^{(past)}(t), \varphi_2^{(past)}(t), ..., \varphi_{I+2}^{(past)}(t)]^T, \end{aligned} \tag{26.29}$$

while M is the mass matrix given by

$$M(t, C(t), C^{(past)}(t)) = (M_{l,m})_{1 \leq l, m \leq I+2}, \tag{26.30}$$

with

$$M_{l,m} = \int_D \left(\hat{\mu}_{xx} \frac{\partial \psi_l}{\partial x} \frac{\partial \psi_m}{\partial x} + \hat{\mu}_{xy} \frac{\partial \psi_l}{\partial x} \frac{\partial \psi_m}{\partial y} + \hat{\mu}_{yx} \frac{\partial \psi_l}{\partial y} \frac{\partial \psi_m}{\partial x} + \hat{\mu}_{yy} \frac{\partial \psi_l}{\partial y} \frac{\partial \psi_m}{\partial y} \right) dx dy, \tag{26.31}$$

and finally the $(I+2)\times 1$-force matrix F, corresponding to the RHS of (26.16), reads

$$F(t) = \frac{d\phi_1}{dt} \cdot [0, 0, ..., 0, 1, 0]^T + \frac{d\phi_2}{dt} \cdot [0, 0, ..., 0, 0, 1]^T, \tag{26.32}$$

where we used (26.19) and (26.23)

case (b)

Now C and $C^{(past)}$ are $I \times 1$-matrices, while M is an $I \times I$-matrix, taking a similar form to that in case (a). Moreover the force matrix now reads

$$F(t) = [F_1(t), F_2(t), ..., F_I(t)]^T, \qquad (26.33)$$

with

$$F_i(t) = -M_{i,I+1}\frac{d}{dt}\varphi_{I+1}(t) - M_{i,I+2}\frac{d}{dt}\varphi_{I+2}(t), 1 \le i \le I. \qquad (26.34)$$

26.5   FULL DISCRETISATION

We construct a suitable finite difference method to solve numerically the IVP (26.27)-(26.28). We need only consider the case of ϕ-excitation, (case (a)), the case of φ-excitation being completely analogous. The analysis proceeds similarly to that in [DVM96].

Let Δt be a time step and $t_k = k \cdot \Delta t$, $(k = 0, 1, 2, ...)$, be the corresponding equidistant time points. We define an approximation $C^{(k)} = [\varphi_1^{(k)}, \varphi_2^{(k)}, ..., \varphi_{I+2}^{(k)}]^T$ of $C(t_k) = [\varphi_1(t_k), \varphi_2(t_k), ..., \varphi_{I+2}(t_k)]^T$, $(k = 1, 2, ...)$, by the following recurrent nonlinear algebraic system

$$M^{(k)} \cdot \frac{C^{(k)} - C^{(k-1)}}{\Delta t} = \frac{F(t_k) + F(t_{k-1})}{2}, \quad k = 1, 2, ... \qquad (26.35)$$

starting from, see (26.28),

$$C^{(0)} = 0. \qquad (26.36)$$

By means of $C^{(k)}$ we could construct an approximation $\varphi_h^{(k)}(x, y)$ of $\varphi_h(x, y; t_k)$, (26.26). As the matrix $M^{(k)}$, arising in (26.35) must depend on the unknown $C^{(k)}$, we set up an iterative procedure, the number of iterations being denoted by n_k. The approximation of $C^{(k)}$ at the l-th iteration level is denoted by $C^{(k),(l)}$. The corresponding approximation of $\varphi_h^{(k)}$ is written as $\varphi_h^{(k),(l)}(x, y)$. We take $\varphi_h^{(k)} := \varphi_h^{(k),(n_k)}$, which is then used as the input at the subsequent time point t_{k+1}.

In the iterative procedure the matrix $M^{(k),(l)}$, approximating the matrix $M^{(k)}$, is generated by an averaging procedure over the interval $[t_{k-1}, t_k]$, as described in detail in [DVM96].

26.6   NUMERICAL RESULTS AND CONCLUDING REMARKS

In this paper we presented a mathematical model and a FEM for the evaluation of the local magnetic field patterns in a 2D tooth region of an electric machine. We reformulated the governing Maxwell equations in terms of the scalar potential and its time derivative. Thus a refined material characterisation could be included in the model, when also an adapted finite difference scheme for the time discretisation was applied. A particular feature of the FEM developed concerns the special set of basis functions constructed to cope with the nonlocal BCs.

To illustrate the feasibility of the present approach for the local magnetic field computations, we considered a test problem with practical relevance, viz the evaluation of the local field patterns in a one tooth region of an asynchronous machine.

The realistic flux patterns enforced through ∂D_1 and ∂D_2 are

$$\phi_j(t) = a_{j,1}cos(2\pi ft + \gamma_{j,1}) + a_{j,15}cos(30\pi ft + \gamma_{j,15})$$

$$+a_{j,17}cos(34\pi ft + \gamma_{j,17}), \; j = 1,2 \tag{26.37}$$

where the amplitudes and phase angles are given in Table 1, with $f = 50Hz$.

Table 1   amplitudes and angles of excitation

	Amplitudes (T)			Phases (degree)		
	1	15	17	1	15	17
ϕ_1	1.262	0.0178	0.0105	+25.	+109.	-36.
ϕ_2	1.268	0.0067	0.0050	+5.9	-155.	+27.

The numerically evaluated B_xH_x-loops and B_yH_y-loops for the points 1 and 2, as indicated in Figure 1 are shown in Figure 3. These loops confirm the uni-directional (point 1) and rotational (point 2) field patterns, which are expected a priori from physical arguments.

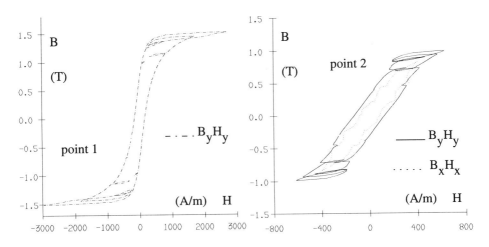

Figure 3 B_xH_x- and B_yH_y-loops in point 1 and point 2

ACKNOWLEDGMENT

We gratefully acknowledge the financial support by the Belgian Government in the frame of the Inter-University Attraction Poles for fundamental research. One of us (R.V.K.) thanks the Belgian National Foundation for Scientific Research (NFWO) for financial support.

References

[Cia78]   Ciarlet P. (1978) *The Finite Element Method of Elliptic Problems.* North-Holland, Amsterdam.

[DVM96] Dupré L., VanKeer R., and Melkebeek J. (1996) On a Numerical model for the Evaluation of Electromagnetic Losses in Electric Machinery. *Int. J. of Num. Meth. in Eng.* 39: 1535–1553.

[HR81]   Halliday and Resnick (1981) *Fundamental of Physics.* John Wiley & Sons, New York.

[May91] Mayergoyz I. (1991) *Mathematical models of hysteresis.* Springer Verlag, New York.

[VDM96] VanKeer R., Dupré L., and Melkebeek J. (1996) On a Numerical model for the Evaluation of Electromagnetic Losses in Electric Machinery. *J. Comp. and Appl. Math.* 72: 179–191.

27

A RP-Adaptivity Scheme for the Finite Element Analysis of Linear Elliptic Problems

J. Ed Akin[1] and J. R. Maddox[2]

[1]Rice University, Houston, TX, USA

[2]Shell Services Co., Houston, TX, USA

27.1   INTRODUCTION

In any adaptive method, the development of internal data representations, system assembly and solution procedures, error estimation techniques, and refinement algorithms is contingent upon the selection of an appropriate element formulation. In traditional r-adaptivity methods, wide latitude is associated with this element selection process. The only real requirement is that a fixed number of degrees of freedom be used. Previous work in this area has opted to employ a fixed number of elements with similar topology to satisfy this requirement. Simple linear triangle and quadrilateral elements are prevalent in the literature.

Traditional h-adaptivity approaches also commonly use simple linear triangle and quadrilateral elements. Control of the element shape after several levels of refinement is an important consideration in these approaches. Use of these elements, coupled with appropriate element bisection techniques, reduces element shape deterioration and its associated numerical conditioning problems. The p-adaptivity schemes typically use hierarchic finite elements where the interpolation functions of order p are a subset of the interpolation functions of order $p + 1$. This allows the element matrices for a refined mesh to be generated by adding new terms to the previous element matrices. The element topology is never affected.

It has been shown that optimal hp-refinement can lead to exponential rates of convergence in the energy norm in terms of the number of degrees of freedom. As a result, work on hybrid adaptivity schemes has focused on the use of hp-adaptivity. These systems typically use (initially) linear quadrilateral elements with hierarchical interpolation functions. They are the most challenging to program.

The Mathematics of Finite Elements and Applications
Edited by J. R. Whiteman © 1997 John Wiley & Sons Ltd.

For the hybrid rp-adaptivity scheme considered in this prototype effort, it was desirable to select an element formulation which would allow the polynomial order of the interpolation functions to be increased and decreased in a conceptually simple manner. Direct manifestations of such an element formulation would be both a relatively simple internal data representation and a straightforward refinement algorithm. This would be a major advantage over the conceptually difficult traditional p-refinement approach associated with hierarchical shape functions, which adds tangential derivatives of the unknown solution as additional degrees of freedom. As a result, heuristics must be developed to ensure interelement continuity and avoid directional conflicts during the assembly process. These implementation difficulties are not trivial problems.

A fully node-based adaptivity scheme could implement the p-refinement by simply adding or deleting new nodes to the elements in question. By requiring these nodes to lie on the element boundary interface, explicit continuity enforcement at the boundaries would not be required.

These considerations suggested the use of serendipity elements. The traditional two-dimensional methodology for generating general serendipity element interpolation functions can be extended to three-dimensions and collapsed to one-dimension. In addition, it is readily implemented as a computer program. The code presented by Zavarise et al. [ZVS91] was converted to Fortran 90 and incorporated into the prototype code, forming the nucleus of a serendipity-based p-refinement methodology extension of the MODEL code [Aki94].

The general serendipity elements do display some troubling characteristics that prove to be somewhat of a hindrance in the nodal gradient recovery and error estimation phase. The lack of polynomial completeness in interpolation functions of elements with edges of quartic or higher order coupled with the presence of parasitic polynomial terms (though not as many as Lagrangian formulations) will result in some procedural differences from the classical gradient recovery techniques [ZZ92a] at that point.

27.2   DATA STRUCTURES

Finite element analysis programs that do not support adaptivity, such as the code described by Akin [Aki94], typically employ a series of sequential list data representations (i.e., arrays). In the rudimentary case, finite element node data is stored in one such structure and element nodal incidences are stored in another. These essentially static structures are inefficient when dynamic alterations to a mesh, other than relatively straightforward r-refinements, are introduced. Sophisticated adaptive refinement algorithms require more flexibility and increased access to information about the mesh than this classical data representation allows.

Early adaptive researchers advocated the use of a more advanced tree structure to implement a localized h-refinement methodology. Conceptually, this form of data representation parallels the recursive element subdivision process, with a simple extension of the tree corresponding to a refinement. This approach is complicated by the frequent introduction of finite element nodes that are not corner nodes of all the undivided neighboring elements. These *irregular* nodes require special algorithmic and data structure considerations such as the addition of extra refinement heuristics, status tags to indicate the regularity or irregularity of a given node, and the addition of constraints to preserve interelement continuity.

Demkowicz et al [DORH89] augment the tree data structure with additional structures in order to implement a hybrid two-dimensional hp-refinement strategy. The crux of these additions is a pseudo-dynamic series of arrays for keeping track of variable degrees of freedom.

The core requirement for a serendipity element-based adaptivity scheme is an underlying data structure that facilitates efficient addition and deletion of finite element nodes while concomitantly keeping memory usage within reasonable limits. Biswas and Strawn [BS93] concluded that dynamically-allocated linked lists addressed these concerns in their three-dimensional unstructured tetrahedron program. That work provides an excellent precedent for the development of data structures to support rp-adaptivity using low order serendipity elements.

After careful consideration, it was decided to employ a series of tiered, dynamically allocated linked lists for the problem at hand. Separate lists are maintained for finite elements and their associated edges and nodes. Unlike sequential data structures such as arrays, where conceptual list items are physically located in memory fixed distances apart, linked representations can have items located at any valid memory locations. In order to maintain the list order, some additional space is used to store a corresponding pointer or memory address to the next list item. This, of course, requires a computer programming language that supports derived types (a.k.a. records or structures), pointers, and dynamic memory features such as Fortran 90.

One principal advantage of employing such list structures pertains to memory usage. Since the lists are dynamically allocated, space is utilized only when it is needed and is returned to the system heap when it is not. The traditional notion of declaring worst case, maximum-sized arrays where storage is frequently wasted is effectively made obsolete. There is some extra space used for the pointers, but this is an acceptable trade-off as this space is only a small percentage of the total usage. Another principal advantage of this list approach is the ease by which data node additions and deletions are made. When the program deletes a finite element node and frees its accompanying space, the pointers are simply rerouted to reflect the change in list order. There is no need to shift large amounts of array space to add or delete data. These macroscopic points are the principal drivers for utilizing the linked list approach. The conceptual idea is to have an element-driven data structure that is flexible enough to give lower level access. Thus, the tiered hierarchy was implemented where each element data node has pointers to its associated edge data nodes, which in turn point to the finite element node data defining the edge. Consequently, if an element identifier is known, all other information related to that element is immediately accessible. The edge data and finite element node data lists can optionally be traversed individually to allow maximum flexibility while maintaining a high degree of hierarchical abstraction.

The element list is singly linked, meaning it is capable of being traversed in only one direction, from the beginning of the list to the end of the list. It is equipped with a header node (as are all the primary lists) which does not represent an item in the list. Its main purpose is to produce a condition where the list is never truly empty. This simplifies the list insertion scheme by removing an algorithmic special case. The data node structure for members of this list contains the following fields:

- *ID*. An integer element identifier.
- *CRNR_SEQ*. An integer array that stores the four corner node identifiers defining the boundary of the element. The order of these corner node identifiers specifies how to define the element normal.
- *EDG*. An array of pointers to specific members of the edge list that correspond to edges of the element. Every element has four edges (for this two-dimensional version).
- *ADJ*. An array of pointers to specific members of the element list that represent adjacent elements in the mesh. Every element will have at least one and at most four adjacent (facing) elements. Pointers that are null or unassociated constitute a situation where the edge is a boundary edge.
- *NEXT*. A pointer to the next member of the element list.

The edge list is also a singly linked list. The data node structure for members of this list contains:

- *ID*. An integer edge identifier.
- *NS_HDR*. A derived type header for a doubly linked circular sublist of element node pointers.
- *NEXT*. A pointer to the next member of the edge list.

The salient feature of this list is that one of its fields is actually a derived type that serves as a header to a doubly linked circular sublist. The principal data items in this sublist are pointers to the particular data nodes in the finite element node list that define the edge. The header node serves a dual purpose in this instance. Besides eliminating an algorithmic special case during list insertion, it serves as a means of determining the list end during traversal. This circular implementation is a direct manifestation of the use of serendipity elements in the adaptivity scheme. Each edge can be shared by at most two elements, with both having a different positive sense. The circular list can be traversed from either end making the computation of element interpolation functions and derivatives a simple task. Each data structure node of this sublist has the following fields:

- *LPTR*. A pointer to the adjacent left-hand member in the sublist.
- *NPTR*. A pointer to a specific member of the finite element node list.
- *RPTR*. A pointer to the adjacent right-hand member in the sublist.

Two of these fields are the right and left directional pointers (or links) that give the list its circular characteristic. The other field is a pointer to a specific member of the finite element node list. The finite element node pointed to is one contained in the serendipity element edge. Thus, all the pertinent nodal information for an edge is readily accessible. One more important item to note is that due to the use of parabolic edges to describe the edge geometry in the prototype, an edge's node sublist will always have at least three members. At least two of these nodes will be used in the analytic interpolation since the analysis always commences using linear edges.

It is also important to note here that due to the nature of *rp*-adaptivity using serendipity elements, the element and edge lists never change after their initial construction. Thus, it was not necessary to implement them using linked representations. The principal advantage of using lists in these cases is to allow a dynamic initial build that does not require *a priori* knowledge about the problem size.

The finite element node list is doubly linked. This means it can be traversed in order from the beginning of the list to the end or vice versa. This implementation enables efficient modification of the list. Every time an edge order is increased or decreased, the doubly linked circular sublist associated with the edge in question is traversed and the purely analytical nodes are deleted and re-defined. Since the entry point into the finite element node list for this operation is arbitrary (i.e., no sequential traversal of the list was performed), deleting the data for a particular node while maintaining list order becomes something of a problem. The doubly linked implementation rectifies this problem by making the deletion algorithmically simple without incurring excessive overhead expense. Each data node contains the following fields:

- $PREV$. A pointer to the previous member in the finite element node list.
- ID. An integer finite element node identifier.
- X. A double precision x-coordinate for the finite element node in question.
- Y. A double precision y-coordinate for the finite element node in question.
- DOF_INFO. An integer flag which serves to classify the analytic or geometric nature of the node.
- $CNSTRNT\_INDICATOR$. An integer code stipulating the constraint types (a.k.a. restraint types) associated with each initial nodal degree of freedom.
- $RELOCATION\_FLAG$. An integer flag specifying the movement constraints on the node during the r-refinement phase.
- $NEXT$. A pointer to the next member in the finite element node list.

The DOF_INFO and $RELOCATION\_FLAG$ fields deserve some further explanation. The prototype code expects parabolic element edges in the geometric interpolation. In the analytic field variable interpolation, any edge order is supported. Consequently, the DOF_INFO field was installed to serve as a mechanism for tracking this information. A zero value for this flag indicates the node serves a purely geometric function. A positive value indicates the node serves a purely analytical function. A negative value indicates the node serves both a geometric and analytical function. After every refinement iteration, the modified finite element node list is traversed and consecutive, progressively increasing (in the absolute value sense) integer values are stored in the field for analytic nodes. These numbers play a critical role in determining the system degree of freedom numbers during the assembly and solution process.

The $RELOCATION\_FLAG$ field was added to give the user more control over the r-refinement process. A zero value for this flag indicates a node is free to relocate. A positive value for this flag indicates a node is restrained from relocating. A negative value for this flag indicates the node is a critical boundary definition node. Such a node must be relocated so as to remain on its original parabolic segment.

The computational efficiency of linked list data structure searches is described mathematically by the upper bound $O(n)$. This means, in the worst case, seek operations on the list are linearly proportional to the number of members n in the list. In this respect, only logarithmic data structures like binary trees are more efficient. Conceptually, list structures are simpler to understand and implement than trees. Additionally, the tiered list structure employed here, especially the doubly linked circular edge sublists, logically correspond to the serendipity element approach. For insertions, the extreme case of sequentially adding list members in ascending order is an $O(n^2)$ operation that requires a complete list traversal every time something is added.

Since there is no requirement that the members be stored sequentially, insertions can be made at the list front to avoid speed problems associated with large models.

27.3   GRADIENT RECOVERY

The error estimation process used in the prototype depends on the accuracy of the recovered gradients. Traditionally, gradient recovery techniques can be grouped into three main categories: (a) direct differentiation of the finite element solution coupled with averaging, (b) local projection coupled with averaging, and (c) global projection. While other miscellaneous techniques do exist, their complex implementation and excessive computational expense preclude them from serious consideration. The obvious and consistent approach to gradient recovery is to directly differentiate the finite element solution and evaluate the resulting functions at the points of interest. For elasticity problems, stresses are calculated by differentiating the displacement solution and using the appropriate constitutive relations. It is well known that this approach yields lower order, discontinous derivatives that are least accurate at the element boundaries and their associated nodes, where accurate gradients are usually desired. The discontinuities are addressed by averaging, but the general inaccuracy of this method makes it undesirable.

The local projection and averaging technique exploits for problems with smooth solutions the existence of locations in many finite element solutions where the usual rate of convergence is exceeded by one order. These optimal sampling points are known as *superconvergent* points. Gradient information at superconvergent points is extrapolated to the nodes and averaged to yield improved gradient accuracy.

The global projection method employs the theory of conjugate approximations to obtain more accurate, continuous gradients across interelement boundaries. This method assumes a C_0 continuous interpolation of the gradients of the same form as that used for the interpolation functions. The nodal values are evaluated using a least squares fit to the consistently evaluated finite element solution. This recovery process is a global or system-level computation that is comparatively expensive. Additionally, this technique yields nodal superconvergence for linear elements only. Although these traditional methods for gradient recovery work in the error estimation process developed by Zienkiewicz and Zhu in most cases, the reliability of the estimator could be adversely affected for quadratic elements. Thus correction factors were introduced. A general purpose, more accurate procedure would eliminate the need for such measures.

The classical superconvergent patch recovery (SPR) technique assumes the mesh in question is constructed of homogeneous, regular elements (i.e., each element utilizes the same interpolation functions). This assumption has two major ramifications. First, it allows the patch definition scheme to be vertex-driven, which means each patch is composed of elements having a particular finite element vertex node in common. Since each element surrounding the vertex node uses the same interpolation functions, this is a valid approach. The assumption also makes the least squares fit polynomial expansion selection process straightforward. The use of regular elements implies there are no areas in the mesh where the interpolation order differs from the order used in each piecewise element domain.

A finite element approach employing p-adaptivity through the use of serendipity elements lends itself to interpolation function irregularities. These irregularities are caused by elements with a varying number of nodes on each edge, yielding interpolation functions that differ in degree and number of polynomial expansion terms throughout the mesh. As a result, a vertex-driven patch determination method becomes a less attractive methodology. Also, selection of the polynomial expansion to use in the nodal gradient recovery process becomes more complicated. These issues necessitate a series of experimental deviations from the classical superconvergent patch recovery technique proposed by Zienkiewicz and Zhu [ZZ92b, ZZ92a].

The prototype code implements an element-driven approach to the gradient recovery process in order to circumvent some of the problems presented by the use of general serendipity elements. Consequently, two element-driven patch definition schemes were studied in the course of this experimental effort.

The first scheme defines the composition of a patch as a particular central element and all elements sharing common edges with it. The central idea behind using such a patch is to have access to as much gradient information in the sphere of influence of the central patch element without using too many elements. For boundary elements having no adjacent elements, the patch is simply constructed using the reduced set of available adjacent elements. There is no reason to employ a special case methodology for boundary corner nodes using this scheme. Using the Zienkiewicz-Zhu vertex-driven approach, corner node patches may produce an underdetermined least squares problem when solving for the coefficients. They remedy this situation by extending the recovery process of interior patches to include boundary node locations, when applicable.

The second scheme defines the composition of a patch as a particular central element and all its surrounding facing elements. This patch has access to a wider array of sampling points during the recovery process (implying improved accuracy), but the trade-off is that it is slightly more involved computationally. Once again, this approach requires no special handling for corner nodes like in the classical vertex-driven scheme. In both methods, the procedure is quite inexpensive. The number of equations to be solved is modest and the actual recovery is performed only on the central patch element's nodes.

One scheme implemented in the prototype code for recovering the nodal gradient values uses a complete polynomial expansion of the central patch element's highest edge degree. As an example, consider a patch of linear and quadratic edges. Since the patch's highest degree edge is quadratic, the complete polynomial expansion $\mathbf{P} = \begin{bmatrix} 1 & x & y & x^2 & xy & y^2 \end{bmatrix}$ is used for the least squares fit. This choice for the gradient expansion is potentially problematic. At any linear-linear edge interface corners, the quadratic fit has x^2 and y^2 terms which are not present in the corner node interpolation functions. Along the quadratic edge, it replaces cubic parasitic terms x^2y and xy^2 with a quadratic term.

A second implemented approach for recovering the nodal gradient values attempts to account for the higher order terms excluded in the previous approach. The complete polynomial expansion of one degree higher than the central patch element's highest edge degree is utilized for this purpose. Thus, for the current patch example, the complete polynomial expansion $\mathbf{P} = \begin{bmatrix} 1 & x & y & x^2 & xy & y^2 & x^3 & x^2y & xy^2 & y^3 \end{bmatrix}$ is used for the least squares fit. This choice for the expansion is also potentially problematic. At the linear-linear edge interface corners, it includes six higher order terms which are not present in the corner node interpolation functions. Along the quadratic edge, it includes four higher order terms which are not present in the interpolation functions.

The last approach tested in the prototype recovers nodal gradient values using the corner node interpolation function for a homogeneous serendipity element having edges of the same order as the central patch element's highest order edge. Thus, for the corner patch example, the polynomial expansion $\mathbf{P} = \begin{bmatrix} 1 & x & y & x^2 & xy & y^2 & x^2y & xy^2 \end{bmatrix}$ is used for the least squares fit. Once again, this choice for the expansion may cause problems. At the linear-linear edge interface corners, it includes four extra higher order terms which are not present in the corner node interpolation functions. Along the quadratic edge, it includes two higher order terms not present in the interpolation functions.

27.4   IMPLEMENTATION CONSIDERATIONS

Implementation of this modified superconvergent patch recovery technique can conceptually be separated into four distinct subprocesses. First, the element patch is defined. Next, sampling point data from the previous solution iteration is used to construct the least squares problem. An appropriate methodology is then used to solve the least squares problem. Lastly, the coefficients returned by the least squares fit are used to evaluate the recovered nodal gradient values. The second step, construction of the least squares problem, requires repetitive access to coordinate information and derivative values of patch sampling points. Any efficient implementation of this process requires a mechanism for storage and efficient access to the sampling point data. Additionally, the solution of the least squares problem can be computed directly from the normal equations. However, this approach is susceptible to roundoff error and sensitive to near-singularities. A singular value decomposition and backsubstitution is the recommended method for performing the least squares fit. For the sake of completeness, these facets of the experimental prototype are addressed below.

In order to facilitate the gradient recovery process, a generalized list structure has been implemented to store pertinent sampling point and derivative data. This gradient recovery data list is singly linked. It is equipped with a header node to facilitate the list insertion action. Each member of this list contains:

- EL_ID. An integer element identifier.
- MAX_NDS_XI. An integer specifying the maximum number of nodes in one parametric direction.
- MAX_NDS_ETA. The maximum number of nodes in the other parametric direction.
- NO_SP. The number of sampling or Gauss points used in the element assembly integration process.
- SD_HDR. Header node to a singly linked sampling point data list.
- $NEXT$. A pointer to the next member of the gradient recovery data list.

The SD_HDR field is a header node to a singly linked list containing information about the sampling points for a particular element. Each data node contains the following fields:

- X. A double precision sampling point x-coordinate value.
- Y. A double precision sampling point y-coordinate value.
- $DERIVS$. A double precision array of sampling point derivative values.
- $NEXT$. A pointer to the next member of the sampling point data sublist.

During the element assembly process of each solution cycle, the element data list is created and the sampling point information is added to the list. At the end of each adaptive cycle, the list is deleted and reinitialized.

The recovery variations discussed are relatively straightforward to implement. Zienkiewicz and Zhu have shown that the classical superconvergent patch recovery yields higher accuracy than previously used gradient recovery methods. Additionally, the recovery process is relatively inexpensive, making it practical for use in error estimators and for solution post-processing. As an example, the recovered velocities from a potential flow example are accurate and smooth. The recovery process has shown a tendency to break down in patches covering a relatively large percentage of the problem domain (i.e., in coarse meshes). Unstable behavior is sometimes apparent in the nodal gradient values on exterior boundaries. This is somehow attributable to large gradient changes across the patch area as well as large differences in polynomial order across the patch area.

27.5 P-REFINEMENT HEURISTICS

The computation of the error, $\|e_\sigma^*\|_i$ for each element allows a refinement indicator to be derived. Indeed, the ratio

$$\xi_i = \frac{\|e_\sigma^*\|_i}{\bar{e}_m} > 1 \tag{27.1}$$

where $\bar{e}_m$ is the mean allowed element error contribution, targets the elements to be refined. Its value can also be used to determine the required polynomial order change to achieve an assumed rate of convergence in the solution. In the prototype, a rate of convergence $O(\Delta^p)$ in the area covered by the element was assumed, where Δ is the current distance between analysis nodes on each edge of the element. Mathematically, Δ is defined as $\Delta = \frac{h}{p}$ where h is the length of the element edge in parametric space and p is the degree of a typical edge in the serendipity element. A new predicted distance between analysis nodes should be

$$\Delta_{new} = \frac{\Delta_{old}}{\xi^{1/p_{old}}} = \frac{h}{p_{old}\xi^{1/p_{old}}} = \frac{h}{p_{new}} \tag{27.2}$$

where $p_{new} = p_{old}\xi^{1/p_{old}}$. Table 1 demonstrates how this computed value for ξ can be used to drive the p-adaptivity process. It is readily apparent that this heuristic can be used to signal that an edge be enriched or degraded. Since the prototype initially assumes that all edges have a linear analysis variation, the degradation operation is rarely encountered in practice. From the implementation perspective, this refinement scheme is relatively straightforward. Once some percentage of the candidate elements (i.e., those having ξ values greater than one) have been selected for refinement, the analysis nodes for each edge are deleted from the edge and finite element node data structures. Procedurally, this means each doubly linked circular edge sublist is

ξ	p_{old}	$\xi^{1/p_{old}}$	p_{new}	Operation	Degree Change
4.00	2.00	2.00	4.00	Enrich	+2
1.00	2.00	1.00	2.00	No Change	NA
0.50	2.00	0.71	1.41	No Change	NA
0.05	2.00	0.22	0.45	Degrade	-1
4.00	3.00	1.59	4.76	Enrich	+2
2.00	3.00	1.26	3.78	Enrich	+1
0.50	3.00	0.79	2.38	No Change	NA
4.00	8.00	1.19	9.51	Enrich	+2
2.00	8.00	1.09	8.72	Enrich	+1

Table 1   P-refinement heuristic ramifications.

traversed, the nodes having nonpositive DOF_INFO flags are deleted from the doubly linked finite element node list, and the edge sublist is reinitialized. The edges are then repopulated with analysis nodes to correctly match the desired polynomial variation on the edge. In the parent space, the new analysis nodes are equally distributed. The quadratic interpolation functions in unit coordinates are used to determine the physical coordinates of the new nodes.

Table 1 reminds us that however the error ratio is employed to predict a new edge degree, that degree estimate must be replaced by an integer. Thus, one is not adapting the element exactly as the current information would allow. To take full advantage of the error indicator we now carry out an additional r-adaption to change the element size to attempt to make up for selecting an integer degree.

27.6   NODE RELOCATION PROCEDURE

In r-refinement, the total number of degrees of freedom in a problem remain unchanged, making it impossible to reduce the approximation error to zero. Therefore, the goal of r-adaptivity is to relocate the nodes in some optimal manner. It has been shown that such an optimal situation occurs when the error is equally distributed between elements. Mathematically, this is expressed $\|e_\sigma^*\|_i = k$ where $\|e_\sigma^*\|_i$ is an individual element's (remaining) error contribution and k is a constant. The idea is to relocate nodes so as to redistribute the finite element gradient errors, decreasing the maximum errors and increasing the minimum errors. In most cases, the average value of the overall error decreases.

Any viable node redistribution scheme must address certain difficulties. It must produce new coordinates that are consistent with the problem boundary domains and with the existing mesh connectivity. It must also stipulate a large enough movement to ensure quick improvements. One computationally efficient approach is implemented in [RA90]. This approach uses the error estimators $\|e_\sigma^*\|_i$ in elements surrounding the node to alter A_i, the area of a particular element. For example, if the element error

$\|\mathbf{e}_\sigma^*\|_i$ is larger than average, the geometric nodes around element i are moved so as to reduce A_i. Since a node is usually shared by several elements, its local movement should be in the direction of the element of largest $\|\mathbf{e}_\sigma^*\|_i$. The magnitude of the movement depends on the relative magnitude of the $\|\mathbf{e}_\sigma^*\|_i$ values around the node. This suggests the idea of assigning a weighted quantity to each element. This weighted value is placed at the geometric center of the element and attracts the nodes surrounding it. Each node is moved towards the larger weight value associated with the elements around it. This movement is described mathematically as

$$ x_n^{\nu+1} = \frac{\sum_{i \,\epsilon\, \{N\}} \bar{x}_i \left(\frac{\|\mathbf{e}_\sigma^*\|_i}{A_i} \right)}{\sum_{i \,\epsilon\, \{N\}} \left(\frac{\|\mathbf{e}_\sigma^*\|_i}{A_i} \right)} \tag{27.3} $$

where $x_n^{\nu+1} \,\epsilon\, R^2$ is the new location of the nth node, $\{N\}$ is the set of all elements that contain x_n, $\bar{x}_i \,\epsilon\, R^2$ is defined as the centroid of the element i, and A_i is the area of element i. The weight assigned to element i is defined as the ratio $\|\mathbf{e}_\sigma^*\|_i/A_i$. It is readily seen that if $\|\mathbf{e}_\sigma^*\|_i$ is large, the weight of i is large and nodes near i are moved towards $\bar{x}_i$. As a result, an element i with $\|\mathbf{e}_\sigma^*\|_i > \|\mathbf{e}_\sigma^*\|_N$ for all neighboors N of i is reduced in size. For nodes on the boundary, the motion can be restricted to be tangent to the boundary or they can be projected back to the boundary after being moved.

27.7   CONCLUSIONS

The prototype rp-adaptivity code [Mad95] developed for this study has proven itself to be a viable methodology for adaptively refining linear elliptic problems. It has been successfully tested on a number of problems. The integration of new data structures and their closely associated assembly, error estimation, and refinement algorithms produces encouraging results. At each step of the solution the r-adaptivity provides a way to compensate for the element error under-correction or over-correction associated with the restriction of selecting an integer polynomial degree on each edge of the element. To properly assess its merit some model problems need to be run to compare error vs. number of unknowns with this rp-scheme as well as with the r-adaptivity turned off.

Many current problem areas still need to be addressed before the code can be considered for general purpose use. First, the ability to handle general boundary conditions within the scope of the refinement schemes must be implemented. Secondly, more investigative work should be done with the p-refinement heuristics to determine a pseudo-optimal method of application. Next, more work should be done investigating the superconvergent patch recovery variations discussed above. Conditions where the methodologies give poor results (e.g., coarse meshes) should be thoroughly understood. Lastly, the actual interpolation form utilized in the highest degree element in the patch should be implemented for the gradient recovery and its viability assessed. That parametric form is easily generated and appears to be a consistent way to fit the superconvergent gradients in the patch.

An area for enhancement activity is the modification of this method to input elements with higher than linear order analysis edges in the initial mesh. This change would give the user more power and flexibility and it would open the way to more robust adaptivity making full use of the program's enrichment and degradation capabilities.

The applicability of this scheme to three-dimensions is something which should be considered. The interpolation function routine can generate interpolation functions for three-dimensional general serendipity elements. The addition of an element face list to the core data structures is an essential component of a three-dimensional version of this code. This list would simply indicate the edges associated with a particular face and would be an additional tier between the element and edge list structures. Other required list modifications would be minor in comparison to this change.

References

[Aki94]   Akin J. E. (1994) *Finite Elements for Analysis and Design*. Academic Press, London.

[BS93]   Biswas R. and Strawn R. (1993) A New Procedure for Dynamic Adaption of Three-Dimensional Unstructured Grids. In *AIAA 31st Aerospace Sciences Meeting and Exhibit*. Reno, NV.

[DORH89] Demkowicz L., Oden J. T., Rachowicz W., and Hardy O. (1989) Toward a Universal h-p Adaptive Finite Element Strategy. *Comp. Methods Appl. Mech. Eng.* 77: 79–112.

[Mad95]   Maddox J. R. (1995) An RP-Adaptivity Scheme for Finite Element Analysis of Linear Elliptic Problems. Master's thesis, Rice University, Houston, TX.

[RA90]   Ramaswamy B. and Akin J. E. (1990) Design of an Optimal Grid for Finite Element Methods in Incompressible Fluid Flow Problems,. *Eng. Comput.* 7: 311–326.

[ZVS91]   Zavarise G., Vitaliani R., and Schrefler B. (1991) An Algorithm for Generation of Shape Functions in Serendipity Elements. *Eng. Comput.* 8: 19–31.

[ZZ92a]   Zienkiewicz O. C. and Zhu J. Z. (1992) The Superconvergent Patch Recovery and "A Posteriori" Error Estimates. Part 1: The Recovery Technique. *Int. J. Num. Meth. Eng.* 33: 1331–1364.

[ZZ92b]   Zienkiewicz O. C. and Zhu J. Z. (1992) The Superconvergent Patch Recovery (SPR) and Adaptive Finite Element Refinement. *Comp. Methods Appl. Mech. Eng.* 101: 207–224.

Index

The index is compiled alphabetically word-by-word.

Italic numbers denote pages on which diagrams appear.